Theory and Design of Enzymatic and Fermentation Processes

The second edition of *Theory and Design of Enzymatic and Fermentation Processes* covers the kinetics and design of enzymatic and fermentation processes both under aerobic and anaerobic conditions. It discusses microorganism biochemistry and metabolism concisely, and covers process design, oxygen and heat transfer and scale-up/scale-down topics systematically. The book includes over a hundred solved examples, questions and problems.

Features:

- Provides exclusive coverage of enzyme kinetics.
- Discusses pertinent aspects of the fermentation process, including kinetics, stoichiometry, mass balances and oxygen transfer.
- Explores oxygen transfer theory and its application to fermentation processes.
- Reviews stoichiometry and kinetics of fermentation processes.
- Includes computer simulations, homework problems, examples, PowerPoint slides and solutions manual.

This book can be a useful resource as a textbook for undergraduate and graduate students in chemical engineering, as well as for process engineers working with enzymatic and fermentation reactions.

Theory and Design of Enzymatic and Fermentation Processes

Second Edition

Davide Dionisi

CRC Press
Taylor & Francis Group
Boca Raton London New York

CRC Press is an imprint of the
Taylor & Francis Group, an **informa** business

Designed cover image: Getty Images

Second edition published 2026
by CRC Press
2385 NW Executive Center Drive, Suite 320, Boca Raton FL 33431

and by CRC Press
4 Park Square, Milton Park, Abingdon, Oxon, OX14 4RN

CRC Press is an imprint of Taylor & Francis Group, LLC

First edition published by CRC Press 2022

ISBN: 9781041027263 (hbk)
ISBN: 9781041061106 (pbk)
ISBN: 9781003633822 (ebk)

DOI: 10.1201/9781003633822

Typeset in Times
by Newgen Publishing UK

Dedication

To Luca and Valerio

Contents

Preface.....xi
About the Author.....xiii

Chapter 1 Enzymatic Processes.....1

1.1 Introduction to Enzymes.....1
1.1.1 Enzyme Nature, Chemistry and Structure.....1
1.1.2 Enzymatic Catalysis and Mechanisms.....2
1.1.3 Enzyme Immobilisation and Materials.....3
1.1.4 Applications of Enzymes.....5
1.2 Enzyme Kinetics with the Michaelis-Menten Model.....7
1.2.1 Derivation of the Michaelis-Menten Kinetics.....7
1.2.2 Analysis of the Michaelis-Menten Kinetics.....10
1.2.3 Kinetic Parameters from Experimental Data.....13
1.2.4 Mass Balances for Enzymatic Reactions using the Michaelis-Menten Kinetics.....32
1.3 Enzyme Kinetics with Effects of Inhibition, pH, Temperature, Immobilisation.....38
1.3.1 Effect of Inhibitors.....38
1.3.2 Effect of pH.....48
1.3.3 Effect of Temperature.....51
1.3.4 Enzyme Kinetics with Immobilised Enzymes.....54
1.4 Design and Economics for Enzymatic Processes.....77

Chapter 2 Introduction to Fermentation Processes.....86

2.1 Fermentation Processes and Microorganisms.....86
2.2 Classification of Microorganisms.....88
2.3 Structure and Components of Microbial Cells.....90
2.4 Stoichiometry and Metabolism of Cell Growth and Product Formation.....95
2.4.1 Metabolic Products.....99
2.4.2 Maintenance and Endogenous Metabolism.....100
2.5 Qualitative Time Profiles of Microbial Growth and Product Formation.....100
2.6 Processes and Reactors Used in Fermentation.....101
2.7 Starting Materials.....106
2.8 Culture Cultivation and Maintenance.....107
2.9 Extraction and Separation of Metabolites.....107
2.10 Sterilisation of Media and Equipment.....108
2.11 Geometry and Materials for Fermentation Vessels.....109

Chapter 3 Fermentation Processes: Kinetics and Stoichiometry 110

3.1 Kinetics of Fermentation Processes 110
3.2 Effect of Inhibitors, Temperature and pH on the Growth Rate 117
3.3 Experimental Measurement of Kinetic Parameters 120
3.4 Yield Coefficients and Growth Stoichiometry 123
3.5 Rate Equations for Secondary Metabolites, Maintenance and Endogenous Metabolism 127

Chapter 4 Fermentation Processes: Mass Balances and Design for Batch, Continuous and Fed-Batch Reactors 139

4.1 General Equation for Mass Balances 139
4.2 Mass Balances for Batch Reactors 140
4.3 Mass Balances for Continuous Reactors 143
4.4 Mass Balances for Fed-Batch Reactors 149
4.5 Modifications of the Mass Balances for Secondary Metabolites, Maintenance and Endogenous Metabolism 154
4.6 Use of Continuous Reactors for the Estimation of Kinetic Parameters 160
4.7 Comparison of Batch, Continuous and Fed-Batch Reactors 162

Chapter 5 Fermentation Processes: Oxygen Transfer 169

5.1 Theory 169
5.2 Measurement of Mass Transfer Coefficients 175
5.3 Correlations for Mass Transfer Coefficients 178
5.3.1 Correlations for Power Draw in Agitated Vessels 182
5.4 Oxygen Balances in Batch Reactors 191
5.5 Oxygen Balance in Continuous Reactors 198
5.6 Considerations for Scale-Up and Scale-Down 203

Chapter 6 Fermentation Processes: Heat Generation and Heat Balances 217

6.1 Enthalpy of Reaction 217
6.2 Rate of Heat Generation of Fermentation Reactions 222
6.3 Heat Balances 228
6.3.1 Heat Balances for Continuous Fermentation 228
6.3.2 Heat Balances for Batch Fermentation 245
6.4 Effect of Variable Temperature on Reaction Rates 250
6.5 Heat Transfer Coefficients in Agitated Vessels 251
6.6 Considerations for Scale-Up and Scale-Down 256

Chapter 7 Downstream Processes 266

7.1 Separation of Microorganisms and Crossflow Filtration 266
7.1.1 Separation of Microorganisms 266
7.1.2 Crossflow Filtration 268

7.2 Ion exchange 274
7.2.1 Introduction 274
7.2.2 Ion Exchange Resins and Column Operation........... 274
7.2.3 Regeneration... 277
7.2.4 Ion Exchange Equilibria....................................... 279
7.2.5 Mass Transfer and Mass Balances in Ion Exchange 283
7.3 Adsorption... 297
7.3.1 Introduction 297
7.3.2 Adsorption Equilibrium....................................... 299
7.3.3 Mass Transfer and Mass Balances in Adsorption..... 300
7.4 Chromatography 307
7.4.1 Introduction 307
7.4.2 Mass Transfer and Mass Balances in Chromatography... 307

Chapter 8 Design Summary and Examples of Industrial Fermentation Processes 316

8.1 Summary of Design Considerations 316
8.2 Bioethanol 317
8.3 Penicillin 322
8.4 Whisky 323
8.5 Citric Acid 324
8.6 L-Glutamic Acid 325
8.7 Cell-based Virus Vaccines 326
8.8 Processes with Undefined Mixed Cultures 328
8.9 Production of Cellulases 330

Appendices 335
Bibliography 423
Index 427

Preface

This book is based on my course EX4011 Biochemical Engineering at University of Aberdeen. The course is for level 4 undergraduate students in chemical engineering. In this second edition, I added two chapters, on enzymatic reactions (Chapter 1) and downstream processes (Chapter 7). I kept the other chapters virtually the same as for the first edition, with minor edits. The book covers the kinetics and design of enzymatic and fermentation processes, defined in a broad sense as any industrial processes that use enzymes or living microorganisms/cells, both under aerobic and anaerobic conditions. Chapter 1 covers enzymatic reaction, with focus on reaction kinetics and mass balances under various conditions (free and immobilised enzymes, effect of temperature, pH and inhibition). Chapters 2–6 cover fermentation processes. After a concise introduction on microorganisms and their metabolism (Chapter 2), rate equations and stoichiometry of fermentation processes are covered in detail (Chapter 3). Chapter 4 shows how to derive and use mass balances for the design of the most common types of fermentation processes. Chapter 5 covers oxygen transfer and oxygen mass balances, going from the fundamental theory to the use of oxygen transfer concepts to design and scale-up/scale-down aerobic fermentations. Chapter 6 does the same for heat transfer and includes heat generation, heat transfer and heat balances, showing how to use these concepts for the design and scale-up/scale-down of fermentation processes. Chapter 7 covers some of the separation and purification processes, which are often carried out after fermentation or enzymatic reactions. The final chapter (Chapter 8) shows some industrially relevant process examples. The book includes over 100 solved examples, questions and problems. The examples are solved in their respective chapters, answers and solutions to questions and problems are provided in the Appendix. The Appendix also covers the solution of differential equations and of systems of equations in Excel.

A key feature of this book is the study of enzymatic and fermentation processes with the same tools used for the design of nonbiological chemical reaction processes, i.e. kinetics and stoichiometry, mass and energy balances, mass and heat transfer. Therefore, this book can be considered an extension of chemical reaction engineering concepts to biochemical engineering processes. The design of biochemical engineering processes shares many similarities with the design of nonbiological processes, however there are also important differences due to peculiarities of biological reactions, e.g. the form of reaction kinetics and the autocatalytic nature of fermentation reactions. This book covers these similarities and differences, showing the specific aspects of process design for biochemical engineering processes.

Like most models used in engineering, the value of the models and equations presented in this book is not that they give us accurate numerical results. In fact, mathematical models of enzymatic and fermentation processes are inevitably great simplifications of very complex biochemical and physical processes and should not be expected to give accurate numerical predictions. As with other areas of engineering, the value of models and equations applied to biochemical engineering processes is that they help us in identifying the effect of operating variables on process performance and so they help us with process understanding, design and optimisation. With this is mind, models and equations are essential tools in the study of enzymatic and fermentation processes but should always be used together with experimental validation.

Compared to other textbooks on biochemical engineering, this book is perhaps less comprehensive but covers mass balances, oxygen and heat transfer and scale-up/ scale-down topics in more detail, more systematically and with more examples and problems.

I believe that this book can be a useful resource as a textbook for undergraduate and graduate students in chemical engineering and can also be useful to process engineers working with fermentation reactions.

I wish to thank my publisher Dr Gagandeep Singh at CRC Press for his support with the publication process over the years. I am greatly indebted to my students of the course Biochemical Engineering at University of Aberdeen for their useful comments over the years on my teaching materials, which are the basis for this book. I also acknowledge the help and assistance of my PhD students Ifeoluwa Bolaji, Chinedu Casmir Etteh, Rita Noelle Moussa, Adamu Rasheed, Igor Silva, Serena Simonetti, Bhattarai Bishnu, who over the years contributed to the teaching of my course and to the revision of this book. A special thanks to Serena Simonetti and Rita Noelle Moussa for the prompt and careful revision of the draft chapters of the first edition. I would also like to thank my family in Italy and my wife Federica for her support during all the weekends that took to complete the writing of this book.

About the Author

Davide Dionisi is Professor of Chemical Engineering at University of Aberdeen, where he teaches biochemical engineering and energy from biomass for undergraduate and postgraduate programmes. He also teaches a CPD course on biological wastewater treatment and contributes to teaching chemical engineering design projects and to project supervision. He is co-ordinator of the MSc programmes in Renewable Energy Engineering and in Sustainable Product, Process and Energy Engineering. Before joining the University of Aberdeen in 2012, Professor Dionisi worked as Lecturer for the School of Mathematical, Physical and Natural Sciences at Sapienza University, Rome, Italy, and as Process Engineer/Principal Process Engineer for Syngenta, Berkshire, UK.

Professor Dionisi's research focuses on biological processes for wastewater treatment and for waste valorisation, e.g. activated sludge, anaerobic fermentation and anaerobic digestion, including new processes to convert the organic matter into sustainable and renewable products, e.g. bioplastics, organic acids, hydrogen. Professor Dionisi has worked in over 20 research projects with a total funding of over £2m. He has published over 70 papers on peer-reviewed journals, two patents, two books as only author and has given invited lectures to many national and international conferences. Overall, Professor Dionisi's scientific outputs have received over 3500 citations. Recently, the focus of Professor Dionisi's research has been about bringing the new processes he has developed together with his colleagues, which combine biological and chemical processes, closer to commercialisation.

Professor Dionisi is a Fellow of the IChemE (Institution of Chemical Engineers) and contributes to the IChemE magazine TCE (The Chemical Engineer). He also writes articles on LinkedIn on teaching and research matters.

1 Enzymatic Processes

This chapter introduces enzymes and their chemical nature, structure and applications. Enzymes are used in many large-scale processes, e.g. bioethanol production, in the food industry and in many other industries. Most of this chapter is focused on the kinetics of enzyme-catalysed reactions. In the end, enzymes are catalysts (biological catalysts) and processes with enzymatic reactions should be designed with the same tools, i.e. reaction kinetics, mass and energy balances, used for the study of any chemical reaction processes. However, these tools need to be adapted to the particular nature of enzymatic reactions, and this will be the main object of this chapter.

1.1 INTRODUCTION TO ENZYMES

This section gives a general introduction about enzymes, with the main concepts that are necessary to understand the modelling of the rate of enzymatic reactions, which will be the main focus of the other sections in this chapter.

1.1.1 Enzyme Nature, Chemistry and Structure

Enzymes are biological molecules, i.e. produced by living organisms, that catalyse chemical reactions. Chemically, enzymes are proteins, i.e. they are polymers of amino acids. Amino acids are described by the general empirical formula $NH_2CHRCOOH$, where R is attached to the CH group and represents a side chain that has a different structure for different amino acids. Table 1.1 shows the R group for the different amino acids. Enzymes are made of hundreds or thousands of amino acids joined together and can have molecular weights (MW) of up to more than 1×10^6 g/mol (or 1×10^6 Da), with complex three-dimensional shapes.

In enzymes, amino acids are linked together in one or more polypeptide chains. The sequence of amino acids in the polypeptides is called the primary structure. The localised chain structure of the polypeptides is called the secondary structure. The term tertiary structure refers to the three-dimensional arrangement of individual polypeptide structures, while the quaternary structure refers to the three-dimensional arrangement of the whole enzyme, composed of one or more polypeptide chains.

Some enzymes are only activated when a cofactor (metal ions, e.g. Zn^{2+}, Ca^{2+}, Mg^{2+}, or organic compounds) attaches to them. Enzymes are usually highly soluble in water although the solubility is dependent on environmental conditions such as pH, temperature and presence of salts.

DOI: 10.1201/9781003633822-1

TABLE 1.1
Side Chain for the Different Amino Acids

Amino acid	Formula of the R group
Arginine	$CH_2CH_2CH_2NHCNH_2NH$
Histidine	$CH_2(C_3H_2NHN)$
Lysine	$CH_2CH_2CH_2NH_2$
Aspartic acid	CH_2COOH
Glutamic acid	CH_2CH_2COOH
Serine	CH_2OH
Threonine	$CHOHCH_3$
Asparagine	CH_2CONH_2
Glutamine	$CH_2CH_2CONH_2$
Cysteine	CH_2SH
Selenocysteine	CH_2SeH
Glycine	H
Proline	CH_2CH_2CH
Alanine	CH_3
Valine	$CHCH_3CH_3$
Isoleucine	$CHCH_3CH_2CH_3$
Leucine	$CH_2CHCH_3CH_3$
Methionine	$CH_2CH_2SCH_3$
Phenylalanine	$CH_2(C_6H_5)$
Tyrosine	$CH_2(C_6H_4)OH$
Tryptophan	$CH_2(C_8H_5NH)$

1.1.2 Enzymatic Catalysis and Mechanisms

Enzymes are biological catalysts, which, like all catalysts, decrease the energy required for the reaction to occur without changing the energy of the reagents and products. Like all catalysts, enzymes modify (increase) the reaction rate (they decrease the activation energy) without changing the thermodynamics of the reaction (the reaction enthalpy, reaction free energy, etc are not affected by the presence of enzymes). Figure 1.1 shows a conceptual scheme of the effect of catalysts on the progression of a reaction. In the presence of a catalyst, the energy level of reagents and products does not change, but the energy barrier to be overcome for the reaction to occur is greatly reduced.

The mechanisms of enzyme catalysis can be complex, however in general the catalytic action occurs on one particular site of the enzyme, called the active site, which binds to the reagent (substrate) molecule forming the enzyme/substrate complex (the enzyme/substrate complex can include more than one substrate if the reaction involves more than one reagent). The general model for enzyme-substrate interaction is often called the lock-and-key model (Figure 1.2).

Once the enzyme/substrate complex is formed, the enzyme can destabilise the substrate(s) by different mechanisms depending on the enzyme/reaction: proximity effect (bringing the substrates or different parts of the same substrate close

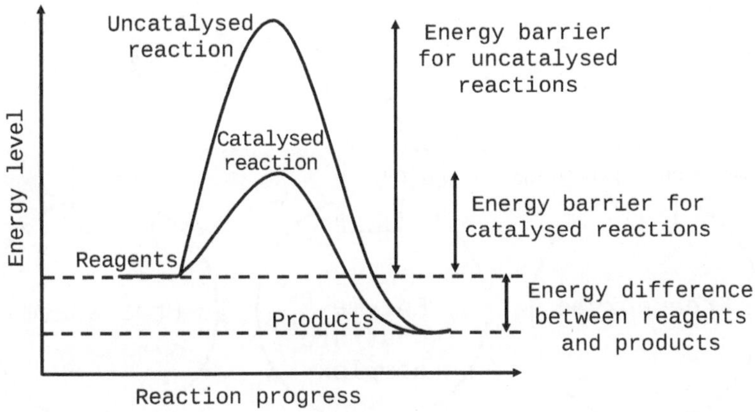

FIGURE 1.1 Conceptual scheme of the effect of catalysts on the reaction progression.

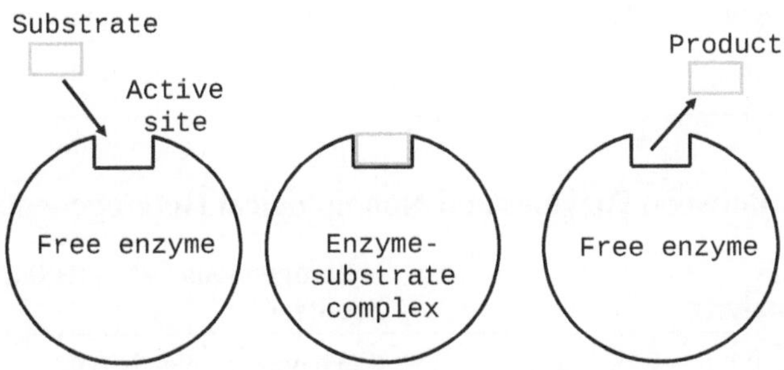

FIGURE 1.2 Conceptual sketch of the lock-and-key model for enzymatic catalysis.

enough to react), orientation effect (binding substrates in favourable position), acid-base catalysis (donating or accepting protons from the substrate(s)) and many other mechanisms. Figure 1.3 shows a model of the mechanism of acid-base catalysis for the enzymatic hydrolysis of a dipeptide.

Compared to nonbiological catalysts, e.g. metals, enzymes are usually more specific (each enzyme only catalyses one reaction or a limited number of reactions), more subject to inhibition by other molecules and work in a narrower temperature range. Table 1.2 summarises the comparison between enzymes and nonbiological catalysts.

1.1.3 Enzyme Immobilisation and Materials

Free enzymes are dissolved in the reaction medium and are frequently used in industrial processes. However, free enzymes have some limitations:

- They cannot be easily separated from the reaction medium, therefore they cannot be re-used, which increases the cost.
- They are mixed with the reaction products and may decrease their purity.

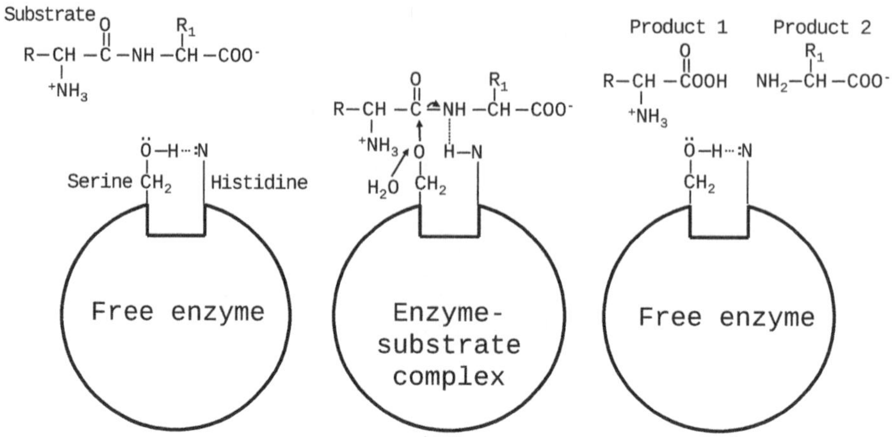

FIGURE 1.3 Conceptual sketch of the acid-base mechanism for the enzymatic hydrolysis of a dipeptide with the enzyme chymotrypsin. (Adapted from J.E. Bailey, D.F. Ollis, *Biochemical Engineering Fundamentals*, McGraw-Hill Chemical Engineering Series, Second Edition, 1986.)

TABLE 1.2
Comparison Between Enzymes and Nonbiological Heterogeneous Catalysts

Enzymatic Catalysts	Heterogeneous Catalysts (e.g. Metals, Zeolites)
Very high selectivity	Can have low selectivity
Work at approximately room temperature	Often need high temperatures
Only work in water-based media	Can work in water and organic solvents
Only work for biological reactions	Can work for any type of reaction

- They are exposed to environmental factors (e.g. inhibitors) that may decrease their activity.
- They cannot be combined with other enzymes for multi-step reactions that require multiple enzymes.

To overcome these limitations of free enzymes, in some applications enzymes are immobilised on solid (insoluble) supports so that they can be retained in the reaction medium and easily separated from the products. Supports can be porous or nonporous. In porous supports, enzymes are attached to the external surface and inside the pores, while in nonporous supports the enzymes are only attached to the external surface of the support. Immobilisation reduces the need for new enzymes and avoids product contamination with the enzymes. Also, immobilised enzymes retain in some cases their activity for longer than free enzymes and can be combined with other enzymes on the same solid to perform multi-step reactions requiring multiple enzymes.

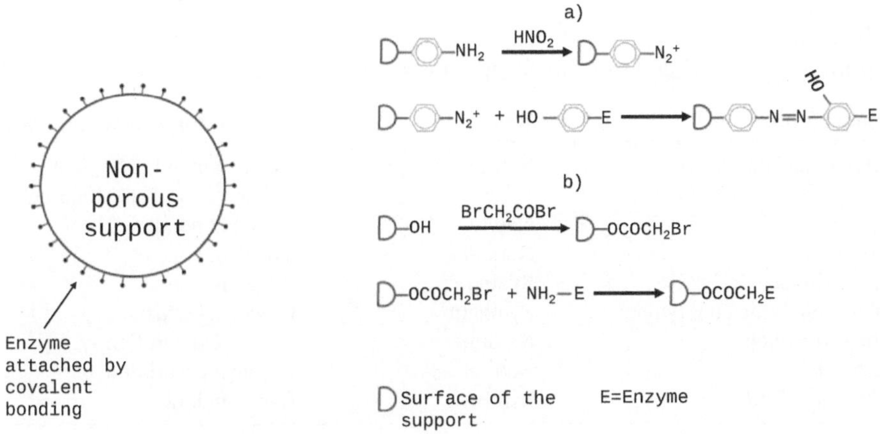

FIGURE 1.4 Conceptual sketch of enzymes attached to the external surface of a nonporous support and examples of covalent bonding: (a) diazo method; (b) alkylation method. (Adapted from J.E. Bailey, D.F. Ollis, *Biochemical Engineering Fundamentals*, McGraw-Hill Chemical Engineering Series, Second Edition, 1986.)

Enzymes can be immobilised using chemical or physical methods. In chemical methods, covalent bonds are formed between the support and the enzyme. In physical methods, the enzyme is adsorbed on the support by means of physical adsorption or ionic binding (or with other mechanisms). Figure 1.4 shows a sketch of enzymes attached by covalent bonding on the external surface of a nonporous support and some chemical mechanisms for covalent bonding between supports and enzymes.

Solid supports can be natural or synthetic materials. Table 1.3 shows some materials used as supports for enzymes and the immobilisation methods. Supports for enzyme immobilisation must be chosen carefully for each enzymatic reaction: supports must be insoluble in the reaction medium (water), must be able to bind effectively to high concentrations of enzyme, must be resistant to changes in pH, temperature and medium composition and must be easily separated from the reaction medium. Laboratory experiments are essential to test the suitability of a particular support for the enzymatic reaction under consideration.

1.1.4 Applications of Enzymes

Enzymes are produced naturally inside cells in living organisms and are essential catalysts in almost all reactions occurring in living cells. In this book we will consider industrial applications of enzymes, i.e. applications of enzymes that have been extracted and purified from the living cells that produced them. Enzymes are used in many industrial sectors, e.g. biofuels (bioethanol production), detergents, food and drink, textiles.

Enzymes are extracted and separated from cells by killing and splitting the cells (e.g. by grinding and/or lysing) and by separating the enzymes from the other cell

TABLE 1.3
Examples of Materials Used for Support for Immobilised Enzymes

Support aterial	Type	Type of Bond with Enzymes
Cellulose and derivatives	Natural	Covalent bonds, physical interaction, ionic binding
Agarose	Natural	Covalent bonds
Dextran	Natural	Covalent bonds
Polyacrylamide derivatives	Synthetic	Covalent bonds
Maleic anhydride copolymers	Synthetic	Covalent bonds
Activated carbon	Natural	Physical adsorption
Silica gel	Natural	Physical adsorption
Amberlite CG50	Synthetic	Ionic binding

TABLE 1.4
Examples of Enzymes and of Their Applications

Enzyme	Source	Reaction	Examples of Application
Amylase	*Bacillus subtilis* and other microorganisms	Starch hydrolysis	Bioethanol production, glucose production, detergents
Cellulase	*Trichoderma viride* and other microorganisms	Cellulose hydrolysis	Textile industry, biofuels, animal feed and other industries
Trypsin	Animal pancreas	Protein hydrolysis	Food industry (e.g. infant formula milk), medical uses
Pepsin	Animal stomach	Protein hydrolysis	Food industry
Glucose oxidase	*Aspergillus niger* and other microorganisms	Glucose removal	Food industry, medical uses (e.g. glucose detection)
Lipase	Many microorganisms, pancreas	Hydrolysis of triglycerides	Food industry, biofuels, detergents, textile and other industries

components. We will see some examples of how enzymes for industrial applications are produced from cells in Chapter 8.

In industry, enzymes are used with different degrees of purity depending on the application. Some enzymes are needed with high purity (e.g. clinical diagnosis) in other cases mixtures of different enzymes and other components are acceptable. Enzyme names usually refer to the function of the enzyme and end with the suffix "-ase", although in some cases they retain the name of the protein (e.g. trypsin). As we have already seen, the reagent of enzymatic and more in general of biological reactions is often called the "substrate". Table 1.4 shows some examples of enzymes and of their applications.

1.2 ENZYME KINETICS WITH THE MICHAELIS-MENTEN MODEL

In this section we will derive an equation for the rate of enzymatic reactions. The rate equation represents a relationship between the rate of the reaction and the main factors that influence it. Like most chemical reactions of any types, the rate of enzymatic reactions is affected by the concentration of the limiting reagent (substrate). Also, the rate is expected to be affected by the concentration of the enzyme and, like all chemical reactions, by the temperature. First, we will consider the effect of the substrate and of the enzyme concentrations using the most common model for the kinetics of enzymatic reactions, the Michaelis-Menten model. This model was developed in the early 1900s.

1.2.1 Derivation of the Michaelis-Menten Kinetics

Let us consider the simplest enzymatic reaction, $S \rightarrow P$, and assume that the reaction mechanism can be described well enough by the lock-and-key model.

We want to find an expression for the substrate removal rate $\left(-r_S\right) = f\left(S, E_0\right)$. The substrate removal rate r_S is negative because the substrate is removed by the reaction, and $\left(-r_S\right)$ is positive. Units of $\left(-r_S\right)$ are, e.g. $\frac{\text{mol}}{\text{L} \cdot \text{min}}$ or more in general $\frac{\text{mass or mol}}{\text{volume} \cdot \text{time}}$.

The Michaelis-Menten model is based on the assumption that the free enzyme (concentration [E]) and the substrate (concentration [S]) react with a reversible reaction (reactions 1 and –1, with kinetic parameters k_1 and k_{-1}) to form the enzyme-substrate complex (concentration [ES]), which then is converted (reaction 2, with kinetic parameter k_2) to the product (concentration [P]) and to the free enzyme in an irreversible reaction. The total enzyme (concentration [E_0]) is the sum of the free enzyme and of the enzyme, which forms the enzyme-substrate complex. Note that in this chapter concentrations will be indicated with or without square brackets [] with the same meaning (e.g. [S] or just S).

This reaction model is shown by equations (1.1) –(1.3):

$$E + S \underset{k_{-1}}{\overset{k_1}{\leftrightarrow}} ES \overset{k_2}{\rightarrow} E + P \tag{1.1}$$

$$\left(-r_S\right) = r_P = k_2\left[ES\right] \tag{1.2}$$

$$\left[E_0\right] = \left[E\right] + \left[ES\right] \tag{1.3}$$

Based on this model, the Michaelis-Menten kinetics can be derived using two different assumptions. What we will call "Assumption 1" states that reactions 1 and –1 are very fast compared to reaction 2 and can be assumed to be at equilibrium. On the other hand, "Assumption 2" states that the concentration of the enzyme/substrate complex ES is constant in all cases. Note that the two assumptions are not the same. According to Assumption 1, the concentration of the enzyme-substrate complex [ES], and the concentrations of free enzyme and substrate [E] and [S] are in equilibrium, therefore [ES] changes as [E] and [S] change. Instead, according to Assumption 2, $[ES]$ does not change during the reaction even though $[E]$ and $[S]$ change.

According to Assumption 1, since we are assuming that reactions 1 and −1 are in equilibrium, we can write the equilibrium constant as

$$\frac{[ES]}{[E][S]} = K_{eq} = \frac{k_1}{k_{-1}} \tag{1.4}$$

Therefore, $[ES] = K_{eq}[E][S]$ and $(-r_S) = r_P = k_2[ES] = k_2K_{eq}[E][S]$.

The concentration of free enzyme $[E]$ is given by the concentration of the total enzyme $[E_0]$ minus the concentration of the enzyme bound to the substrate, therefore, $[E] = [E_0] - [ES]$. By substituting the equation for $[ES]$ derived above, we can write

$$[E] = [E_0] - K_{eq}[E][S] \tag{1.5}$$

which can be re-arranged as

$$[E] = \frac{[E_0]}{1 + K_{eq}[S]} \tag{1.6}$$

Equation (1.6) can be substituted into the equation for $(-r_S)$ derived above to give

$$(-r_S) = k_2K_{eq}\frac{[E_0]}{1 + K_{eq}[S]}[S] \tag{1.7}$$

which can be re-arranged as

$$(-r_S) = k_2\frac{[E_0][S]}{\frac{1}{K_{eq}} + [S]} \tag{1.8}$$

Equation (1.8) in turn becomes

$$(-r_S) = k_2\frac{[E_0][S]}{K_m + [S]} \tag{1.9}$$

where $K_m = \frac{1}{K_{eq}} = \frac{k_{-1}}{k_1}$.

Equation (1.9) represents the relationship between the rate of the enzymatic reaction and the concentration of enzyme and substrate that we were looking for. Equation (1.9) was derived according to Assumption 1. We will now see how the model represented by equations (1.1)–(1.3) can lead to a rate equation by using Assumption 2.

According to Assumption 2, the assumption of constant [*ES*] means that [*ES*] does not change with time, i.e.

$$\frac{d[ES]}{dt} = k_1[E][S] - k_{-1}[ES] - k_2[ES] = 0 \tag{1.10}$$

Equation (1.10) can be re-arranged as $[ES] = \frac{k_1[E][S]}{k_{-1}+k_2}$. Therefore, we can write $(-r_S) = r_P = k_2[ES] = k_2\frac{k_1[E][S]}{k_{-1}+k_2}$. As done for Assumption 1, we can still assume that $[E] = [E_0] - [ES]$, which in this case can be re-arranged as

$$[E] = [E_0] - \frac{k_1[E][S]}{k_{-1}+k_2} \tag{1.11}$$

and then as

$$[E] = \frac{[E_0]}{1+[S]\frac{k_1}{k_{-1}+k_2}} \tag{1.12}$$

Therefore, according to Assumption 2, $(-r_S)$ can be written and re-arranged as

$$(-r_S) = k_2\frac{k_1[S]}{k_{-1}+k_2}\frac{[E_0]}{1+[S]\frac{k_1}{k_{-1}+k_2}} = k_2\frac{k_1[S]}{k_{-1}+k_2}\frac{[E_0](k_{-1}+k_2)}{k_{-1}+k_2+[S]k_1} = k_2\frac{[E_0][S]}{\frac{k_{-1}+k_2}{k_1}+[S]} \tag{1.13}$$

which, can also be written as

$$(-r_S) = k_2\frac{[E_0][S]}{\frac{k_{-1}+k_2}{k_1}+[S]} = k_2\frac{[E_0][S]}{K_m+[S]} \tag{1.14}$$

where $K_m = \frac{k_{-1}+k_2}{k_1}$.

Equations (1.9) and (1.14) are the same equation, which is called the Michaelis-Menten kinetics. This shows that Assumptions 1 and 2, although they are different, in practice give the same result. Note that the meaning of the parameter K_m is different according to Assumptions 1 and 2. According to Assumption 1, $K_m = \frac{k_{-1}}{k_1}$ instead, according to Assumption 2, $K_m = \frac{k_{-1} + k_2}{k_1}$. However, in practical engineering cases this difference does not matter. In practice, K_m is a parameter to be determined from experimental data, regardless of its chemical meaning. The two derivations of the same equation according to two different assumptions indicate that we cannot determine which one of Assumptions 1 or 2 applies to an enzymatic reaction, which follows the Michaelis-Menten kinetics. This principle applies much more widely in the field of chemical reaction kinetics. The fact that the rate of a reaction follows a certain kinetic model derived according to a certain mechanism, is not in itself enough proof that the reaction follows that particular mechanism. Indeed, the same or a very similar (and practically indistinguishable due to the inevitable dispersion of the experimental data) equation could be obtained with a different mechanism or with different assumptions. On the other hand, if the rate of a reaction does not follow the equation derived under a certain mechanism and assumptions, it is a definite prove that the hypothesised mechanism and/or assumptions are not applicable to that reaction. In order to prove that a reaction mechanism is the correct one for a particular reaction, in addition to the experimental data fitting the derived equation other evidence is needed, e.g. measurement of the time profiles of the intermediates, advanced analytical techniques. However, for many engineering problems, it is sufficient to have an equation that describes the rate of the reaction under a sufficiently wide range of substrate (reagent) concentration and temperature, without having definite evidence of the reaction mechanism. In this section we will follow this approach, and we will now analyse the Michaelis-Menten rate equation without further considerations about the reaction mechanism.

1.2.2 Analysis of the Michaelis-Menten Kinetics

We have seen that the rate equation $(-r_S) = k_2 \frac{[E_0][S]}{K_m + [S]}$ is called the Michaelis-Menten kinetics and is used to describe the rate of enzymatic reactions. In this rate equation, $[S]$ is the concentration of substrate, expressed in mol/L, g/L or other equivalent units; $[E_0]$ is the concentration of the total enzyme, expressed in mol/L, g/L or other equivalent units. Note that the mass of enzymes is sometimes expressed in units of activity, simply referred as units, (rather than in mol or g), because of the difficulty of isolating pure enzymes and in knowing their molecular weight. One unit of an enzyme is defined as the amount of enzyme that gives a specified reaction rate under certain conditions. For example, for the enzyme glucoamylase, one unit can be defined as the mass of enzyme that produces 1 mmol of glucose per minute starting from a 4% solution of soluble starch at 60 ^{0}C and pH 4.5 (other definitions of one unit for the same enzyme, glucoamylase, are also possible).

k_2 and K_m are parameters to be determined, for the particular enzyme and for the particular reaction and reaction conditions, from experimental data. The units of K_m have to be the same as the units of S, while the units of k_2 depend on the units used for $\left[E_0\right]$ and have to be consistent with the desired units for r_S. For example, if r_S is in $\frac{\text{g}}{\text{L} \cdot \text{min}}$ and E_0 is in g/L, then k_2 will be in $\frac{\text{g}}{\text{g}_{\text{enzyme}} \cdot \text{min}}$.

According to the Michaelis-Menten kinetics, if the substrate concentration is very large, i.e $\left[S\right] \gg K_m$ then $\left(-r_S\right) \cong k_2\left[E_0\right]$. The maximum reaction rate $k_2\left[E_0\right]$ when the substrate is in excess is also called V_{max}. Under these conditions, the substrate removal rate is independent (zero order) of the substrate concentration. On the other hand, if the substrate concentration is very low, $\left[S\right] \ll K_m$, then $\left(-r_S\right) \cong \frac{V_{max}\left[S\right]}{K_m}$, i.e. the substrate rate is first order with the substrate concentration. Numerically, the parameter K_m coincides with the value of the substrate concentration for which the substrate rate is half of its maximum value. Indeed, if $\left[S\right] = K_m$ then $\left(-r_S\right) = \frac{V_{max}}{2}$.

The typical profile of $\left(-r_S\right)$ vs $\left[S\right]$ corresponding to the Michaelis-Menten kinetics is shown in Figure 1.5.

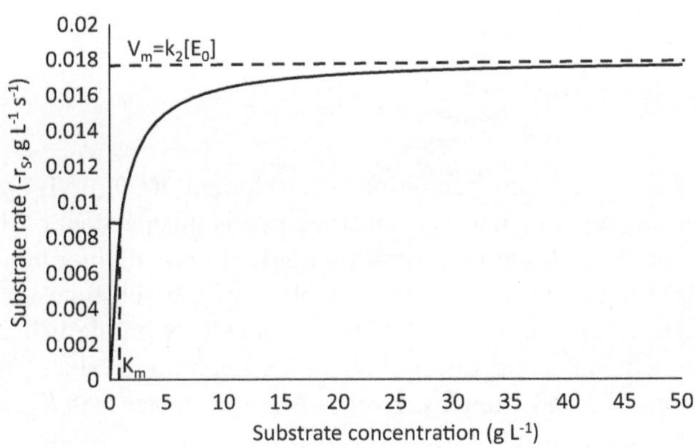

FIGURE 1.5 Typical profile of the substrate rate vs the substrate concentration for the Michaelis-Menten kinetics.

Example 1.1

Consider an enzymatic reaction, which follows the Michaelis-Menten kinetics with k_2 =1.3 min^{-1} and K_m=150 mg/L. The reaction is carried out with E_0 =100 mg/L. Calculate V_{max} and the substrate removal rate for S = 4000, 2000, 150, 5, 1 mg/L.

Solution

The substrate removal rate is in this case (with [S] expressed in mg/L)

$$(-r_S) = 1.3\,\text{min}^{-1}\,\frac{100\text{ mg L}^{-1}[S]}{50\text{ mg L}^{-1} + [S]}$$

V_{max} is the maximum rate of substrate removal when the substrate is in excess:

$$V_{max} = k_2 E_0 = 1.3\,\text{min}^{-1} \cdot 100\text{ mg L}^{-1} = 130\,\frac{\text{mg}}{\text{L min}}$$

For the given values of S:

$S = 4000$ mg/L, $(-r_S) = 125.30\,\frac{\text{mg}}{\text{L min}}$

$S = 2000$ mg/L, $(-r_S) = 120.93\,\frac{\text{mg}}{\cdot\text{L min}}$

$S = 150$ mg/L, $(-r_S) = 65.00\,\frac{\text{mg}}{\text{L min}}$

$S = 5$ mg/L, $(-r_S) = 4.19\,\frac{\text{mg}}{\text{L min}}$

$S = 1$ mg/L, $(-r_S) = 0.86\,\frac{\text{mg}}{\text{L min}}$

Note that the substrate concentrations of 4000 and 2000 mg/L are very high and close to the region where the substrate rate is independent of the substrate concentration. Indeed, the rate for 4000 mg/L is only slightly higher than the rate for 2000 mg/L, in spite of the doubling of the substrate concentration. The substrate concentration of 150 mg/L coincides with the value of K_m and, as expected, the substrate rate is half of its maximum value. The substrate concentrations of 5 and 1 mg/L are very low, much lower than K_m, and the substrate rate is approximately linearly proportional to the substrate concentration (the rate of 4.19 mg L^{-1} min^{-1} for S = 5 mg/L is 4.87 times higher than the rate of 0.86 mg L^{-1} min^{-1} for S = 1 mg/L).

1.2.3 Kinetic Parameters from Experimental Data

We have seen that the kinetic parameters k_2 and K_m of the Michaelis-Menten equation $\left(-r_S\right)=k_2\frac{[E_0][S]}{K_m+[S]}$ are usually derived from experimental data. There are several methods to obtain these kinetic parameters from experimental data.

1.2.3.1 Lineweaver-Burk Plot

In this method, the enzymatic reaction is carried out in a batch reactor with known enzyme concentration $\left[E_0\right]$ and known initial substrate concentration S_{IN} (or S_0). The substrate concentration $[S]$ is measured as a function of the time t during the experiment. From the data of $[S]$ vs t, the reaction rate at any time step can be calculated as the difference between substrate concentration at two consecutive time steps divided by the length of the time step. This is shown in Table 1.5. Note that the substrate concentration difference ΔS is defined as $\Delta S=S\left(t=t_n\right)-S\left(t=t_{n-1}\right)$ and is negative, therefore $-\Delta S$ is positive.

In the Lineweaver-Burk method, the Michaelis-Menten kinetics is rearranged as

$$\frac{1}{\left(-r_S\right)}=\frac{K_m}{k_2\left[E_0\right]}\frac{1}{[S]}+\frac{1}{k_2\left[E_0\right]} \tag{1.16}$$

From equation (1.16), it follows that if we plot $\frac{1}{\left(-r_S\right)}$ vs $\frac{1}{[S]}$ (Lineweaver-Burk plot) we should obtain a line with slope equal to $\frac{K_m}{k_2\left[E_0\right]}$ and intercept equal to $\frac{1}{k_2\left[E_0\right]}$, from which we calculate k_2 and K_m. This is shown in Figure 1.6.

TABLE 1.5
Data Analysis Required for the Application of the Lineweaver-Burk Method

Time	Substrate Concentration	Substrate Rate $\left(-r_S\right)=\frac{-\Delta S}{\Delta t}$
0	S_{IN}	$\frac{S_{IN}-S_1}{t_1}$
t_1	S_1	$\frac{S_1-S_2}{t_2-t_1}$
t_2	S_2	$\frac{S_2-S_3}{t_3-t_2}$
t_3	S_3	…
…	…	…

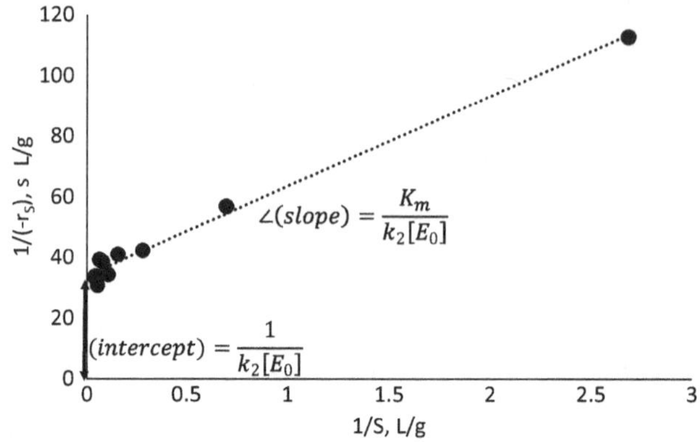

FIGURE 1.6 Lineweaver-Burk plot for the determination of the parameters of the Michalis-Menten equation.

The limitation of the Lineweaver-Burk method is that the value of the slope is strongly influenced by the data points at high $1/S$, i.e. at low S. The points at low S are often the most difficult to determine, and this can lead to inaccurate parameter estimation.

Example 1.2

A batch experiment is carried out with an enzyme concentration of 15 mg/L. The concentration of the substrate vs time is reported in Table 1.6. Determine the parameters of the Michaelis-Menten kinetics using the Lineweaver-Burk method.

TABLE 1.6
Experimental Data for Example 1.2

Time (h)	Substrate Concentration (mg/L)
0	199.86
0.1	149.32
0.2	102.02
0.3	62.65
0.4	33.95
0.5	15.22
0.6	5.73
0.7	2.15
0.8	0.79
0.9	0.27
1	0.09

Solution
To use the Lineweaver-burk plot, the experimental data must be re-arranged as $\frac{1}{\left(-r_S\right)}$ vs $\frac{1}{[S]}$. The substrate removal rate $\left(-r_S\right)$ should be calculated as in Table 1.5, obtaining Table 1.7. Note that, the way the rate $\left(-r_S\right)$ has been calculated according to Table 1.5, for the last data point the rate cannot be calculated.

From Table 1.7, the plot of $\frac{1}{\left(-r_S\right)}$ vs $\frac{1}{[S]}$ is shown in Figure 1.7.

The plot of $\frac{1}{\left(-r_S\right)}$ vs $\frac{1}{[S]}$ gives a slope of 0.14977 h and an intercept of 0.00154 L h mg^{-1}. We have therefore, according to the Lineweaver-Burk linearisation

TABLE 1.7
Re-Arrangements of the Experimental Data for the Lineweaver-Burk Plot

Time (h)	S (mg/L)	1/S (L/mg)	$-r_S$ (mg L^{-1} h^{-1})	1/($-r_S$) (h L mg^{-1})
0	199.86	0.00500	505.4	0.00198
0.1	149.32	0.00670	473	0.00211
0.2	102.02	0.00980	393.7	0.00254
0.3	62.65	0.01596	287	0.00348
0.4	33.95	0.02946	187.3	0.00534
0.5	15.22	0.06570	94.9	0.01054
0.6	5.73	0.17452	35.8	0.02793
0.7	2.15	0.46512	13.6	0.07353
0.8	0.79	1.26582	5.2	0.19231
0.9	0.27	3.70370	1.8	0.55556
1	0.09	11.11111		

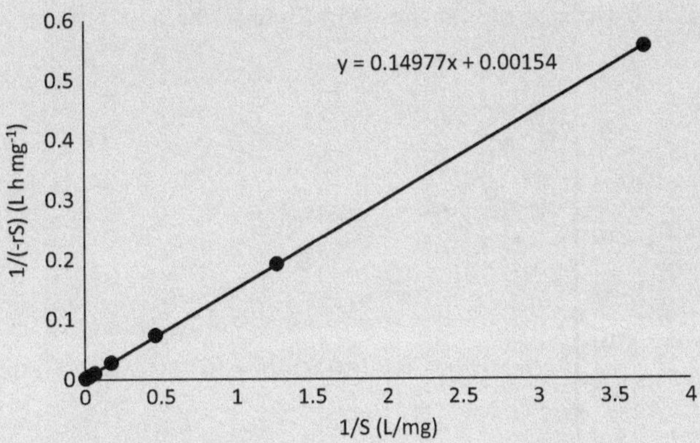

FIGURE 1.7 Lineweaver-Burk plot for Example 1.2.

$$\frac{1}{k_2[E_0]} = 0.00154\,\mathrm{L\ h\ mg^{-1}} \rightarrow k_2 = \frac{1}{0.00154[E_0]}$$

$$= \frac{1}{0.00154\,\mathrm{L\,h\,mg^{-1}} \cdot 15\,\mathrm{mg\ L^{-1}}} = 43.4\,\mathrm{h^{-1}}$$

and

$$\frac{K_m}{k_2[E_0]} = 0.14977h \rightarrow K_m = 0.14977h \cdot k_2[E_0]$$

$$= 0.14977\,\mathrm{h} \cdot 43.4\,\mathrm{h^{-1}} \cdot 15\,\mathrm{mg\,L^{-1}}$$

$$= 97.4\,\mathrm{mg\,L^{-1}}$$

1.2.3.2 Eadie-Hofstee Plot

The Eadie-Hofstee method is based on the same experimental procedure and data seen for the Lineweaver-Burk plot (batch reactor with measurement of S vs time) but uses a different linearisation procedure. In this method, the Michaelis-Menten kinetics is linearised as follows:

$$(-r_S)K_m + (-r_S)S = k_2[E_0][S] \Rightarrow (-r_S) = k_2[E_0] - \frac{(-r_S)K_m}{[S]} \tag{1.17}$$

From equation (1.17), it follows that if we plot $(-r_S)$ vs $\frac{(-r_S)}{[S]}$ we should obtain a line with slope equal to $-K_m$ and intercept equal to $k_2[E_0]$, from which we calculate k_2 and K_m. The Eadie-Hofstee plot is shown in Figure 1.8.

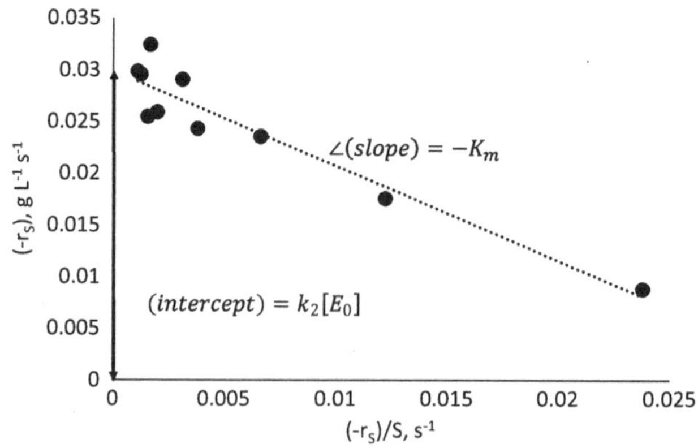

FIGURE 1.8 Eadie-Hofstee plot for the determination of the parameters of the Michalis-Menten equation.

Compared to the Lineweaver-Burk plot, in the Eadie-Hofstee plot the slope is mainly affected by the points at high S (low $1/S$), which is good as they are often the most accurate. However, the substrate rate, which is affected by a relatively large error, appears in both the x and y axes. This can reduce the accuracy of parameter determination.

Example 1.3

Using the experimental data from Example 1.2 and Table 1.6, determine the parameters k_2 and K_m of the enzymatic reaction with the Eadie-Hofstee method.

Solution

We need to re-arrange the experimental data from Table 1.6 to include $\left(-r_S\right)$ and $\frac{\left(-r_S\right)}{[S]}$, this is shown in Table 1.8.

The plot of $\left(-r_S\right)$ vs $\frac{\left(-r_S\right)}{[S]}$ according to the data in Table 1.8 is shown in Figure 1.9.

The plot of $\left(-r_S\right)$ vs $\frac{\left(-r_S\right)}{[S]}$ gives a slope of –130.19 mg/L and an intercept of 873.01 mg L^{-1} h^{-1}. We have therefore, according to the Eadie-Hofstee linearisation

TABLE 1.8
Re-Arrangements of the Experimental Data for the Eadie-Hofstee Plot

Time (h)	S (mg/L)	$-r_S$ (mg L^{-1} h^{-1})	$(-r_S)/S$ (h^{-1})
0	199.86	505.4	2.5288
0.1	149.32	473	3.1677
0.2	102.02	393.7	3.8590
0.3	62.65	287	4.5810
0.4	33.95	187.3	5.5169
0.5	15.22	94.9	6.2352
0.6	5.73	35.8	6.2478
0.7	2.15	13.6	6.3256
0.8	0.79	5.2	6.5823
0.9	0.27	1.8	6.6667
1	0.09		

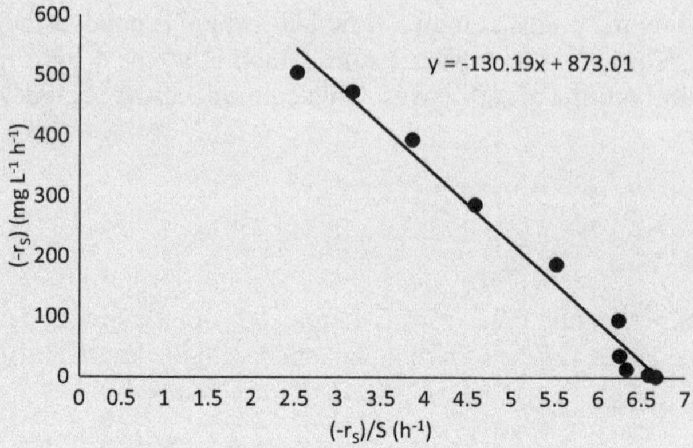

FIGURE 1.9 Eadie-Hofstee plot for Example 1.3.

$$k_2\left[E_0\right] = 873.01\,\text{mg}\,\text{L}^{-1}\text{h}^{-1} \rightarrow k_2 = \frac{873.01\,\text{mg}\,\text{L}^{-1}\text{h}^{-1}}{\left[E_0\right]}$$

$$= \frac{873.01\,\text{mgL}^{-1}\text{h}^{-1}}{15\,\text{mgL}^{-1}} = 58.2\text{h}^{-1}$$

and

$$K_m = -\left(-130.19\,\text{mg}\,\text{L}^{-1}\right) = 130.19\,\text{mg}\,\text{L}^{-1}$$

Note that the values obtained for k_2 and K_m are somewhat different from the values obtained in Example 1.2 with the Lineweaver-Burk method, with the same experimental data.

1.2.3.3 Hanes-Woolf Plot

The Hanes-Woolf method is based on the same experimental procedure and data seen for the Lineweaver-Burk plot but uses a different linearisation method. The re-arrangement of the Michaelis-Menten kinetics is in this case the following:

$$\frac{\left(-r_S\right)}{\left[S\right]} = \frac{k_2\left[E_0\right]}{K_m + \left[S\right]} \Rightarrow \frac{\left[S\right]}{\left(-r_S\right)} = \frac{K_m}{k_2\left[E_0\right]} + \frac{\left[S\right]}{k_2\left[E_0\right]} \tag{1.18}$$

From this re-arrangement, it follows that if we plot $\frac{\left[S\right]}{\left(-r_S\right)}$ vs $\left[S\right]$ we should obtain a linear trend with slope equal to $\frac{1}{k_2\left[E_0\right]}$ and intercept equal to $\frac{K_m}{k_2\left[E_0\right]}$, from which we calculate k_2 and K_m. The Eadie-Hofstee plot is shown in Figure 1.10.

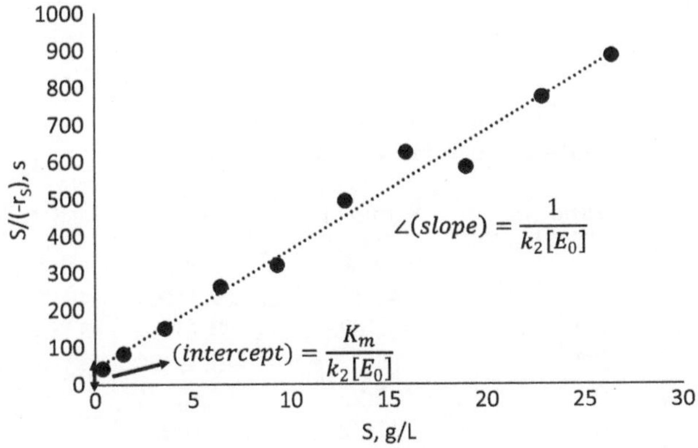

FIGURE 1.10 Hanes-Woolf plot for the determination of the parameters of the Michalis-Menten equation.

This method is often accurate in the determination of k_2, since this parameter is obtained directly from the slope, which is mainly affected by the points at high S, which are usually the least affected by error. However, this method can lead to inaccurate estimates of K_m as this parameter is obtained from the intercept, which is mainly affected by the points at low S.

Example 1.4

Using the experimental data from Example 1.2 and Table 1.6, determine the parameters k_2 and K_m of the enzymatic reaction with the Hanes-Woolf method.

Solution

We need to re-arrange the experimental data from Table 1.6 to include, according to equation (1.18), $\frac{[S]}{(-r_S)}$, this is shown in Table 1.9.

The plot of $\frac{[S]}{(-r_S)}$ vs $[S]$ according to the data in Table 1.9 is shown in Figure 1.11.

The plot of $\frac{[S]}{(-r_S)}$ vs $[S]$ gives a slope of 0.00118 h L mg^{-1} and an intercept of 0.14769 h. We have therefore, according to the Hanes-Woolf linearisation

$$\frac{1}{k_2[E_0]} = 0.00118\,\text{h}\,\text{L}\,\text{mg}^{-1} \rightarrow k_2 = \frac{1}{0.00118\,\text{h}\,\text{L}\,\text{mg}^{-1}[E_0]}$$

$$= \frac{1}{0.00118\,\text{h}\,\text{L}\,\text{mg}^{-1} \cdot 15\,\text{mg}\,\text{L}^{-1}} = 56.6\,\text{h}^{-1}$$

TABLE 1.9
Re-Arrangements of the Experimental Data for the Hanes-Woolf Plot

Time (h)	S (mg/L)	$S/(-r_S)$ (h)
0	199.86	0.39545
0.1	149.32	0.31569
0.2	102.02	0.25913
0.3	62.65	0.21829
0.4	33.95	0.18126
0.5	15.22	0.16038
0.6	5.73	0.16006
0.7	2.15	0.15809
0.8	0.79	0.15192
0.9	0.27	0.15000
1	0.09	

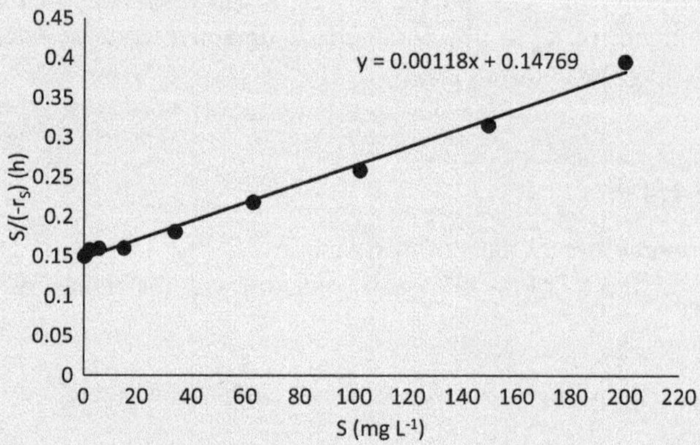

FIGURE 1.11 Hanes-Woolf plot for Example 1.4.

and

$$\frac{K_m}{k_2\left[E_0\right]} = 0.14769\,\text{h} \rightarrow K_m = 0.14769\,\text{h}\cdot k_2\left[E_0\right]$$
$$= 0.14769\,\text{h}\cdot 56.6\,\text{h}^{-1}\cdot 15\,\text{mg}\,\text{L}^{-1} = 125.4\,\text{mg}\,\text{L}^{-1}$$

Note that the values obtained for k_2 and K_m are somewhat different from the values obtained with other methods, with the same experimental data.

1.2.3.4 Integration Method

The integration method uses the same type of experiments used by the Lineaweaver-Burk, Eadie Hofstee and Hanes-Woolf methods, but uses a different linearisation method. In this method, the re-arrangement of the Michaelis-Menten kinetics is the following:

$$\frac{dS}{dt} = r_S = -\frac{k_2[E_0]}{K_m + [S]} \Rightarrow (\text{integrating}) - \frac{dS}{[S]}(K_m + [S])$$
$$= k_2[E_0]dt \Rightarrow \int_{S_{IN}}^{S} -\frac{dS}{S}(K_m + S) \tag{1.19}$$
$$= k_2 E_0 \int_0^t dt \Rightarrow -K_m ln\frac{S}{S_{IN}} - (S - S_{IN}) = k_2 E_0 t$$

The final form of the previous equation can be re-arranged as

$$\frac{1}{t} ln\frac{S_{IN}}{S} = \frac{k_2 E_0}{K_m} - \frac{(S_{IN} - S)}{t}\frac{1}{K_m} \tag{1.20}$$

This equation indicates that if we plot $\frac{1}{t} ln\frac{S_{IN}}{S}$ vs $\frac{(S_{IN} - S)}{t}$ we should obtain a linear trend with a slope equal to $-\frac{1}{K_m}$ and intercept equal to $\frac{k_2 E_0}{K_m}$, from which we calculate k_2 and K_m. The integration method is shown in Figure 1.12.

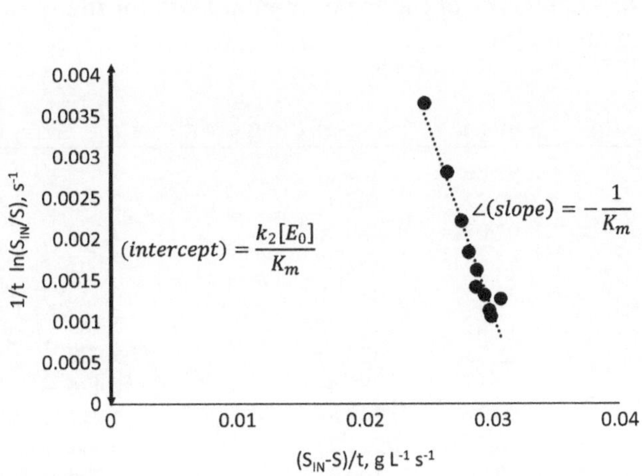

FIGURE 1.12 Plot of the integration method for the determination of the parameters of the Michalis-Menten equation.

This method is often more accurate than the previous ones as it does not require the calculation of $(-r_S)$, which is often less accurate. However, this method can be inaccurate if the experiment is very fast with small t values.

Example 1.5

Using the experimental data from Example 1.2 and Table 1.6, determine the parameters k_2 and K_M of the enzymatic reaction with the integration method.

Solution

We need to re-arrange the experimental data from Table 1.6 to include, according to equation (1.20), $\frac{1}{t}\ln\frac{S_{IN}}{S}$ and $\frac{(S_{IN}-S)}{t}$. This is shown in Table 1.10.

The plot of $\frac{1}{t}ln\frac{S_{IN}}{S}$ vs $\frac{(S_{IN}-S)}{t}$ according to the data in Table 1.10 is shown in Figure 1.13.

The plot of $\frac{1}{t}ln\frac{S_{IN}}{S}$ vs $\frac{(S_{IN}-S)}{t}$ gives a slope of –0.0152 L mg^{-1} and an intercept of 10.7618 h^{-1}. We have therefore, according to the integration method

$$-\frac{1}{K_m} = -0.00152\,\mathrm{L\,mg^{-1}} \rightarrow K_m = \frac{1}{0.0052\,\mathrm{L\,mg^{-1}}} = 65.7\,\mathrm{mg\,L^{-1}}$$

TABLE 1.10
Re-Arrangements of the Experimental Data for the Integration Plot

Time (h)	S (mg/L)	$(S_{IN}$-S)/t (mg L^{-1} h^{-1})	$\frac{1}{t}ln\frac{S_{IN}}{S}$
0	199.86		
0.1	149.32	505.40	2.915
0.2	102.02	489.20	3.362
0.3	62.65	457.37	3.867
0.4	33.95	414.78	4.432
0.5	15.22	369.28	5.150
0.6	5.73	323.55	5.920
0.7	2.15	282.44	6.474
0.8	0.79	248.84	6.917
0.9	0.27	221.77	7.341
1	0.09	199.77	7.706

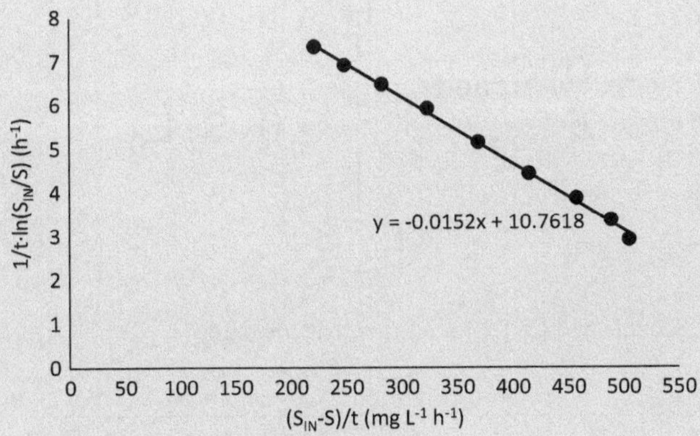

FIGURE 1.13 Integration plot for Example 1.5.

and

$$\frac{k_2 E_0}{K_m} = 10.7618\,\mathrm{h}^{-1} \rightarrow k_2 = \frac{10.7618\,\mathrm{h}^{-1}\cdot K_m}{[E_0]}$$

$$= \frac{10.7618\,\mathrm{h}^{-1}\cdot 65.7\,\mathrm{mg\,L}^{-1}}{15\,\mathrm{mg\,L}^{-1}} = 47.1\,\mathrm{h}^{-1}$$

Note that the values obtained for k_2 and K_m are somewhat different from the values obtained with other methods, with the same experimental data.

1.2.3.5 CSTR Method

Differently from previous methods in sections 1.2.3.1–1.2.3.4, this method uses experiments carried out in continuous-flow reactors (CSTRs) rather than in batch.

In CSTR experiments (shown in Figure 1.14), the CSTR receives a feed of flow rate Q (units of, e.g. L/min) with substrate concentration S_{feed} (units of, e.g. g/L) and enzyme concentration E_{feed} (units of, e.g. g/L). The concentration of the enzyme in the reactor and in the feed is the same ($E_0 = E_{feed}$, the enzyme is a catalyst and is not converted or generated in the reactor). The substrate is converted in the reactor, the substrate concentration in the reactor and in the effluent is S. The residence time in the reactor is $\tau = V/Q$. The CSTR method requires doing various experiments at different residence times, with each experiment lasting long enough to achieve the steady state, i.e. the state where the substrate concentration in the reactor (S) does not change with time. For each experiment, the substrate concentration at steady state should be measured and a table with the substrate concentration vs the residence time should be generated (e.g. Table 1.11).

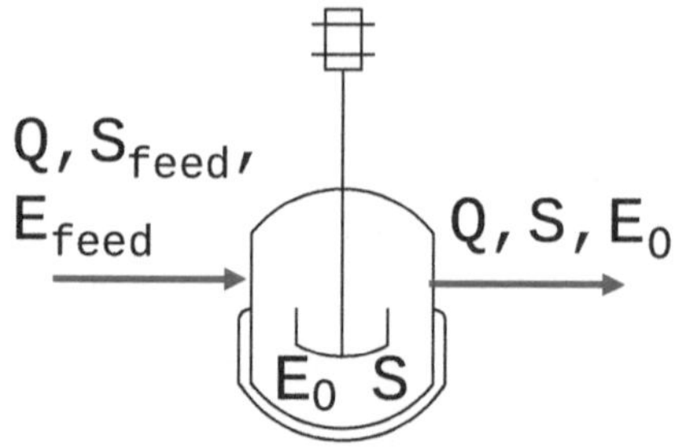

FIGURE 1.14 Scheme of a CSTR for an enzymatic reaction.

TABLE 1.11
Example of Experimental Data Collected in CSTR for the Determination of the Michaelis-Menten Parameters

Residence Time τ	Substrate Concentration S in the Effluent (or in the Reactor)
τ_1	S_1
τ_2	S_2
τ_3	S_3
…	…

The mass balance for the substrate in this CSTR at steady state is

$$QS_{feed} = (-r_S)V + QS \Rightarrow QS_{feed} = k_2 \frac{E_0 S}{K_m + S} V + QS \tag{1.21}$$

Introducing the residence time and re-arranging

$$\frac{S_{feed} - S}{\tau} = k_2 \frac{E_0 S}{K_m + S} \Rightarrow \frac{\tau}{S_{feed} - S} = \frac{K_m}{k_2 [E_0]} \frac{1}{[S]} + \frac{1}{k_2 [E_0]} \tag{1.22}$$

It follows that if we plot $\frac{\tau}{S_{feed} - S}$ vs $\frac{1}{[S]}$ we should obtain a linear trend with slope equal to $\frac{K_m}{k_2 [E_0]}$ and intercept equal to $\frac{1}{k_2 [E_0]}$ (Figure 1.15).

This method is very similar to the Lineweaver-Burk plot, however it uses data collected in CSTR rather than in batch. Data collected in CSTR can be more accurate

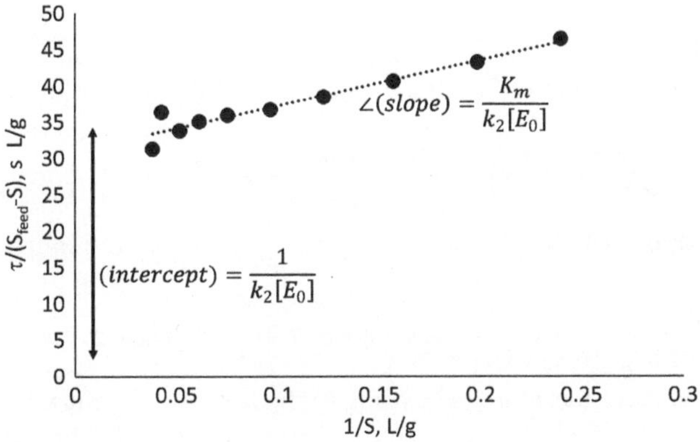

FIGURE 1.15 Plot of the CSTR method for the determination of the parameters of the Michalis-Menten equation.

than those collected in batch (each data point is collected at a steady state and so is less affected by concentration changes during sampling/measuring), however CSTR experiments are more complex to run than batch experiments.

Example 1.6

An enzymatic reaction is carried out in a CSTR at different residence times. In all cases the substrate and enzyme concentrations in the feed are 200 and 15 mg/L, respectively. The concentration of the substrate in the effluent vs the residence time τ is reported in Table 1.12. Determine the parameters of the Michaelis-Menten kinetics.

TABLE 1.12
Experimental Data for Example 1.6

Residence Time (τ, *h*)	Substrate Concentration (mg/L)
0.1	151.93
0.2	111.34
0.3	82.87
0.4	63.98
0.5	49.31
0.6	40.48
0.7	35.62
0.8	29.89
0.9	25.75
1	22.94

Solution

According to Equation (1.22), the experimental data in Table 1.12 must be re-arranged as $\frac{\tau}{S_{feed}-S}$ vs $\frac{1}{[S]}$. This is shown in Table 1.13.

From Table 1.13, the plot of $\frac{\tau}{S_{feed}-S}$ vs $\frac{1}{[S]}$ is shown in Figure 1.16.

The plot of $\frac{\tau}{S_{feed}-S}$ vs $\frac{1}{[S]}$ gives a slope of 0.09778 h and an intercept of .00140 L h mg^{-1}. We have therefore, according to the Lineweaver-Burk linearisation:

TABLE 1.13
Re-Arrangements of the CSTR Experimental Data to Obtain a Linearised Plot

Time (h)	*S* (mg/L)	1/*S* (L/mg)	$\frac{\tau}{S_{feed}-S}$ (h mg^{-1} L^{-1})
0.1	151.93	0.00658	0.002080
0.2	111.34	0.00898	0.002256
0.3	82.87	0.01207	0.002561
0.4	63.98	0.01563	0.002941
0.5	49.31	0.02028	0.003318
0.6	40.48	0.02470	0.003761
0.7	35.62	0.02807	0.004258
0.8	29.89	0.03346	0.004703
0.9	25.75	0.03883	0.005165
1	22.94	0.04359	0.005648

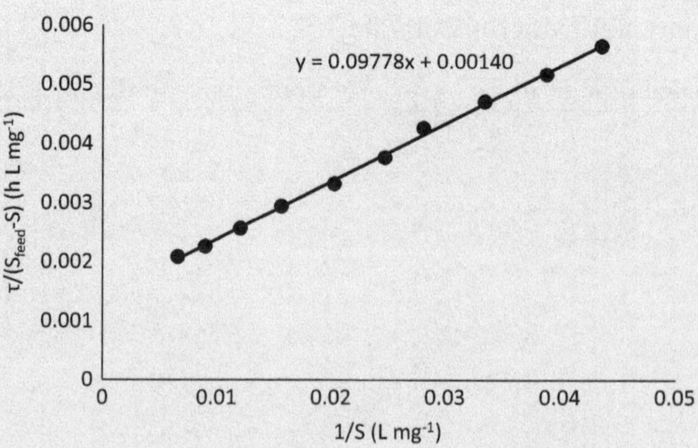

FIGURE 1.16 Plot of the CSTR method for Example 1.6.

$$\frac{1}{k_2[E_0]} = 0.00140\,\mathrm{L\,h\,mg^{-1}} \rightarrow k_2 = \frac{1}{0.00140[E_0]}$$
$$= \frac{1}{0.00140\,\mathrm{L\,h\,mg^{-1}} \cdot 15\,\mathrm{mg\,L^{-1}}} = 47.7\,\mathrm{h^{-1}}$$

and

$$\frac{K_m}{k_2[E_0]} = 0.09778\,\mathrm{h} \rightarrow K_m = 0.09778\,\mathrm{h} \cdot k_2[E_0]$$
$$= 0.09778\,\mathrm{h} \cdot 47.7\,\mathrm{h^{-1}} \cdot 15\,\mathrm{mg\,L^{-1}} = 69.9\,\mathrm{mg\,L^{-1}}$$

1.2.3.6 Nonlinear Parameter Fitting Method

This method uses data collected during batch reactions, exactly the same type of data of S vs t used in Sections 1.2.3.1–1.2.3.4. However, in this case the calculation of the substrate rate is not necessary, and the method uses nonlinear fitting of the k_2 and K_m parameters based on the experimental data.

The first step in this method is the solution of the differential equation that describes the substrate profile. The mass balance for the substrate in a batch reactor is

$$\frac{dS}{dt} = r_S = -k_2 \frac{E_0 S}{K_m + S} \tag{1.23}$$

This equation can be solved analytically or with any suitable numerical methods, e.g. with the method (finite differences) described in Appendix A. By using this method, the mass balance for the substrate is solved numerically by using the finite differences:

$$S(t + \Delta t) = S(t) + \frac{dS}{dt}(t) \cdot \Delta t \tag{1.24}$$

The use of this method is shown in Table 1.14. Initially, the method requires setting some arbitrary values of k_2 and K_m, in order to have an initial profile of S vs t according to Table 1.14.

The initial substrate profile calculated according to Table 1.14 with the arbitrary values of k_2 and K_m will be of course different from the experimental data. In order to match the calculated profile with the experimental one, the parameters k_2 and K_m need to be adjusted to minimise the difference between the calculated and experimental data. This difference is represented by the objective function:

$$\sum_i \left(\mathrm{Exp}\ S(t_i) - \mathrm{Calc}\ S(t_i)\right)^2 \tag{1.25}$$

TABLE 1.14
Calculation of the Time Profile of the Substrate Concentration in a Batch Reactor

Time	$\frac{dS}{dt}$	S
0	$r_S(S = S_{IN})$	S_{IN}
$t_1 = \Delta t$	$r_S(S = S_1)$	$S_1 = S_{IN} + \frac{dS}{dt}(t = 0) \cdot \Delta t$
$t_2 = 2\Delta t$	$r_S(S = S_2)$	$S_2 = S_1 + \frac{dS}{dt}(t = t_1) \cdot \Delta t$
$t_3 = 3\Delta t$	$r_S(S = S_3)$	$S_3 = S_2 + \frac{dS}{dt}(t = t_2) \cdot \Delta t$
...	...	...

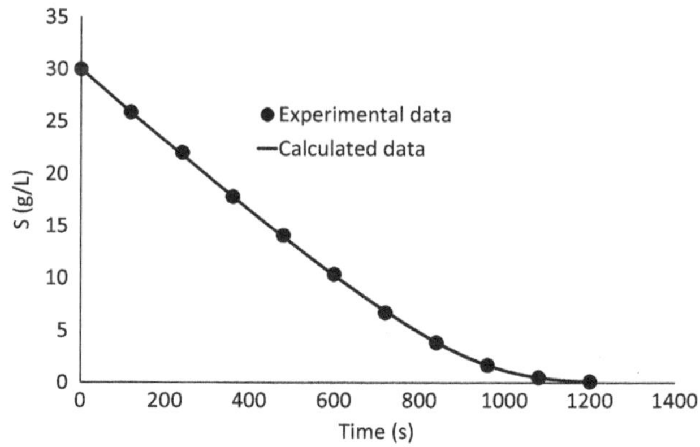

FIGURE 1.17 Comparison of experimental data and data calculated with the optimised k_2 and K_m parameters according to the nonlinear parameter fitting method.

where Exp $S(t_i)$ and Calc $S(t_i)$ represent the experimental and calculated concentration of substrate at time t_i, respectively. Equation (1.25) represents the sum of the deviations between the calculated and experimental concentrations over the length of the batch run. The difference $(\text{Exp } S(t_i) - \text{Calc } S(t_i))$ is squared to eliminate the risk that large negative and positive deviations may cancel each other, giving incorrect parameters. According to this method, the parameters k_2 and K_m are those parameters for which the objective function (1.25) is minimised. Minimisation of the objective function can be done with various methods and with various solvers, e.g. by using the Solver add-in in Excel. An example of the use of the nonlinear parameter fitting method is shown in Figure 1.17.

The method is often the most accurate, however it is also the most complex. This method suffers from the limitation that the points at higher substrate concentration are usually fitted better than the points at lower concentration, because they give a higher contribution to the objective function.

Example 1.7

Using the experimental data from Example 1.2 and Table 1.6, determine the parameters k_2 and K_m using the nonlinear parameter fitting method. Also, compare the experimental data in Table 1.6 with the data calculated with the parameters from each of the methods in Examples 1.2–1.5.

Solution

The first step is to generate a table like Table 1.14 with arbitrary values of k_2 and K_m, obtaining a calculated profile of S vs time. As an example, here values of k_2 and K_m of 100 h^{-1} and 100 mg L^{-1} are chosen. The accuracy of the method depends on using a Δt value which is small enough, therefore a $\Delta t = 0.01$ h is chosen in this case. Table 1.15 shows the first data points calculated in this way (the whole table spans from time 0 to 1 h consisting of 101 data points).

The second step is to generate an objective function as equation (1.25), with the difference between the values in Table 1.15 (calculated) and in Table 1.6 (experimental):

$$\text{Objective function} = (149.32 - 108.82)^2 + (102.02 - 44.4)^2 + \ldots$$

where 149.32, 102.02 are the experimental data (Table 1.6) at time 0.1 and 0.2 h and 108.82, 44.4 are the calculated values (Table 1.15) at the same times. The objective function will also include the difference between experimental and calculated data for all data points in Table 1.6.

With the assumed arbitrary values of k_2 and K_m the objective function has a value of 8680.59 $(mg/L)^2$, which is very high (large discrepancy between experimental and calculated data).

The final step is to determine the parameters k_2 and K_m, which minimise the value of the objective function, i.e. bring the calculated data as close as possible to the experimental data. By using the Solver add-in in Excel, the objective function is minimised to 1.079 $(mg/L)^2$ with the optimum values of the parameters:

$$k_2 = 47.6\,\text{h}^{-1}$$

and

$$K_m = 68.7\,\text{mg}\,\text{L}^{-1}$$

With these optimum values of k_2 and K_m, the comparison between the experimental and the calculated data is shown in Figure 1.18.

TABLE 1.15
Calculation of S as a Function of Time for Arbitrary Values of k_2 and K_m

Time (h)	$\frac{dS}{dt}$	S
0	-999.767	199.86
0.01	-982.513	189.86
0.02	-964.357	180.03
0.03	-945.253	170.39
0.04	-925.158	160.94
0.05	-904.028	151.69
0.06	-881.824	142.65
0.07	-858.511	133.83
0.08	-834.061	125.25
0.09	-808.454	116.91
0.10	-781.681	108.82
0.11	-753.746	101.00
0.12	-724.672	93.47
0.13	-694.500	86.22
0.14	-663.296	79.27
0.15	-631.149	72.64
0.16	-598.18	66.33
0.17	-564.538	60.35
0.18	-530.401	54.70
0.19	-495.978	49.40
0.20	-461.502	44.44
...	...	...

Note: The first 20 data points only are shown.

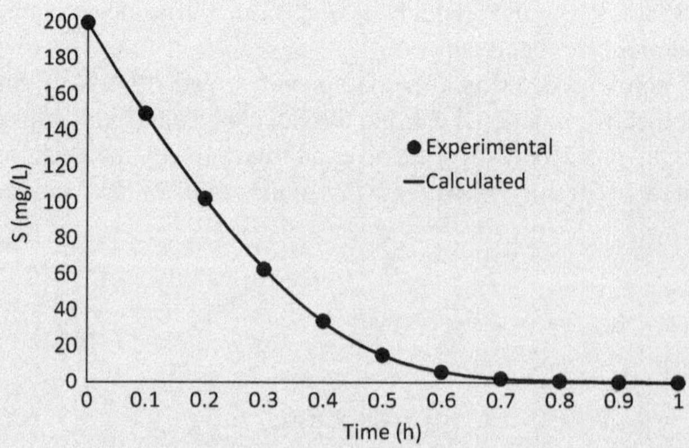

FIGURE 1.18 Comparison of the experimental data in Table 1.6 with the calculated data after parameter fitting with the nonlinear method.

TABLE 1.16
Summary of the Michaelis-Menten Parameters Obtained with the Experimental Data in Table 1.6 Obtained with the Different Methods for Parameter Estimation

Method	k_2 (h^{-1})	K_m (mg/L)
Lineweaver-Burk	43.4	97.4
Eadie-Hofstee	58.2	130.2
Hanes-Woolf	56.6	125.4
Integration	47.1	65.7
Nonlinear	47.6	68.7

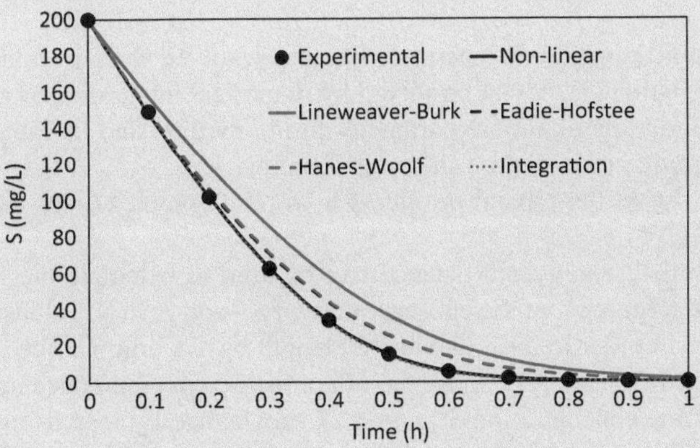

FIGURE 1.19 Comparison of the fitting of the experimental data with the different models.

In summary, the same experimental data in Table 1.6 led to the following estimations of the parameters with the different methods (Table 1.16).

For each set of parameters, the time profiles of S vs time can be calculated by numerical integration of the differential equation of the batch reactor, according to the method in Table 1.14. The resulting plots of S vs time are shown and compared in Figure 1.19.

Figure 1.19 shows that the best fitting with the experimental data is obtained by the nonlinear and integration methods, which give virtually identical profiles. The Hanes-Woolf and the Eadie-Hofstee methods also give almost identical profiles but with worse fitting than the nonlinear and integration methods. The worst fitting is obtained by the Lineweaver-Burk method.

1.2.4 Mass Balances for Enzymatic Reactions using the Michaelis-Menten Kinetics

In this section, we will cover how to write mass balances for enzymatic reactions and how to use them to simulate these reactions and to identify the effect of the most important variables on the reaction performance. Here we assume that the kinetic parameters k_2 and K_m for the reaction of interest have been determined and are known.

We will consider separately the case of batch reactors and of continuous-flow reactors.

1.2.4.1 Mass Balances for Batch Reactors

The mass balance for the substrate in the batch reactor has already been seen:

$$\frac{dS}{dt} = r_S = -k_2 \frac{E_0 S}{K_m + S} \tag{1.26}$$

With k_2, K_m and E_0 known, this mass balance can be solved with the condition that at $t = 0$ $S = S_{IN}$. Equation (1.26) can be solved by numerical integration as shown earlier in the section about the nonlinear parameter fitting method and in Table 1.14, or by analytical integration (this will be shown in Example 1.8).

Figure 1.20 shows the typical profile of S vs time obtained from the solution of equation (1.26).

The solution of the substrate balance can be used to calculate the length of the batch process as a function of the chosen values of S_{IN} and E_0. The values of S_{IN} and E_0 can be modified in order to have the desired length of reaction, i.e. the desired productivity. Figures 1.21 and 1.22 show the effect of the enzyme concentration and of the initial substrate concentration. Figure 1.21 shows that, by increasing the enzyme

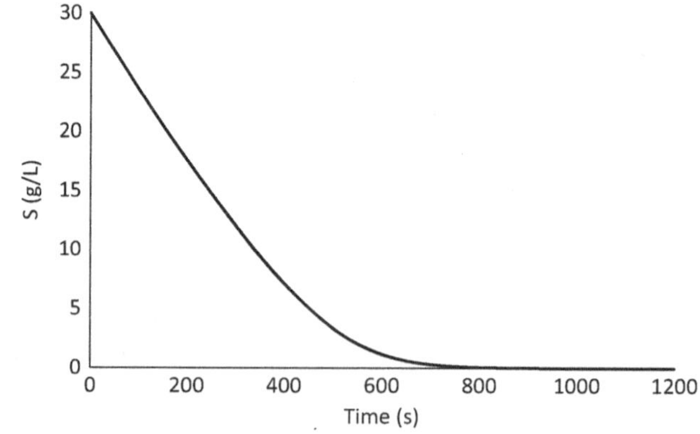

FIGURE 1.20 Example of the substrate profile in a batch reactor.

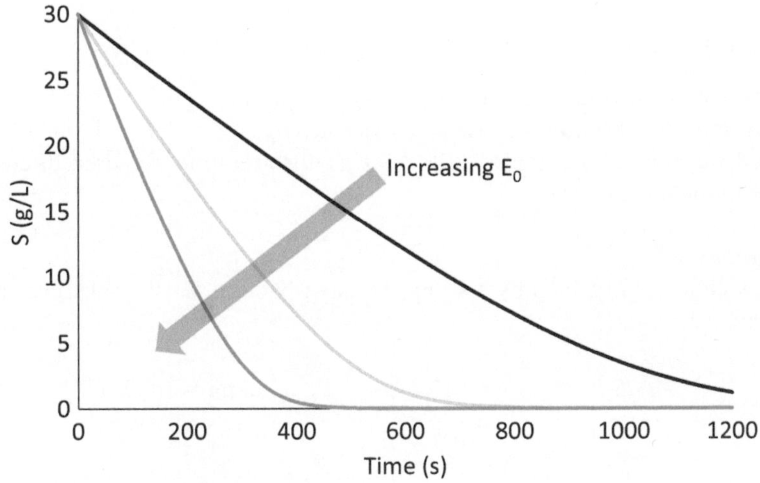

FIGURE 1.21 Effect of the enzyme concentration on the substrate profile in a batch reactor.

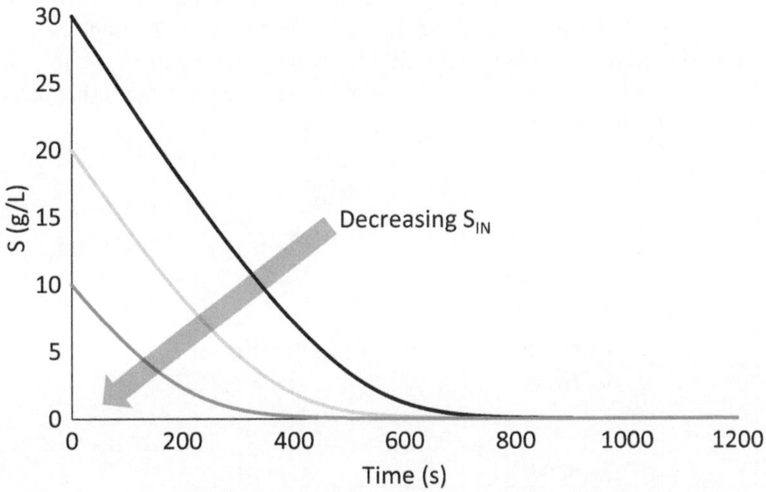

FIGURE 1.22 Effect of the initial substrate concentration on the substrate profile in a batch reactor.

concentration, the reaction becomes faster, with higher productivity. However, the higher cost of the enzymes must be considered. Figure 1.22 shows that, by decreasing the initial substrate concentration, the reaction time decreases, as expected, however the average reaction rate is lower because of the lower average substrate concentration, leading to lower productivity.

Example 1.8

Consider an enzymatic reaction with $k_2 = 2.5 \cdot 10^{-4}$ mol unit^{-1} s^{-1} and $K_m = 0.15$ M. Assume the reaction is carried out in a batch reactor with $E_0 = 10$ units/L and initial substrate concentration 0.80 M. Calculate the time required to convert 99% of the initial substrate.

Solution

The solution is obtained by solving the mass balance for the substrate in the batch reactor:

$$\frac{dS}{dt} = r_S = -k_2 \frac{E_0 S}{K_m + S} = -2.5 \cdot 10^{-4} \frac{10 \text{ units L}^{-1} S}{0.15 \text{ M} + S}$$

with the substrate expressed in M.

The substrate profile can be calculated by discretising the time domain in small Δt intervals. As an example, the initial calculation is shown in Table 1.17.

99% conversion of the substrate is obtained when the substrate concentration is reduced to 1% of the initial value, i.e. to 0.008 M. From the table of S vs t, it can be observed that $S = 0.008$ after 592 s, i.e. the time required is 9 min 52 s.

Another method to solve this problem is to integrate the substrate mass balance, i.e. to integrate

$$\frac{dS}{dt} = -k_2 \frac{E_0 S}{K_m + S}$$

TABLE 1.17
Initial Calculations of the Profiles of *S* vs Time

Time (s)	S	$\frac{dS}{dt}$
0	0.8000	-0.002105
1	0.7979	-0.002104
2	0.7958	-0.002104
3	0.7937	-0.002103
4	0.7916	-0.002102
5	0.7895	-0.002101
6	0.7874	-0.002100
7	0.7853	-0.002099
8	0.7832	-0.002098
9	0.7811	-0.002097
10	0.7790	-0.002096
…	…	…

between time 0 and t_{final} (the time required for 99% conversion) and between S_{IN} (0.80 M) and S_{final} (0.008 M).

The integration can be done by re-arranging and by separation of the variables:

$$-\frac{(K_m+S)}{k_2E_0}\frac{dS}{S}=dt\rightarrow-\frac{K_m}{k_2E_0}\frac{dS}{S}-\frac{1}{k_2E_0}dS=dt\rightarrow-\frac{K_m}{k_2E_0}\int_{S_{IN}}^{S_{final}}\frac{dS}{S}-\frac{1}{k_2E_0}\int_{S_{IN}}^{S_{final}}dS$$

$$=\int_0^{t_{final}}dt\rightarrow t_{final}=\frac{1}{k_2E_0}\left(K_m\ln\frac{S_{IN}}{S}+S_{IN}-S\right)$$

$$=\frac{1}{2.5\cdot10^{-4}\text{ mol unit}^{-1}\text{s}^{-1}10\,\text{unit L}^{-1}}$$

$$\left(0.15\,\text{mol L}^{-1}\ln\frac{0.8}{0.008}+0.8\,\text{mol L}^{-1}-0.008\text{mol L}^{-1}\right)$$

$$=593\,\text{s}$$

The result is virtually the same as the one obtained above with the discretisation method, the small difference due to the inevitable numerical error in the discretisation of differential equations by finite differences. It can be verified that, by using a smaller discretisation in Table 1.17, e.g. 0.1 rather than 1 s, the time to reach $S = 0.008$ becomes 593 s as obtained by integration.

1.2.4.2 Mass Balances for Continuous-Flow Reactors

If the enzymatic reaction is carried out in a CSTR (continuous reactor) rather than in a batch reactor, the mass balance for the substrate can be used to calculate the substrate concentration at steady state as a function of the residence time $\tau = V/Q$.

The mass balance for the substrate in a CSTR has already been seen

$$QS_{feed}=k_2\frac{E_0S}{K_m+S}V+QS \tag{1.27}$$

Equation (1.27) can be re-arranged introducing the residence time $\tau = V/Q$:

$$S_{feed}=k_2\frac{E_0S}{K_m+S}\tau+S \tag{1.28}$$

This equation has the only unknown S and can be solved with various analytical or numerical methods. If using Excel and Solver, the following procedure can be used to calculate S:

1. The equation is written as $S_{feed} - k_2 \frac{E_0 S}{K_m + S}\tau - S = 0$
2. For any value of τ, Solver is used to find the value of S for which the left-hand side of the equation is equal to zero. This is the solution of the equation.
3. Step 2 is repeated for any values of τ for which we want to calculate S.

The steps above allows the calculation of S as a function of the residence time τ, for given values of the kinetic parameters, of the enzyme concentration and of the substrate concentration in the feed. Figure 1.23 shows a typical profile of S vs τ for a CSTR.

The substrate concentration in the reactor and in the effluent is lower for longer residence times, as expected. Note that when S is low, only small S reductions are obtained by increasing the residence time.

Figure 1.24 shows the effect of increasing the enzyme concentration. By increasing the enzyme concentration, the substrate concentration in the reactor and in the effluent decreases for a given residence time (faster reaction rate). However, higher enzyme concentration increases the process costs.

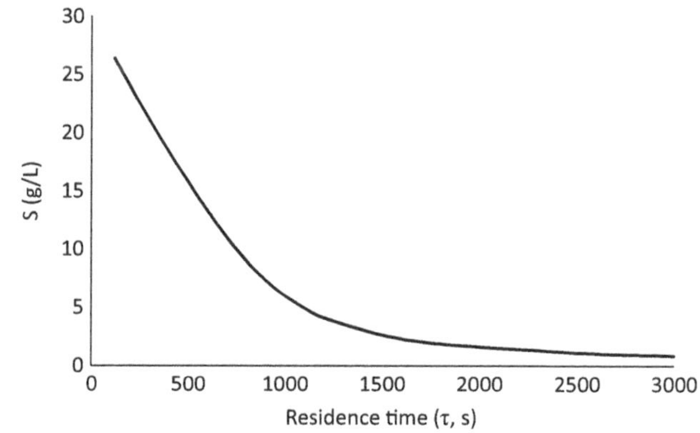

FIGURE 1.23 Typical profile of S vs residence time for a CSTR.

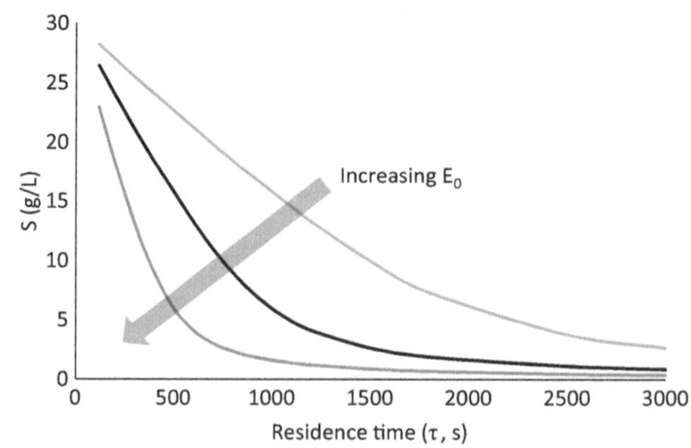

FIGURE 1.24 Effect of enzyme concentration on substrate profile vs residence time for a CSTR.

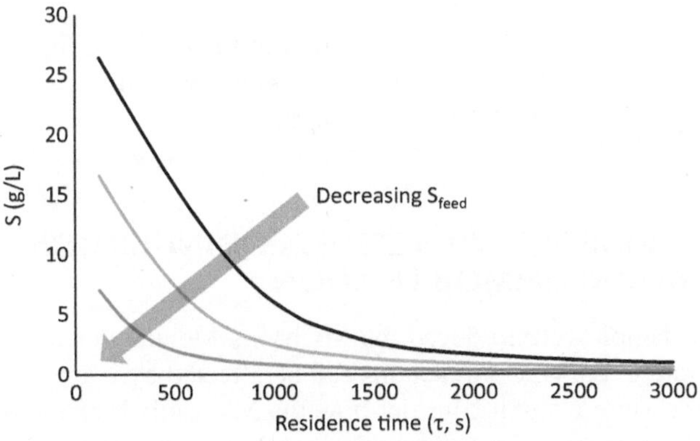

FIGURE 1.25 Effect of substrate concentration in the feed on substrate profile vs residence time for a CSTR.

Figure 1.25 shows the effect of the substrate concentration in the feed. Lower substrate concentrations in the feed give lower effluent substrate concentration, however the reaction rate is also lower (the reaction rate is proportional to the substrate concentration).

Example 1.9

Consider the enzymatic reaction of Example 1.8. This reaction is now carried out in a CSTR, with enzyme and substrate concentration in the feed of 10 units/L and 0.8 M, respectively. Calculate the residence time required in the CSTR to achieve a 99% conversion of the substrate. Compare with the time in a batch reactor calculated in Example 1.8.

Solution

The required residence time τ is obtained by re-arranging equation (1.28) and solving for τ:

$$\tau = \frac{S_{feed} - S}{k_2 E_0 S}\left(K_m + S\right)$$

$$= \frac{0.8\,\text{mol}\,\text{L}^{-1} - 0.008\,\text{mol}\,\text{L}^{-1}}{2.5\cdot 10^{-4}\,\text{mol}\,\text{unit}^{-1}\text{s}^{-1}\,10\,\text{unit}\,\text{L}^{-1}\,0.008\,\text{mol}\,\text{L}^{-1}}\left(0.15 + 0.008\right)\text{mol}\,\text{L}^{-1} = 6257\text{s}$$

The residence time required in a CSTR to reduce the substrate concentration from 0.8 (feed concentration) to 0.008 M (effluent concentration corresponding to 99% conversion) is 6257 s, i.e. 1 h 44 min 17 s. This time is much longer than the time calculated for the batch reactor in Example 1.8. As explained earlier, this is due to the fact that in CSTRs enzymes are exposed to one single concentration of substrate (the concentration in the reactor/effluent), while in batch processes the substrate concentration varies over time from high to low values, giving on average a faster substrate rate.

1.3 ENZYME KINETICS WITH EFFECTS OF INHIBITION, PH, TEMPERATURE, IMMOBILISATION

In the previous section we introduced the Michaelis-Menten kinetics for enzymatic reactions. The rate of enzymatic reactions can be affected by many properties of the reaction medium. Here we will consider how the Michaelis-Menten kinetics can be modified to account for the effect of inhibitors, pH and temperature. For simplicity's sake, we will always refer to the simple reaction $S \rightarrow P$.

1.3.1 Effect of Inhibitors

Inhibition of enzymatic reactions is due to the presence in reaction media of substances that interfere with the normal process of the reactions and therefore decrease the reaction rate. For examples, the polypeptide aprotinin inhibits protein hydrolysis catalysed by certain protease enzymes, such as trypsin and chymotrypsin. Another example is the inhibition of amylase enzymes by certain polyphenols, blocking starch hydrolysis. Enzyme inhibition is often undesirable, and care is taken to remove any inhibitors from the reaction media. However, in some cases enzyme inhibitors are added on purpose to prevent the occurrence of undesired enzymatic reactions, for example enzyme inhibitors are used in food processing to extend shelf life and prevent spoilage.

We will consider two types of inhibition: competitive and noncompetitive inhibition (Figure 1.26, other types of inhibition mechanisms are also possible).

Competitive inhibition occurs when the inhibitor has a similar shape, size, MW, as the substrate and binds to the active site in place of the substrate, without leading to the desired product *P*. Noncompetitive inhibition occurs when the inhibitor, usually

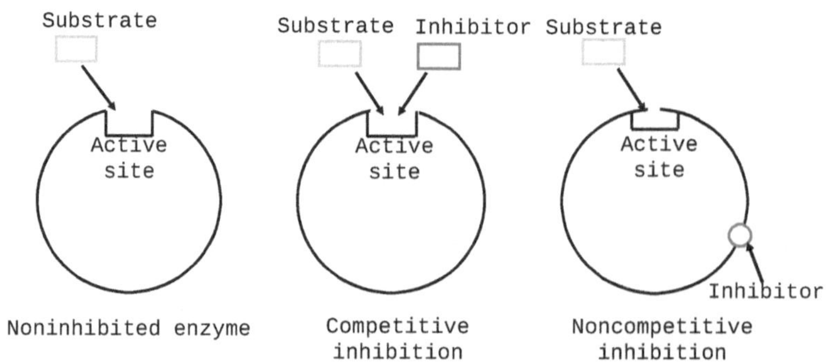

FIGURE 1.26 Conceptual scheme of competitive and noncompetitive inhibition.

of a different shape than the substrate, binds to other parts of the enzyme (not to the active site), making the enzyme inactive.

1.3.1.1 Kinetics with Competitive Inhibitors

We will model competitive inhibition with this reaction scheme:

$$\begin{array}{l} E+S \leftrightarrow ES \rightarrow E+P \\ + \\ I \\ \updownarrow \\ EI \end{array} \tag{1.29}$$

We want now to derive a modified Michaelis-Menten kinetics for competitive inhibition. The rate of substrate removal is still given by $(-r_S) = r_P = k_2[ES]$. Also, it is still valid that $\frac{[E][S]}{[ES]} = K_m \Rightarrow [ES] = \frac{[E][S]}{K_m}$ (this follows from either of Assumptions 1 or 2 in the base Michaelis-Menten derived earlier).

We assume that the inhibitor's binding to the active site is at equilibrium with (here K_I is defined in the same way as K_m):

$$\frac{[E][I]}{[EI]} = K_I \Rightarrow [EI] = \frac{[E][I]}{K_I} \tag{1.30}$$

From the previous equations, it follows that

$$(-r_S) = k_2[ES] = k_2\frac{[E][S]}{K_m} \tag{1.31}$$

The total enzyme that we add to the reaction medium (E_0) is present as free enzyme (E), as enzyme/substrate complex (ES) or as enzyme/inhibitor complex (EI):

$$[E_0] = [E] + [ES] + [EI] \tag{1.31}$$

By substituting the equations for $[ES]$ and $[EI]$ that we have seen earlier:

$$[E_0] = [E] + \frac{[E][S]}{K_m} + \frac{[E][I]}{K_I} \Rightarrow [E] = \frac{[E_0]}{1 + \frac{[S]}{K_m} + \frac{[I]}{K_I}} \tag{1.32}$$

Therefore, $\left(-r_S\right)$ can be written as

$$\left(-r_S\right)=k_2\frac{[S]}{K_m}\frac{[E_0]}{1+\frac{[S]}{K_m}+\frac{[I]}{K_I}}\Rightarrow\left(-r_S\right)=k_2\frac{[E_0][S]}{K_m\left(1+\frac{[I]}{K_I}\right)+[S]} \tag{1.33}$$

This is the rate equation for competitive inhibition.

In general, the rate for competitive inhibition is, as expected, lower than the noninhibited rate $\left(-r_S\right)=k_2\frac{[E_0][S]}{K_m+[S]}$.

If the substrate concentration is very low, i.e. $[S]\ll K_m\left(1+\frac{[I]}{K_I}\right)$ (low substrate concentrations) then the inhibited rate becomes $\left(-r_S\right)=k_2\frac{[E_0][S]}{K_m\left(1+\frac{[I]}{K_I}\right)}$, which is lower than the noninhibited rate for low substrate concentrations: $\left(-r_S\right)=k_2\frac{[E_0][S]}{K_m}$.

If the substrate concentration is very large, i.e. $[S]\gg K_m\left(1+\frac{[I]}{K_I}\right)$ (high substrate concentrations) then the inhibited rate becomes $\left(-r_S\right)=k_2\left[E_0\right]=V_{max}$, which is the same as the noninhibited rate for high concentrations of S.

This shows that at high substrate concentration, competitive inhibition has no effect. In other words, competitive inhibition can be compensated by increasing the substrate concentration. This can be explained by looking at the conceptual mechanism of competitive inhibition (Figure 1.26). When present at high enough concentrations, the substrate can outcompete the inhibitor for access to the active site. In these cases, the inhibitor has no effect on the reaction rate.

Example 1.10

Consider an enzymatic reaction with V_{max} = 25 mg L^{-1}min^{-1} and K_m = *15* mg L^{-1}. Calculate the reaction rate (substrate conversion rate) in the presence of a competitive inhibitor at concentration 10 mg L^{-1} for substrate concentrations of 100 and 5000 mg L^{-1}. Assume K_I = 1 mg/L.

Solution
The reaction rate is given by

$$(-r_S) = k_2 \frac{[E_0][S]}{K_m\left(1+\frac{[I]}{K_I}\right)+[S]} = \frac{25\,\text{mg L}^{-1}\text{min}^{-1}[S]}{15\,\text{mg L}^{-1}\left(1+\frac{10\,\text{mg L}^{-1}}{1\,\text{mg L}^{-1}}\right)+[S]}$$

For S = 100 mg/L, $(-r_S) = 9.44$ mg L^{-1}min^{-1}. Note that this reaction rate is severely reduced by the presence of the inhibitor, as the noninhibited reaction rate for S = 100 mg/L would be $(-r_S) = 21.74$ mg L^{-1}min^{-1}.
For S = 5000 mg/L, $(-r_S) = 24.20$ mg L^{-1}min^{-1}. This rate is very close to V_{max}. This shows that competitive inhibition can be compensated by increasing the substrate concentration, obtaining the maximum noninhibited rate.

1.3.1.2 Kinetics with Noncompetitive Inhibitors

We will model noncompetitive inhibition with this reaction scheme:

$$\begin{array}{ccc} E+S & \leftrightarrow & ES \rightarrow E+P \\ + & & + \\ I & & I \\ \updownarrow & & \updownarrow \\ EI+S & \leftrightarrow & ESI \end{array} \tag{1.34}$$

With this mechanism, we assume that the inhibitor can bind to both E and ES and that EI can react with S giving ESI but not the product P.

We assume that the value of K_m is the same for the reactions of E and EI with S, i.e. that $K_m = \frac{[E][S]}{[ES]} = \frac{[EI][S]}{[ESI]}$ from which $[ES] = \frac{[E][S]}{K_m}$ and $[ESI] = \frac{[EI][S]}{K_m}$. Similarly, we also assume that the value of K_I is the same for the reactions of E and ES with I, i.e. that $K_I = \frac{[E][I]}{[EI]} = \frac{[ES][I]}{[ESI]}$ from which $[EI] = \frac{[E][I]}{K_I}$ and $[ESI] = \frac{[ES][I]}{K_I} = \frac{[E][S][I]}{K_m K_I}$.

The enzyme balance is in this case

$$[E_0] = [E]+[ES]+[EI]+[ESI] \Rightarrow [E_0] = [E]+\frac{[E][S]}{K_m}+\frac{[E][I]}{K_I}+\frac{[E][S][I]}{K_m K_I} \tag{1.35}$$

From the previous equations, it follows that

$$\left(-r_S\right)=k_2\left[ES\right]=k_2\frac{[E][S]}{K_m}=k_2\frac{[S]}{K_m}\left(\frac{\left[E_0\right]}{1+\frac{[S]}{K_m}+\frac{[I]}{K_I}+\frac{[S][I]}{K_mK_I}}\right) \tag{1.36}$$

The obtained rate equation can be re-arranged as

$$\left(-r_S\right)=k_2\frac{[S]\left[E_0\right]}{K_m+[S]+\frac{K_m[I]}{K_I}+\frac{[S][I]}{K_I}}=k_2\frac{[S]\left[E_0\right]}{K_m\left(1+\frac{[I]}{K_I}\right)+[S]\left(1+\frac{[I]}{K_I}\right)} \tag{1.37}$$

which can be finally rearranged as

$$\left(-r_S\right)=k_2\frac{[S]\left[E_0\right]}{\left(K_m+[S]\right)\left(1+\frac{[I]}{K_I}\right)} \tag{1.38}$$

This is the rate equation for noncompetitive inhibition.

The rate for noncompetitive inhibition is in all cases lower than the noninhibited rate $\left(-r_S\right)=k_2\frac{\left[E_0\right][S]}{K_m+[S]}$.

Indeed, if the substrate concentration is very large, i.e. $[S]\gg K_m$ (high substrate concentrations) then the inhibited rate becomes $\left(-r_S\right)=\frac{k_2\left[E_0\right]}{\left(1+\frac{[I]}{K_I}\right)}=\frac{V_{max}}{\left(1+\frac{[I]}{K_I}\right)}$, which is lower than the noninhibited rate for high concentrations of S: $\left(-r_S\right)=k_2\left[E_0\right]=V_{max}$. In other words, noncompetitive inhibition decreases the maximum substrate rate obtained when the substrate is in excess. On the contrary, as we have seen competitive inhibition does not affect the maximum substrate rate.

Differently from competitive inhibition, noncompetitive inhibition cannot be compensated by increasing the substrate concentration. This can be explained by looking at the conceptual mechanism of noncompetitive inhibition (Figure 1.26). The noncompetitive inhibitor binds on a different site than the substrate and it affects the enzyme at any substrate concentration.

Example 1.11

Consider the enzymatic reaction of Example 1.10, with the same kinetic parameters and concentration of inhibitor but assuming that the inhibitor is noncompetitive. Calculate the substrate rate for substrate concentrations of 100 and 5000 mg/L.

Solution

In this case the inhibitor is noncompetitive, and the reaction rate is given by

$$(-r_S) = k_2 \frac{[S][E_0]}{(K_m + [S])\left(1 + \frac{[I]}{K_I}\right)} = \frac{25\,\text{mg L}^{-1}\,\text{min}^{-1}[S]}{(15\,\text{mg L}^{-1} + [S])\left(1 + \frac{10\text{ mg L}^{-1}}{1\text{ mg L}^{-1}}\right)}$$

For S = 100 mg/L, $(-r_S) = 1.98\text{ mg L}^{-1}\text{ min}^{-1}$. Note that the severe inhibition compared to the noninhibited reaction rate for S = 100 mg/L which would be $(-r_S) = 21.74\text{ mg L}^{-1}\text{ min}^{-1}$.

For S = 5000 mg/L, $(-r_S) = 2.27\,\text{mg L}^{-1}\,\text{min}^{-1}$. Note that this rate is still much lower than the V_{max} in the absence of inhibition, in spite of the very high substrate concentration. This shows that noncompetitive inhibition, differently from competitive inhibition, cannot be compensated by increasing the substrate concentration.

1.3.1.3 Obtaining Kinetic Parameters with Inhibitors

In previous sections we have seen the modified Michaelis-Menten kinetics in the presence of inhibitors. In this section, we will see how experimental data can be used to determine if a substance is an inhibitor or not, whether the inhibition is competitive or noncompetitive and how to determine the kinetic parameters of the inhibition.

The method includes running batch experiments at different concentrations of the (suspected) inhibitor concentration $[I]$, including the case where $[I] = 0$. For each experiment, we obtain values of $(-r_S)$ vs $[S]$. By measuring and comparing the substrate rate at different values of $[I]$ we can determine the nature of inhibition and the parameter K_I.

The type of linearisation of the experimental data is different for competitive and noncompetitive inhibition.

For competitive inhibition, the rate equation can be linearised as follows:

$$\frac{1}{(-r_S)} = \frac{K_m\left(1 + \frac{[I]}{K_I}\right)}{k_2[E_0]}\frac{1}{[S]} + \frac{1}{k_2[E_0]} \quad (1.39)$$

This linearised equation shows that, for a given concentration $[I]$ of a competitive inhibitor, by plotting $\frac{1}{(-r_S)}$ vs $\frac{1}{[S]}$ (Lineweaver-Burk plot), we should obtain a straight line with slope equal to $\frac{K_m\left(1+\frac{[I]}{K_I}\right)}{k_2[E_0]}$ and intercept equal to $\frac{1}{k_2[E_0]}$. For a competitive inhibitor, by increasing the inhibitor concentration $[I]$, the slope will increase while the intercept will not change (Figure 1.27).

For noncompetitive inhibition, the rate equation can be linearised as follows:

$$\frac{1}{(-r_S)} = \frac{K_m\left(1+\frac{[I]}{K_I}\right)}{k_2[E_0]}\frac{1}{[S]} + \frac{\left(1+\frac{[I]}{K_I}\right)}{k_2[E_0]} \tag{1.40}$$

For a given concentration $[I]$ of a noncompetitive inhibitor, by plotting $\frac{1}{(-r_S)}$ vs $\frac{1}{[S]}$ (Lineweaver-Burk plot), we should obtain a straight line with slope equal to $\frac{K_m\left(1+\frac{[I]}{K_I}\right)}{k_2[E_0]}$ (same slope as for competitive inhibition) and intercept equal to $\frac{\left(1+\frac{[I]}{K_I}\right)}{k_2[E_0]}$ (different than for competitive inhibition). For a noncompetitive inhibitor,

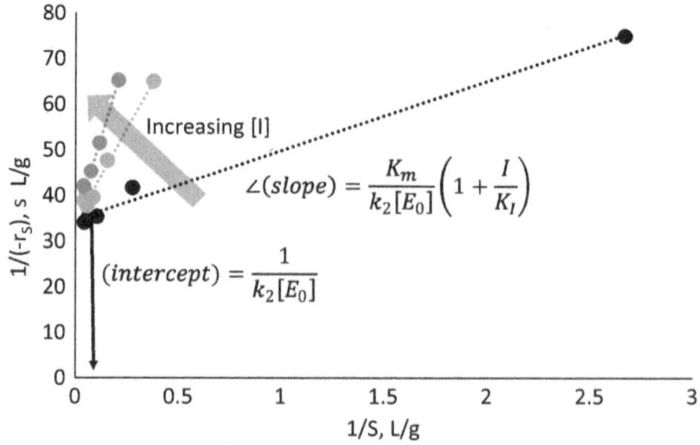

FIGURE 1.27 Linearisation of experimental data for a competitive inhibitor.

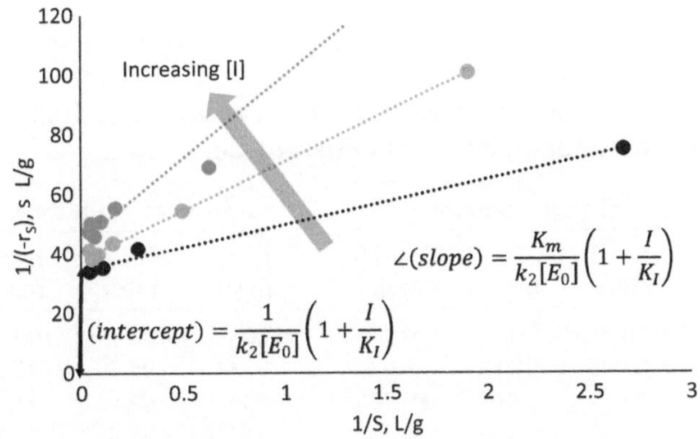

FIGURE 1.28 Linearisation of experimental data for a noncompetitive inhibitor.

by increasing the inhibitor concentration [I], both the slope and the intercept will increase (Figure 1.28).

In summary, by comparing the plots of $\frac{1}{(-r_S)}$ vs $\frac{1}{[S]}$ for different values of [I], we can determine the nature and the parameters of the inhibition.

Example 1.12

It is known that for a certain enzymatic reaction with a given enzyme concentration the maximum noninhibited rate is V_{max} = 7.0 mg L^{-1} min^{-1} and the K_m without inhibition is 10 mg L^{-1}. Two substances are suspect inhibitors, and they are tested at different concentrations obtaining the experimental data in Table 1.18. Determine whether these substances are inhibitors, the type of inhibition and the K_I for each substance.

Solution

From a first look at the experimental data we see that the reaction rate decreases for increasing concentrations of both I_1 and I_2, which indicates that both substances are inhibitors.

Considering I_1, the data collected at the three concentrations can be converted into reaction rates and then into plots of $\frac{1}{(-r_S)}$ vs $\frac{1}{[S]}$ with the same method used for noninhibited data in Example 1.2. The resulting plot is shown in Figure 1.29.

From Figure 1.29, it is evident that the slope changes while the intercept remains approximately the same for the lines obtained at the different I_1 concentrations. This indicates that I_1 is a competitive inhibitor. The inhibition parameter K_I can be calculated from the slope of the regression line, by rearranging Equation (1.39):

TABLE 1.18
Experimental Data with Substrate Concentration vs Time for Suspect Inhibitors I_1 and I_2 at Different Concentrations

	Suspect inhibitor I_1			Suspect inhibitor I_2		
t (s)	$I_1 = 5$ mg/L	$I_1 = 10$ mg/L	$I_1 = 20$ mg/L	$I_2 = 5$ mg/L	$I_2 = 10$ mg/L	$I_2 = 20$ mg/L
0	100	100	100	100	100	100
3	85.98	88.82	91.03	93.56	95.79	97.18
6	68.45	73.62	80.42	83.23	89.65	94.45
9	56.39	64.95	73.28	75.89	85.48	91.49
12	43.71	52.74	63.95	69.34	81.01	88.62
15	32.12	43.52	55.43	60.52	75.34	85.89
18	23.91	35.09	48.65	53.42	70.76	83.01
21	14.97	27.97	43.01	47.54	65.45	80.32
24	9.34	20.83	37.21	39.62	60.65	77.49
27	5.85	16.12	32.76	33.96	56.65	74.76
30	3.19	12.11	27.71	26.55	51.98	71.99

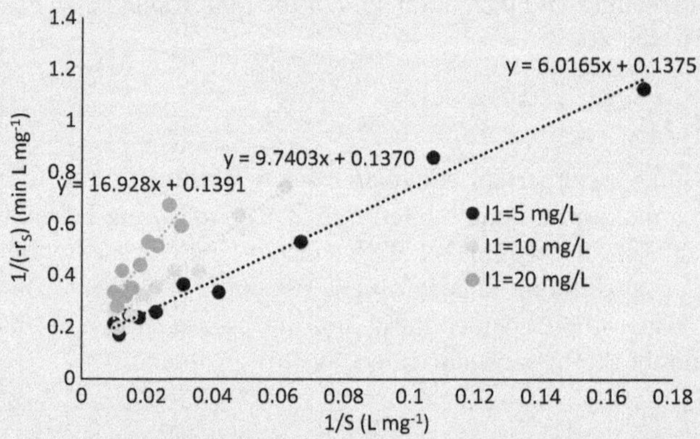

FIGURE 1.29 Plot of $\frac{1}{(-r_S)}$ vs $\frac{1}{[S]}$ for I_1 from experimental data in Table 1.18.

$$\frac{K_m\left(1+\frac{[I]}{K_I}\right)}{V_{max}} = \text{slope} \rightarrow K_I = \frac{[I]}{\text{slope}\frac{V_{max}}{K_m} - 1}$$

$$= \frac{[I]}{\text{slope}\frac{7\frac{\text{mg}}{\text{L}\cdot\text{min}}}{10\frac{\text{mg}}{\text{L}}} - 1} = \frac{[I]}{\text{slope}\cdot 0.7\text{min}^{-1} - 1}$$

By replacing the values of the slope from Figure 1.29, 6.0165, 9.7403, 16.928 min corresponding to concentrations of inhibitor of 5, 10, 20 mg/L, respectively, we obtain three values for K_I: 1.56, 1.72, 1.84 mg/L. The values are similar, the difference being due to the scatter of the experimental data. These values can be averaged to obtain K_I = 1.72 mg/L, which is the best estimate for substance I_1.

We can repeat the same analysis for substance I_2. The plot of $\frac{1}{(-r_S)}$ vs $\frac{1}{[S]}$ from the experimental data for substance I_2 is shown in Figure 1.30.

Despite the scatter in the experimental data, it is evident that the plots in Figure 1.30 have different slope and difference intercept. This indicates that I_2 is a noncompetitive inhibitor. The inhibition parameter K_I can be calculated either from the intercept or from the slope of the plots by re-arranging Equation (1.40). If using the slope, we obtain the same equation obtained for competitive inhibitors:

$$K_I = \frac{[I]}{\text{slope}\frac{V_{max}}{K_m} - 1} = \frac{[I]}{\text{slope} \cdot 0.7\text{min}^{-1} - 1}$$

From the slopes of Figure 1.30, we obtain the three values of K_I (in mg/L) of 3.45, 3.36, 3.59.

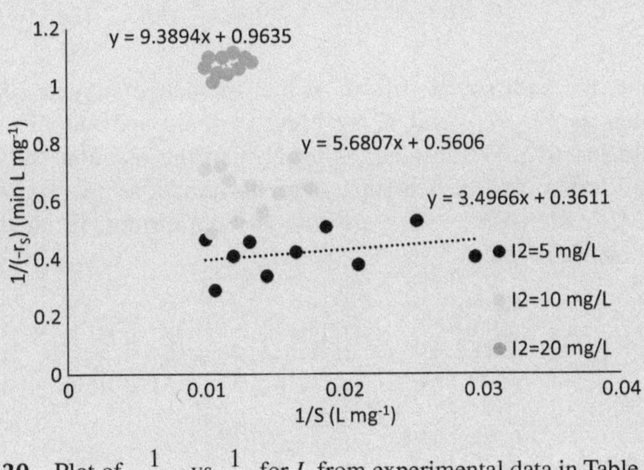

FIGURE 1.30 Plot of $\frac{1}{(-r_S)}$ vs $\frac{1}{[S]}$ for I_2 from experimental data in Table 1.18.

If using the intercept, the re-arrangement of equation (1.40) gives

$$K_I = \frac{[I]}{\text{intercept}\, V_{max} - 1} = \frac{[I]}{\text{intercept} \cdot 7\,\text{min}^{-1} - 1}$$

From the intercepts of Figure 1.30, we obtain the three values of K_I (in mg/L) of 3.27, 3.42, 3.48.

The six values just obtained for K_I can be averaged to obtain $K_I = 3.43$ mg/L, which is the best estimate for K_I for substance I_2.

1.3.2 Effect of pH

Enzymes are only active in a certain pH range. One of the reasons for the pH effect is that certain enzymes have ionic groups on their active site and the enzyme is only active when the ionic group is in a certain form.

We will use the model mechanism below, which takes into account the fact that pH influences the reactions with protons' exchange in water:

$$\begin{array}{l} E^- + H^+ \\ \updownarrow \\ EH + S \leftrightarrow EHS \rightarrow EH + P \\ + \\ H^+ \\ \updownarrow \\ EH_2^+ \end{array} \tag{1.41}$$

EH represents the undissociated form of the free active enzyme (that so far we have just written as *E*), EH_2^+ and *E* represent cationic and anionic forms of the enzyme that are inactive. We assume, as usual, that the reaction between *EH* and *EHS* is in equilibrium. We also assume that the reactions of protons' exchange ($EH + H^+ \leftrightarrow EH_2^+$,$EH \leftrightarrow E^- + H^+$) are also in equilibrium. From these equilibrium assumptions, we define

$$K_m = \frac{[E][S]}{[ES]}, K_1 = \frac{[EH][H^+]}{[EH_2^+]}, K_2 = \frac{[E^-][H^+]}{[EH]} \tag{1.42}$$

The substrate rate can be written as

$$(-r_S) = k_2[EHS] = k_2 \frac{[EH][S]}{K_m} \tag{1.43}$$

The total enzyme concentration $[E_0]$ can be written as

$$\begin{aligned}[E_0] &= [EH]+[EHS]+[E^-]+[EH_2^+] \Rightarrow [E_0] \\ &= [EH]+\frac{[EH][S]}{K_m}+\frac{K_2[EH]}{[H^+]}+\frac{[EH][H^+]}{K_1}\end{aligned} \tag{1.44}$$

Solving for $[EH]$ we obtain

$$[EH] = \frac{[E_0]}{1+\frac{[S]}{K_m}+\frac{K_2}{[H^+]}+\frac{[H^+]}{K_1}} \tag{1.45}$$

This equation for $[EH]$ can now be inserted into the rate equation:

$$(-r_S) = k_2\frac{[EH][S]}{K_m} = k_2\frac{[S]}{K_m}\frac{[E_0]}{1+\frac{[S]}{K_m}+\frac{K_2}{[H^+]}+\frac{[H^+]}{K_1}} = \frac{k_2[E_0][S]}{K_m+[S]+\frac{K_mK_2}{[H^+]}+\frac{K_m[H^+]}{K_1}} \tag{1.46}$$

and finally

$$(-r_S) = \frac{k_2[E_0][S]}{K_m\left(1+\frac{K_2}{[H^+]}+\frac{[H^+]}{K_1}\right)+[S]} \tag{1.47}$$

We see that, according to the considered model, pH (i.e. the concentration of H^+ ions, $pH = -\log[H^+]$ and $[H^+] = 10^{-pH}$) affects K_m.

When the term $1+\frac{K_2}{[H^+]}+\frac{[H^+]}{K_1} \cong 1$ (optimum pH), we obtain the reaction rate for the optimum pH $(-r_S) = \frac{k_2[E_0][S]}{K_m+[S]}$.

If the pH is not optimum, the term $1+\frac{K_2}{[H^+]}+\frac{[H^+]}{K_1} > 1$ and the reaction rate decreases, approaching zero if the pH is far from the optimum value with $1+\frac{K_2}{[H^+]}+\frac{[H^+]}{K_1} \gg 1$.

The plot of equation (1.47) vs pH is shown in Figure 1.31.

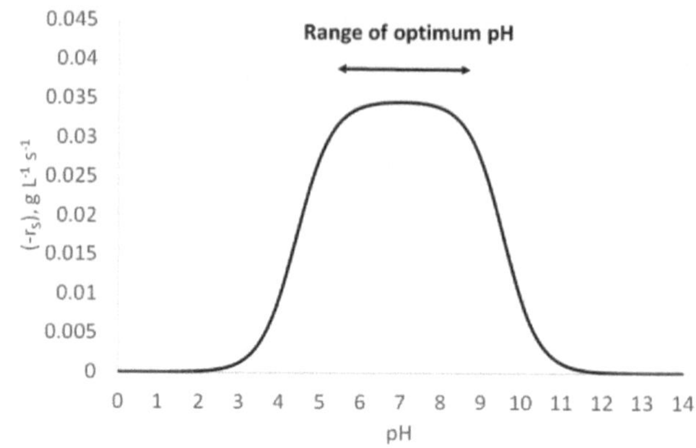

FIGURE 1.31 Example of plot of reaction rate vs pH according to Equation (1.47).

Example 1.13

Consider an enzymatic reaction with $K_m = 10$ mg/L. The reaction is affected by pH with parameters $K_1 = 1 \cdot 10^{-6}$ and $K_2 = 1 \cdot 10^{-8}$. For the substrate concentration of 20 mg/L, calculate the reduction in the reaction rate at pH 5.0 compared to pH 7.0. Also, calculate which substrate concentration is required to have at pH 5.0 the same reaction rate as for $S = 20$ mg/L at pH 7.0.

Solution

The ratio between the actual reaction rate and the maximum reaction rate (for the optimum pH and with substrate in excess) is obtained by re-arranging equation (1.47).

$$\frac{(-r_S)}{V_{max}} = \frac{[S]}{K_m\left(1+\frac{K_2}{[H^+]}+\frac{[H^+]}{K_1}\right)+[S]}$$

With the numerical values of this example

$$\frac{(-r_S)}{V_{max}} = \frac{20\,\text{mg L}^{-1}}{10\,\text{mg L}^{-1}\left(1+\frac{1\cdot 10^{-8}}{10^{-\text{pH}}}+\frac{10^{-\text{pH}}}{1\cdot 10^{-6}}\right)+20\,\text{mg L}^{-1}}$$

For pH = 7.0 we obtain

$$\frac{(-r_S)}{V_{max}} = 0.66$$

Note that pH = 7 is within the optimum pH range for this reaction, however the reaction rate is lower than V_{max} because the substrate concentration is not in excess.

For pH = 5.0 we obtain

$$\frac{(-r_S)}{V_{max}} = 0.50$$

Therefore, at pH 5.0 the reaction rate is reduced by 0.16/0.66 = 24% compared to the value at pH 7.0. In order to obtain $\frac{(-r_S)}{V_{max}} = 0.66$ at pH 5.0, we need to increase the substrate concentration to 39 mg/L, as it can be obtained by trial and error from the equation for $\frac{(-r_S)}{V_{max}}$ above.

1.3.3 Effect of Temperature

Enzymes have an optimum temperature range where their activity is the highest and where they give the highest reaction rates.

Like all chemical reactions, as temperature increases the reaction rate increases, within a certain temperature range. This can be described by using for the reaction parameter k_2 the usual Arrhenius model with activation energy:

$$(-r_S) = k_2 \frac{[E_0][S]}{K_m + [S]} \text{ with } k_2 = Ae^{-\frac{E_a}{RT}} \tag{1.48}$$

This equation gives k_2 increasing with temperature

However, as temperature increases, enzyme deactivation (or denaturation) also occurs. Deactivation is the alteration of enzyme structure due to high temperature, with loss of enzyme activity and decrease in the reaction rate.

Enzyme deactivation can be described by using a simple first-order model:

$$\frac{d[E_0]}{dt} = -k_d[E_0] \tag{1.49}$$

where k_d is the deactivation constant which can be assumed to increase exponentially with the temperature T:

$$k_d = A_d e^{-\frac{E_d}{RT}} \tag{1.50}$$

where A_d and E_d are the frequency factor and the activation energy for the deactivation reaction.

Using the first-order model above, we can calculate the decrease in the concentration of the active enzyme over time, calling $[E_{0IN}]$ the initial concentration of the active enzyme:

$$\frac{d\left[E_0\right]}{\left[E_0\right]} = -k_d dt \Rightarrow \int_{\left[E_{0IN}\right]}^{\left[E_0\right]} \frac{d\left[E_0\right]}{\left[E_0\right]} = -k_d \int_0^t dt \Rightarrow \ln\frac{\left[E_0\right]}{\left[E_{0IN}\right]} = -k_d t \Rightarrow \left[E_0\right] = \left[E_{0IN}\right] e^{-k_d t} \tag{1.51}$$

This equation describes how the concentration of the active enzyme $[E_0]$ decreases with time.

Overall, the temperature effect on the Michealis-Menten kinetics can be included as follows:

$$\left(-r_S\right) = k_2 \frac{\left[E_0\right]\left[S\right]}{K_m + \left[S\right]} = \frac{Ae^{-\frac{E_a}{RT}}\left[E_{0IN}\right]e^{-k_d t}\left[S\right]}{K_m + \left[S\right]} \tag{1.52}$$

In this equation, the term $e^{-\frac{E_a}{RT}}$ increases as the temperature increases, while the term $e^{-k_d t}$ decreases as the temperature increases (k_d increases with temperature).

Therefore, the reaction rate $(-r_S)$ shows an optimum value as a function of the temperature. The value of $-r_S$ is also affected by the time length for which the enzyme is exposed at the given temperature (term $e^{-k_d t}$). The plot of the reaction rate vs temperature according to equation (1.51) is shown in Figure 1.32.

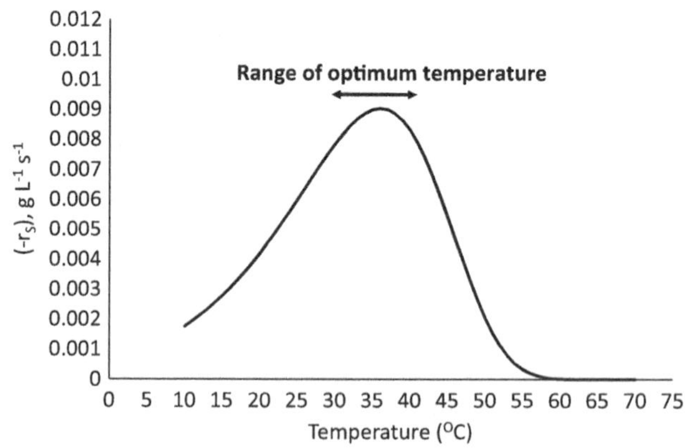

FIGURE 1.32 Plot of reaction rate vs temperature according to Equation (1.52).

Example 1.14

Consider an enzymatic reaction affected by temperature and characterised by the kinetic parameters below according to equations (1.50) and (1.52). Calculate the fraction of the initial enzyme that is still active after spending 60 s at $T = 70$ °C, and calculate the reaction rate if the initial enzyme concentration is 30 mg/L and the substrate concentration is 50 mg/L.

$$A = 4.8 \cdot 10^8 \,\mathrm{mg\, mg_{enzyme}^{-1}\, s^{-1}}$$

$$E_a = 58{,}000 \frac{\mathrm{J}}{\mathrm{mol}}$$

$$A_d = 1.2 \cdot 10^{11} \,\mathrm{s^{-1}}$$

$$E_d = 82{,}000 \frac{\mathrm{J}}{\mathrm{mol}}$$

$$K_m = 15 \,\mathrm{mg\, L^{-1}}$$

Solution

The fraction of the initial enzyme that is still active during denaturation is obtained by re-arranging equation (1.51):

$$\frac{[E_0]}{[E_{0IN}]} = e^{-k_d t}$$

The denaturation constant is given by equation (1.50):

$$k_d = A_d e^{-\frac{E_d}{RT}} = 1.2 \cdot 10^{11}\,\mathrm{s^{-1}}\, e^{-\frac{82{,}000 \frac{\mathrm{J}}{\mathrm{mol}}}{8.31 \frac{\mathrm{J}}{\mathrm{mol \cdot K}}(273.15+70)}} = 0.039\,\mathrm{s^{-1}}$$

Therefore,

$$\frac{[E_0]}{[E_{0IN}]} = e^{-0.039\,\mathrm{s^{-1}} \cdot 60\,\mathrm{s}} = 0.097$$

So, 9.7% of the initial enzyme is still active after exposure to 70 °C for 60 s. Under these conditions, the reaction rate for initial enzyme concentration of 30 mg/L and for substrate concentration of 50 mg/L is obtained by adapting equation (1.52):

$$(-r_S) = \frac{4.8\cdot 10^8\,\mathrm{mg\,mg_{enzyme}^{-1}\,s^{-1}}\,e^{-\frac{58{,}000\frac{J}{mol}}{8.31\frac{J}{mol\cdot K}(273.15+70)}}\,0.097\cdot 30\frac{mg}{L}\cdot 50\frac{mg}{L}}{15\frac{mg}{L}+50\frac{mg}{L}} = 1.57\frac{mg}{L\cdot s}$$

1.3.4 Enzyme Kinetics with Immobilised Enzymes

As we have seen, enzymes can be immobilised on supports to retain them in the reactor or to protect them from environmental effects. In this section, we will see how the Michaelis-Menten kinetics, which in its simple form refers to free enzymes, can be modified to account for immobilisation effects. We will consider first the case of nonporous and then of porous supports.

1.3.4.1 Nonporous Supports

We will use the conceptual model in Figure 1.33 to develop a kinetic model for enzymatic reactions on nonporous supports.

We assume that the substrate has a concentration in the bulk S_b (this is the concentration that so far we have just called S), which decreases to S_s on the enzyme and support surface due to mass transfer limitations. Note that the concentration S_S, although is taken in proximity of the support surface, is still referred to the liquid volume and has the same units as S_b (e.g. g or mol per L). We assume that mass transfer occurs inside a liquid film (actual thickness is in the µm range), which separates the bulk solution from the enzyme-support surface.

The rate of substrate transfer (flux, J_S units of, e.g. mol m^{-2} s^{-1}) across the liquid film to the surface of the support with the immobilised enzyme can be written as

$$J_S = -D_S \frac{dS}{dx} \tag{1.53}$$

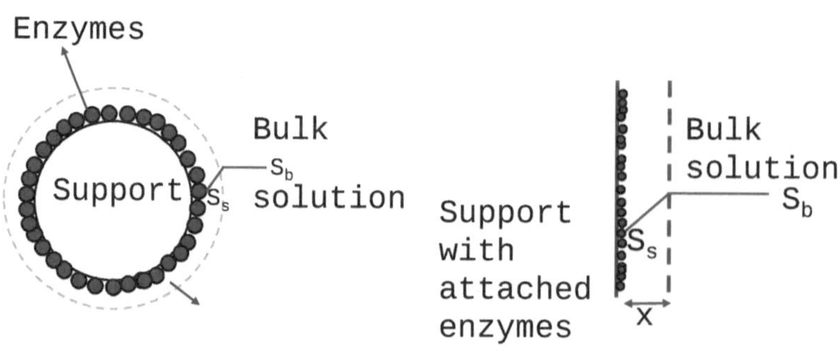

FIGURE 1.33 Conceptual model for enzymatic reactions on nonporous supports.

This is the rate of substrate transfer per unit of support surface. This transfer occurs in the liquid film. D_S (units of, e.g. m^2 s^{-1}) is the diffusivity of the substrate in the film.

Assuming that the substrate profile in the liquid film is linear, the flux can be written as (where x is the thickness of the film):

$$J_S = k_L\left(S_b - S_s\right) \tag{1.54}$$

where

$$k_L = \frac{D_S}{x} \tag{1.55}$$

k_L is a mass transfer coefficient, units of, e.g. m/s.
The rate of substrate reaction on the support surface is

$$\left(-r_S\right)_{surf} = \frac{k_2 E_{surf} S_s}{K_m + S_s} \tag{1.56}$$

where $\left(-r_S\right)_{surf}$ is the substrate rate per unit of support surface (rather than per unit of reactor volume) with units of, e.g. mol m^{-2} s^{-1} (same units as the flux). E_{surf} is the enzyme concentration referred to the support surface in units of, e.g. g/m^2.

At steady state, the rate of substrate transfer across the film and the rate of substrate reaction on the enzyme must be the same, otherwise the substrate would accumulate on the support surface.
Therefore,

$$k_L\left(S_b - S_s\right) = \frac{k_2 E_{surf} S_s}{K_m + S_s} \tag{1.57}$$

Assuming the kinetic parameters k_L, k_2, K_m, the enzyme concentration and the bulk substrate concentration are known, equation (1.57) allows to calculate S_s.

Note that if the mass transfer rate is very large (k_L very large) or the reaction is very slow ($k_2 E_{surf}$ very small), then $S_s \to S_b$, i.e. the substrate concentration on the support surface is approximately equal to the bulk concentration. Instead, if the mass transfer is very slow (k_L very small) or the reaction is very fast ($k_2 E_{surf}$ very large), $S_s \to 0$.

Graphically, S_S can be found as the intersection point of the plots of $J_S = k_L\left(S_b - S_s\right)$ and $\left(-r_S\right)_{surf} = \dfrac{k_2 E_{surf} S_s}{K_m + S_s}$, as shown Figure 1.34.

This graphical representation helps in understanding the effect of the mass transfer rate and of the intrinsic reaction rate (described by the parameter k_2) on the substrate

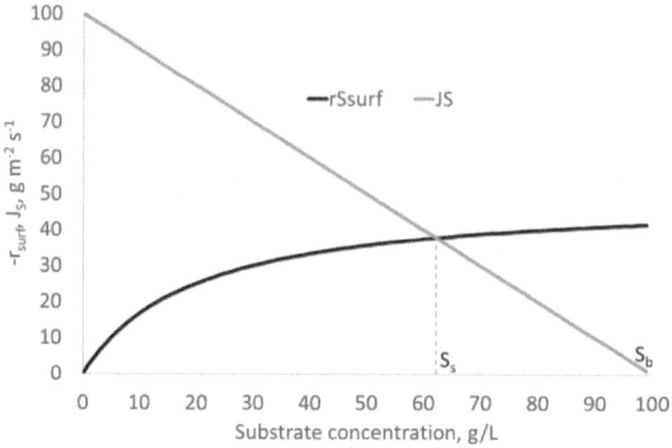

FIGURE 1.34 Graphical determination of the substrate concentration on the surface of a nonporous support (S_s).

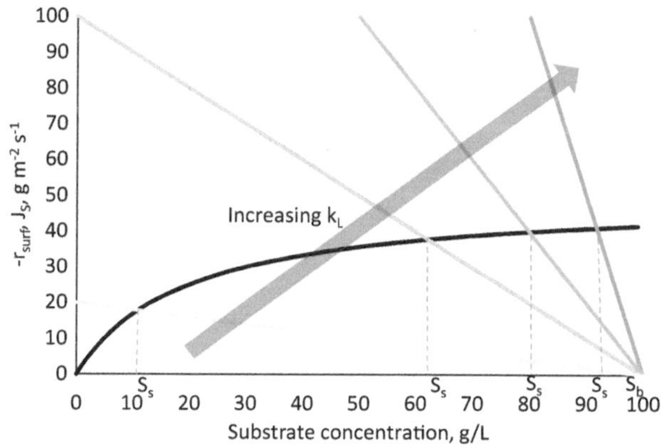

FIGURE 1.35 Effect of the mass transfer coefficient k_L on the substrate concentration on the surface of a nonporous support (S_s).

concentration on the support surface and therefore on the overall substrate removal rate with the immobilised enzyme.

Figure 1.35 shows that, by increasing the mass transfer rate (which is obtained by increasing k_L), S_s increases and becomes closer to S_b. As a consequence of the increase in S_s, the reaction rate also increases.

Figure 1.36 shows that, by increasing the reaction parameter k_2 for fixed mass transfer parameter k_L, S_s decreases and becomes closer to 0. As a consequence of the decrease in S_s, the mass transfer rate ($J_S = k_L\left(S_b - S_s\right)$) increases and, overall, the reaction rate increases.

We have seen that the reaction rate increases if k_L (the mass transfer coefficient) increases.

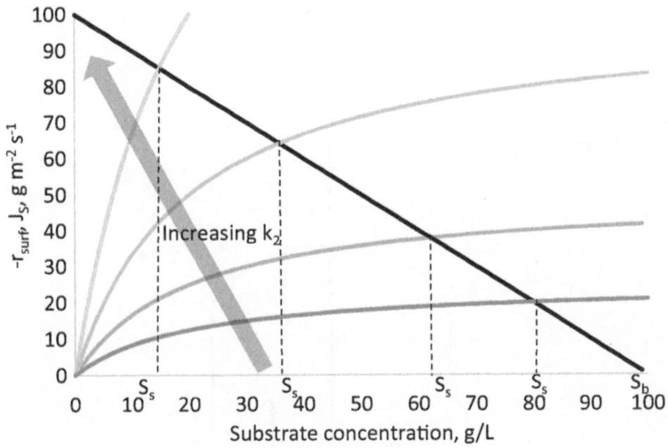

FIGURE 1.36 Effect of the reaction parameter k_2 on the substrate concentration on the surface of a nonporous support (S_s).

Several parameters contribute to the value of k_L (according to equation (1.55), $k_L = \frac{D_S}{x}$): the nature of the substrate and of the reaction medium, temperature and agitation rate.

The nature of the substrate and temperature affect the diffusivity D_S, while the agitation rate affects the film thickness x.

Usually, the agitation rate is the most important effect and a significant increase in the mass transfer, and therefore of the reaction rate, can be obtained by increasing the agitation speed.

So far, we have expressed the reaction rate for nonporous supports per unit of support area, $\left(-r_S\right)_{surf} = \frac{k_2 E_{surf} S_s}{K_m + S_s}$ where $\left(-r_S\right)_{surf}$ has units of, e.g. mol m^{-2} s^{-1} . In order to use the reaction rate in mass balances for batch or continuous reactors, we need to have the reaction rate $\left(-r_S\right)$per unit volume of reactor, e.g. with units of mol m^{-3} s^{-1} .

To obtain the volumetric reaction rate from the surface reaction rate, we need to multiply the latter by the total area of the support per unit volume of reactor:

$$\left(-r_S\right) = \left(-r_S\right)_{surf} \frac{\text{Total area of the support}}{\text{Reactor volume}} = \frac{k_2 E_{surf} S_s}{K_m + S_s} \frac{N_{beads} \pi D_{beads}^2}{V} \quad (1.58)$$

where we are assuming the support is made of spheric beads of diameter D_{beads} with a number of beads N_{beads} in the reactor volume V (working volume). We will refer to the

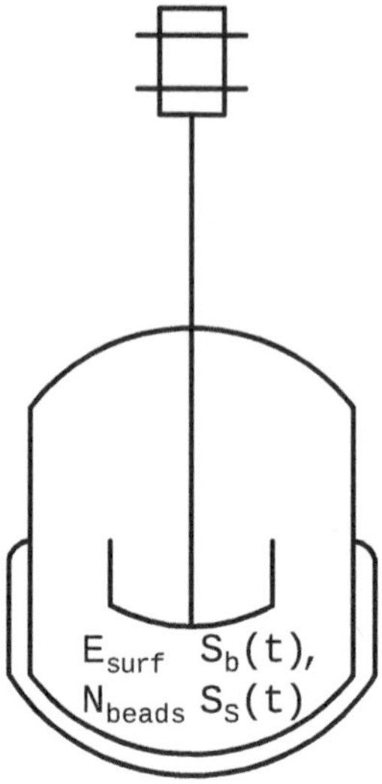

FIGURE 1.37 Conceptual scheme of a batch reactor with enzymes immobilised on a nonporous support.

ratio between the total area of the support and the reactor volume as $\sigma = \frac{N_{beads}\pi D_{beads}^2}{V}$.

Note that in all the cases that we will consider, it will be $\frac{k_2 E_{surf} S_s}{K_m + S_s} = k_L\left(S_b - S_s\right)$.

Once we have the rate equations for immobilised enzymes on nonporous supports, we can include them in the mass balances for the substrate in a batch or continuous-flow reactor with immobilised enzymes.

The conceptual scheme of a batch reactor with enzymes immobilised on a nonporous support is shown in Figure 1.37.

For batch reactors, the mass balance for the substrate in the bulk liquid (S_b) is the following, coupled with the equation stating that the mass transfer and the reaction rates are equal:

$$\frac{dS_b}{dt} = -k_L\left(S_b - S_s\right)a \tag{1.59}$$

$$k_L\left(S_b - S_s\right) = \frac{k_2 E_{surf} S_s}{K_m + S_s} \tag{1.57}$$

The simultaneous solution of equations (1.57) and (1.59) gives the profiles of S_b and S_s over time during the batch reaction. In order to solve the mass balances, the values of the reaction kinetic and mass transfer parameters, of the enzyme concentration on the support surface and of the total area of the support per unit volume of the reactor must be known. In addition, the initial values of S_b and S_s must also be known. In practice, in the solution of the two equations in Excel, the differential equation with $\frac{dS_b}{dt}$ can be discretised as shown in previous sections, with S_s calculated at each time step as a function of S_b by analytically solving the second-order equation $k_L\left(S_b - S_s\right) = \frac{k_2 E_{surf} S_s}{K_m + S_s}$ with S_s as unknown.

The typical profiles of S_b and S_s in batch reactors are shown in Figure 1.38.

The mass balance equations for a batch reactor shown above can also be used to show the effect of changes in the mass transfer and in the reaction rates on the substrate profiles over time.

Figure 1.39 (effect of the mass transfer rate) shows that by increasing k_L (i.e. the mass transfer rate), the concentration S_b decreases faster, i.e. the overall reaction rate

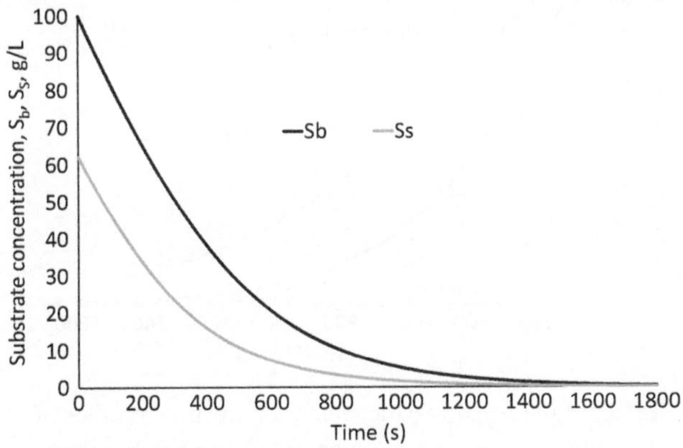

FIGURE 1.38 Typical profiles of the substrate in the bulk (S_b) and on the surface of a nonporous support (S_s) in a batch reactor.

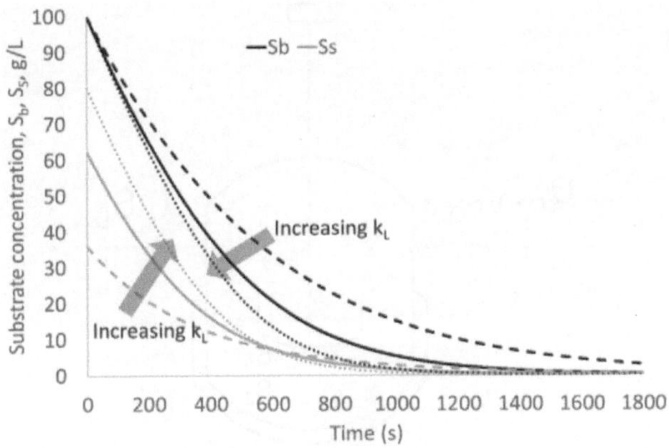

FIGURE 1.39 Effect of the mass transfer coefficient (k_L) on the time profiles of the substrate in the bulk (S_b) and on the surface of a nonporous support (S_s) in a batch reactor.

increases as expected. By increasing k_L, S_s increases, this is what makes the reaction rate to increase.

Figure 1.40 (effect of the intrinsic reaction rate) shows that, by increasing k_2 (i.e. the reaction rate) the concentration S_b decreases faster, i.e. the overall reaction rate increases as expected. By increasing k_2, S_s decreases because the faster reaction brings down the substrate concentration on the support surface.

If the reaction with enzymes immobilised on nonporous supports is carried out in a CSTR, rather than in a batch reactor, the conceptual scheme is shown in Figure 1.41. For CSTRs, we assume that the support with the enzyme is retained in the reactor and so it does not need to be fed continuously. Therefore, we assume that the feed only contains the substrate at concentration S_{bfeed}.

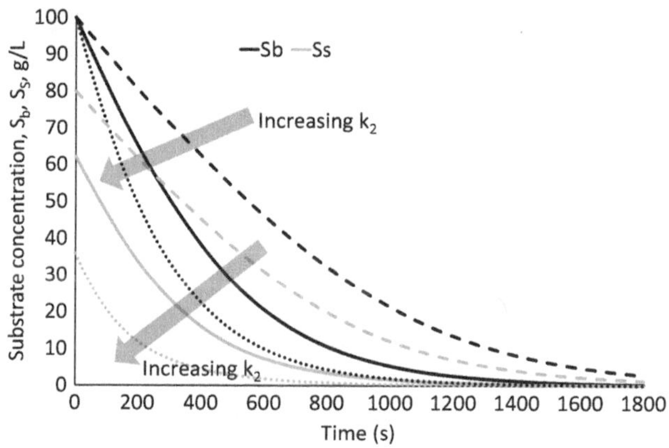

FIGURE 1.40 Effect of the reaction parameter k_2 on the time profiles of the substrate in the bulk (S_b) and on the surface of a nonporous support (S_s) in a batch reactor.

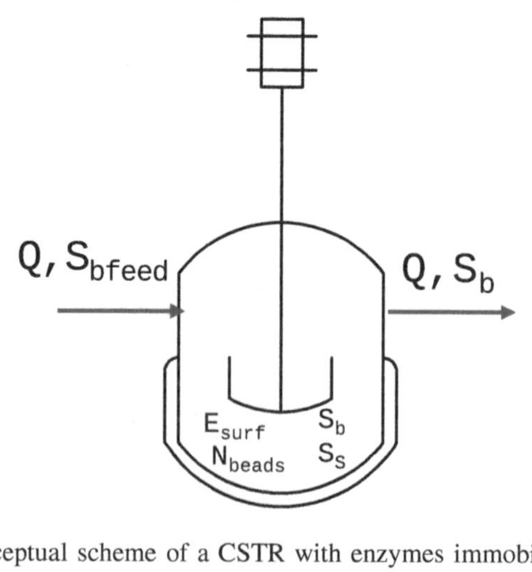

FIGURE 1.41 Conceptual scheme of a CSTR with enzymes immobilised on a nonporous support.

The mass balances for the substrate in the bulk liquid (S_b) and on the support surface (S_s) are the following:

$$QS_{bfeed} = (-r_S)V + QS_b \Rightarrow QS_{bfeed} = k_L(S_b - S_s)aV + QS_b \Rightarrow S_{bfeed} = \frac{k_2 E_{surf} S_s}{K_m + S_s} a\tau + S_b \tag{1.60}$$

$$k_L(S_b - S_s) = \frac{k_2 E_{surf} S_s}{K_m + S_s} \tag{1.57}$$

Note that S_s does not enter or leave the reactor as the enzymes are confined in the reactor.

The simultaneous solution of equations (1.57) and (1.60) gives S_b and S_s at steady state. In order to solve the mass balances, the values of substrate concentration in the feed, of the reaction kinetic and mass transfer parameters, of the enzyme concentration on the support surface and of the total area of the support per unit of reactor volume must be known. In practice, if using Excel these two equations can be solved by isolating S_b from the mass balance for S_b and substituting it into the mass balance for S_s, which will give a second order equation in the unknown S_s, which can be solved analytically.

Figure 1.42 shows the typical profiles of S_b and S_s vs the residence time in CSTRs.

The effect of the mass transfer rate and of the intrinsic reaction rate in CSTRs is analogous to their effect in batch reactors and is shown in Figures 1.43 and 1.44.

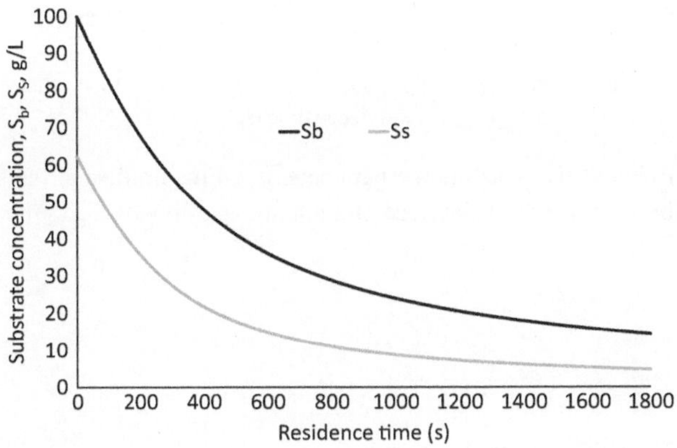

FIGURE 1.42 Typical profiles of the substrate in the bulk (S_b) and on the surface of a nonporous support (S_s) vs the residence time in a CSTR.

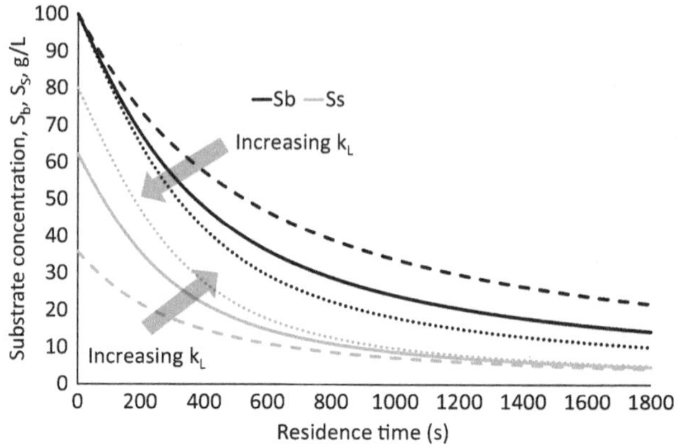

FIGURE 1.43 Effect of the mass transfer coefficient (k_L) on the profiles of the substrate in the bulk (S_b) and of the substrate on the surface of a nonporous support (S_s) vs the residence time in a CSTR.

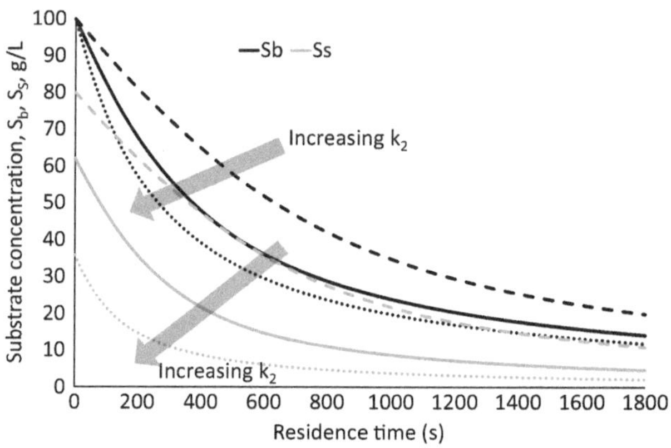

FIGURE 1.44 Effect of the reaction rate parameter k_2 on the profiles of the substrate in the bulk (S_b) and of the substrate on the surface of a nonporous support (S_s) vs the residence time in a CSTR.

Example 1.15

Assume an enzymatic reaction is carried out with the enzyme immobilised on a nonporous support. The parameters for this system are reported below. Calculate the substrate conversion rate per unit area of support.

$$k_L = 0.012\,\mathrm{m\,s^{-1}}$$

$$S_b = 400\,\text{mg L}^{-1}$$

$$k_2 = 450\,\text{mg g}_{\text{enzyme}}^{-1}\ \text{s}^{-1}$$

$$E_{surf} = 5\,\text{g m}^{-2}$$

$$K_m = 20\,\text{mg L}^{-1}$$

Solution

It is required to calculate the substrate concentration on the surface of the support, S_s. This should be done using equation (1.57), which can be solved graphically or analytically.

The graphical method relies on calculating the two plots $k_L\left(S_b - S_s\right)$ and $\dfrac{k_2 E_{surf} S_s}{K_m + S_s}$ as a function of S_s, in the range of S_s from 0 to 400 mg/L (400 mg/L is the substrate in the bulk S_b, which represents the highest possible value for S_s in this example).

With the numerical values for this example (note the conversion of the substrate concentration from mg/L to g/m^3 for consistency of units):

$$k_L\left(S_b - S_s\right) = 0.012\text{m}\,\text{s}^{-1}\left(400 - S_s\right)\text{gm}^{-3}$$

$$\frac{k_2 E_{surf} S_s}{K_m + S_s} = \frac{0.45\,\text{gg}_{\text{enzyme}}^{-1}\text{s}^{-1}5\,\text{g}\,\text{m}^{-2} S_s}{20\,\text{g}\,\text{m}^{-3} + S_s}$$

The plots of $k_L\left(S_b - S_s\right)$ and $\dfrac{k_2 E_{surf} S_s}{K_m + S_s}$ vs substrate concentration is shown in Figure 1.45. S_s is the value at the intersection of the two plots, i.e. $S_s = 227.6\text{gm}^{-3}$. As an alternative to the graphical method, S_s can be determined analytically by rearranging equation (1.57) as

$$\left(-k_L\right)S_s^{\,2} + \left(k_L S_b - k_2 E_{surf} - k_L K_m\right)S_s + k_L S_b K_m = 0$$

This is a second-order equation in the unknown S_s, which can be solved analytically:

$$S_s = \frac{-\left(k_L S_b - k_2 E_{surf} - k_L K_m\right) \pm \sqrt{\left(k_L S_b - k_2 E_{surf} - k_L K_m\right)^2 + 4k_L^{\,2} S_b K_m}}{-2k_L}$$

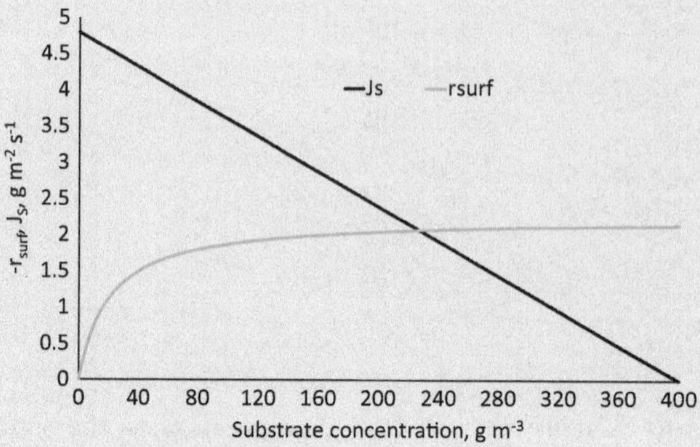

FIGURE 1.45 Graphical determination of S_s for Example 1.15.

With the values of this example, only the solution with the negative sign makes physical sense, and this gives $S_s = 227.6\,\text{g}\,\text{m}^{-3}$ as found with the graphical method.

The substrate conversion rate is therefore:

$$k_L(S_b - S_s) = 0.012\,\text{ms}^{-1}(400 - S_s)\text{gm}^{-3} = \frac{k_2 E_{surf} S_s}{K_m + S_s}$$
$$= \frac{0.45\text{gg}^{-1}_{\text{enzyme}}\text{s}^{-1} 5\,\text{gm}^{-2} S_s}{20\,\text{gm}^{-3} + S_s} = 2.07\text{gm}^{-2}\text{s}^{-1}$$

Example 1.16

Consider the data from Example 1.15, calculate the minimum mass transfer coefficient k_L that is required to achieve a reaction rate not lower than 80% of the maximum possible.

Solution

With the data from Example 1.16, the maximum possible reaction rate is $k_2 E_{surf} = 2.25\,\text{gm}^{-2}\text{s}^{-1}$. 80% of this rate is $1.80\,\text{gm}^{-2}\text{s}^{-1}$. The substrate concentration S_s that corresponds to this rate is obtained from solving:

$$\frac{k_2 E_{surf} S_s}{K_m + S_s} = 1.80\text{gm}^{-2}\text{s}^{-1} \rightarrow S_s = \frac{1.80 \cdot K_m}{k_2 E_{surf} - 1.80}$$
$$= \frac{1.80\,\text{gm}^{-2}\text{s}^{-1} \cdot 20\,\text{gm}^{-3}}{2.25\,\text{gm}^{-2}\text{s}^{-1} - 1.80\,\text{gm}^{-2}\text{s}^{-1}} = 80\,\text{gm}^{-3}$$

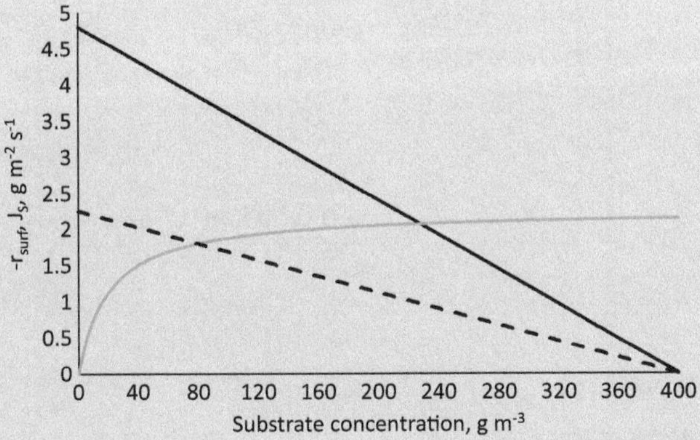

FIGURE 1.46 Graphical visualisation of the reaction rate with the k_L of Example 1.15 (solid black line) and of Example 1.16 (dashed black line).

The k_L that is required to have $S_s = 80$ g m^{-3} can be obtained graphically or analytically by solving equation (1.56) for k_L:

$$k_L = \frac{k_2 E_{surf} S_s}{S_b\left(K_m + S_s\right) - K_m S_s - S_s^2}$$
$$= \frac{2.25\,\text{gm}^{-2}\text{s}^{-1} \cdot 80\,\text{gm}^{-3}}{400\,\text{gm}^{-3} \cdot (20+80)\,\text{gm}^{-3} - 20\,\text{gm}^{-3} \cdot 80\,\text{gm}^{-3} - 80^2\left(\text{gm}^{-3}\right)^2} = 0.00563\,\frac{\text{m}}{\text{s}}$$

The graphical comparison of S_s and reaction rate with the k_L of Example 1.15 and with the lower k_L from this example is shown in Figure 1.46.

Example 1.17

With the data of Example 1.15, calculate the substrate conversion rate per unit of reactor volume in a reactor of volume 1 m^3 where there are 20,000 spherical support particles of diameter 0.5 cm.

Solution

The substrate conversion rate per unit of reactor volume is given by Equation (1.58):

$$\left(-r_S\right) = \frac{k_2 E_{surf} S_s}{K_m + S_s} \frac{N_{beads} \pi D_{beads}^2}{V}$$

With the data of Example 1.15 we have seen that the substrate rate per unit of external area of the support is

$$\frac{k_2 E_{surf} S_s}{K_m + S_s} = 2.07\,\text{gm}^{-2}\text{s}^{-1}$$

Therefore, we obtain

$$(-r_S) = 2.07\text{gm}^{-2}\text{s}^{-1}\,\frac{20{,}000\pi 0.005^2\,\text{m}^2}{1\text{m}^3} = 3.25\text{gm}^{-3}\text{s}^{-1}$$

Example 1.18

Consider a CSTR where an enzymatic reaction is carried out. The enzyme is attached to nonporous spherical support particles. The operating parameters of the CSTR are listed below.

$$k_L = 0.0060\,\text{ms}^{-1}$$

$$S_{bfeed} = 1.2\,\text{gL}^{-1}$$

$$k_2 = 0.75\,\text{g}\,\text{g}_{\text{enzyme}}^{-1}\,\text{s}^{-1}$$

$$E_{surf} = 10\,\text{gm}^{-2}$$

$$K_m = 50\,\text{mg L}^{-1}$$

$$a = 0.7\frac{\text{m}^2}{\text{m}^3}$$

$$\tau = 1.5\text{h}$$

$$V = 500\,\text{L}$$

Calculate the substrate conversion rate in this reactor in kg d^{-1}.

Solution

It is required to calculate S_b and S_s in the CSTR using the mass balances for the substrate in this type of reactor:

$$S_{bfeed} = \frac{k_2 E_{surf} S_s}{K_m + S_s} a\tau + S_b$$

$$k_L\left(S_b - S_s\right) = \frac{k_2 E_{surf} S_s}{K_m + S_s}$$

S_b can be isolated from the first equation:

$$S_b = S_{bfeed} - \frac{k_2 E_{surf} S_s}{K_m + S_s} a\tau$$

and substituted into the second equation

$$k_L\left(S_{bfeed} - \frac{k_2 E_{surf} S_s}{K_m + S_s} a\tau - S_s\right) = \frac{k_2 E_{surf} S_s}{K_m + S_s}$$

which can be re-arranged into a second-order equation in the only unknown S_s:

$$\left(-k_L\right) S_s^2 + \left(k_L S_{bfeed} - k_L k_2 E_{surf} a\tau - k_L K_m - k_2 E_{surf}\right) S_s + k_L S_{bfeed} K_m = 0$$

This equation can be written in the form:

$$a_1 S_s^2 + b_1 S_s + c_1 = 0$$

where

$$a_1 = -k_L = -0.0060\,\text{ms}^{-1}$$

$$\begin{aligned} b_1 &= k_L S_{bfeed} - k_L k_2 E_{surf} a\tau - k_L K_m - k_2 E_{surf} \\ &= 0.0060\,\text{ms}^{-1} \cdot 1{,}200\,\text{gm}^{-3} - 0.0060\,\text{ms}^{-1} 0.75\,\text{gg}_{\text{enzyme}}^{-1} \text{s}^{-1} 10\,\text{gm}^{-2} 0.7 \frac{\text{m}^2}{\text{m}^3} 5400\text{s} \\ &\quad - 0.0060\text{ms}^{-1}\,50\,\text{gm}^{-3} - 0.75\,\text{gg}_{\text{enzyme}}^{-1} \text{s}^{-1} 10\,\text{gm}^{-2} \\ &= -170.7\,\text{gm}^{-2}\text{s}^{-1} \end{aligned}$$

$$c_1 = k_L S_{bfeed} K_m = 0.0060\,\text{ms}^{-1} \cdot 1200\,\text{gm}^{-3}\,50\,\text{gm}^{-3} = 360\,\text{g}^2\text{m}^{-5}\text{s}^{-1}$$

By solving for S_s

$$\begin{aligned} S_s &= \frac{-b_1 \pm \sqrt{b_1^2 - 4a_1c_1}}{2a_1} \\ &= \frac{170.7\ \text{gm}^{-2}\text{s}^{-1} \pm \sqrt{170.7^2\,\text{g}^2\text{m}^{-4}\text{s}^{-2} + 4 0.0060\text{m}\,\text{s}^{-1}\,360\,\text{g}^2\text{m}^{-5}\,\text{s}^{-1}}}{-2 \cdot 0.0060\text{m}\,\text{s}^{-1}} \\ &= 2.11\text{gm}^{-3} \end{aligned}$$

(the negative root is the only one with physical sense).

Correspondingly,

$$S_b = 1200\,\mathrm{gm^{-3}} - 0.75\,\mathrm{gg^{-1}_{enzyme}}\ \mathrm{s^{-1}}10\,\mathrm{gm^{-2}}0.7\frac{\mathrm{m^2}}{\mathrm{m^3}}5{,}400\,s\frac{2.11\,\mathrm{g\ m^{-3}}}{50\,\mathrm{g\ m^{-3}} + 2.11\,\mathrm{g\ m^{-3}}}$$
$$= 52.70\,\mathrm{gm^{-3}}$$

Note the large extent of substrate conversion which is reduced from 1200 to 52.70 g m^{-3}. Also note the significant mass transfer limitations with S_s being considerably lower than S_b.

Since the residence time is 1.5 h and the reactor volume is 500 L, the influent flow rate must be

$$Q = \frac{V}{\tau} = \frac{500\ \mathrm{L}}{1.5\ \mathrm{h}} = 333.3\frac{\mathrm{L}}{\mathrm{h}} = 8.0\frac{m^3}{d}$$

The required substrate conversion rate in kg/d is therefore:

$$Q\left(S_{bfeed} - S_b\right) = 8.0\frac{m^3}{d}(1{,}200 - 52.70)\,\mathrm{gm^{-3}} = 9.2\,\mathrm{kgd^{-1}}$$

1.3.4.2 Porous Supports

In porous supports, enzymes are immobilized inside the pores of the support, i.e. we have enzymes inside the volume of the support in addition to enzymes on the external surface. The conceptual scheme of enzymes immobilised on porous supports is shown in Figure 1.47.

We want now to develop an equation for the rate of enzymatic reactions for enzymes immobilised in porous supports.

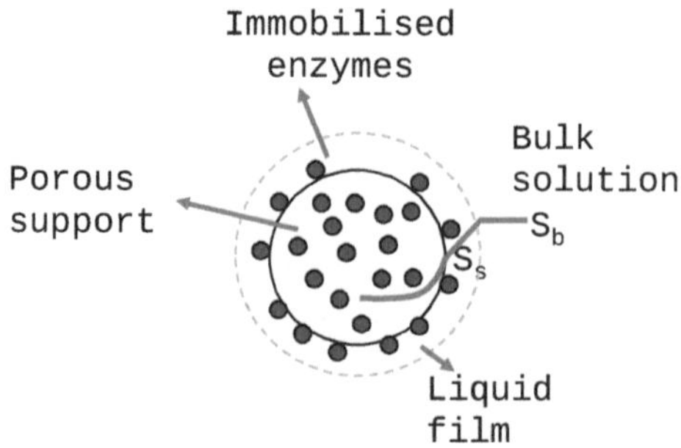

FIGURE 1.47 Conceptual model for enzymatic reactions on porous supports.

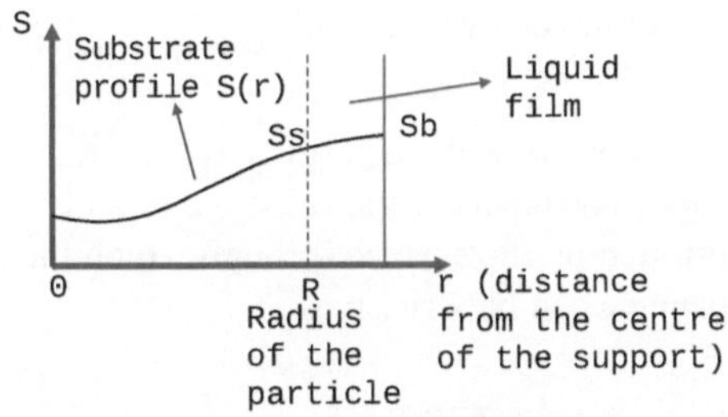

FIGURE 1.48 Conceptual scheme for the substrate profile inside the pores of a porous support.

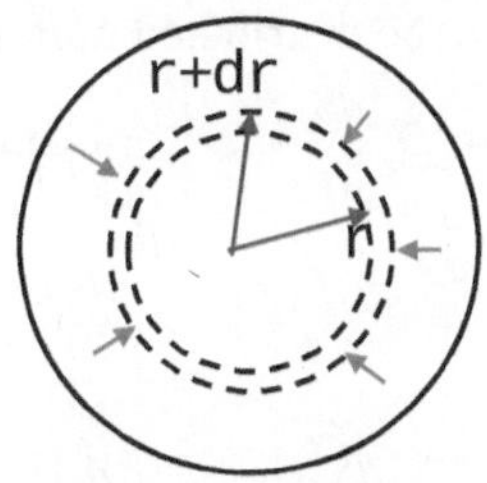

FIGURE 1.49 Conceptual scheme for mass balances inside a porous support.

In porous supports, the substrate needs to transfer from the bulk solution to the external surface of the support and then inside the pores. The substrate concentration decreases inside the pores because of the reaction on the enzymes.

In order to derive an equation of the rate of reaction with enzymes immobilised in porous support, we need to calculate the substrate concentration inside the support.

The conceptual model for the substrate profile inside a porous support is shown in Figure 1.48.

The conceptual scheme that we will use here to develop the substrate mass balance inside a porous support is shown in Figure 1.49.

We assume that, inside the pores of the support the substrate moves by diffusion with diffusivity D_e (D_e stands for effective diffusivity). Although the use of D_e is the same as the molecular diffusivity, D_e is not a physical property of the solute and liquid but depends on the geometry of the particle and pores.

The flux of the substrate per unit of spherical surface is (Fick's law of diffusion): $D_e \frac{dS}{dr}$ (in mol m^{-2} s^{-1} or equivalent units) where the sign is positive (instead of the usual negative sign) because the substrate moves towards the inside of the particle, in the opposite direction to the orientation of the r axis.

The steady state mass balance for the substrate in the volume defined by the radius values r and $r+dr$ is

$$\text{Subs. in}\big|_{r+dr} = \text{Subs.removed by reaction}\big|_{\text{Volume between } r \text{ and } r+dr} + \text{Subs. out}\big|_{r} \quad (1.61)$$

Like in all mass balances, each of these terms has the units of mass (or mol) per unit time, e.g. mol/s or kg/s. The terms of the substrate in and out are therefore equal to the substrate flux multiplied by the surface through which the substrate flows.

Therefore the mass balance can be written as

$$D_e \frac{dS}{dr} 4\pi r^2 \bigg|_{r+dr} = \frac{k_2 E_0 S}{K_m + S} 4\pi r^2 dr + D_e \frac{dS}{dr} 4\pi r^2 \bigg|_{r} \quad (1.62)$$

Here E_0 represents the volumetric concentration of enzyme inside the support (e.g. in kg/m^3), The enzyme is assumed to be distributed uniformly inside the support.

In the equation above, the term $D_e \frac{dS}{dr} 4\pi r^2 \Big|_{r+dr}$ can be written as (note that in general $f(x+dx)$ can be approximated by $f(x) + \frac{df}{dx} dx$):

$$D_e \frac{dS}{dr} 4\pi r^2 \bigg|_{r+dr} = D_e \frac{dS}{dr} 4\pi r^2 \bigg|_{r} + \frac{d}{dr}\left[D_e \frac{dS}{dr} 4\pi r^2 \right] dr \quad (1.63)$$

with:

$$\frac{d}{dr}\left[D_e \frac{dS}{dr} 4\pi r^2 \right] dr = D_e 8\pi r \frac{dS}{dr} dr + D_e \frac{d^2 S}{dr^2} 4\pi r^2 dr \quad (1.64)$$

Therefore, the substrate mass balance can be rewritten as

$$D_e 8\pi r \frac{dS}{dr} dr + D_e \frac{d^2 S}{dr^2} 4\pi r^2 dr = \frac{k_2 E_0 S}{K_m + S} 4\pi r^2 dr \quad (1.65)$$

which can be simplified as

$$D_e \frac{2}{r} \frac{dS}{dr} + D_e \frac{d^2 S}{dr^2} = \frac{k_2 E_0 S}{K_m + S} \quad (1.66)$$

The substrate mass balance (1.66) is a second-order differential equation that can be solved to obtain the profile of S inside the pores at steady state (i.e. with S not changing with time), $S(r)$. In equation (1.66), $r = 0$ is the centre and $r = R$ is the external surface of the support. The profile $S(r)$ determines the reaction rate for immobilised enzymes. This second-order differential equation can be solved with the boundary conditions:

$r = 0 \quad \dfrac{dS}{dr} = 0$ (the substrate profile should be symmetrical, no change at the centre)

$r = R \quad S = S_s$ (at the outer surface of the particle the concentration is equal to S_s)

To solve the differential equation (1.66), the values of D_e, k_2, K_m, E_0 must also be known. The equation can be solved in Excel or with other software. If using Excel, the equation can be solved by discretisation of the r domain, in a similar way as done earlier. Note that, because of the presence of the term $\frac{2}{r}$ in equation (1.66), the numerical solution cannot be calculated for $r = 0$ but must be approximated by the solution for $r = \Delta r$, which is the starting point of the solution. Table 1.19 shows the numerical solution of the substrate mass balance on a porous support.

The equation must be integrated until $r = R$. The value of $S|_{r=r1}$ should be found as the value for which the boundary condition of $S = S_s$ for $r = R$ is satisfied (this condition can be imposed using Solver in Excel).

The typical profile of the substrate inside a porous support in shown in Figure 1.50.

The substrate concentration is the highest at the external surface of the support ($S = S_s$ for $r = R$) and the lowest at the centre of the support ($r = 0$).

In reaction processes limited by mass transfer, the extent of mass transfer limitations is often expressed by the effectiveness factor (η) defined as

TABLE 1.19
Numerical Solution of the Substrate Mass Balance Inside the Pores of a Porous Support

r	$\dfrac{dS}{dr}$	$\dfrac{d^2S}{dr^2}$	S
$r_1 = \Delta r$	0	$\dfrac{1}{D_e}\left(\dfrac{k_2E_0S}{K_m+S} - D_e\dfrac{2}{r}\dfrac{dS}{dr}\right)\Bigg\|_{r=r1}$	$S\|_{r=r1}$
$r_2 = 2\Delta r$	$\dfrac{dS}{dr}\Big\|_{r=r1} + \dfrac{d^2S}{dr^2}\Big\|_{r=r1} \cdot \Delta r$	$\dfrac{1}{D_e}\left(\dfrac{k_2E_0S}{K_m+S} - D_e\dfrac{2}{r}\dfrac{dS}{dr}\right)\Bigg\|_{r=r2}$	$S\|_{r=r2} = S\|_{r=r1} + \dfrac{dS}{dr}\Big\|_{r=r1} \cdot \Delta r$
$r_3 = 3\Delta r$	$\dfrac{dS}{dr}\Big\|_{r=r2} + \dfrac{d^2S}{dr^2}\Big\|_{r=r2} \cdot \Delta r$	$\dfrac{1}{D_e}\left(\dfrac{k_2E_0S}{K_m+S} - D_e\dfrac{2}{r}\dfrac{dS}{dr}\right)\Bigg\|_{r=r3}$	$S\|_{r=r3} = S\|_{r=r2} + \dfrac{dS}{dr}\Big\|_{r=r2} \cdot \Delta r$
…	…	…	…

$$\eta = \frac{\text{actual reaction rate}\left(\dfrac{\text{mol or mass}}{\text{time}}\right)}{\text{reaction rate without mass transfer limitations}} \tag{1.67}$$

For enzyme reactions inside porous supports η can be calculated as

$$\eta = \frac{D_e \left.\dfrac{dS}{dr}\right|_{r=R} 4\pi R^2}{\dfrac{k_2 E_0 S_s}{K_m + S_s}\dfrac{4}{3}\pi R^3} \tag{1.68}$$

For example, for the case shown in Figure 1.50, η is 0.92.

The effect of the mass transfer rate is accounted for in the parameter D_e. Figure 1.51 shows the effect of D_e (i.e. of the mass transfer rate) on the substrate profile inside

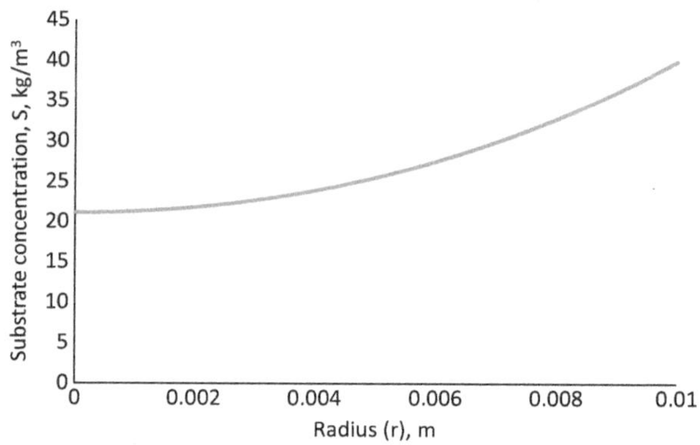

FIGURE 1.50 Typical profile of the substrate inside a porous support, calculated by solving Equation (1.66).

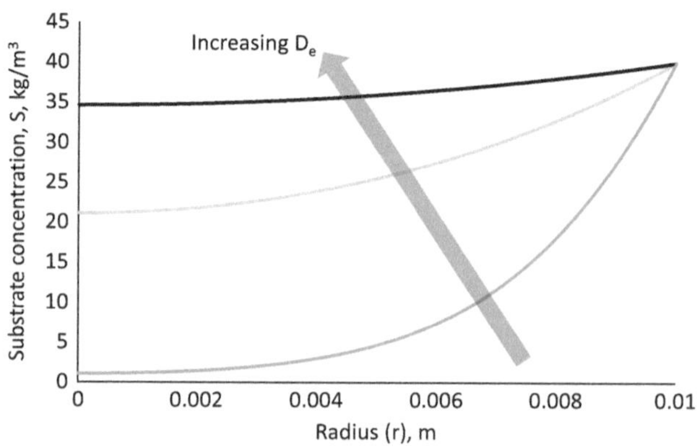

FIGURE 1.51 Effect of D_e on the substrate profile inside a porous support.

the pores. As D_e increases, the substrate concentration inside the pores remains closer to S_s. The effectiveness factors for these simulations were 0.65 (lowest D_e), 0.92 and 0.98 (highest D_e).

The effect of the intrinsic reaction rate (represented by the parameter k_2) on the substrate profile inside the pores is shown in Figure 1.52. As k_2 increases, the substrate concentration inside the pores decreases because the faster reaction consumes the substrate faster. The lower substrate concentration in the pores gives a lower effectiveness factors, which for the simulation in this graph is 0.50 (highest k_2), 0.92 and 0.99 (lowest k_2).

The same effect as k_2 can be obtained by changing the enzyme concentration E_0.

The substrate profiles inside the pores are also greatly affected by the size of the support particles. Larger particles cause larger substrate gradients inside the pores, due to the longer path that the substrate molecules need to travel along. This causes a reduction in the effectiveness factor. For example, for the simulations in Figure 1.53 the effectiveness factor is 0.81 (largest particles), 0.92 and 0.98 (smallest particles).

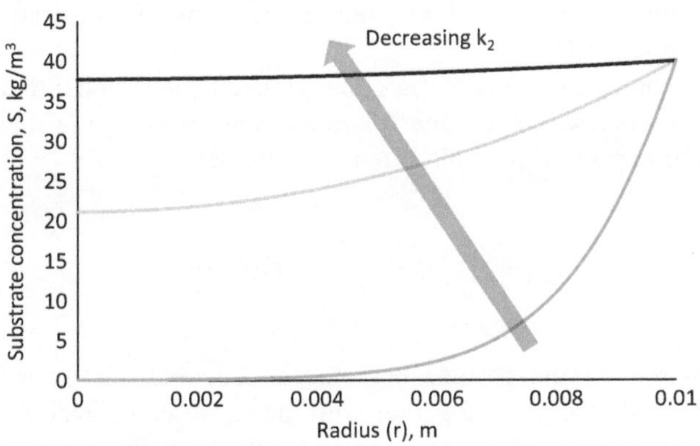

FIGURE 1.52 Effect of the intrinsic reaction rate (represented by the parameter k_2) on the substrate profile inside a porous support.

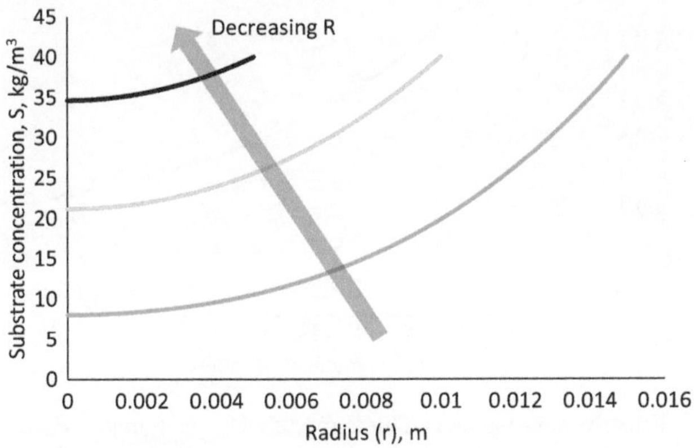

FIGURE 1.53 Effect of the size of the particle on the substrate profile inside a porous support.

The effectiveness factor η is often expressed as a function of the Thiele modulus (Φ), defined as (note that the Thiele modulus is adimensional):

$$\Phi = R\sqrt{\frac{\frac{k_2E_0}{K_m}}{D_e}} \tag{1.69}$$

The Thiele modulus represents the ratio between the reaction and the mass transfer rates. The effectiveness factor can be shown to depend only on Φ and on the ratio K_m/S_s. Low values of Φ indicate low reaction rate and/or fast mass transfer, under these conditions the effectiveness factor is high. High values of Φ indicate high reaction rate and/or slow mass transfer, under these conditions the effectiveness factor is low. Low values of K_m/S_s represent reaction kinetics which are less affected by the substrate concentration, these conditions lead to high η. High values of K_m/S_s represent reaction kinetics, which are more affected by the substrate concentration, these conditions lead to low η.

Figure 1.54 shows the effect of the Thiele modulus on the effectiveness factor.

Similarly to what we have done for nonporous support, the rate of enzymatic reactions can be expressed per unit of reactor volume:

$$\left(-r_S\right) = \eta\frac{k_2E_0S_s}{K_m + S_s}\frac{4}{3}\pi R^3\frac{N_{beads}}{V} \tag{1.70}$$

The concentration on the particle surface can be obtained by solving the substrate balance inside the pores together with the substrate balance in the liquid film surrounding the particle, where the rate is $k_L\left(S_b - S_s\right)$. For simplicity's sake, and also considering the fact that usually in porous particles the internal mass transfer

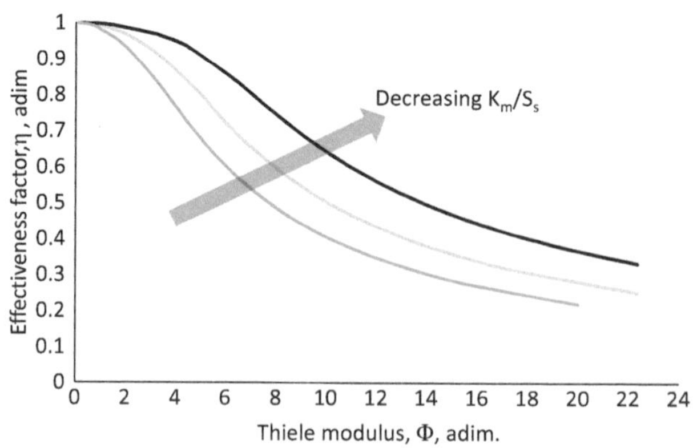

FIGURE 1.54 Effectiveness factor vs Thiele modulus for enzymatic reactions inside porous supports.

limitations are more important than external limitations, we will always assume that $S_s = S_b$.

The mass balances for the substrate when the enzyme is immobilised on porous supports can be written for batch and continuous-flow reactors in the same way as we have seen for nonporous supports.

Substrate mass balance for a batch reactor:

$\frac{dS_b}{dt} = r_S = -\eta \frac{k_2 E_0 S_b}{K_m + S_b} \frac{4}{3} \pi R^3 \frac{N_{beads}}{V}$ (where η should be calculated as a function of S_b at every time step)

Substrate mass balance for a CSTR:

$$QS_{bfeed} = (-r_S)V + QS_b \tag{1.71}$$

with r_S having the same expression used for the batch reactor above, η should be calculated as a function of S_b.

The mass balances for the substrate in the batch reactor and in the CSTR can be solved in the same way described earlier and can give the profile of $S_b(t)$ or S_b as a function of the residence time.

Example 1.19

Calculate the substrate profile inside the pores for an enzymatic reaction on a porous support with the parameters below. Also calculate the effectiveness factor and the Thiele modulus.

$$S_s = 10\,\text{g L}^{-1}$$

$$E_0 = 500\,\text{g m}^{-3}$$

$$k_2 = 7.2\,\text{gg}^{-1}\text{s}^{-1}$$

$$K_m = 5\,\text{g L}^{-1}$$

$$D_e = 2.3 \cdot 10^{-6}\,\text{m}^2\text{s}^{-1}$$

$$R = 1\,\text{cm}$$

Solution

It is required to solve the equation:

$$D_e \frac{2}{r}\frac{dS}{dr} + D_e \frac{d^2S}{dr^2} = \frac{k_2 E_0 S}{K_m + S}$$

which becomes, with the numerical values of this example and after unit conversion, with S expressed in g m^{-3}:

$$2.3\cdot10^{-6}\,m^2s^{-1}\frac{2}{r}\frac{dS}{dr} + 2.3\cdot10^{-6}\,m^2s^{-1}\frac{d^2S}{dr^2} = \frac{7.2gg^{-1}s^{-1}\cdot500\,gm^{-3}S}{5000\,gm^{-3} + S}$$

This should be solved by finite differences as shown in Table 1.19. Table 1.20 reports as an example the first ten calculated rows, obtained by using a Δr of 0.0001 m.

The plot of the substrate concentration as a function of the radius r is shown in Figure 1.55.

From the profile of S, the effectiveness factor and the Thiele modulus can be calculated:

$$\eta = \frac{D_e \left.\frac{dS}{dr}\right|_{r=R} 4\pi R^2}{\frac{k_2 E_0 S_s}{K_m + S_s}\frac{4}{3}\pi R^3} = \frac{2.3\cdot10^{-6}\,m^2s^{-1}\cdot2.61\cdot10^6\,gm^{-4}4\pi0.01^2m^2}{\frac{7.2\,gg^{-1}s^{-1}\cdot500\,gm^{-3}10{,}000\,gm^{-3}}{5000gm^{-3} + 10{,}000\,gm^{-3}}\frac{4}{3}\pi0.01^3m^3} = 0.75$$

TABLE 1.20
Example of Calculations for Example 1.19

r	$\frac{dS}{dr}$	$\frac{d^2S}{dr^2}$	S
0.0001	0.00E+00	2.44E+08	9.24E+02
0.0002	2.44E+04	2.44E+08	9.27E+02
0.0003	3.26E+04	8.19E+07	9.30E+02
0.0004	4.09E+04	8.24E+07	9.34E+02
0.0005	4.91E+04	8.30E+07	9.39E+02
0.0006	5.75E+04	8.36E+07	9.45E+02
0.0007	6.60E+04	8.44E+07	9.51E+02
0.0008	7.45E+04	8.53E+07	9.59E+02
0.0009	8.31E+04	8.63E+07	9.67E+02
0.001	9.19E+04	8.74E+07	9.76E+02
…	…	…	…

Note: The first 10 rows only are shown. Units are m for r and g m-3 for S

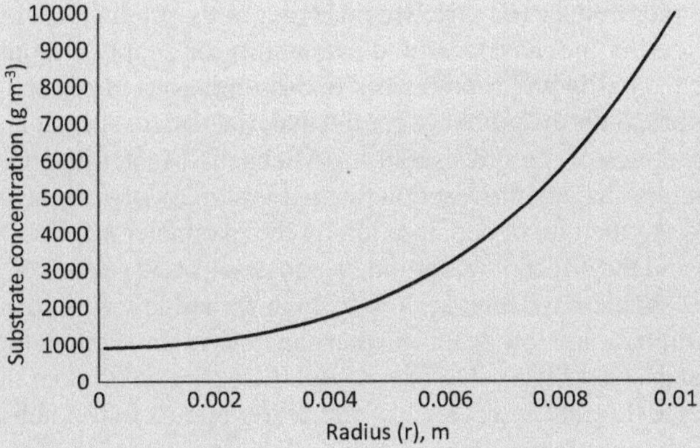

FIGURE 1.55 Plot of S vs the support radius r for Example 1.19.

$$\Phi = 0.01\text{m}\sqrt{\frac{\dfrac{7.2\,\text{gg}^{-1}\text{s}^{-1}\cdot 500\text{gm}^{-3}}{5000\text{gm}^{-3}}}{2.3\cdot 10^{-6}\,\text{m}^2\text{s}^{-1}}} = 5.6$$

Example 1.20

With the data of Example 1.19, calculate the reaction rate per unit volume of reactor if 2000 support particles are present in a reactor of 250 L.

Solution

The substrate conversion rate per unit of reactor volume is

$$\begin{aligned}(-r_S) &= \eta\frac{k_2E_0S_s}{K_m + S_s}\frac{4}{3}\pi R^3\frac{N_{beads}}{V}\\ &= 0.75\frac{7.2\text{gg}^{-1}\text{s}^{-1}\cdot 500\text{gm}^{-3}10{,}000\,\text{gm}^{-3}}{5000\text{gm}^{-3} + 10{,}000\,\text{gm}^{-3}}\pi 0.01^3\,\text{m}^3\frac{2000}{0.250\text{m}^3}\\ &= 45.2\,\text{gm}^{-3}\text{s}^{-1}\end{aligned}$$

1.4 DESIGN AND ECONOMICS FOR ENZYMATIC PROCESSES

In the design of enzymatic processes, the first step is the identification of the most appropriate types of enzymes to be used for the process under consideration. Ideally, the best enzymes should give high reaction rate per unit of enzyme, high selectivity towards the desired product (no or low conversion to by-products), should be easily available (considering purchase cost and availability of supply), should not inactivate or be inhibited easily and should work under conditions of pH and temperature, which are appropriate for the products, for the materials of construction of the vessels and for the economics of the process. In the Michaelis-Menten model and with the notation used in this chapter, the reaction rate per unit mass of enzyme is represented by the parameter k_2, the rate of denaturation by the parameter k_d, the effect of pH by equation (1.47) and the effect of inhibition by equations (1.33) and (1.38). Ideally, the chosen enzymes should have high k_2, low k_d, high K_1 and low K_2 in the pH equation (1.47, note that high K_1 and low K_2 ensure that enzymes maintain their highest activity in a wide pH range) and high value of K_I (inhibition parameter, note that high K_I in equations (1.33) and (1.38) indicates low inhibition effect) for all substances likely to be present in the reaction medium. Clearly, also considering the requirements for low cost and availability, it is usually not possible to find the "ideal" enzyme and the choice of the best enzyme for any given process is always a compromise.

The cost of enzymes is highly variable even for enzymes that catalyse the same reaction, depending on purity, activity, supply chain, etc. The importance of the careful choice of the enzymes and of their supplier is evident if we consider that enzymes and enzyme preparations can be found on the market for prices in the wide price range 1–100 \$/kg and sometimes outside this range. Before purchasing bulk quantities of enzymes, it is recommended to test the suitability of any proposed enzymes at lab scale under conditions comparable to full-scale conditions. The performance of enzymes in a range of temperatures, the resistance to inhibition and to denaturation, the optimum pH range and any important process performance should be tested in the lab before going ahead with scale up and commercialisation.

The concentration of enzyme is another parameter to be chosen during process design. As we have seen in this chapter, higher enzyme concentrations give faster reaction rates, however they also increase the amount of enzyme required with a corresponding increase in the costs. Once the kinetic parameters of the enzymatic reactions under appropriate environmental conditions have been determined, simulations can be done to identify the optimum concentration of enzyme taking reaction rate and cost considerations into account. Higher reaction rate (obtained, e.g. by using higher concentration of enzymes) allows a higher product production per unit of reactor volume and per unit of time. These benefits should be weighted against the higher enzyme costs.

In the design of enzymatic processes, another parameter to be chosen is the substrate concentration. We have seen that for low substrate concentration, higher substrate concentration gives higher reaction rates, however when the substrate concentration is high enough, the reaction rate reaches its maximum value (limited by the concentration of enzyme) with no further increase for higher substrate concentrations. In general, it is desirable to design enzymatic processes running with an initial substrate concentration (or with a substrate concentration in the feed for continuous-flow processes) that is high enough to reach the highest possible substrate conversion rate. However, as for other design choices, the substrate concentration is also a compromise considering substrate availability, solubility and costs, desired product concentration and other practical considerations.

Another design consideration is process type: batch or continuous-flow. As we have seen in this chapter, for enzymatic reactions the average rate of reaction is higher for a batch than for a continuous flow process, because of the higher average substrate concentration. The extent of this kinetic effect depends on the nature of the reaction and on the substrate concentration, in particular on the substrate concentration and on the parameter K_m. In addition to the faster reaction rate, batch processes have also the advantage of easy operation and traceability of the products and feeds. On the other hand, batch processes require downtime between batches for product extraction, filling and cleaning, which reduces the overall productivity. Continuous-flow processes can operate without downtime and this may more than compensate, depending on the process nature and conditions, for the lower reaction rates. In general, continuous-flow processes are used when high product production rates are required while batch processes are preferred when traceability of the product is required.

Another design choice is the use of free or of immobilised enzymes. Free enzymes have the advantages of avoiding the cost of the supports (including the research costs and laboratory tests to find the right support) and of avoiding any mass transfer limitations. However, the use of free enzymes requires higher amounts of enzymes as fresh enzymes need to be added in every batch or continuously for a continuous-flow process. In addition, free enzymes need to be separated from reaction media, which may be expensive, to avoid contamination of the products. Immobilised enzymes have the advantage of being retained in the reaction media, which brings lower requirements for fresh enzymes and lower separation costs. However, finding a suitable support may be challenging and the reaction rate for immobilised enzymes is limited by mass transfer. In general, enzyme immobilisation has to be preferred for large volume processes or for processes where inhibition effects are important with inhibition being attenuated by immobilisation. As usual, it is recommended to test immobilised enzymes in the laboratory, considering the effectiveness of immobilisation and the effect of mass transfer on the reaction rate, before scale up and commercialisation.

The cost analysis of enzymatic processes can be done in the same way as any other processes in the chemical industry. The overall process cost is determined by capital and operating costs. Capital costs include the purchase cost of equipment (e.g. reactors, agitators, separation units, pumps), land, installation, civil infrastructure, electrical connections, control systems and others. Operating costs include enzymes, substrates, other reagents, chemicals for pH control, labour and energy costs. The market price of the product and its production rate must be such to compensate the capital and operating costs and generate a profit for the investors. The payback time, return on investment and other indicators of the economic viability of an enzymatic process can be calculated with the same tools used for other chemical processes, as described in textbooks listed in the Bibliography section.

QUESTIONS AND PROBLEMS

Note that some of the problems in this chapter have been adapted from M.L. Shuler, F. Kargi, M. DeLisa. Bioprocess Engineering. Basic Concepts. Third Edition. Prentice Hall, USA, 2017.

1.1 Determine which are the units of the parameters k_2 and K_m in the Michaelis-Menten equation in the following cases:
(a) $(-r_S)$ expressed in mol L^{-1} s^{-1}, E_0 in mol L^{-1} and S in kg m^{-3};
(b) $(-r_S)$ expressed in g L^{-1} h^{-1}, E_0 in mol L^{-1} and S in mol m^{-3};
(c) $(-r_S)$ expressed in g L^{-1} h^{-1}, E_0 in units L^{-1} and S in kmol m^{-3};
(d) $(-r_S)$ expressed in kg m^{-3} d^{-1}, E_0 in units m^{-3} and S in g L^{-1}

1.2 A definition of the activity of the enzyme lipase is the following (adapted from Enzyme Assay Units, Deerland Probiotics and Enzymes, www.deerland.com/uploads): "One unit of enzyme activity is defined as that quantity of a standard lipase preparation that liberates the equivalent of 1 μmol of fatty acid per minute from the substrate emulsion under the described assay conditions (pH 7.00 and 37°C). The assay is based on the measurement of the amount of free fatty acids formed from an olive oil emulsion in the presence of sodium taurocholate over a fixed time interval."

Assume that the enzyme lipase is sold as a powder with 200 units/g. Assume for simplicity's sake that the produced fatty acid is palmitic acid (in reality it is a mixture of fatty acids).

Calculate the mass of enzyme powder that is needed in the following cases:
(a) For a rate of 10 mmol/L of fatty acid per hour in a 2 L vessel;
(b) For a rate of 5 mmol/L of fatty acid per minute in a 200 mL flask;
(c) For a rate of 2 kg/m^3 of fatty acid per hour in a 5 m^3 vessel

1.3 The enzyme lactase catalyses the hydrolysis of the disaccharide lactose into the sugars – galactose and glucose.

Calculate the production rate of galactose and of glucose in the following cases:
(a) Lactose is being converted at a rate of 10 mol L^{-1} h^{-1}. Express the production rate of galactose and of glucose in mol L^{-1} h^{-1}
(b) Lactose is being converted at a rate of 10 g L^{-1} h^{-1}. Express the production rate of galactose and of glucose in g L^{-1} h^{-1}

1.4 Sucrose is hydrolysed by the catalytic action of the enzyme sucrose as follows:

$$sucrose \xrightarrow{sucrase} product$$

Starting with a sucrose concentration S_0 = 1.0 mmol/L and an enzyme concentration E_0 = 0.01 mmol/L the following kinetic data are obtained in a batch reactor:

t, h	*S*, mmol/L
1	0.84
2	0.68
3	0.53
4	0.38
5	0.27
6	0.16
7	0.09
8	0.04
9	0.018
10	0.006
11	0.0025

Determine the kinetic parameters k_2 and K_m of the Michaelis-Menten model using each of the applicable methods seen in the lectures. Compare and briefly discuss the obtained values of the parameters.

1.5 Substrate *S* and enzyme *E* flow through a CSTR ($V = 6$ L). From the entering and leaving concentrations and flow rate find a rate equation to represent the action of the enzyme on the substrate.

E_0, mol/L	S_0, mol/L	*S*, mol/L	*Q*, L/hr
0.02	0.2	0.04	3.0
0.01	0.3	0.15	4.0
0.001	0.69	0.60	1.2

1.6 Consider the hydrolysis of the triglyceride tripalmitin into palmitic acid and glycerol. Assuming validity of the Michaelis-Menten kinetics, calculate the time profiles of tripalmitin, palmitic acid and glycerol in a batch reactor, expressing the concentrations in g/L.

Kinetic parameters: $k_2 = 2$ g g_{enzyme}^{-1} min^{-1}; $K_m = 10$ g/L.

Rection conditions: $S_0 = 80$ g/L, $E_0 = 3$ g/L. No products present at the start. Reaction length: 4 h.

1.7 Consider the same reaction as Problem 6 but this time in a CSTR, assuming that the substrate concentration in the feed is the same as the initial concentration in Problem 1.6. Calculate the profiles of tripalmitin, palmitic acid and glycerol as a function of the residence time in the reactor in the range 0–4 h.

1.8 A pesticide inhibits the activity of a particular enzyme *E*.

(a) In the laboratory, the initial rate data shown below are obtained with an enzyme concentration of 0.01 mol/l. Is the pesticide a competitive or non-competitive inhibitor? Find the values of the kinetic parameters k_2, K_m and K_I.

S, mol/L	**$(-r_S)$, mol/(L min)·10^6** No inhibitor	**10^{-5} M inhibitor**
$3.3 \cdot 10^{-4}$	56	37
$5.0 \cdot 10^{-4}$	71	47
$6.7 \cdot 10^{-4}$	88	61
$1.65 \cdot 10^{-3}$	129	103
$2.21 \cdot 10^{-3}$	149	125

(c) It is desired to determine the concentration of the pesticide in an unknown sample. 50 ml of an enzyme solution with 0.01 mol enzyme/L are mixed with 50 mL of a solution $8 \cdot 10^{-4}$ M of substrate and with 25 mL of the unknown sample containing the pesticide. The initial rate observed is 18 µmol/(min L). What is the pesticide concentration in the sample assuming no other inhibitors or substrates are present in the sample?

1.9 The enzymatic reaction $S \rightarrow 2P$ is carried out in a CSTR. There are three feed streams to the reactor:

- one stream containing the enzyme at concentration 0.01 mol/L, with a flow rate of 0.2 L/s;
- one stream containing the substrate at a concentration 2 mol/L, with a flow rate of 0.1 L/s;
- one stream containing all the salts and pH buffer required for the reaction. Unfortunately, this stream is contaminated with an inhibitor I (competitive inhibitor) at a concentration 0.1 mol/l. The flow rate of this stream is 0.5 L/s. The reactor has a single effluent stream, where the product concentration is 0.2 mol/L. What is the volume of the reactor?

Kinetic data:
Michaelis-Menten kinetics with competitive inhibition. Assume that the inhibitor concentration I is not affected by the inhibition reaction

$$k_2 = 0.3 \text{ mol S/(mol E·s)}$$

$$K_m = 0.1 \text{ mol S/l}$$

$$K_I = 0.1 \text{ mol I/l}$$

1.10 An enzyme ATPase has a molecular weight of 5×10^4 Da, K_m of 10^{-4} M, and $k_2 = 104$ mol ATP min^{-1} mol enzyme^{-1} at 37 °C. The reaction catalysed is the following:

$$\text{ATP} \xrightarrow{\text{ATPase}} \text{ADP} + P_i$$

This can also be represented as

$$E + S \leftrightarrow ES \xrightarrow{k_2} E + P$$

where S is the ATP. The enzyme at this temperature is unstable. The enzyme inactivation kinetics are first order:

$$\left[E_0\right] = \left[E_{0IN}\right] e^{-k_d t}$$

where E_{0IN} is the initial enzyme concentration and k_d = 0.1 min^{-1}. In an experiment with a partially pure enzyme preparation, 10 mg of total crude protein (containing enzyme) is added to a 1 mL reaction mixture containing 0.02 M ATP and incubated at 37 °C. After 12 h the reaction ends and the inorganic phosphate (Pi) concentration is found to be 0.002 M, which was initially zero. What fraction of the crude enzyme preparation was the enzyme?

Hint: Assume that throughout all the reaction time $[S] >> K_m$.

1.11 An enzyme catalyses the following reaction: $S \rightarrow 2P +$ other products

A batch experiment is carried out using the free enzyme, at initial enzyme concentration E_0 and the plot in Figure 1.56 is obtained.

In this plot, $Y\left(\frac{l \cdot \min}{g}\right) = \left(\frac{1}{-r_s}\right)$ and $X\left(\frac{1}{g}\right) = \frac{1}{S}$

Commercially it is desired to immobilise the enzyme on a nonporous support, in order to facilitate downstream separation. The support is spherical and the diameter of the beads is D_{bead}. The surface density of the enzyme

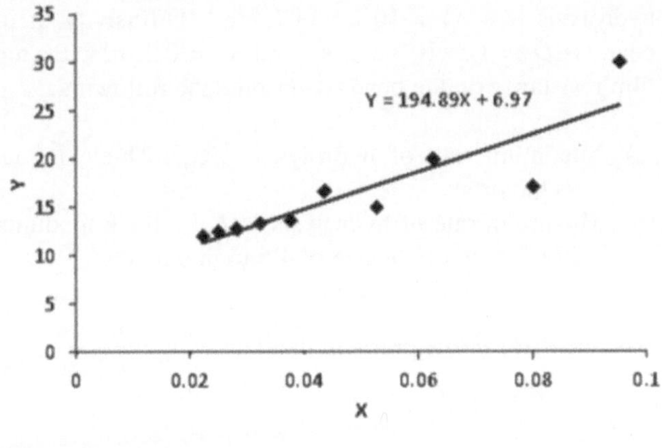

FIGURE 1.56 Plot for Problem 1.11.

on the support is $E_{surface}$. It is desired to run the process in batch, with an initial substrate concentration S_0. It is desired to produce a product concentration P_{out} in a time t_{batch}.

Assuming that there are no diffusion limitations for the immobilised enzyme, calculate the number of support beads per unit volume that are necessary in the batch reactor to achieve the desired performance.

Data:

E_0 = 0.1 g/l (enzyme concentration in the batch reactor in the free-enzyme experiment)
D_{bead} = 1 mm (diameter of the support beads)
$E_{surface}$ = $1\cdot10^{-2}$ g enzyme/cm^2 support (amount of enzyme bound to the surface of the support)
S_0 = 50 g/l (initial concentration of substrate in the commercial process)
P_{out} = 1 mol/l (desired product concentration in the commercial process)
t_{batch} = 2 h (time required to obtain Pout in the commercial process)
MW_S = 70 (molecular weight of species *S*)

1.12 Urea dissolved in aqueous solution is degraded to ammonia and CO_2 by the enzyme urease immobilised on surfaces of nonporous polymeric beads. The conversion rate is limited by the transfer of urea to the surface of the beads through liquid film, and the conversion takes place on the surfaces of the beads. The following parameters are given for the system:

k_L = 0.2 cm/s; K_m = 200 mg/L; $k_2 \times E_{surf}$ = 0.1 mg urea cm^{-2} support surface s^{-1}; S_b = 1000 mg urea/L.

(a) Determine the surface concentration of urea
(b) Determine the rate of urea degradation

1.13 The enzyme urease is immobilised in Ca-alginate beads 2 mm in diameter. When the urea concentration in the bulk liquid is 0.5 mM, the rate of urea hydrolysis is $(-r_S)$ = 10 mmol L^{-1} h^{-1}. Diffusivity of urea in Ca-alginate beads is $D_e = 1.5\times10^{-5}$ cm^2 s^{-1} and K_m = 0.2 mM. By neglecting the liquid film resistance on the beads determine the following:

(a) Maximum rate of hydrolysis $k_2[E_0]$, Thiele modulus and effectiveness factor
(b) Maximum rate of hydrolysis $k_2[E_0]$, Thiele modulus and effectiveness factor for a particle size of 4 mm in diameter.

Assume that for this problem in all cases $\eta = \dfrac{3}{\phi}$

1.14 Calculate the time profile in a batch reactor of the substrate (both substrate in the bulk and attached to the support) for an enzymatic reaction with enzymes attached to a nonporous support.
Parameters: $k_2 = 0.002$ g g_{enzyme} s^{-1}; $K_m = 5$ g/L; $E_{surf} = 25$ g/m^2; $k_L = 0.0008$ m/s; $a = 0.5$ m^2/m^3. Initial concentration of the bulk substrate (S_b) = 50 g/L. Length of the simulation: 6 h.

1.15 Using the same parameters of Problem 1.14, calculate the effect of the residence time in a CSTR, in the range 0–6 h, for the same enzymatic reaction on a nonporous support. Assume that S_{feed} in the CSTR is the same, 50 g/L, as the initial concentration of the bulk substrate in Problem 1.14.

1.16 Consider an enzymatic reaction with the enzyme confined inside porous support particles. Calculate the profile of the substrate S inside the porous particle as a function of the distance from the centre.

Parameters: $k_2 = 0.05$ g g_{enzyme} s^{-1}; $K_m = 10$ g/L; $E_0 = 40$ g/m^3; $D_e = 3.0\times10^{-9}$ m^2/s; $S_b = 70$ g/L; $R = 2$ mm

Assume negligible mass transfer resistance in the liquid film surrounding the particles. Initial concentration of the bulk substrate (S_b) = 70 g/L.

Also, calculate the effectiveness factor and the Thiele modulus.

2 Introduction to Fermentation Processes

This chapter introduces the main concepts on fermentation processes used in this book. Humans have been using fermentation processes for thousands of years. There is evidence of the use of fermented alcoholic beverages in Neolithic China, and the fermentation of milk to make yogurt was probably discovered even before then. Nowadays industrial fermentation processes are carried out at very large scales, for example bioethanol is produced from starchy or sugary feedstocks using yeasts at rates of millions of cubic meters per year. Commercial-scale fermentation processes need therefore to be designed and optimised taking into consideration their kinetics and stoichiometry, mass and energy balances, mass and heat transfer, which are the main focus of this book.

2.1 FERMENTATION PROCESSES AND MICROORGANISMS

In this book we will use the term "fermentation" in the broadest sense to indicate any reactions that involve the growth of living microorganisms. In a stricter sense, the term fermentation is often used to indicate anaerobic reactions, which produce a compound of interest in the liquid phase (e.g. alcoholic fermentation), but in this book we will refer to both aerobic and anaerobic processes. In Chapter 1 we studied enzymes, which although are produced from living organisms, are not themselves "living" and do not grow. Instead, from this chapter onwards, we will consider living microorganisms, which have their own metabolism and grow.

Microorganisms (Figure 2.1) are unicellular (mostly) or multicellular organisms, in the order of micrometers in size, which grow (duplicate) on a carbon source and mineral nutrients.

For engineering purposes, in general terms the metabolism of any microorganisms can be schematised in a simple way (Figure 2.2). The microorganisms live in water and absorb from the external environment the carbon source (often called the "substrate") and the mineral elements that they need to grow (increase in number). The growth of microorganisms occurs by duplication. When they grow, microorganisms generate products that are released into the environment. Depending on the type

DOI: 10.1201/9781003633822-2

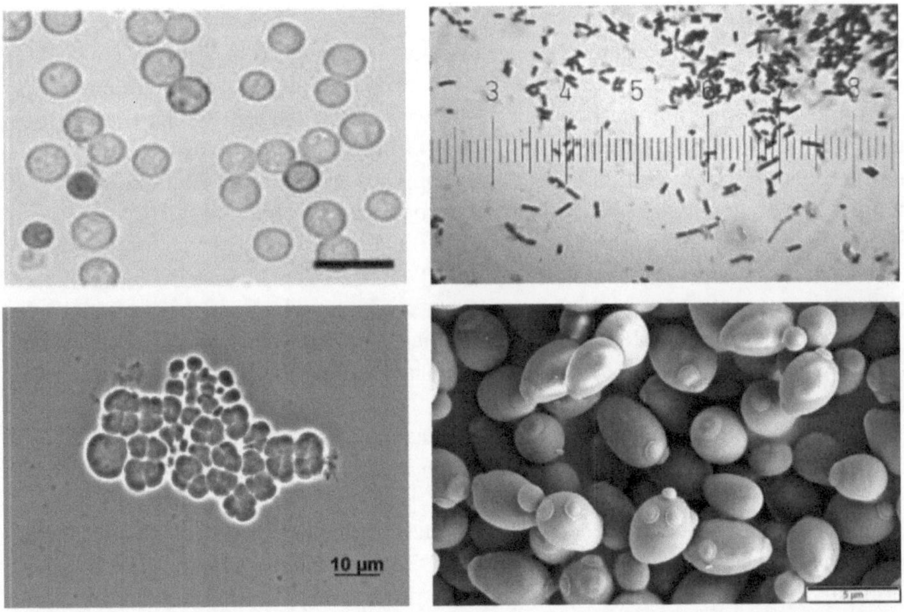

FIGURE 2.1 Examples of microorganisms. Top left: *Aspergillus Niger* conidia (van Leeuwen et al., 2013, Creative Commons license). Bar is 10 µm; Top right: *Lactobacillus acidophilus*. (Bob Blaylock, CC BY-SA, https://creativecommons.org/licenses/by-sa/3.0, https://commons.wikimedia.org/wiki/File:20101212_200110_LactobacillusAcidophilus.jpg.) Numbered ticks are 11 µm apart; Bottom left: *Methanosarcina mazei* (https://bacdive.dsmz.de/strain/7096, Copyright Leibniz-Institut DSMZ-Deutsche Sammlung von Mikroorganismen und Zellkulturen GmbH); Bottom right: *Saccharomyces cerevisiae*, SEM image. (Mogana Das Murtey and Patchamuthu Ramasamy, CC BY-SA 3.0, https://commons.wikimedia.org/w/index.php?curid=52254246.)

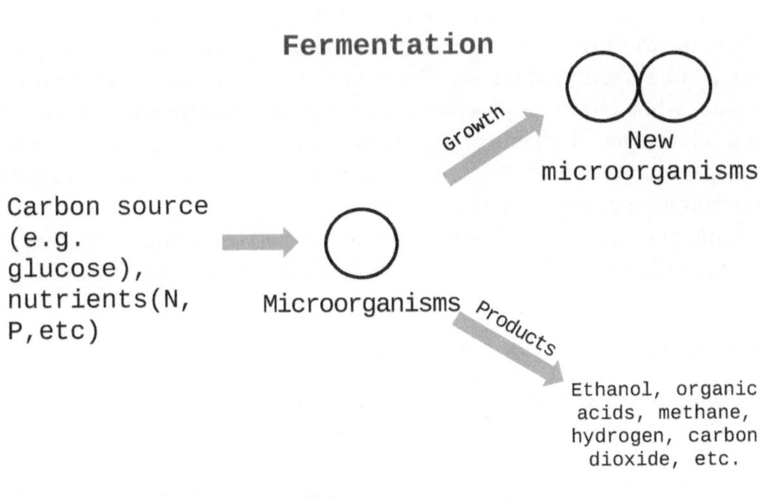

FIGURE 2.2 Conceptual scheme of microbial metabolism.

of microorganisms and the environmental conditions, different products can be generated by the microorganisms, e.g. carbon dioxide (which is produced in most cases), organic acids, methane and many others. In this book, we will generally use "substrate" to express the carbon source and we will use the terms "microorganisms" and "biomass" with the same meaning. In addition to microorganisms, the theories and models presented in this book can be applied to cultures of animal cells, e.g. mammalian or stem cells, which are used in medical and other applications.
Fermentation processes are carried out for two main reasons:

- We are interested in the fermentation products, which are of industrial relevance, for example ethanol, antibiotics, enzymes. In medical applications, viral vaccines and therapeutic proteins are among the products of the fermentation of mammalian cells;
- We want to remove polluting organic substances from the water phase. This is what happens in biological wastewater treatment processes.

When they are used to produce some products, fermentation processes are an alternative to processes that use chemical synthesis. Some products can be produced by both fermentation and chemical processes and the process chosen to be used at industrial scale depends on the cost and availability of the feedstocks. For example, lactic acid (Figure 2.3) can be produced by fermentation of carbohydrates by microorganisms of *Lactobacillus* or other species or by chemical synthesis from acetaldehyde, which is derived from coal or petroleum. Glutamic acid is another example of a substance that can be produced by either fermentation or chemical synthesis, the fermentation process being currently the preferred one in the industry. Generally, fermentation processes have the advantage over chemical synthesis processes, of requiring fewer reaction stages. When fermentation processes are available to produce a certain chemical, microorganisms may be able to convert the feedstock into the desired chemical in just one stage. On the other hand, chemical synthesis may require many reaction stages to obtain the same chemical. Another advantage of fermentation processes, is that they are able to work with wet feedstocks (many organic materials are present in nature with a large fraction of moisture) and in some cases with mixed feedstocks as organic waste, while chemical processes usually require purified feedstocks. On the other hand, advantages of chemical processes are the faster reaction rates and often the easier separation of the products, due to the absence of water and the consequent higher concentration of the products.

Fermentation products are also very important in medical applications, for example production of antibiotics and vaccines (examples of these processes are in Chapter 8).

2.2 CLASSIFICATION OF MICROORGANISMS

Microorganisms can be classified in many ways. From the phylogenetic point of view, microorganisms can be classified into one of the three domains Bacteria, Archaea and Eukarya. Bacteria and Archaea are unicellular microorganisms without a membrane-bound nucleus (prokaryotes). They have a length typically in the range 0.5–5.0

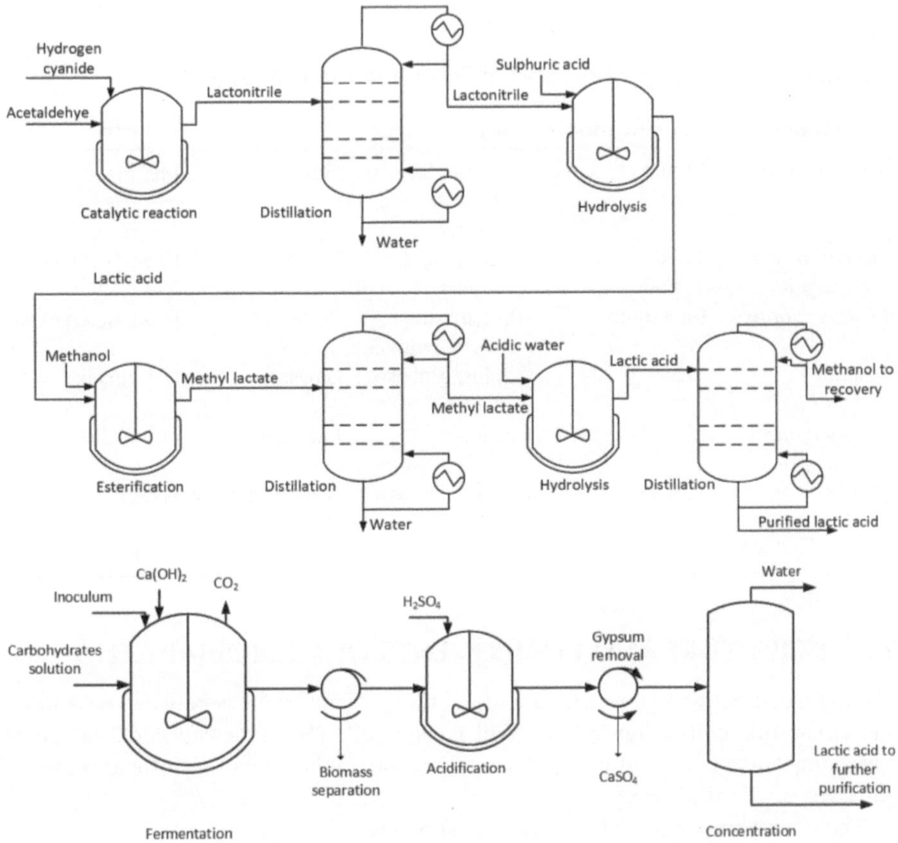

FIGURE 2.3 Process scheme for lactic acid production using chemical synthesis (top) and fermentation processes (bottom).

micrometres and can be of various shapes, e.g. spherical, or rod-shaped. Bacteria typically reproduce by binary fission. Microorganisms in the Eukarya domain are characterised by having a membrane-bound nucleus and membrane-bound organelles. Eukarya microorganisms are much larger than Bacteria or Archaea. Most of the Eukarya microorganisms, used in fermentation processes, belong to the groups of yeasts and moulds. Yeast reproduces by budding. In budding, a small cell is formed from a parent cell. The nucleus of the parent cell splits into a daughter nucleus and migrates into the daughter cell. The bud continues to grow until it separates from the parent cell, forming a new cell. Moulds grow as multicellular filaments called hyphae and reproduce by generating large number of small spores.

Microorganisms are named according to their species, which is a binary combination of the name of the genus followed by a single specific epithet, which indicates some properties or characteristics of the microorganisms, e.g. *Saccharomyces cerevisiae*. Table 2.1 reports some examples of microorganisms used in industrial fermentations.

TABLE 2.1
Examples of Microorganisms and of Fermentation Products

Microorganism	Classification	Carbon Source	Product	Industry
Penicillium spp.	Mould	Sucrose, lactose, glucose, and others	Penicillin	Pharmaceutical
Saccharomyces cerevisiae	Yeast	Glucose and others	Ethanol	Energy, chemical
Bacillus subtilis	Bacterium	Proteins and carbohydrates	Proteases	Food, detergents
Candida antarctica	Yeast	Lipids, glucose, lactose	Lipases	Detergents
Lactobacillus delbruekii	Bacterium	Lactose	Lactic acid	Food, plastics
Various	Various	Sugars, molasses	Amino acids	Food, pharmaceutical, chemical

2.3 STRUCTURE AND COMPONENTS OF MICROBIAL CELLS

The main distinction in the structure of microorganisms' cells is between prokaryotic and eukaryotic cells (Figure 2.4). Prokaryotic cells (Bacteria and Archaea) do not have a nucleus or internal compartmentalised organelles, that are instead present in eukaryotic cells (Eukarya).

The main components of cells are listed below.

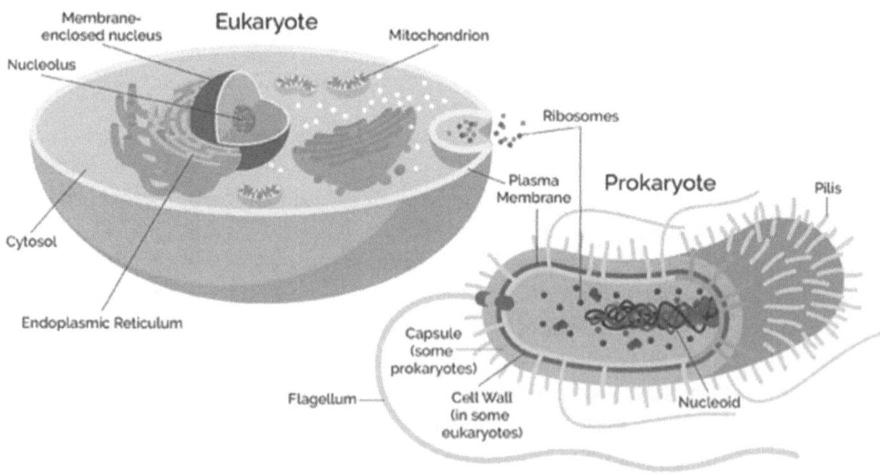

FIGURE 2.4 Differences between eukaryotic and prokaryotic cells. (Adapted from N. Gleitchmann, 2020, www.technologynetworks.com/cell-science/articles/prokaryotes-vs-eukaryotes-what-are-the-key-differences-336095.)

Cell membrane. The cell membrane (Figure 2.5) separates the cell from the external environment, allowing the flux of the molecules in and out of the cell. The cell membrane is selectively permeable, allowing only the in and out flow of substances, which are useful for cellular metabolism. Cell membranes are mainly composed of lipids and proteins. The most abundant lipids are phospholipids, but also glycolipids and sterols are present. Phospholipids are characterised by the presence of fatty acids and phosphoric groups.

Cytoplasm. Cytoplasm includes all the materials inside the cell. Cytoplasm is composed of water and of all the organic and inorganic constituents (enzymes, DNA, RNA, organelles) of the cell. Approximately 80% of the cytoplasm is made of water.

Organelles. Organelles (Figure 2.5) are internal cell compartments separated by the rest of the cytoplasm by membranes. Organelles are present almost exclusively

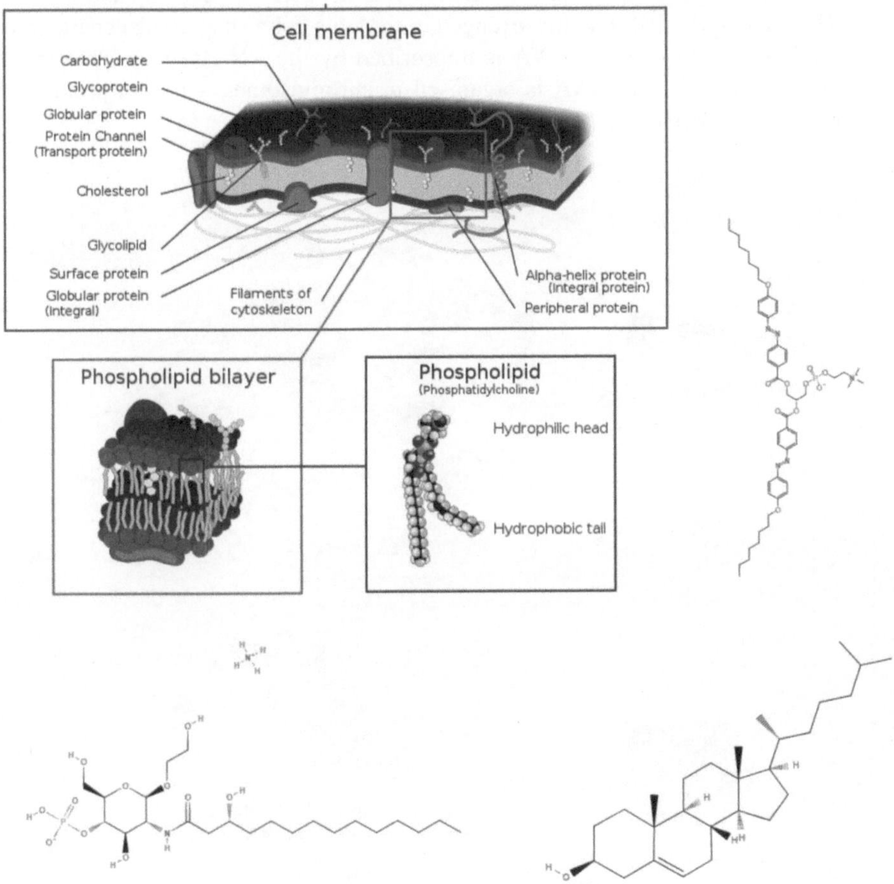

FIGURE 2.5 Cell membrane (top left) and some of its main constituents, phospholipids (top right, Cdpc-phospholipid), glycolipids (bottom left, Hedht-glycolipid), sterols (bottom right, cholesterol). Images from https://commons.wikimedia.org/wiki/File:Cell_membrane_detailed_diagram_4.svg (cell membrane) and from Molview.org (chemical structures).

in eukaryotic cells. Examples of organelles are the nucleus, which contains the DNA, mitochondria, which control the energy generation, and Golgi apparatus, which controls the processing and modifications of proteins.

Enzymes. Enzymes (Figure 2.6), covered in Chapter 1 when used as standalone catalysts extracted from cells, are proteins that increase the rate of the metabolic reactions inside the cell. Enzymes are biological catalysts that decrease the activation energy of the reaction. Enzymatic reactions occur on the active sites of the enzymes, where the substrate(s) bind and is destabilised, allowing for new bonds to form. Enzymes are named after their substrate or after the chemical reaction they catalyse, with the word ending in "-ase".

DNA (deoxyribonucleic acid). DNA (Figure 2.7) contains the genetic information that regulates the growth and metabolism of all living organisms. DNA is composed of nucleotides, which are made of a sugar called deoxyribose, a phosphate group and one of the four bases, cytosine, guanine, adenine and thymine. The nucleotides are spatially arranged in a double helix shape. The genetic information contained in the DNA is transcribed by the RNA to form proteins. In eukaryotic cells, the DNA is organised in chromosomes, while in prokaryotic cells it is dispersed in the cytoplasm.

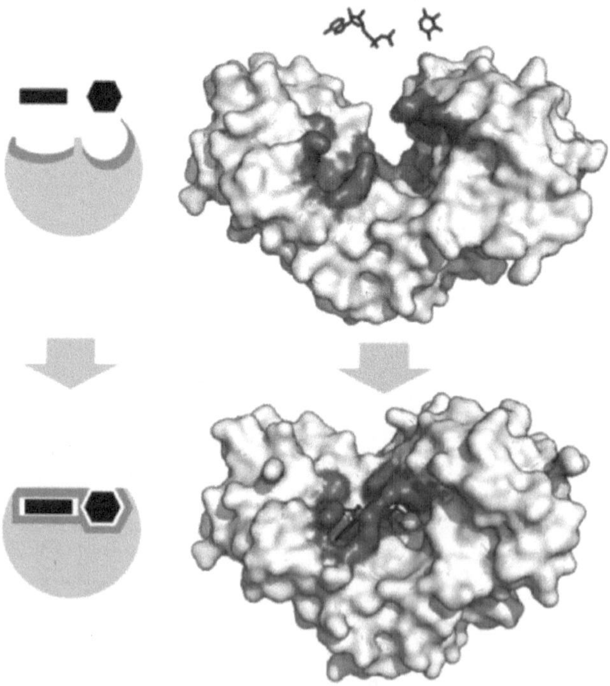

FIGURE 2.6 Scheme of an enzymatic reaction (hexokinase enzyme that binds adenosine triphosphate and glucose) (https://commons.wikimedia.org/wiki/File:Hexokinase_induced_fit.svg).

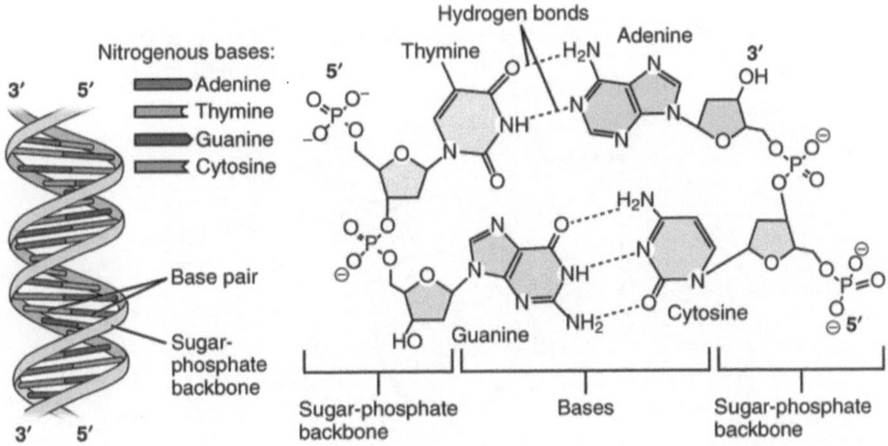

FIGURE 2.7 DNA double helix and the four nucleotides, joined by their phosphate groups and by double bonds (https://en.wikipedia.org/wiki/Nucleotide#/media/File:0322_DNA_Nucleotides.jpg).

RNA (ribonucleic acid). RNA is made of nucleotides (Figure 2.8). Cytosine, guanine and adenine are the same as in DNA, but the nucleotide uracil replaces thymine. The sugar in RNA is ribose, while it is deoxyribose in DNA. Differently from DNA, RNA is usually arranged in a single filament, rather than as a double helix. There are three main types of RNA. Messenger RNA (mRNA) translates the genetic information in the DNA and conveys it to ribosomes. Transfer RNA (tRNA) delivers amino acids to ribosomes, while ribosomal RNA is the main constituent in ribosomes, which assemble the amino acids into proteins using instructions from the mRNA.

ATP and ADP (adenosine triphosphate, adenosine di-phosphate). ATP and ADP (Figure 2.9) are the molecules used by cells as energy carriers. They are made of an adenine base (the same used in DNA and RNA), the sugar ribose and two (ADP) or three (ATP) phosphoric groups. ATP is synthesised from ADP by adding one phosphate molecule. This process requires energy, which is obtained from the catabolic (see Section 2.4) reactions. ATP is an energy-rich molecule which transfers energy to the reactions that produce cell materials (anabolic reactions, Section 2.4). When energy is required, ATP transfers one of its phosphate groups to the molecule that needs energy. When this molecule releases the phosphate group, energy is generated, and the biosynthetic reaction proceeds. The reactions that produce and consume ATP are schematised as follows:

$$\text{ADP} + P_i \rightarrow \text{ATP} \tag{2.1}$$

$$\text{ATP} \rightarrow \text{ADP} + \text{P}_i \tag{2.2}$$

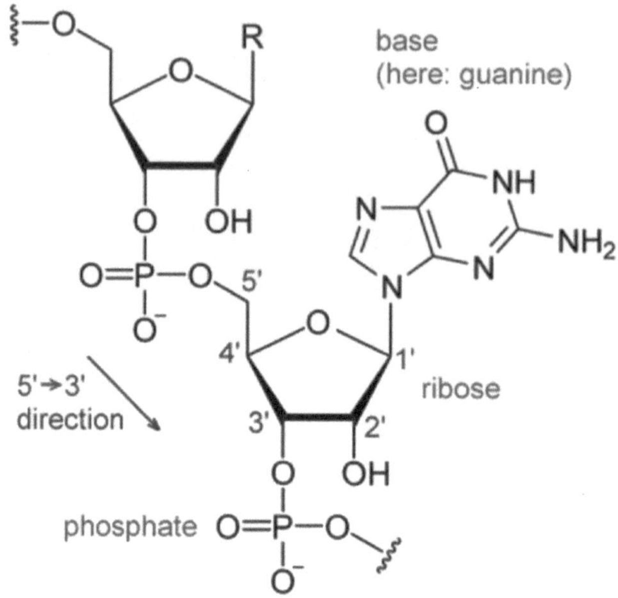

FIGURE 2.8 Scheme (https://commons.wikimedia.org/wiki/File:RNA_chemical_structure.GIF) of a RNA nucleotide with the ribose sugar, a base (here: guanine) and a phosphate group which joins two consecutive nucleotides.

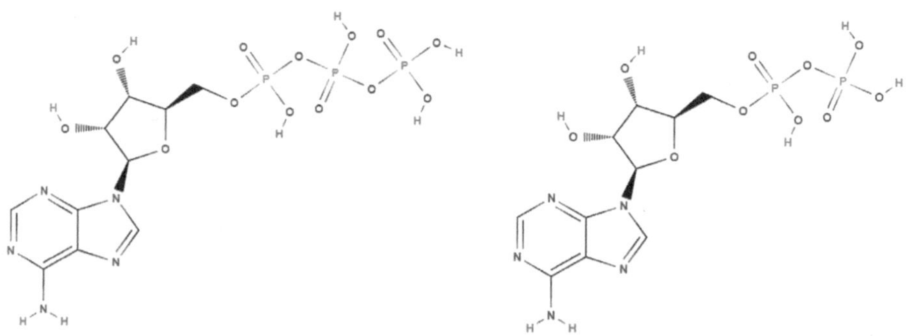

FIGURE 2.9 Adenosine triphosphate (ATP, left) and adenosine diphosphate (ADP, left). Structures from molview.org.

In these reactions, P_i stands for the phosphate group PO_4^{3-}. For simplicity, equations (2.1) and (2.2) and other metabolic reactions in this chapter are shown without the water molecules and the protons, which also take part in the reactions.

NAD⁺, NADH (nicotinamide adenine dinucleotide). NAD^+ and NADH (Figure 2.10, NAD^+ is the oxidised form, NADH is the reduced form) are used in oxidation and reduction reactions in the cells. These molecules are made of two nucleotides

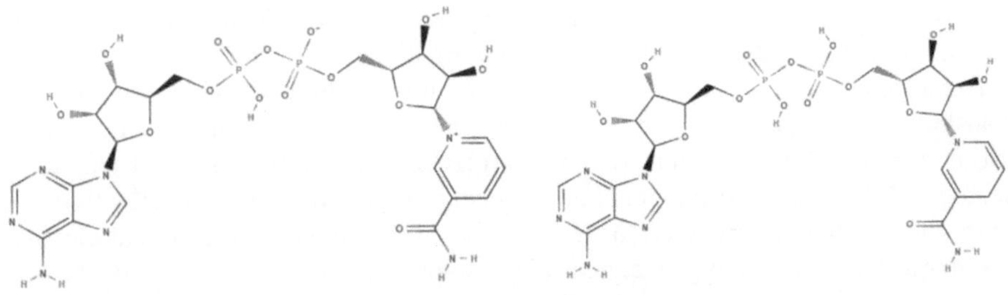

FIGURE 2.10 NAD^+ (left) and NADH (left). Structures from molview.org.

(adenine and nicotinamide) joined through their phosphate groups. Each molecule of NAD^+ removes two hydrogen atoms from the biological molecule to be oxidised. One of these hydrogen atoms is removed as H^+ (proton), the other as H^- (hydride ion, hydrogen atom with one proton and two electrons). The hydride ion is transferred to NAD^+, which becomes NADH, the proton is released into the solution. The reverse reaction occurs when NADH is converted back to NAD^+: the hydride ion from NADH and a proton from the solution are transferred to the molecule to be reduced, regenerating the NAD^+. As alternatives to NAD^+/NADH, the nucleotides FAD/$FADH_2$ (flavin adenine dinucleotide) are also used by cells with the same functions.

2.4 STOICHIOMETRY AND METABOLISM OF CELL GROWTH AND PRODUCT FORMATION

Overall, the growth and product generation by microorganisms can be schematised by the following general chemical stoichiometry:

$$\begin{aligned} &\text{microorganisms} + \text{carbon source} + \text{other elements} \\ &+ (\text{electron acceptor}) \rightarrow \text{new microorganisms} \\ &+ (\text{carbon products}) + (CO_2) + (H_2O) \end{aligned} \tag{2.3}$$

In equation (2.3) the species in brackets are not always produced or used. Electron acceptors are molecules that remove electrons from the carbon source. The most common electron acceptor in fermentation processes is oxygen. Other electron acceptors can be nitrate (NO_3^-), sulphate (SO_4^{2-}) and others.

To understand the concept of electron acceptor, let us consider, for example, the aerobic fermentation of glucose into carbon dioxide (CO_2). In the carbon dioxide molecule, the two electrons in each bond between C and O atoms are more attracted by the O atom than by the C atom. Therefore, we say that each O atom in the CO_2 molecule has oxidation state "–2". The C atom in the CO_2 molecule has oxidation state "+4" so that overall, the CO_2 molecule has oxidation state "0". In the O_2 molecule, the electrons are equally shared between the two O atoms, so the oxidation state of oxygen is 0. Therefore, in the aerobic fermentation of glucose into carbon dioxide each oxygen atom gains two electrons, so oxygen is called "electron acceptor". The substrate (glucose in this case) is called "electron donor".

To understand the cellular metabolism better, it is useful to introduce the concepts of anabolism and catabolism. Anabolism includes all the reactions that lead to the generation of new cell components, e.g. new proteins, cell membranes, etc. Catabolism includes all the reactions that generate the energy required by anabolic reactions. In aerobic fermentation processes (or in processes where there is an external electron acceptor different from oxygen, e.g. nitrate or sulphate), energy is obtained by the oxidation of the substrate to carbon dioxide and water. In aerobic processes, the substrate is the electron donor, and oxygen is the electron acceptor. In anaerobic processes, on the other hand, there is no external electron acceptor. In these cases, in the catabolic reactions a fraction of the substrate is used as electron acceptor and another fraction is used as electron donor, generating the required energy. The fraction of the substrate that is used as the electron acceptor is converted to reduced compounds (e.g. methane or hydrogen) while the fraction of the substrate that is used as the electron donor is converted to oxidised compounds (carbon dioxide). In any cases, transfer of energy from catabolic to anabolic reactions happens through ATP molecules.

As an example of anabolic and catabolic processes, let us consider the metabolism of glucose by microorganisms. The first reactions in this metabolism (Figure 2.11) are the same for aerobic and anaerobic microorganisms and involve the conversion of glucose into pyruvic acid (or pyruvate). This metabolic pathway forms, per each molecule of glucose converted, two molecules of ATP and two molecules of NADH. The stoichiometry for the conversion of glucose into pyruvate (called glycolysis) is

$$\text{Glucose} + 2\text{NAD}^+ + 2\text{ADP} + 2\text{P}_i \rightarrow 2\text{pyruvate} + 2\text{NADH} + 2\text{ATP} \tag{2.4}$$

Part of the metabolites produced in this pathway, instead of being converted to pyruvate, can be used for anabolic pathways for the formation of the various cellular components. For example, the first reaction product, glucose 6-phosphate can be used for the formation of nucleotides (DNA and RNA) and polysaccharides for cell walls. P-glycerate can be used for the formation of amino acids and other cellular components.

While the pathways to produce pyruvate from glucose are the same in aerobic and anaerobic microorganisms, the fate of pyruvate depends on whether we have an electron acceptor or not. In the presence of an electron acceptor, pyruvate is converted to acetyl-coenzyme A, which is then oxidised to carbon dioxide in a sequence of reactions called the tricarboxylic acid cycle (also called the Krebs cycle, Figure 2.12). The oxidants in the tricarboxylic acid cycle are NAD^+ and FAD molecules, which are converted to NADH and $FADH^+$. The stoichiometry of pyruvate oxidation to acetyl-CoA and of the tricarboxylic acid cycle are reported in equations (2.5) and (2.6) respectively:

$$\text{Pyruvate} + \text{CoA} + \text{NAD}^+ \rightarrow \text{acetyl-CoA} + \text{CO}_2 + \text{NADH} \tag{2.5}$$

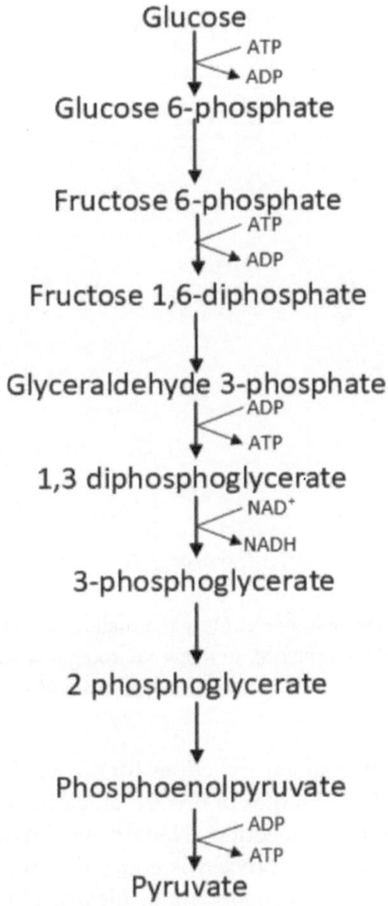

FIGURE 2.11 Steps of the conversion of glucose into pyruvate (glycolysis) according to the Embden-Meyerhof-Parnas pathway. From one molecule of fructose 1,6-diphosphate two molecules of glyceraldehyde 3-phosphate are produced, which both go through the subsequent steps, therefore overall glycolysis produces 2 moles of ATP and 2 moles of NADH per mole of glucose. The molecules in this pathway can also be used for the anabolism to produce the biosynthetic components.

$$\text{acetyl-CoA} + 3\text{NAD}^+ + \text{FAD} + \text{ADP} \rightarrow \text{CoA} + 2\text{CO}_2 + 3\text{NADH} + \text{FADH}^+ + \text{ATP} \tag{2.6}$$

Overall, in aerobic conditions, the oxidation of pyruvate to acetyl-CoA and the oxidation of acetyl-CoA in the tricarboxylic acid cycle yield, per mole of pyruvate, 4 moles of NADH, 1 mol of $FADH^+$ and 1 mol of ATP. Considering that 1 mole of glucose is converted to pyruvate yields 2 moles of pyruvate, 2 moles of NADH and

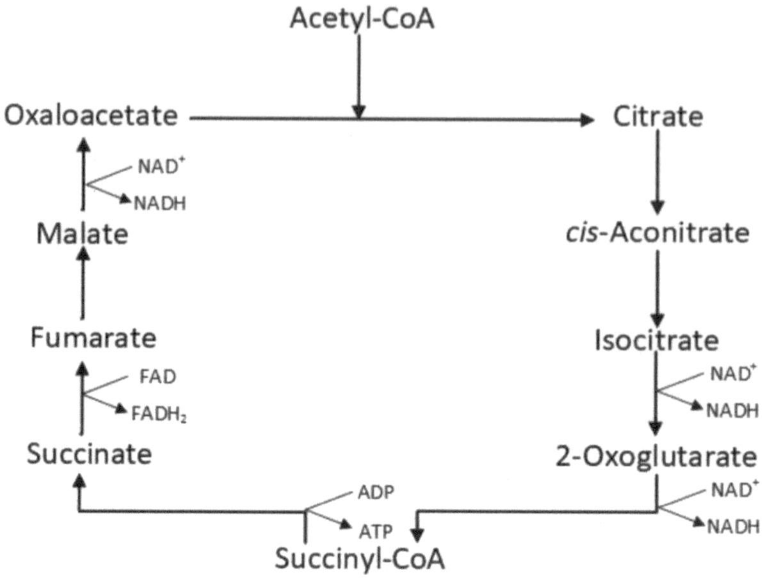

FIGURE 2.12 Tricarboxylic acid cycle. Only the main compounds in the cycle are shown. In this cycle, ADP and ATP are replaced, in many microorganisms, by guanosine diphosphate (GDP) and guanosine triphosphate (GTP) with the same functions.

2 moles of ATP, the overall catabolic reactions for the aerobic metabolism of glucose produce 10 moles of NADH, 2 moles of $FADH^+$ and 4 moles of ATP.

In aerobic conditions the nucleotides NADH and $FADH^+$ are oxidised back to NAD^+ and FAD by oxygen. This process is coupled to the generation of ATP from ADP and is called oxidative phosphorylation. Ideally, although there is variability for different microorganisms and different environmental conditions, oxidative phosphorylation produces 3 moles of ATP per mol of NADH oxidised and 2 moles of ATP per mole of $FADH_2$ oxidised (equations (2.7) and (2.8)):

$$NADH + 3ADP + 3P + 0.5O_2 \rightarrow NAD^+ + 3ATP + H_2O \tag{2.7}$$

$$FADH_2 + 2ADP + 2P + 0.5O_2 \rightarrow FAD + 2ATP + H_2O \tag{2.8}$$

Therefore, the aerobic catabolism of glucose can produce up to 38 moles of ATP per mole of glucose. Of these 38 moles of ATP, 2 come from the glycolysis, 2 from the tricarboxylic acid cycle and 34 from the oxidative phosphorylation (of which 30 come from NADH and four from $FADH_2$). These ATP molecules are used as energy vectors for the anabolic reactions that produce the cell components.

Under anaerobic conditions there is no external electron acceptor that oxidises the reduced nucleotides NADH to their NAD^+ form. Therefore, under anaerobic conditions the NADH, produced by the glycolysis reaction, are oxidised to NAD^+ by the glucose itself or by other metabolic intermediates, with the overall result that reduced products are generated and released into the medium. These products are organic acids (acetic, propionic, butyric and others), hydrogen, methane, alcohols. These products have often considerable value and may be the reason for carrying out the fermentation. It is important to note that under anaerobic conditions the only metabolism that generates energy (in the form of ATP) is the glycolysis, in the absence of the pyruvate oxidation to acetyl-CoA and of the tricarboxylic acid cycle. As we have seen, the glycolysis generates 2 moles of ATP per mole of glucose. Therefore, the amount of energy that can be generated per mole of glucose converted in the catabolic pathways is much lower under anaerobic conditions (2 moles of ATP) than under aerobic conditions (up to 38 moles of ATP). Considering that the energy requirements for anabolic reactions are approximately the same under aerobic and anaerobic conditions, it follows that under anaerobic conditions a much larger fraction of the glucose needs to be used for catabolism than under aerobic conditions. Therefore, the number of microorganisms produced per unit mass of glucose consumed is usually much lower under anaerobic conditions than under aerobic conditions. Consequently, under anaerobic conditions a large fraction of the glucose is converted to the reduced products. In most cases, these reduced products and not the microorganisms are the desired products of the fermentation, therefore the low growth yield of anaerobic microorganisms is a favourable outcome. However, the low generation of microorganisms under anaerobic conditions also makes the fermentation slower, which is usually a negative consideration.

2.4.1 Metabolic Products

Products are produced during the cell metabolism for different reasons. One important class of products are those produced during cell growth (primary metabolites). Primary metabolites are essential products of cell metabolism and are produced during the growth of microorganisms. Primary metabolites can be produced during anaerobic or aerobic growth. Production of primary metabolites under anaerobic conditions is more common. As we have seen in Section 2.4, under anaerobic conditions there are no external electron acceptors and therefore the substrate acts both as electron acceptor and electron donor, producing oxidised (carbon dioxide) and reduced (organic acids, ethanol, methane) compounds. Under aerobic conditions, oxygen is the external electron acceptor and normally the only final metabolic products are carbon dioxide and water. However, in some cases, e.g. lack of certain enzymes, primary metabolites may be produced under aerobic conditions.

Another important class of products are called secondary metabolites. These metabolites are normally not produced during normal cell metabolism and during growth and are mainly produced when cell growth has stopped or has slowed down considerably. Secondary metabolites are produced when microorganisms are stressed by environmental factors, such as deficiency in any nutrients, trace elements or an unbalanced mineral medium. For example, polyhydroxyalkanoates (PHAs) are biological polymers that are produced when the supply of carbon is more than the supply of nitrogen or phosphorus (or of other elements) or when the metabolic apparatus is not ready for a fast growth on the available carbon source. As another example, production of citric acid is carried out by the aerobic microorganisms *Aspergillus Niger* and is favoured using iron-deficient media, because iron is needed for the enzymes that convert citric acid. Many secondary metabolites have important applications, for example penicillin and L-glutamic acid (Chapter 6). L-glutamic acid is produced when the mineral medium is unbalanced because of vitamin (biotin) limitation, because of the presence of surfactants or for other reasons.

2.4.2 Maintenance and Endogenous Metabolism

Equation (2.3) describes the use of the substrate to produce new microorganisms ("growth" process). In addition to growth, other types of metabolism take place in fermentation processes. One of these metabolisms is called maintenance. Maintenance metabolism is the use of substrate not related to microbial growth. This metabolism includes substrate use to provide the energy needed to maintain the internal osmotic pressure, to repair cellular components, etc. The substrate used for maintenance is oxidised to generate energy for these processes. Another process, which is nongrowth-related is endogenous metabolism. Endogenous metabolism includes the self-oxidation by the cells of internal components (e.g. enzymes), which are no longer required for the metabolism. This later metabolism does not involve the use of the substrate, but it causes a reduction in the concentration of microorganisms. Therefore, endogenous metabolism can be assimilated to a death process for the microorganisms.

2.5 QUALITATIVE TIME PROFILES OF MICROBIAL GROWTH AND PRODUCT FORMATION

Let us assume that we have a small number of microorganisms of the same species in suspension in water. We assume that all the elements necessary for growth (e.g. N, P, metals, etc) and one biodegradable carbon source (e.g. glucose, ethanol, etc) are present in the water solution. We also assume that the pH and temperature of the medium are adequate for the growth of microorganisms. What will happen to the microorganisms and substrate concentration over time in this system? Initially, the microorganisms will need to adapt to the carbon source (acclimation phase). In the acclimation phase, the microorganisms will develop the internal enzymes and the metabolic apparatus needed to degrade the substrate. The length of the acclimation phase might vary from several hours to days, depending on the microorganisms and on the substrate. If the microorganisms are already acclimated to the substrate, there will be no acclimation phase.

At the end of the acclimation phase, the microorganisms are adapted to the substrate and start growing on it. During the growth phase, the concentration of microorganisms increases and the substrate concentration decreases. When the substrate becomes limiting, the growth rate decreases until it stops completely when the substrate is completely removed. Figure 2.13 shows, qualitatively, the profiles of substrate and biomass during this experiment (a process like this, with no continuous feeding of the substrate is called a batch process, see Section 2.6).

As we have seen, in many cases, fermentation processes are carried out to produce a certain product of interest. As we have seen in Section 2.4.1, products can be

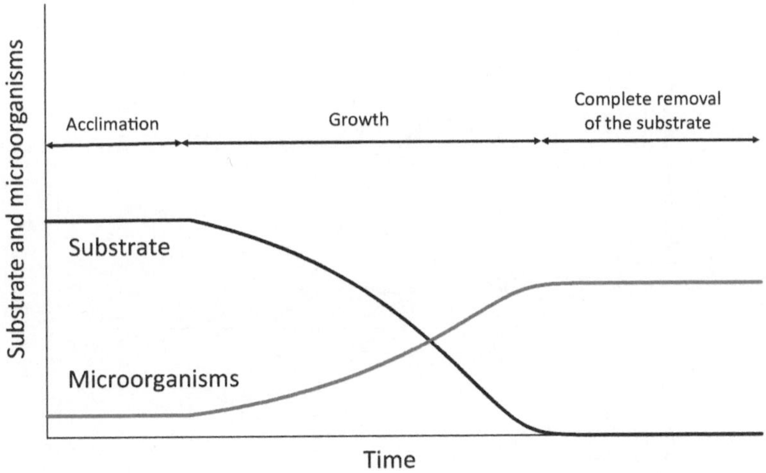

FIGURE 2.13 Qualitative profiles of microorganisms and substrate during batch growth.

produced either during biomass growth on the substrate (primary metabolites), or at the end of biomass growth when the substrate has been (almost) completely removed (secondary metabolites). Figure 2.14 shows the qualitative profiles of substrate, microorganisms, and products during batch growth.

2.6 PROCESSES AND REACTORS USED IN FERMENTATION

The three main fermentation processes used in the industry are batch, continuous and fed-batch. In batch fermentations, the batch reactor is inoculated with microorganisms, a substrate is added, microorganisms grow (and possibly produce a product) and when the maximum biomass concentration is reached the reactor is emptied to start a new batch. Batch fermentations are relatively easy to manage and to keep free of microbial contaminants as there is no continuous addition of the medium. Disadvantages of batch fermentations are the downtime required for cleaning and emptying, the need to add a fresh inoculum for each batch, and the low initial reactor rates, due to the low biomass concentration at the start of the batch.

In continuous fermentation the substrate and nutrients are continuously added to the reactor, and the medium which contains microorganisms and products is continuously removed. Continuous fermenters with a growth-limiting substrate are also called chemostats or, when the concentration of microorganisms is measured from the turbidity of the culture and controlled by manipulation of the residence time, turbidostats. Ideally, the continuous reactor's feed is sterile (does not contain microorganisms) and microorganisms are added only at the start-up (inoculation) of the reactor. At the end of the start-up, typically continuous reactors reach a steady state, i.e., a condition where the all the measured parameters (concentrations and temperature) are constant with time. Compared to batch processes, continuous processes have the advantages of having no downtime and higher biomass concentrations (because at

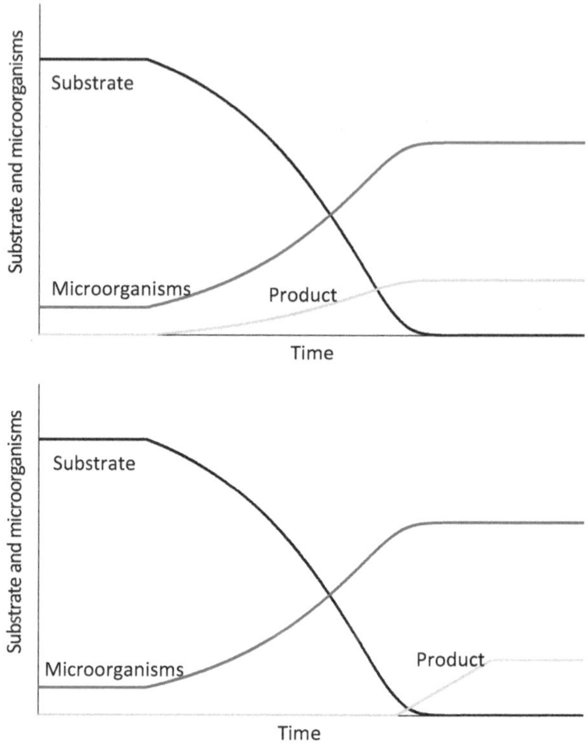

FIGURE 2.14 Qualitative profiles of substrate, microorganisms, and product during batch growth. Top: product produced during biomass growth (primary metabolite). Bottom: product produced at the end of biomass growth (secondary metabolite).

steady state typically most of the substrate is removed and the biomass concentration is the highest), with consequent higher reaction rates and higher productivities. In addition, the fact that continuous processes are operated at steady state with constant values of substrate, biomass, and nutrients concentration, makes it easier to study the effect of these parameters on the fermentation. However, continuous processes need careful control of the residence time and are not easy to use to produce secondary metabolites, which require a sequence of growth and non-growth conditions. Furthermore, continuous processes are sensitive to contamination because of the continuous feeding and more at risk of culture mutations for long run times.

Another type of fermentation process is the fed-batch. In fed-batch processes the volume of the reaction medium varies with time. The fed-batch cycle is started with a low volume of medium, residual from the previous cycle. The sterile feed, containing substrates and nutrients, is then added slowly over the whole length of the cycle, until the liquid medium reaches the maximum volume. Afterwards, the reaction medium, containing microorganisms and products, is removed from the reactor, leaving a small residual volume in the reactor and adding the feed, starting a new cycle. Compared to batch processes, fed-batch fermentations give usually higher productivities but are more at risk of contamination because of the continuous feeding. Fed-batch processes give lower volumetric productivities than continuous processes, because of the downtime (required for withdrawal of the liquid medium) and because part of the volume of the reactor is not withdrawn and does not contribute to the productivity. However, fed-batch processes are highly flexible because they allow changing of the fermentation conditions and of the medium composition during feeding, which is particularly suitable to produce secondary metabolites. Furthermore, the concentration of microorganisms during feeding is comparable to the concentration of continuous processes and higher than at the beginning of batch processes, therefore fed-batch processes are a good compromise between batch and continuous processes and are widely used in the industry.

As far as reactor types are concerned, several types are available for fermentation reactions. The most common are the stirred tank reactor, the fluidised bed reactor, and the packed bed reactor.

Stirred tank reactors (Figure 2.15) are characterised by the presence of one or more agitators supported by a central shaft. Typically, these vessels have baffles to improve the fluid dynamics and avoid vortexing. In aerobic fermentations, air or oxygen are typically supplied using spargers located in the bottom part of the vessel. Various types of agitators are available, depending on the characteristics of the fermentation medium and on the main tasks required by the agitation system. If the main

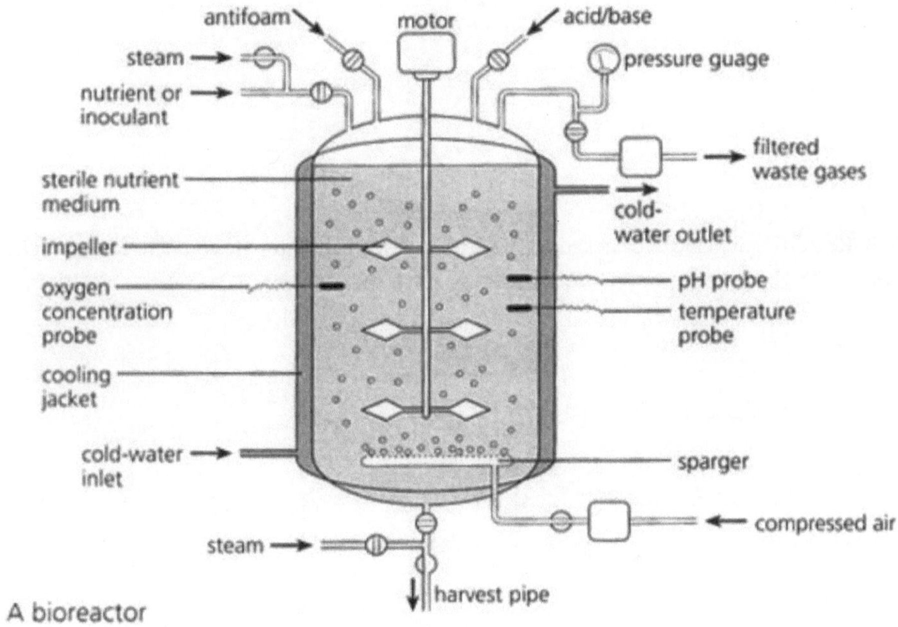

FIGURE 2.15 Scheme of the main components of a stirred tank fermenter (http://2010.igem.org/Team:UCL_London/Fermenter_Mechanics).

task is dispersing and breaking the gas bubbles, ensuring high mass transfer coefficient for oxygen, then small diameter agitators such as Ruston turbines are used. These agitators can rotate at high-speed ensuring good gas dispersion in relatively low viscosity fermentation media. For fermentation media with higher viscosity and higher cell concentrations, large low-speed agitators are used, such as the hydrofoil agitators. Ideally it can be assumed that agitated vessels are perfectly mixed, i.e. that the concentration of substrate, microorganisms and products is uniform in any points of the reactor.

Packed bed reactors are characterised by microorganisms growing attached to support media. The feed containing the substrate and nutrients flows downwards from the top of the bed. Microorganisms are retained on the support and, ideally, are not present in the liquid effluent, containing the products, which is collected from the bottom of the reactor. In packed bed reactors, the concentration of substrates and products varies along the vertical length of the column, however the concentration gradients are lower at higher flow rates because of the increased mixing due to the higher turbulence (Figure 2.16).

Fluidised bed fermenters (Figure 2.17) are similar to packed beds, however in this case the microorganisms, immobilised on support materials, are kept in suspension by circulation of the gas or of the liquid phase. As in packed bed fermenters, the liquid phase flows out of the reactor while the immobilised microorganisms are retained.

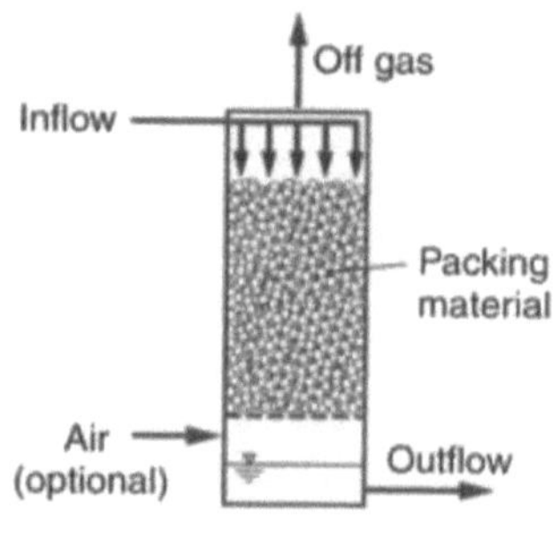

FIGURE 2.16 Scheme of a packed bed fermenter (https://sites.google.com/site/fermentersin/types).

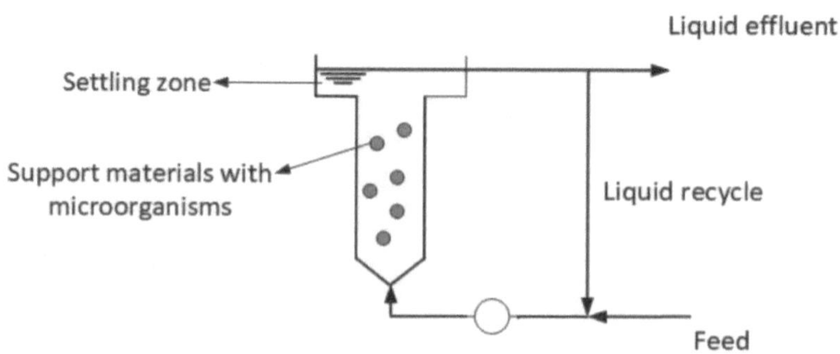

FIGURE 2.17 Scheme of a fluidised bed fermenter.

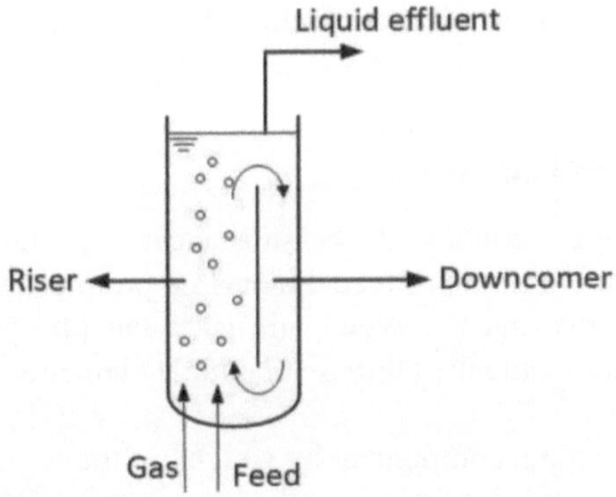

FIGURE 2.18 Scheme of an airlift fermenter.

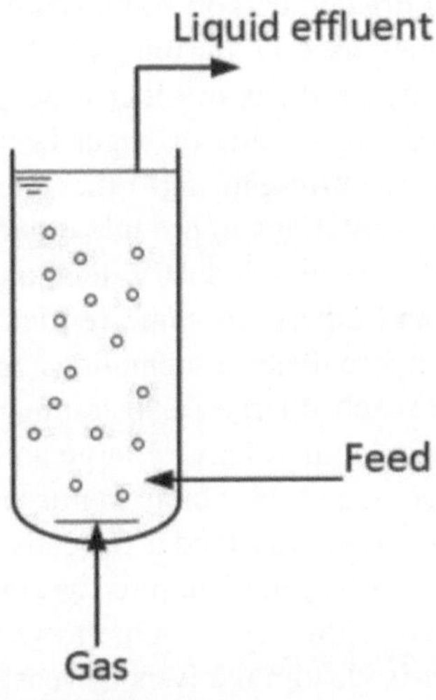

FIGURE 2.19 Scheme of a bubble column fermenter.

Airlift fermenters (Figure 2.18) are divided into two zones, the riser, and the downcomer. In the riser, a gas is fed decreasing the density of the mixture compared to the downcomer, where there is no gas. The higher density of the fluid in the downcomer causes fluid movement and recirculation between the two zones. Airlift reactors have the advantage over stirred tank fermenters, because of lower shear rates (due to the absence of mechanical aerators) and are therefore suitable for shear-sensitive cultures.

The bubble column reactor (Figure 2.19) is the simplest type of fermenter. It is conceptually similar to stirred tank reactors, but the mixing is only provided by the flow of the gas phase, in the absence of mechanical agitators. The mass and heat transfer rates obtained in bubble column fermenters are lower, due to the absence of agitation, than the rates for stirred tank reactors, however bubble columns are suitable for shear-sensitive cultures.

2.7 STARTING MATERIALS

Fermentation media must contain all the substances and nutrients required for the microorganisms to grow and to produce the desired products. Since microorganisms are made of carbon, hydrogen, oxygen, nitrogen and phosphorus and of smaller amounts of many other elements, all these chemical elements must be present in the fermentation medium.

Water is usually the main component, by weight, of the fermentation medium and, in most cases must be purified, with removal of suspended solids and microorganisms before use. If the water has high c salts or calcium carbonate concentration (hard water), these salts must also be removed.

For most fermentations, the carbon source is organic and supplied at concentrations of 10 g/L or higher. Typical organic carbon sources are sugars, such as glucose and sucrose. Sugars are sometimes used as purified substrates, however in most cases, for cost reasons, sugar-rich by-products of other processes are used, e.g. molasses. Molasses are a by-product of sugar cane or sugar beet production and are rich in carbohydrates, mainly sucrose, nitrogen and other elements and vitamins. Other starting materials used in fermentations to provide organic carbon sources are whey, a lactose-rich by-product of the dairy industry, methanol, typically produced from natural gas, vegetable oils and others. In some fermentations the carbon source is inorganic, CO_2, e.g. in the microalgae fermentations to produce biofuels, however these processes have not yet reached large-scale commercialisation.

Nitrogen must also be supplied in relatively large amounts, as the nitrogen is present in proteins and nucleic acids and its content in microorganisms is 10–15% (w/w). Nitrogen concentration in fermentation media depends on the type of fermentation and is often higher than 1 g/L. Nitrogen is supplied as ammonium salts or as nitrogen-rich substances, such as urea, corn steep liquor (a by-product of starch extraction from corn), yeast extract (derived from the lysis of yeast cells) or others.

Phosphorus, present inside the cells in nucleic acids, ATP, NADP and other substances, is generally provided as inorganic phosphate ions, which are also used as pH buffers. In most fermentations, the concentration of phosphorus required for microbial metabolism, excluding the needs for pH buffering, is not higher than 100 mg/L.

Providing sulphur is essential as sulphur is contained in several amino acids and some vitamins. Sulphur, if not present enough in the material used to supply the carbon source, is usually supplied as inorganic sulphate or sulphide salt, at concentrations often in the order of 20 mg/L.

Other elements that need to be present at lower concentrations in the fermentation medium are calcium, iron, potassium, sodium, magnesium (supplied at concentrations typically in the order of 10 mg/L or lower) and trace elements, such as nickel, cobalt, copper, manganese, selenium, zinc and molybdenum (typical concentrations in fermentation media 0.1–1 mg/L). These elements are required for certain enzymes and for a variety of functions in the cells and are often present in the required amount in the water or in the material used to supply the carbon source. If they are not present at high enough concentrations, they need to be added with their salts.

Other substances such as vitamins, growth factors and precursors can also be added to the fermentation medium if required by specific microorganisms and processes.

2.8 CULTURE CULTIVATION AND MAINTENANCE

Regardless of the type of process, in many cases, fermentations are started in very small reactors where the microorganisms are inoculated from frozen stock cultures. In these reactors, microorganisms are supplied with the carbon source and with all the required nutrients and are left to acclimate to the substrate and to grow in numbers. Once microorganisms are acclimated and have reached a high enough concentration, they are inoculated in a larger reactor, in order to increase the cell mass and numbers. The process is repeated several times in larger reactors until the cell mass is large enough to be inoculated in the full-scale fermentation vessel.

The progress of the fermentation is usually monitored by measuring the growth of the microorganisms using direct or indirect methods. Direct methods include direct microscopic counts, electronic cell counters and plate-counting techniques. Indirect methods include turbidimetric techniques, which measure the turbidity of the suspension, which is proportional to the concentration of microorganisms, and dry weight measurement. ATP can also be measured using a bioluminometry technique, giving an estimation of the concentration of viable microorganisms.

During the run of fermentation processes, the medium conditions need to be monitored to ensure maintenance of the optimum values of pH, dissolved oxygen and temperature. pH is maintained at the desired value by addition to the medium of a buffer and/or by controlled addition of acid or base in response to the readings of a pH sensor. Temperature is controlled by sensors in the medium that usually control the flow rate of a cooling or heating fluid in the vessel's jacket or coils. The concentration of dissolved oxygen is measured with dissolved oxygen sensors and is controlled by manipulating either the air (or oxygen) flow rate or the agitation speed. During fermentation processes, foam production can also be a problem, especially for aerobic fermentations. Foam formation is due to the generation by the microorganisms of surface-active agents, e.g. proteins. Foam is controlled by automatic addition of antifoam agents, e.g. vegetable oils or synthetic antifoams. Alternatively, the medium composition can be modified to avoid foam generation.

2.9 EXTRACTION AND SEPARATION OF METABOLITES

After the fermentation, the target product needs in most cases to be separated from the fermentation medium and purified. The sequence of unit operations required for the separation and purification of the product is called "downstream processing" (covered in more detail in Chapter 7). The processes used in downstream processing are different for different fermentations and different products and depend on the physical and chemical properties of the product and on its concentration.

In general, the first step of downstream processing is the separation of the microorganisms from the fermentation medium. This is achieved by a solid-liquid separation stage, which can be sedimentation, centrifugation or filtration.

The processes used to separate and purify the target product depend on whether the product is intracellular or extracellular. For intracellular products (e.g. many enzymes), the cells need to be disrupted to allow the release of the product. Cell disruption can be achieved with mechanical or nonmechanical methods. Mechanical methods include the use of pressure and liquid shear (high-pressure homogenizers), of high-speed bead mills or of ultrasounds. Nonmechanical methods cause cell disruption by permeabilization of the cell membrane, which can be achieved by autolysis, osmotic or heat shock. When the cell membrane is made permeable by one of these methods, the products are released into the medium and can be purified.

Extracellular products are present in the liquid medium after separation of the microorganisms. These products can be separated and purified using a range of unit operations, some of them widely used in other areas of chemical engineering, others more typical of fermentation processes. Chromatography is a separation process used for the recovery of high-value products, e.g. proteins. In chromatography, products are separated based on their different retention when they flow through a stationary phase. The separation of the products in chromatography can occur via different mechanisms, e.g. size exclusion, ion-exchange. Other separation and purification processes include solvent extraction, evaporation, distillation, crystallisation and drying. In solvent extraction, separation is achieved using organic solvents to extract the product from the water-based fermentation medium. Evaporation and distillation require bringing the fermentation medium to the boiling point by supplying heat, using the different volatility of the product and of the other components of the medium. When crystallisation is used, the product is recovered as solid crystals, which can be obtained from the fermentation medium by decreasing the temperature or by adding a chemical that reacts with the product decreasing its solubility. Drying, when used, is the final purification stage and it involves the removal of any residual solvent or water by application of heat.

2.10 STERILISATION OF MEDIA AND EQUIPMENT

In many fermentation processes, it is necessary to avoid the growth of undesired microorganisms. Indeed, undesired microorganisms may compete with the desired species for the substrate, reducing the substrate available for the desired process. In addition, undesired microorganisms may metabolise the target product or contaminate it with other undesired metabolic products, whose removal in downstream processing may be difficult or expensive.

Undesired microorganisms are removed before the start of the fermentation by sterilising the reactor and the medium. Sterilisation is usually carried out by using heat because microorganisms cannot survive at high temperatures. In many cases, the vessel is heated using pressurised steam in the jacket or coils at temperatures up to 121 °C, for an appropriate length. In some cases, the steam is injected into the head space of the vessel above the medium. The medium can also be sterilised by steam heating, however some medium components, e.g. vitamins and glucose, can degrade when exposed to heat. In these cases, the medium is sterilised by filtration, rather than by heating. As an alternative, for heat-sensitive media heat can be applied for very short times at higher temperatures (e.g. flowing the media through heat exchangers) to avoid degradation of the components. Since microorganisms are also present in air, all the air inlets to the fermentation vessels need to be equipped with filters to remove undesired microorganisms, if sterile conditions are required. In addition, if contamination of the environment by the microorganisms in the fermentation is a concern, e.g. in the case of the cultivation of pathogenic microorganisms, the gas exhaust ports in the fermenter need also to be equipped with filters to remove any microorganisms from the outlet gases.

Sterilisation of the vessel and medium can also be achieved by means other than heat, e.g. irradiation with microwave, UV or X rays or with high voltage electric fields. Sterilisation by irradiation is used frequently to sterilise small fermentation accessories, e.g. Petri dishes.

2.11 GEOMETRY AND MATERIALS FOR FERMENTATION VESSELS

Fermentation vessels are typically of cylindrical shape, equipped, in case of agitated vessels, with agitators in the centre. Agitators are commercially available in many different types and manufacturers follow guidelines on the geometrical ratios (e.g. ratio between agitator and vessel diameter). Small lab-scale vessels (up to a few litres) are usually made of glass, while pilot and production scale vessels are made of mild or stainless steel. The construction materials should be nontoxic, resistant to corrosion and able to withstand repeated cycles of sterilisation, i.e. they need to be designed for pressures larger than the atmospheric due to the use of pressurised steam. The pipework should also be designed to be sterilisable and to avoid undesired microorganisms entering the process. For this reason, welded joints are usually preferable to flanges, as flanged joints may loosen due to the repeated sterilisation cycles with heating and cooling, favouring the contamination with air-borne microorganisms.

QUESTIONS AND PROBLEMS

2.1 Describe the main components in cells and their function.

2.2 Explain the difference between primary and secondary products produced by microorganisms.

2.3 Explain why the mass of microorganisms produced per unit mass of substrate removed is usually much larger for aerobic reactions than for anaerobic ones.

2.4. Explain the advantages and disadvantages of the main types of fermentation processes: batch, continuous and fed-batch.

2.5 Explain the advantages and disadvantages of the main types of reactors for fermentation, stirred tank reactor, packed column, fluidised bed, airlift and bubble column.

3 Fermentation Processes
Kinetics and Stoichiometry

3.1 KINETICS OF FERMENTATION PROCESSES

In order to design fermentation processes, it is important to know how to express the rate of the various processes, i.e. biomass growth, substrate removal and product generation.

We will start considering the rate of biomass growth. In quantitative terms, growth is defined as the increase in the number and mass of the microorganisms present in the medium. It is expected that the rate of biomass growth will be proportional to the number, or mass, of microorganisms. Indeed, in any given period of time, two microorganisms will produce more microorganisms than one microorganism, four microorganisms will produce more microorganisms than two of them, and so on. Conceptually, the same happens for the human population (and indeed for the population of any other living species), the more people are there on the planet, the faster the population will grow.

Based on these considerations, we will use the following rate equation for microorganism growth:

$$r_X\left(\frac{\text{kg}}{\text{m}^3\text{d}}\right) = \mu \cdot X \tag{3.1}$$

r_X is the microorganism growth rate, and X is the microorganism concentration. Here we report the variables with their typical units, e.g. $\frac{\text{kg}}{\text{m}^3\text{d}}$ for r_X, but equivalent units (e.g. $\frac{\text{g}}{\text{L}\cdot\text{h}}$) are of course also possible. Typically, the microorganism concentration is expressed as mass, rather than number, so the units of X are mass/volume, e.g. kg/m^3 or equivalent. Note that the unit of time for the rate of fermentation processes is often taken as days (d) or hours (h), rather than seconds, which are more typical for nonfermentation reactions. Clearly, the microorganism growth rate does not depend only on the microorganism concentration, otherwise microorganisms would never stop growing. There must be something that limits microorganism growth. If all the other elements and nutrients that the

DOI: 10.1201/9781003633822-3

cells need are in excess of the requirements for biomass growth, growth of the microorganisms will only be limited by the concentration of the substrate. The effect of the substrate concentration on the growth rate is included in the term μ, which is the specific growth rate (it is called "specific" because it is the growth rate per unit mass of microorganisms). In general, we expect, and this has been experimentally observed in a great number of cases, that substrate concentration only limits biomass growth when the substrate concentration is low. Instead, when the substrate concentration is very high, changes in the substrate concentration do not affect the growth rate, which then reaches its maximum value for the given substrate and type of microorganisms. These considerations lead to the very frequent use of the equation for μ represented in equation (3.2):

$$\mu\left(\mathrm{d}^{-1}\right)=\frac{\mu_{max}S}{K_S+S} \tag{3.2}$$

S is the substrate concentration, which is often expressed as kg/m^3, even though mol/ m^3 or other equivalent units can also be used. μ_{max} (maximum growth rate, d^{-1} or equivalent units, indicating the maximum possible growth rate for the considered microorganism under the environmental conditions) and K_S (half-saturation constant, same units as the substrate) are parameters, which depend on the species of microorganism and on the substrate. They also depend on the environmental conditions, mainly temperature and pH. Equation (3.2) is called the "Monod" equation, from the French researcher who first proposed it.

Combining equations (3.1) and (3.2) the biomass growth rate is

$$r_X\left(\frac{\mathrm{kg}}{\mathrm{m^3d}}\right)=\frac{\mu_{max}S}{K_S+S}X \tag{3.3}$$

Equation (3.2) is a shifting order rate equation. When the substrate concentration is very large, i.e. $S>>K_S$, the specific growth rate does not depend on the substrate concentration. When the substrate concentration is very low, i.e. $S<<K_S$, $\mu=\frac{\mu_{max}}{K_S}$, the specific growth rate shows a first-order dependence on the substrate concentration. Note that numerically the value of K_S coincides with the substrate concentration for which $\mu=\frac{\mu_{max}}{2}$. The profile of μ vs S given by the Monod equation is shown in Figure 3.1.

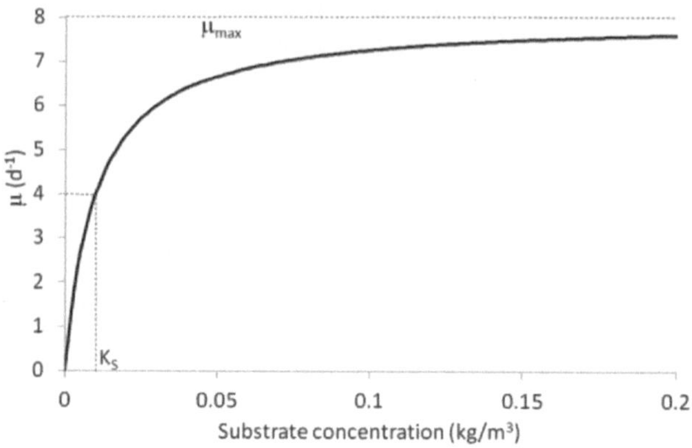

FIGURE 3.1 Typical shape of the rate equation for microbial growth on the limiting substrate S. This curve has been obtained with μ_{max}= 8 d^{-1} and K_S = 0.010 kg/m^3.

Example 3.1

The kinetic parameters of a microorganism on a certain substrate are μ_{max}=4.9 d^{-1} and K_S = 0.1 kg/m^3.

Calculate

(a) the specific growth rate m when the substrate concentration is 10, 5, 0.002, 0.001 kg/m^3;
(b) the growth rate r_X when the substrate concentration is 5 kg/m^3 and the microorganisms concentration is 2.5 kg/m^3.

Solution.
(a) We use equation (3.2). For S = 10 kg/m^3, $\mu = 4.85\,\text{d}^{-1}$; for S = 5 kg/m^3, μ = 4.80 d^{-1}; for S = 0.002 kg/m^3, μ = 0.096 d^{-1}; for S = 0.001 kg/m^3, μ = 0.049 d^{-1}.

Note that for the high concentrations of substrate, 10 and 5 kg/m^3, μ is very close to μ_{max} and almost independent on substrate concentration, while for very low concentrations μ is approximately linearly dependent on S;

(b) For 5 kg/m^3 and $\mu = 4.80\,\text{d}^{-1}$ and using equation (3.1), $r_X = 12.0\dfrac{\text{kg}}{\text{m}^3\text{d}}$

The Monod equation shown by equation (3.2) assumes that the specific growth rate is only limited by the concentration of the carbon source. This is true if all the other elements and nutrients which the microorganisms need are in excess in the growth medium. If any other elements are limiting, equation (3.2) can be modified to include limitation by multiple substances, e.g.

$$\mu(\text{d}^{-1}) = \mu_{max}\frac{S_1}{K_{S1}+S_1}\cdot\frac{S_2}{K_{S2}+S_2}\cdot\ldots \tag{3.4}$$

In equation (3.4), S_1, S_2, etc are the concentration of the limiting substances, e.g. substrate, oxygen, nitrogen, etc, and K_{S1}, K_{S2}, etc are the corresponding half-saturation constants.

Example 3.2

With the kinetic parameters of Example 3.1, assume that there is a second limiting substrate, i.e. oxygen, for which $K_{SO_2} = 0.0005$ kg/m³. Calculate the specific growth rate when $S = 5$ kg/m³ and $O_2 = 0.001$ kg/m³ or 0.008 kg/m³

Solution

Using equation (3.4), for $O_2 = 0.001$ kg/m³ $\mu = 3.20\ d^{-1}$, for $O_2 = 0.008$ kg/m³ $\mu = 4.52\,d^{-1}$.

It is important to observe that equations (3.1)–(3.3) do not describe the acclimation phase, but only apply to completely acclimated microorganisms.

Once we have an equation for the rate biomass growth, we can write the rate equation for the substrate removal and product generation. We expect the rate of substrate removal to be proportional to the rate of biomass growth. Indeed, biomass grows by removing the substrate from the liquid medium, and therefore, the faster the microorganisms grow, the faster the substrate is removed from the medium. Based on these considerations, the rate of substrate removal is often expressed by equation (3.5):

$$r_S\left(\frac{\text{kg}}{\text{m}^3\text{d}}\right) = -\frac{\mu \cdot X}{Y_{X/S}} = -\frac{\mu_{max} S}{K_S + S}\frac{X}{Y_{X/S}} \tag{3.5}$$

r_S is the rate of substrate removal, the minus sign indicating that the substrate is being removed from the medium. The parameter $Y_{X/S}$ is the so-called growth yield, i.e. the biomass produced per unit of substrate removed. $Y_{X/S}$ is defined as follows:

TABLE 3.1
μ_{max} Values Obtained for Various Microorganisms

Microorganism	Limiting substrate	μ_{max} (d^{-1})	K_s (kg/m³)
Pseudomonas sp.	Benzoate	10.4	$4.5 \cdot 10^{-4}$
Pseudomonas putida F1	Toluene	20.6	$1.4 \cdot 10^{-2}$
Pseudomonas putida F1	Benzene	17.5	$1.2 \cdot 10^{-4}$
Mixed culture (activated sludge)	Sodium dodecylsulfate	22.2	$9.6 \cdot 10^{-2}$
Streptococcus zooepidemicus	Glucose	10.8-19.2	-

TABLE 3.2
$Y_{X/S}$ Values Obtained for Various Microorganisms

Microorganism	Substrate	$Y_{X/S}$ (kg/kg)
Klebsiella sp.	Methanol	0.38
Enterobacter aerogenes	Maltose	0.46
Candida utilis	Glucose	0.51
Pseudomonas fluorescens	Acetate	0.28
Pseudomonas methanica	Methane	0.56

$$Y_{X/S}\left(\frac{\text{kg}}{\text{kg}}\right)=\frac{\text{biomass produced}}{\text{substrate removed}}=\frac{r_X}{\left(-r_S\right)} \tag{3.6}$$

The parameter $Y_{X/S}$ depends on the species of microorganism, on the nature of the substrate and also on the environmental conditions, e.g. temperature and pH.

As far as the rate of product production is concerned, we need to distinguish whether the product is a primary or secondary metabolite. If the product is a primary metabolite (kinetics for secondary metabolites production will be covered in section 3.6), i.e. it is produced during biomass growth, then the rate of product production will be proportional to the rate of biomass growth:

$$r_P\left(\frac{\text{kg}}{\text{m}^3\text{d}}\right)=Y_{P/X}r_X=Y_{P/X}\mu\cdot X \tag{3.7}$$

$Y_{P/X}$ is the product yield, i.e. the product produced per unit mass of biomass produced:

$$Y_{P/X}\left(\frac{\text{kg}}{\text{kg}}\right)=\frac{\text{product produced}}{\text{biomass produced}}=\frac{r_P}{r_X} \tag{3.8}$$

Instead of using the parameter $Y_{P/X}$, the parameter $Y_{P/S}$ can be introduced, with the following definition:

$$Y_{P/S}\left(\frac{\text{kg}}{\text{kg}}\right)=\frac{\text{product produced}}{\text{substrate removed}}=\frac{\Delta P}{\left(-\Delta S\right)}=\frac{r_P}{\left(-r_S\right)} \tag{3.9}$$

By using $Y_{P/S}$, the rate of product production can be expressed as

$$r_P\left(\frac{\text{kg}}{\text{m}^3\text{d}}\right)=Y_{P/S}\left(-r_S\right)=Y_{P/S}\frac{\mu X}{Y_{X/S}} \tag{3.10}$$

Comparing equations (3.6) and (3.9) it is also evident that

$$Y_{P/X} = \frac{Y_{P/S}}{Y_{X/S}} \tag{3.11}$$

We have seen that the parameters $Y_{X/S}$, $Y_{P/S}$, $Y_{P/X}$ link biomass growth, substrate consumption and product generation. In fermentations it is often also important to express the rate of oxygen and nutrients consumption by the microorganisms. These can be expressed by using the same approach, e.g. the biomass produced per unit of oxygen consumed can be expressed by the parameter $Y_{X/O2}$:

$$Y_{X/O2}\left(\frac{\text{kg}}{\text{kg}}\right) = \frac{\text{Biomass produced}}{\text{Oxygen consumed}} = \frac{r_X}{\left(-r_{O_{2biom}}\right)} \tag{3.12}$$

In equation (3.12) $r_{O_{2biom}}$ is the rate of oxygen consumption by the microorganisms per unit volume of the reactor. $r_{O_{2biom}}$ is negative, because oxygen is consumed, so in the definition of the parameter Y_{X/O_2}, which is positive, the term $-r_{O_{2biom}}$ is used. From equation (3.12) the equation for the calculation of the rate of oxygen consumption follows immediately:

$$r_{O_{2biom}}\left(\frac{\text{kg}}{\text{m}^3\text{d}}\right) = -\frac{r_X}{Y_{X/O_2}} = -\frac{\mu_{max}S}{K_S+S}\frac{X}{Y_{X/O_2}} \tag{3.13}$$

As an example of the rate of nutrients consumption, let us write an equation for the rate of nitrogen uptake by the microorganisms. This can be written in the same way as done for oxygen, by defining the parameter $Y_{X/N}$, which represents the biomass produced per unit of nitrogen consumed:

$$Y_{X/N}\left(\frac{\text{kg}}{\text{kg}}\right) = \frac{\text{Biomass produced}}{\text{Nitrogen consumed}} = \frac{r_X}{\left(-r_N\right)} \tag{3.14}$$

In equation (3.14) r_N is the volumetric rate of nitrogen consumption by the microorganisms. r_N is negative, because nitrogen is consumed, so in the definition of the parameter $Y_{X/N}$, which is positive, the term $-r_N$ is used. From equation (3.14) the equation for the calculation of the rate of oxygen consumption follows immediately:

$$r_N\left(\frac{\text{kg N}}{\text{m}^3\text{d}}\right) = -\frac{r_X}{Y_{X/N}} = -\frac{\mu_{max}S}{K_S+S}\frac{X}{Y_{X/N}} \tag{3.15}$$

The same approach can be used to express and calculate the production and consumption rate of any products and metabolites generated or consumed during biomass growth.

TABLE 3.3
$Y_{P/S}$ Values Obtained for Various Microorganisms

Microorganism	Substrate	Product	$Y_{P/S}$ (kg/kg)
Saccharomyces cerevisiae	Pre-treated rice husks	Ethanol	0.30–0.40
Scheffersomyces stipitis	Pre-treated rice husks	Ethanol	0.24–0.36
Yeast culture	Pre-treated orange peels	Ethanol	0.36
Streptococcus zooepidemicus	Glucose	Ialuronic acid	0.070-0.085

Example 3.3

A certain species of microorganisms (assume empirical formula $C_5H_7O_2N$) is growing on glucose ($C_6H_{12}O_6$) at a rate of 0.1 kg m^{-3} min^{-1}. The rate of substrate removal is 0.5 kg m^{-3} min^{-1}. Calculate the growth yield $Y_{X/S}$ in units of kg/kg and mol/mol.

Solution
From the definition of $Y_{X/S}$:

$$Y_{X/S} = \frac{r_X}{(-r_S)} = \frac{0.1\dfrac{\text{kg}}{\text{m}^3\text{min}}}{0.5\dfrac{\text{kg}}{\text{m}^3\text{min}}} = 0.2\frac{\text{kg}}{\text{kg}}$$

On a molar basis, 0.1 kg of microorganisms corresponds to 0.88 mol and 0.5 kg of glucose corresponds to 2.78 mol. So, on a molar basis the growth yield is

$$Y_{X/S} = \frac{r_X}{(-r_S)} = \frac{0.88\dfrac{\text{mol}}{\text{m}^3\text{min}}}{2.77\dfrac{\text{mol}}{\text{m}^3\text{min}}} = 0.32\frac{\text{mol}}{\text{mol}}$$

This example shows that it is important to express the growth yield $Y_{X/S}$ with the appropriate units (kg/kg or mol/mol) rather than in the adimensional form without specifying any units.

Example 3.4

A microorganism grows on a certain substrate at a rate of 1.5 $\frac{kg}{m^3h}$. The yield parameters are $Y_{X/S} = 0.3$ kg/kg, $Y_{P/S} = 0.2$ kg/kg, $Y_{X/O_2} = 0.5$ kg/kg, $Y_{X/N} = 10$ kg/kg. Calculate the rate of substrate, oxygen and nitrogen removal and product production. Also, calculate the parameter $Y_{P/X}$.

Solution

$$r_S = -\frac{r_X}{Y_{X/S}} = -\frac{1.5\frac{kg}{m^3\,h}}{0.3\frac{kg}{kg}} = -5\frac{kg}{m^3\,h}$$

$$r_{O_{2biom}} = -\frac{r_X}{Y_{X/O_2}} = -3\frac{kg}{m^3\,h}$$

$$r_N = -\frac{r_X}{Y_{X/N}} = -0.15\frac{kg\ N}{m^3\,h}$$

$$Y_{P/X} = \frac{Y_{P/S}}{Y_{X/S}} = 0.67\frac{kg}{kg}$$

3.2 EFFECT OF INHIBITORS, TEMPERATURE AND PH ON THE GROWTH RATE

The rate of microorganism growth and the kinetic parameters used to describe it depend on many environmental factors, the most important of which are the presence of inhibitors, temperature and pH.

Inhibition may be due to the substrate, to the fermentation products or to other chemicals in the fermentation medium. The Haldane or Andrews equation (3.16) includes substrate inhibition:

$$\mu = \frac{\mu_{max} S}{K_S + S + \frac{S^2}{K_I}} \tag{3.16}$$

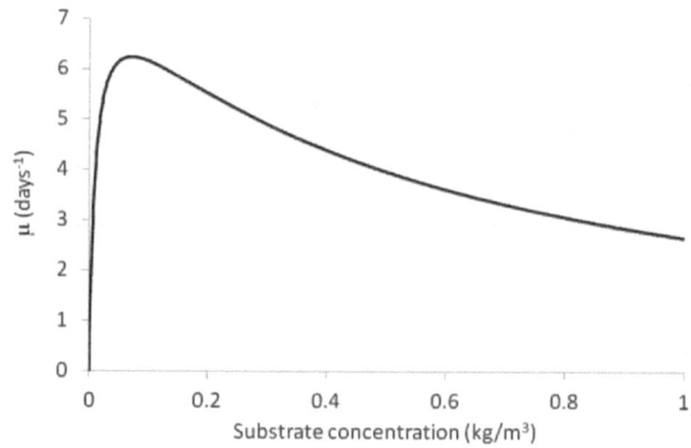

FIGURE 3.2 Typical shape of the rate equation for microbial growth on an inhibiting substrate S. This curve has been obtained with $\mu_{max} = 8.0\ d^{-1}$, $K_S = 0.01\ kg/m^3$ and $K_I = 0.5\ kg/m^3$.

Equation (3.16) is valid for those substrates that inhibit biomass growth when they are present above a certain concentration. In equation (3.16) in addition to μ_{max} and K_S we have an additional parameter K_I. The lower the K_I value, the larger the inhibiting effect of the substrate or the lower the substrate concentration, which is required to give a decrease in μ. A typical plot of the Haldane equation is shown in Figure 3.2.

If microorganisms are growing on substrate S but are inhibited by another substance I (which may be either a product of the fermentation or another chemical), the specific growth rate can have the following form:

$$\mu = \frac{\mu_{max} S}{K_S + S} \frac{1}{1 + \frac{I}{K_I}} \tag{3.17}$$

Another important effect on the specific growth rate is pH. Microorganisms are typically only able to grow in a restricted pH range and, within this range, there is typically a pH value for which the specific growth rate is maximum. The effect of pH on the specific growth rate is often described by equations such as (3.18)

$$\mu = \frac{\mu_{max} S}{K_S + S} \frac{1 + 2 \cdot 10^{0.5(a-b)}}{1 + 10^{(\mathrm{pH}-b)} + 10^{(a-\mathrm{pH})}} \tag{3.18}$$

where a and b are two parameters, which represent, respectively, the low and high extremes of pH for which there is still microbial activity. A plot of the pH inhibition factor, $\frac{1 + 2 \cdot 10^{0.5(a-b)}}{1 + 10^{(\mathrm{pH}-b)} + 10^{(a-\mathrm{pH})}}$, is shown in Figure 3.3.

Another important effect on microbial growth rate is the effect of temperature. Typically, the rate of microbial growth increases with temperature when the temperature

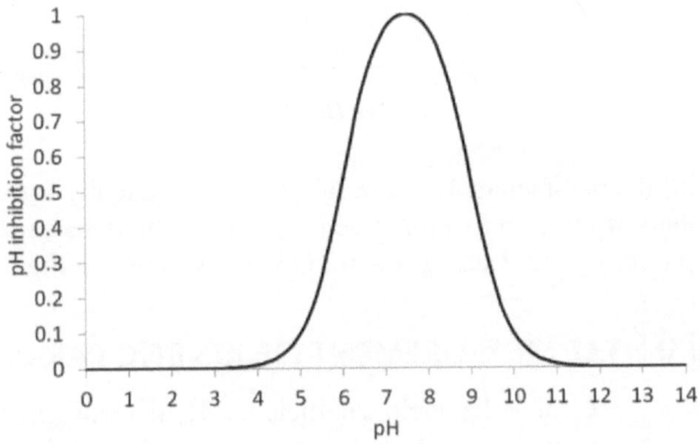

FIGURE 3.3 Typical shape of the pH inhibition factor. This curve has been obtained with $a = 6$, $b = 9$.

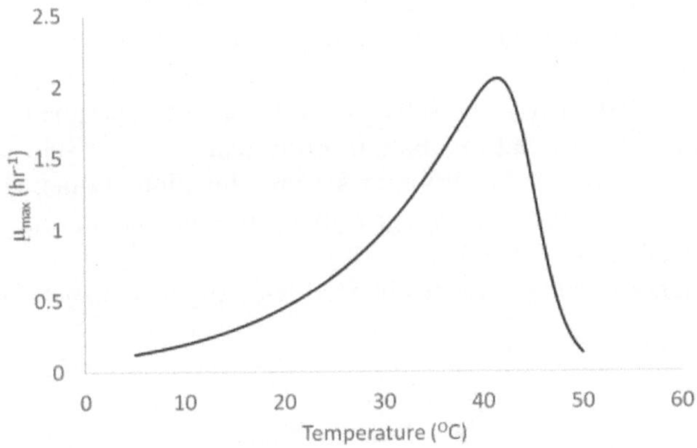

FIGURE 3.4 Plot of μ_{max} vs T according to equation (3.20), using $A = 10^{10}$ h^{-1}, $B = 3.0{\cdot}10^{90}$, $Ea = 58{,}000$ J/mol, $\Delta G_d = 550{,}000$ J/mol.

increases from room temperature to approximately 35–40 °C, even though many microbial species may have an optimum temperature outside this range. At higher temperatures, denaturation of the cellular enzymes may start to occur, with a consequent decrease in microbial activity and in the growth rate. In the region where the reaction rate increases with temperature, it is usually assumed that the temperature dependence can be included in the μ_{max} parameter, e.g. with a conventional Arrhenius equation such as

$$\mu_{max} = Ae^{-\frac{Ea}{RT}} \tag{3.19}$$

An equation that accounts for the effect of temperature in a wider range, including both the increase and decrease in microbial activity is the following:

$$\mu_{max} = \frac{Ae^{-\frac{Ea}{RT}}}{1 + Be^{-\frac{\Delta G_d}{RT}}} \tag{3.20}$$

In equation (3.20) the parameters A, B, Ea, ΔG_d (which physically corresponds to the free energy change of the deactivation reaction) have to be determined experimentally. A typical plot of μ_{max} vs T using equation (3.20) is shown in Figure 3.4.

3.3 EXPERIMENTAL MEASUREMENT OF KINETIC PARAMETERS

The parameters μ_{max}, K_S and the yield coefficients $Y_{i/j}$ depend on the species of microorganisms, on the substrate and on the environmental conditions, mainly temperature and pH. The values of these parameters need to be determined experimentally. A typical procedure to determine these parameters is described here, but other procedures, involving batch or continuous reactors, have also been reported.

We assume we have a batch reactor, where microorganisms, substrate and nutrients are provided. We assume that the only limiting substrate is the organic substrate, and all the nutrients are provided in excess. We also assume that microorganisms are already completely adapted to the substrate. We start the experiment at the time of substrate addition and we call this time 0. From that time, we start measuring the concentration of substrate and microorganisms as a function of time. From each data point, we calculate the rate of biomass growth, r_X, as a function of time. The data will be arranged in a table like Table 3.4.

The parameters μ_{max} and K_S can be obtained re-arranging equation (3.3) as

$$\frac{X}{r_X} = \frac{K_S}{\mu_{max}}\frac{1}{S} + \frac{1}{\mu_{max}} \tag{3.21}$$

TABLE 3.4
Example of Experimental Data to be Collected for the Determination of Kinetic Parameters

Time	X	S	r_X
0	X_0	S_0	
t_1	X_1	S_1	$\frac{X_1 - X_0}{t_1 - t_0}$
t_2	X_2	S_2	$\frac{X_2 - X_1}{t_2 - t_1}$
…	… .	…	…

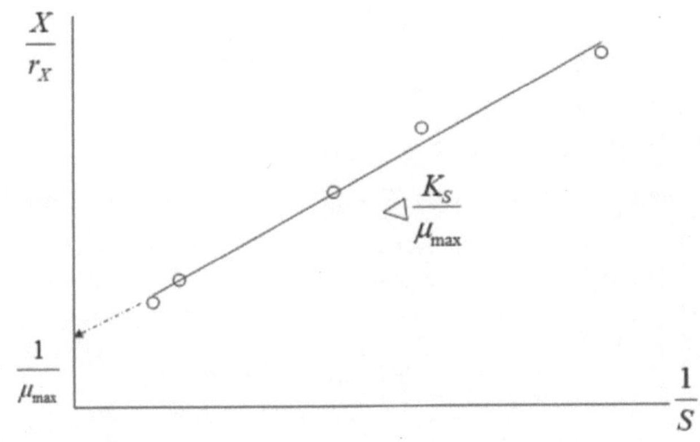

FIGURE 3.5 Plot of $\frac{X}{r_X}$ vs $\frac{1}{S}$ for the calculation of the parameters μ_{max} and K_S.

Therefore, the experimental data collected in Table 3.3 can be re-arranged to plot $\frac{X}{r_X}$ vs $\frac{1}{S}$. The intercept of the graph corresponds to $\frac{1}{\mu_{max}}$ and the slope to $\frac{K_S}{\mu_{max}}$, allowing therefore the calculation of μ_{max} and K_S (Figure 3.5).

From the same experiment, the parameter $Y_{X/S}$ can also be obtained, by calculating the ratio between the biomass produced, $\Delta X = X_{final} - X_0$, and the substrate removed, $(-\Delta S) = -\left[S_{final} - S_0\right]$, over the whole length of the experiment. Alternatively, $Y_{X/S}$ can be calculated as the average of the ratios $\frac{\Delta X}{(-\Delta S)}$ over the length of the experiment.

Example 3.5

The following data (Table 3.5) have been obtained during the batch growth of a certain microorganism on an organic substrate.
Calculate μ_{max}, K_S and $Y_{X/S}$.

Solution
The plot of the experimental profiles of *S* and *X* vs time is given in Figure 3.6.

In order to calculate μ_{max} and K_S, we need to generate the plot of X/r_X vs $1/S$. This plot is reported in Figure 3.7. Note that in the plot of Figure 3.7 we have excluded the last datapoint, at $t = 22$ h, because at that time there was no substrate left and we do not know at what time the substrate had been totally removed.

The intercept of the X/r_X vs $1/S$ graph is equal to 10.164 h. This intercept is equal to $1/\mu_{max}$ (equation (3.21)). This means that

TABLE 3.5
Data for Example 3.5

Time (h)	Substrate (*S*, mg/L)	Biomass (*X*, mg/L)
0	2000	100
2	1930	122
4	1840	148
6	1730	180
8	1600	219
10	1450	265
12	1260	321
14	1040	388
16	780	466
18	480	556
20	180	646
22	0	696

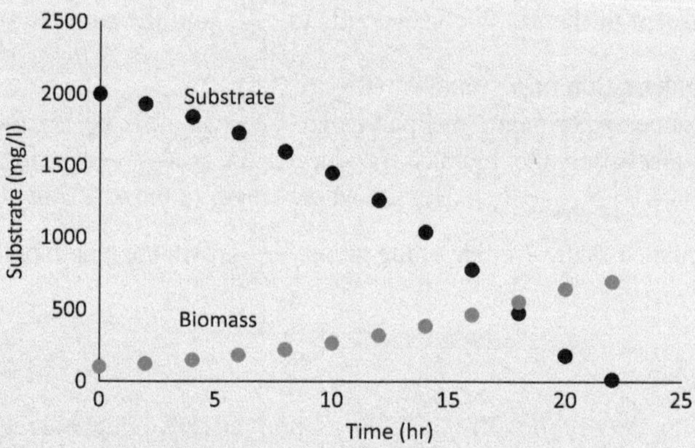

FIGURE 3.6 Plot of experimental data for Example 3.5.

$$\mu_{max} = 0.098\,\text{h}^{-1}$$

The intercept is 1929.9 mg/L.h. This corresponds to $\dfrac{K_S}{\mu_{max}}$. With the known value of μ_{max}, we obtain

$$K_S = 189.1\frac{\text{mg}}{\text{L}}$$

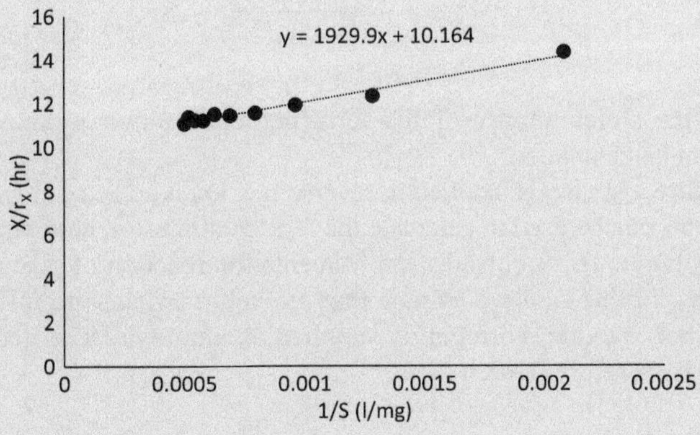

FIGURE 3.7 Linearisation of experimental data for Example 3.5.

To calculate $Y_{X/S}$ we can use the total biomass formation and substrate removal during the experiment:

$$Y_{X/S} = \frac{(696-100)\,\mathrm{mg/l}}{2000\,\mathrm{mg/l}} = 0.3\frac{\mathrm{kg}}{\mathrm{kg}}$$

3.4 YIELD COEFFICIENTS AND GROWTH STOICHIOMETRY

The yield coefficients introduced in the previous sections are linked by the overall stoichiometry that describes the fermentation reaction. For example, let us consider the generic stoichiometry for growth of microorganisms on acetic acid (CH_3COOH):

$$CH_3COOH + aO_2 + bNH_3 \rightarrow cC_5H_7O_2N + dCO_2 + eH_2O \tag{3.22}$$

In this book, we will generally assume that $C_5H_7O_2N$ is the empirical formula for the dry mass of the microorganisms. From equation (3.22), assuming the stoichiometric coefficients *a*, *b*, *c*, *d* and *e* are known, the coefficients $Y_{X/S}$, $Y_{X/O2}$ and $Y_{X/N}$ can be calculated (MW and AW stand for molecular and atomic weight, respectively):

$$Y_{X/S} = \frac{c \cdot MW_{C_5H_7O_2N}}{MW_{CH_3COOH}} \tag{3.23}$$

$$Y_{X/O2} = \frac{c \cdot MW_{C_5H_7O_2N}}{a \cdot MW_{O2}} \tag{3.24}$$

$$Y_{X/N} = \frac{c \cdot MW_{C_5H_7O_2N}}{b \cdot AW_N} \tag{3.25}$$

Therefore, if the stoichiometry of the fermentation reaction is known, the yield coefficients can be calculated.

The procedure can also be applied in reverse, i.e. knowledge of at least one of the yield coefficients can be used to calculate the fermentation stoichiometry if this is not known. For example, let us consider the fermentation reaction of glucose under aerobic conditions. In this case we assume that the only products are microorganisms, water and carbon dioxide. Nitrogen is supplied as ammonia. The generic reaction stoichiometry is therefore written as

$$C_6H_{12}O_6 + aO_2 + bNH_3 \rightarrow cC_5H_7O_2N + dCO_2 + eH_2O \tag{3.26}$$

How can we calculate the stoichiometric coefficients a, b, c, d and e? These coefficients are linked by the elemental balances for C, H, N and O. These constitute four equations, however a fifth equation is needed to calculate the five coefficients. The fifth equation can be obtained by knowledge of an experimental value for one of the yield coefficients. For example, if we assume the growth yield $Y_{X/S}$ is known, the fifth equation is

$$Y_{X/S} = \frac{c \cdot 113}{180} \tag{3.27}$$

In equation (3.27) 113 and 180 are the molecular weights of microorganisms and glucose, respectively. Equation (3.27) can be coupled with the elemental balances for C, O, N and H to calculate all the stoichiometric coefficients. In summary, the five equations that give the stoichiometric coefficients are:

$$\begin{cases} C - balance: 5c + d = 6 \\ O - balance: 2c + 2d + e = 2a + 6 \\ H - balance: 7c + 2e = 3b + 12 \\ N - balance\,c: b = c \\ Y_{X/S} = \dfrac{c \cdot 113}{180} \end{cases} \tag{3.28}$$

Equation (3.28) constitute a system of five equations in the five unknowns a, b, c, d and e.

The concepts expressed in this section can be applied in the same way if the fermentation produces a product of interest. For example, let us consider the generic stoichiometry for ethanol production from glucose under anaerobic conditions:

$$C_6H_{12}O_6 + aNH_3 \rightarrow bC_5H_7O_2N + cC_2H_5OH + dCO_2 + eH_2O \tag{3.29}$$

If the coefficients a, b, c, d, e are known, the product yields can be calculated:

$$Y_{P/X} = \frac{\text{kg product}}{\text{kg biomass}} = \frac{cMW_{C_2H_5OH}}{bMW_{C_5H_7O_2N}} \tag{3.30}$$

$$Y_{P/S} = \frac{\text{kg product}}{\text{kg substrate}} = \frac{cMW_{C_2H_5OH}}{MW_{C_6H_{12}O_6}} \tag{3.31}$$

Similarly, if one of the yield coefficients $Y_{X/S}$, $Y_{X/P}$, $Y_{P/S}$, is known, the stoichiometric coefficients for equation (3.29) can be calculated by combining the equation for the known coefficient with the elemental balances for C, H, N and O.

Example 3.6

Consider the production of citric acid by the aerobic fermentation of glucose by the microorganism *Aspergillus niger*. Assume the stoichiometry of the reaction reported below:

$$C_6H_{12}O_6 + 2.35O_2 + 0.1NH_3 \rightarrow 0.1C_5H_7O_2N + 0.7C_6H_8O_7 + 1.3CO_2 + 3H_2O$$

Calculate the yields $Y_{X/S}$, $Y_{P/X}$, $Y_{P/S}$, $Y_{X/O2}$, $Y_{X/N}$.

Solution

$$Y_{X/S} = \frac{0.1 \cdot MW_{C_5H_7O_2N}}{MW_{C_6H_{12}O_6}} = 0.063\frac{\text{kg}}{\text{kg}}$$

$$Y_{P/X} = \frac{0.7 \cdot MW_{C_6H_8O_7}}{0.1 \cdot MW_{C_5H_7O_2N}} = 11.89\frac{\text{kg}}{\text{kg}}$$

$$Y_{P/S} = \frac{0.7 \cdot MW_{C_6H_8O_7}}{MW_{C_6H_{12}O_6}} = 0.75\frac{\text{kg}}{\text{kg}}$$

$$Y_{X/O2} = \frac{0.1 \cdot MW_{C_5H_7O_2N}}{2.35 \cdot MW_{O2}} = 0.15\frac{\text{kg}}{\text{kg}}$$

$$Y_{X/N} = \frac{0.1 \cdot MW_{C_5H_7O_2N}}{0.1 \cdot AW_N} = 8.07\frac{\text{kg}}{\text{kg}}$$

Example 3.7

Consider the fermentation of glucose ($C_6H_{12}O_6$) to produce ethanol (C_2H_6O) by an anaerobic microorganism. The reaction consumes glucose and ammonia and produces ethanol, water, carbon dioxide and microorganisms. It is known that 0.10 kg of microorganisms are produced per kg of glucose. Determine the stoichiometry of the fermentation reaction.

Solution

The general form of the unbalanced stoichiometry is

$$C_6H_{12}O_6 + a\mathrm{NH}_3 \rightarrow bC_5H_7O_2N + cC_2H_6O + dCO_2 + eH_2O$$

There are five stoichiometric coefficients to determine. We have the four elemental balances for C, H, N and O and the information on the product yield:

$$Y_{X/S} = \frac{b \cdot MW_{C_5H_7O_2N}}{MW_{C_6H_{12}O_6}} = 0.627b = 0.10\frac{\mathrm{kg}}{\mathrm{kg}}$$

From this equation we get that $b = 0.16$.
The elemental balances are

$$\begin{cases} C-balance: 6 = 5b + 2c + d \\ O-balance: 6 = 2b + c + 2d + e \\ H-balance: 12 + 3a = 7b + 6c + 2e \\ N-balance: a = b \end{cases}$$

From the N balance

$$a = b = 0.16$$

From the carbon balance

$$c = \frac{6 - 5b - d}{2} = 2.6 - 0.5d$$

From the oxygen balance

$$e = 6 - 2 \cdot 0.16 - 2.6 + 0.5d - 2d = 3.08 - 1.5d$$

From the H balance

$$12 + 3 \cdot 0.16 = 7 \cdot 0.16 + 6 \cdot (2.6 - 0.5d) + 2 \cdot (3.08 - 1.5d)$$

which gives
$d = 1.73$
Therefore, $e = 0.49$, $c = 1.73$
And the balanced stoichiometry is

$$C_6H_{12}O_6 + 0.16NH_3 \rightarrow 0.16C_5H_7O_2N + 1.73C_2H_6O + 1.73CO_2 + 0.49H_2O$$

3.5 RATE EQUATIONS FOR SECONDARY METABOLITES, MAINTENANCE AND ENDOGENOUS METABOLISM

The rate equations developed in previous sections apply to microorganism growth on the substrate with simultaneous production of the product (primary metabolite). However, in some cases, the product is not produced during microbial growth but rather when the substrate is (almost) completely removed and growth has stopped (secondary metabolite). For secondary metabolites, the rate of product production is no longer proportional to the growth rate according to equation (3.7), but instead it will be proportional to the biomass concentration:

$$r_P = \alpha X \tag{3.32}$$

In equation (3.32), α is a parameter to be determined from experiments. In the most general case, the rate of product production can be expressed as a function of both the biomass growth rate and the biomass concentration:

$$r_P = Y_{P/X} r_X + \alpha X \tag{3.33}$$

During the production of secondary metabolites, there will be some oxygen consumption and change in biomass concentration, however these are usually small and can be ignored in the development of kinetic models.

Other modifications of the growth kinetics are used to account for the phenomena of maintenance and endogenous metabolism (Chapter 2). Maintenance (the use of substrate for non-growth purposes) gives a consumption of the substrate in the absence of growth and is usually described by the following rate equation:

$$r_{Sm} = -m_S X \tag{3.34}$$

In equation (3.34), r_{Sm} is the rate of substrate removal due to the maintenance metabolism. The total rate of substrate consumption is the sum of the substrate removed due to growth and due to maintenance:

TABLE 3.6
m_s Values Obtained for Various Microorganisms

Microorganism	Substrate	m_s (d^{-1})
Candida utilis	Glucose	0.74
Escherichia coli	Maltose	1.4
Penicillium chrysogenium	Glucose	0.50
Klebsiella aerogenes	Glycerol	1.78

$$r_S\left(\frac{\text{kg substrate}}{\text{m}^3\ \text{d}}\right) = -\frac{\mu \cdot X}{Y_{X/S}} - m_S X \tag{3.35}$$

The substrate used for maintenance is oxidised for energy generation, therefore there is an oxygen consumption associated to maintenance ($r_{O_{2\,biom,m}}$):

$$r_{O_{2\,biom,m}} = \frac{-m_S X}{Y_{S/O_{2,m}}} \tag{3.36}$$

The coefficient $Y_{S/O_{2,m}}$ can be calculated from the stoichiometry of total oxidation of the substrate. The total oxygen consumption is the sum of the consumption due to growth and to maintenance:

$$r_{O_{2\,biom,m}} = -\frac{r_X}{Y_{X/O_2}} - \frac{m_S X}{Y_{S/O_{2,m}}} \tag{3.37}$$

Table 3.6 reports values of the maintenance coefficient measured for various microorganisms.

In summary, if we assume that biomass growth and substrate removal are described by the maintenance model, the equations that describe them are equations (3.3) and (3.38):

$$r_X = \frac{\mu_{max} S}{K_S + S} X \tag{3.3}$$

$$r_S = -\frac{\mu_{max} S}{\left(K_S + S\right)} \frac{X}{Y_{X/S}} - m_S X \tag{3.38}$$

In order to determine the values of the kinetic parameters μ_{max}, K_S, $Y_{X/S}$ and m_S, we can use the same experiment as described in Section 3.4, with measurement of the concentration of microorganisms and substrate over time in a batch experiment. The parameters μ_{max} and K_S can be determined by the same procedure described in section 3.4, by plotting $\frac{X}{r_X}$ vs $\frac{1}{S}$ (equation (3.21)). $Y_{X/S}$ and m_S can be determined from the ratio $\frac{r_S}{r_X}$:

$$\frac{r_S}{r_X} = \frac{-\frac{\mu_{max} S \cdot X}{(K_S + S) Y_{X/S}} - m_S X}{\frac{\mu_{max} S \cdot X}{(K_S + S)}} = -\frac{1}{Y_{X/S}} - \frac{m_S}{\mu_{max}} - \frac{m_S K_S}{\mu_{max}} \frac{1}{S} \tag{3.39}$$

Therefore, according to equation (3.39), plotting of $\frac{r_S}{r_X}$ vs $\frac{1}{S}$ should give a straight line with slope equal to $-\frac{m_S K_S}{\mu_{max}}$ and intercept equal to $-\frac{1}{Y_{X/S}} - \frac{m_S}{\mu_{max}}$. Once μ_{max} and K_S are known, the slope of the line $\frac{r_S}{r_X}$ vs $\frac{1}{S}$ will give m_S and the intercept will give $Y_{X/S}$.

Example 3.10

The following data (Table 3.7) have been obtained during the batch growth of a certain microorganism on an organic substrate. Using the Monod model with maintenance, calculate μ_{max}, K_S, $Y_{X/S}$ and m_S (Figure 3.8).

Solution

TABLE 3.7
Data for Example 3.10

Time (h)	Substrate (S, mg/L)	Biomass (X, mg/L)
0	3,000	120
2	2885	152
4	2744	193
6	2561	245
8	2333	310
10	2048	391
12	1688	493
14	1247	619
16	719	769
18	175	921
20	0	966

According to equations (3.21) and (3.39), the parameters μ_{max}, K_S, $Y_{X/S}$ and m_S can be calculated by plotting $\frac{X}{r_X}$ vs $\frac{1}{S}$ and $\frac{r_S}{r_X}$ vs $\frac{1}{S}$. From the intercept of the $\frac{X}{r_X}$ vs $\frac{1}{S}$ plot we calculate μ_{max}:

$$\mu_{max} = \frac{1}{8.85} = 0.11\text{h}^{-1}$$

From the slope of the plot we calculate K_S:

$$K_S = 1712.3 \cdot \mu_{max} = 188\frac{\text{mg}}{\text{L}}$$

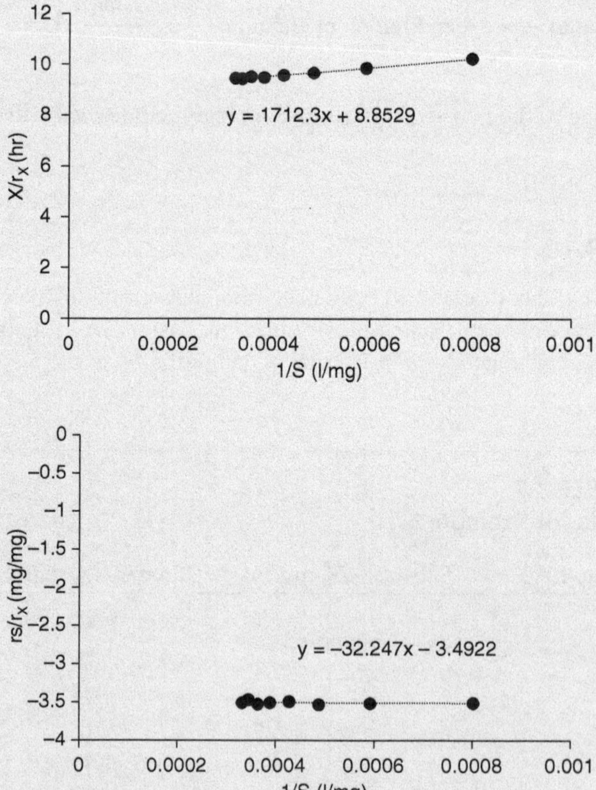

FIGURE 3.8 Linearisation of experimental data for Example 3.10.

From the slope of the $\frac{r_S}{r_X}$ vs $\frac{1}{S}$ plot we calculate m_S:

$$m_S = \frac{\mu_{max}}{K_S} 32.25 = 0.019\,\text{h}^{-1}$$

From the intercept of the plot we calculate $Y_{X/S}$:

$$Y_{X/S} = \frac{1}{3.49 - \frac{m_S}{\mu_{max}}} = 0.30 \frac{\text{kg}}{\text{kg}}$$

Example 3.11

Calculate the coefficient $Y_{S/O_{2,m}}$ which gives the ratio between substrate and oxygen consumed during maintenance metabolism, if the substrate is glucose.

Solution

The coefficient $Y_{S/O_{2,m}}$ can be calculated from the complete oxidation of glucose into carbon dioxide and water:

$$C_6H_{12}O_6 + 6O_2 \rightarrow 6CO_2 + 6H_2O$$

From this stoichiometry: $Y_{S/O_{2,m}} = \frac{180}{6 \cdot 32} = 0.94 \frac{\text{kg}}{\text{kg}}$.

Another phenomenon that occurs during fermentations is endogenous metabolism. Endogenous metabolism is the consumption of cellular components by the microorganisms. Endogenous metabolism does not directly affect the rate of substrate removal, but affects the rate of biomass growth:

$$r_{end}\left(\frac{\text{kg}}{\text{m}^3\,\text{d}}\right) = -bX \tag{3.40}$$

In equation (3.40) r_{end} is the rate of endogenous metabolism, i.e. the rate of consumption of the cellular components (which can be considered a death rate of the microorganisms), and b is an empirical parameter, dependent on temperature and on the type of microorganisms. The net rate of biomass growth is the algebraic sum of the biomass produced due to growth and the biomass consumed due to endogenous metabolism:

$$r_X = \mu X - bX \tag{3.41}$$

Under aerobic conditions, we can assume that endogenous metabolism gives the total oxidation of the biomass to carbon dioxide and water. Therefore, under aerobic conditions endogenous metabolism gives an oxygen consumption, written as:

$$r_{O_{2\,biom,end}} = \frac{-bX}{Y_{X/O_{2,end}}} \tag{3.42}$$

The coefficient $Y_{X/O_{2,end}}$ can be calculated from the stoichiometry of the total oxidation of the microorganisms. The total oxygen consumption is therefore given as the sum of oxygen consumption due to growth and due to endogenous metabolism:

$$r_{O_{2\,biom}} = -\frac{\mu X}{Y_{X/O_2}} - \frac{bX}{Y_{X/O_{2,end}}} \tag{3.43}$$

In summary, if we assume that biomass growth and substrate removal are described by the endogenous metabolism model, the rate equations are

$$r_X = \frac{\mu_{max} S}{K_S + S} X - bX \tag{3.44}$$

$$r_S = -\frac{\mu_{max} S}{\left(K_S + S\right)} \frac{X}{Y_{X/S}} \tag{3.5}$$

The kinetic parameters to be determined are now μ_{max}, K_S, $Y_{X/S}$ and b. They can be obtained by linearization of $\frac{X}{r_S}$ vs $\frac{1}{S}$ and $\frac{r_X}{r_S}$ vs $\frac{1}{S}$.

The linearization of $\frac{X}{r_S}$ vs $\frac{1}{S}$ gives

$$\frac{X}{r_S} = -\frac{Y_{X/S}}{\mu_{max}} - \frac{Y_{X/S}}{\mu_{max}} K_S \frac{1}{S} \tag{3.45}$$

Plotting $\frac{X}{r_S}$ vs $\frac{1}{S}$ will give the values of $\frac{Y_{X/S}}{\mu_{max}}$ (intercept of the plot) and K_S (from the slope).

The linearization of $\frac{r_X}{r_S}$ vs $\frac{1}{S}$ gives

$$\frac{r_X}{r_S} = -Y_{X/S} + b\frac{Y_{X/S}}{\mu_{max}} + \frac{bK_S Y_{X/S}}{\mu_{max}}\frac{1}{S} \quad (3.46)$$

Once the values of $\frac{Y_{X/S}}{\mu_{max}}$ and K_S are known, the plot of $\frac{r_X}{r_S}$ vs $\frac{1}{S}$ will give the values of b (from the slope) and $Y_{X/S}$ (from the intercept). μ_{max} can then be calculated from $\frac{Y_{X/S}}{\mu_{max}}$.

Example 3.12

The following data have been obtained during the batch growth of a certain microorganism on an organic substrate. Using the Monod model with endogenous metabolism, calculate μ_{max}, K_S, $Y_{X/S}$ and b.

Solution

According to equations (3.45) and (3.46), the parameters μ_{max}, K_S, $Y_{X/S}$ and b can be calculated by plotting $\frac{X}{r_S}$ vs $\frac{1}{S}$ and $\frac{r_X}{r_S}$ vs $\frac{1}{S}$. The plots are shown in Figure 3.9.

From the intercept and slope of the $\frac{X}{r_S}$ vs $\frac{1}{S}$ plot we obtain

TABLE 3.8
Data for Example 3.12

Time (h)	Substrate (S, mg/L)	Biomass (X, mg/L)
0	2500	150
2	2382	192
4	2230	246
6	2032	314
8	1788	401
10	1476	510
12	1086	647
14	611	812
16	113	979
18	0	985

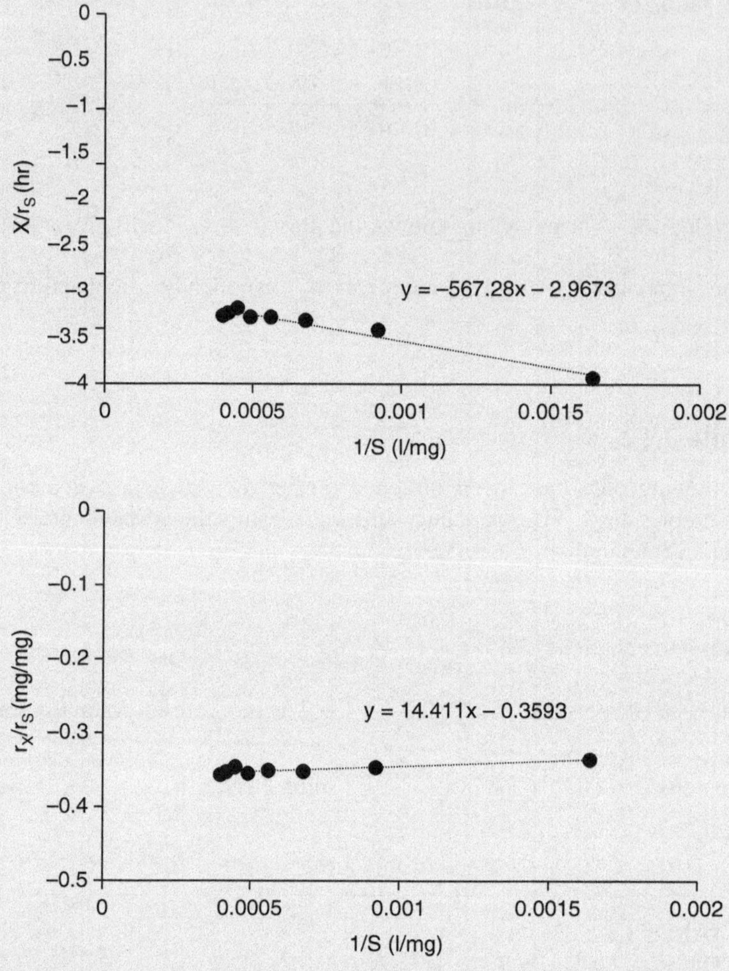

FIGURE 3.9 Linearisation of experimental data for Example 3.11.

$$\frac{Y_{X/S}}{\mu_{max}} = 2.967\frac{\text{mg}}{\text{mg}\cdot\text{h}} \text{ and } \frac{Y_{X/S}}{\mu_{max}}K_s = 567.28\,\frac{\text{mg}}{\text{L}\cdot\text{h}}$$

From the intercept and slope of the $\frac{r_X}{r_S}$ vs $\frac{1}{S}$ plot we obtain

$$\frac{bK_S Y_{X/S}}{\mu_{max}} = 14.411 \text{ and } -Y_{X/S} + b\frac{Y_{X/S}}{\mu_{max}} = -0.359$$

Solving these four equations in the four unknowns μ_{max}, K_S, $Y_{X/S}$ and b we obtain

$$K_S = 191.2\,\text{mg/L} \quad b = 0.025\ \text{h}^{-1} \quad Y_{X/S} = 0.43\ \text{kg/kg} \quad \mu_{max} 0.14\ \text{h}^{-1}$$

Example 3.13

Calculate the coefficient $Y_{X/O_{2,end}}$, which gives the ratio between biomass and oxygen consumed during endogenous metabolism. Assume that the empirical formula for the microorganisms is $C_5H_7O_2N$.

Solution

The coefficient $Y_{X/O_{2,end}}$ can be calculated from the complete oxidation of microorganisms into carbon dioxide and water:

$$C_5H_7O_2N + 5O_2 \rightarrow 5CO_2 + 2H_2O + NH_3$$

From this stoichiometry: $Y_{X/O_{2,end}} = \frac{113}{5 \cdot 32} = 0.71\frac{\text{kg}}{\text{kg}}$

QUESTIONS AND PROBLEMS

3.1 A microorganism grows aerobically on ethanol to produce microorganisms, water and carbon dioxide. The microorganism uses oxygen and ammonia for its metabolism. It is known that the reaction produces 0.20 kg of microorganisms per 1 kg of ethanol consumed. Calculate the stoichiometry of the fermentation reaction.

3.2 Consider the aerobic fermentation of glucose to produce the amino acid L-glutamic acid ($C_5H_9O_4N$) by the microorganisms *Corynebactrium glutamicum*:

$$C_6H_{12}O_6 + 2.215O_2 + 0.83NH_3 \rightarrow 0.10C_5H_7O_2N + 0.73C_5H_9O_4N + 2.61H_2O + 1.85CO_2$$

A plant processes 30,000 t/year of glucose, of which 95% are converted. Calculate, on a yearly basis (t/year):

- the production of L-glutamic acid;
- the production of carbon dioxide;
- the consumption of oxygen

3.3 Citric acid ($C_6H_8O_7$) is produced industrially by the fermentation of glucose using the microorganism *Aspergillus Niger*. A simplified stoichiometry of the fermentation reaction is reported below:

$$C_6H_{12}O_6 + 2.35O_2 + 0.1NH_3 \rightarrow 0.1C_5H_7O_2N + 0.7C_6H_8O_7 \\ + 1.3CO_2 + 3H_2O$$

It is desired to produce 10,000 tonnes per year of citric acid.

- Assuming that ammonia is provided as ammonium chloride (NH_4Cl), calculate the amount of ammonium chloride required per year.
- Calculate the amount of oxygen consumed by the microorganisms per year.
- Calculate the amount of carbon dioxide produced per year.

3.4 The microorganism *Bacillus subtilis* is used for the production of the enzyme protease ($C_3H_5O_2N$). The fermentation reaction is aerobic and uses glucose as substrate and ammonia as nitrogen source. The products are microorganisms, the enzyme, water and carbon dioxide. It is known that 60 g of carbon dioxide and 20 g of microorganisms are produced per 100 g of glucose consumed. Calculate the stoichiometry of the fermentation.

3.5 The growth of an anaerobic microorganisms which produces ethanol on glucose is described by the following stoichiometry:

$$C_6H_{12}O_6 + 0.15NH_3 \rightarrow 0.15C_5H_7O_2N + xC_2H_6O + 2.1CO_2 + 3H_2O$$

Determine the stoichiometric coefficient for ethanol, x.

3.6 A microorganism produces lactic acid ($C_3H_6O_3$) from the sugar lactose ($C_{12}H_{22}O_{11}$) present in liquid cheese whey (the fermentation is anaerobic). Assume that cheese whey contains 50 kg/m^3 of lactose and that the fermentation is started with 80% of the working volume made of cheese way (the remaining 20% is for the inoculum and nutrients). During the fermentation, when 70% of the initial lactose has been consumed, 2 g/L of microorganisms have been produced. Calculate the rate of rate of lactic acid production when the rate of lactose removal is 0.4 g L^{-1} h^{-1}. Assume that the reactants and products in this reaction are lactose, ammonia, microorganisms, lactic acid and water with no carbon dioxide production.

3.7 Assuming that, in the considered temperature range, the temperature dependency of μ_{max} is described by equation (3.19), describe a series of experiments to determine the values of the parameters *A* and *Ea* in equation (3.19), showing the equation used to obtain the parameter values.

3.8 The growth of a microorganism on a substrate is described by the Monod model with maintenance metabolism, with $Y_{X/S} = 0.2$ kg/kg and $m_S = 0.4$ d^{-1}. Calculate the rate of substrate removal when the biomass concentration is 2.5 g/L and the rate of biomass formation is 1.7 g/L.h.

3.9 It is suspected that a substance present in the fermentation medium may inhibit the growth of the microorganisms. Describe a set of experiments to determine the K_I parameter in equation (3.17). Assume that the parameters μ_{max}, K_S and $Y_{X/S}$ are known.

3.10 The maximum growth rate of a microorganisms on a substrate has been measured at different temperatures (Table 3.9). Assuming that, in this temperature range, μ_{max} depends on temperature according to equation (3.19), calculate the parameters *A* and *Ea* and calculate the μ_{max} when the microorganism is at 28 °C.

3.11 The initial rate of substrate removal and microorganisms' growth was measured (Table 3.10) at various initial concentrations. Assuming that the microorganisms' growth can be described by the Monod model with

TABLE 3.9
Data for Problem 3.10

Temperature (°C)	μ_{max} (d^{-1})
20	4.5
25	7.5
30	12.8
35	22.1

TABLE 3.10
Data for Problem 3.11

Initial *X* (kg/m³)	Initial *S* (kg/m³)	Initial r_X (kg m^{-3} d^{-1})	Initial ($-r_S$) (kg m^{-3} d^{-1})
1.0	0.5	6.1	17.9
1.5	0.8	9.5	27.9
2.0	1.2	12.8	37.7
2.5	0.2	13.5	39.7
1.5	0.2	8.1	23.8
1.5	0.05	5.1	15.2
1.5	0.01	1.6	5.6
0.5	1.0	3.2	9.3

endogenous metabolism, use the data to calculate the paramaters μ_{max}, K_S, $Y_{X/S}$, b.

3.12 It is known that the growth of a microorganism on a substrate can be described by the Monod model with endogenous metabolism. Calculate the total oxygen consumption by the microorganisms when the biomass concentration is 1.2 kg/m^3 and the net growth rate is 0.3 g L^{-1} h^{-1}. Assume that the coefficient $Y_{X/O_2} = 0.8$ kg/kg, $b = 0.15$ d^{-1} and $Y_{X/O_{2,end}} = 0.71$ kg/kg.

4 Fermentation Processes

Mass Balances and Design for Batch, Continuous and Fed-Batch Reactors

In this chapter we will focus on the design of batch, continuous and fed-batch fermentation processes. We will cover how to calculate the volume or residence time required for fermentation processes, and how to calculate the composition of the fermentation effluents. Mass balances are the key equations for the design of these processes.

Many processes can be carried out in any fermentation mode: batch, continuous or fed-batch. Examples of batch processes are the fermentation to produce penicillin and other antibiotics and the production of citric acid by *Aspergillus* or *Candida* microorganisms. An example of continuous fermentation is the production of bioethanol by *Saccharomyces cerevisiae* strains, while the fermentative production of amino acids uses most commonly fed-batch processes (e.g. production of L-lysine with *Escherichia coli*).

4.1 GENERAL EQUATION FOR MASS BALANCES

Mass balances have the general form:

$$\text{Accumulation} = \text{Input-Output} + \text{Generation-Consumption} \quad (4.1)$$

Mass balances can be written for each component in the reaction medium or in the system under consideration. All the terms in the mass balance should have the units of a mass flow rate (e.g. kg/d). Accumulation represents the rate at which the component's mass in the system changes. Input and Output represent the mass flow rate of the component that enters or leaves the system. Generation represents the rate of production of the component in the system and Consumption the rate at which the component is consumed in the system. With the signs used in equation (4.1), all the terms will be taken as positive.

This general equation takes different forms depending on whether we are considering a batch or a continuous reactor.

DOI: 10.1201/9781003633822-4

4.2 MASS BALANCES FOR BATCH REACTORS

In batch reactors (Figure 4.1) the inoculum of microorganisms, the substrate and the required nutrients are added at time 0. No continuous addition of feed and continuous removal of the products is carried out. The only exception is oxygen, in the case of aerobic fermentations, which is continuously sparged in the reactor.

For the batch reactor, the general equation for mass balances (4.1), becomes

$$\text{Accumulation} = \text{Generation-Consumption} \tag{4.2}$$

With the rate equations introduced in section 3.1, the mass balance for microorganisms, substrate and products in a batch reactor are

$$\frac{dX}{dt}\left(\frac{\text{kg}}{\text{m}^3\text{d}}\right) = r_X = \mu \cdot X = \frac{\mu_{max} S}{K_s + S} X \tag{4.3}$$

$$\frac{dS}{dt}\left(\frac{\text{kg}}{\text{m}^3\text{d}}\right) = r_S = -\frac{\mu \cdot X}{Y_{X/S}} = -\frac{\mu_{max} S}{K_S + S}\frac{X}{Y_{X/S}} \tag{4.4}$$

$$\frac{dP}{dt}\left(\frac{\text{kg}}{\text{m}^3\text{ d}}\right) = r_P = \mu X \cdot Y_{P/X} = \frac{\mu_{max} S}{K_S + S} X \cdot Y_{P/X} \tag{4.5}$$

Equations (4.3)–(4.5) represent a system of three differential equations in the three unknowns *S(t)*, *X(t)*, *P(t)* and need to be solved simultaneously in order to calculate the profiles of microorganisms, substrate and products in a batch fermentation. In order to solve these equations, the parameters μ_{max}, K_S, $Y_{X/S}$ and $Y_{P/X}$ need to be known and the initial conditions $X = X_0$, $S = S_0$ and $P = P_0$ need to be specified, where X_0, S_0 and P_0 are the concentrations of microorganisms, substrate and products at the start of the fermentation.

The general method to solve systems of differential equations using Microsoft Excel is shown in Appendix B.

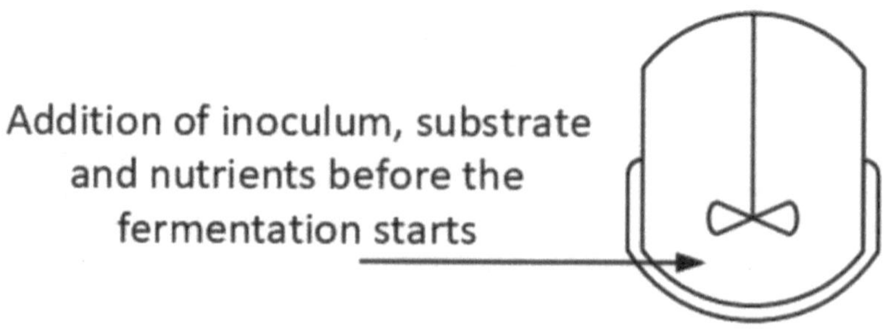

FIGURE 4.1 Batch reactor.

Example 4.1

Calculate the time profiles of substrate, microorganisms and product for a fermentation process characterised by the following parameters:

$$\mu_{max} = 5.2\ d^{-1};\ K_S = 0.05\ kg/m^3;\ Y_{X/S} = 0.15\ kg/kg;\ Y_{P/X} = 2.70\ kg/kg$$

The initial biomass, substrate and product concentrations are 0.1 kg/m^3, 10 kg/m^3 and 0 kg/m^3, respectively.

Solution

In this case equations (4.2)–(4.5) become (using units of kg for microorganisms, substrate and product, m^3 for volume and d for time):

$$\frac{dX}{dt} = \frac{5.2S}{0.05+S}X$$

$$\frac{dS}{dt} = -\frac{5.2S}{0.05+S}\frac{X}{0.15}$$

$$\frac{dP}{dt} = \frac{5.2S}{0.05+S}X \cdot 2.70$$

The solution of these differential equations with the given initial conditions is reported in Figure 4.2

In a batch process we can also use the rate equations and mass balances to calculate the doubling time of microorganisms. The doubling time is the time the microorganisms need to double their concentration. The doubling time only refers to

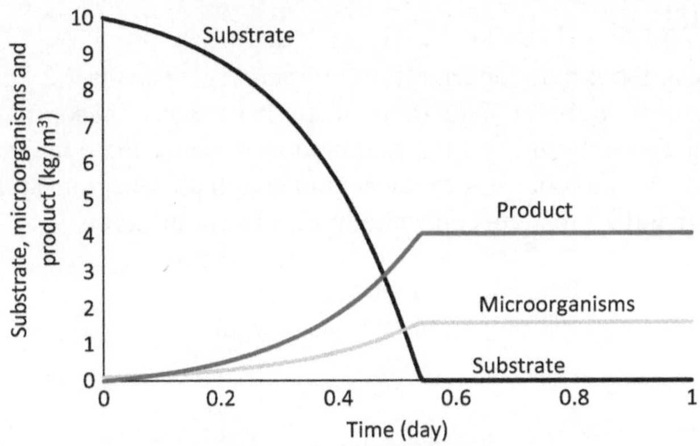

FIGURE 4.2 Solution of Example 4.1.

the condition where biomass is completely acclimated to the substrate and the substrate is in excess (i.e. $S>>K_S$, i.e. $\mu=\mu_{max}$). Under these hypotheses, the mass balance for the microorganisms is

$$\frac{dX}{dt} = \mu_{max} X \tag{4.6}$$

Equation (4.6) can be integrated between X_0 and X and between 0 and t:

$$\int_{X_0}^{X} \frac{dX}{X} = \mu_{max} \int_0^t dt \rightarrow \ln\left(\frac{X}{X_0}\right) = \mu_{max} t \tag{4.7}$$

The doubling time (t_d) can be calculated from equation (4.7), by imposing that $X = 2X_0$:

$$t_d = \frac{\ln 2}{\mu_{max}} \tag{4.8}$$

Example 4.2

A certain microorganism has a doubling time on a substrate of 2.5 h. Calculate the maximum growth rate of the microorganisms under these conditions.

Solution

Re-arranging equation (4.8):

$$\mu_{\max} = \frac{\ln 2}{t_d} = 0.28h^{-1}$$

Example 4.3

In a batch test, the initial concentration of microorganisms is 0.2 kg/m^3. After 12 hours, the concentration of microorganisms has reached 0.8 kg/m^3. Calculate the maximum growth rate and the doubling time under the conditions of the test. Assume that microorganisms are acclimated to the substrate at the start of the test and that the substrate concentration is always in excess.

Solution

Under the hypotheses of this problem (substrate in excess and no acclimation phase):

$$\ln\left(\frac{X}{X_0}\right) = \mu_{max} t$$

Therefore,

$$\mu_{max} = \frac{1}{t}\ln\left(\frac{X}{X_0}\right) = \frac{1}{12}\ln\left(\frac{0.8}{0.2}\right) = 0.1155\ \text{h}^{-1}$$

and

$$t_d = \frac{\ln 2}{\mu_{max}} = 6\,\text{h}$$

4.3 MASS BALANCES FOR CONTINUOUS REACTORS

Continuous reactors are characterised by a continuous flow of feed and effluent. Continuous reactors are only inoculated at the start-up, but they require no further addition of microorganisms (the feed is sterile). Similarly as for batch reactors, we assume that continuous reactors are perfectly mixed (CSTR). This also means that the composition of the effluent is the same as inside the reactor (Figure 4.3).

Assuming we know the values of all the kinetic parameters that we need, in the design of continuous fermenters the main task is to determine the residence time that is required to achieve the desired conversion of the substrate and the desired concentration of microorganisms and products. The calculation of the residence time can be obtained by solving the mass balances for substrate, microorganisms and products.

Assuming that the fermenter is at steady state, the general form of the mass balance (4.1) takes the form

$$\text{Output-Input} = \text{Generation-Consumption} \tag{4.9}$$

Assuming the rate equations for substrate removal, microorganisms and product production seen in Chapter 3, the mass balances in a continuous fermenter are

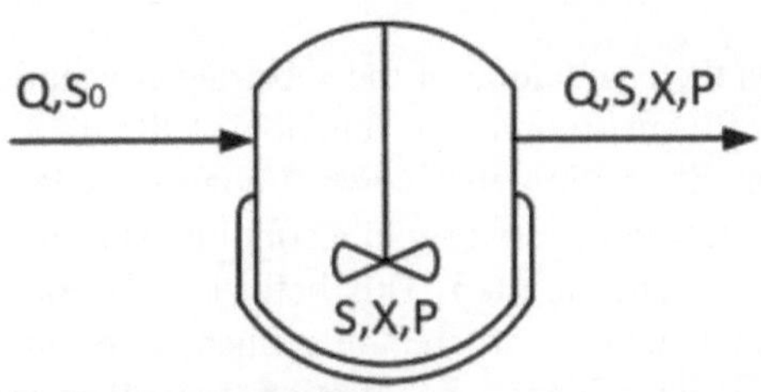

FIGURE 4.3 Scheme of a continuous fermenter.

$$r_X \cdot V = Q \cdot X\left(\text{biomass balance}\right) \tag{4.10}$$

$$Q \cdot S_0 = Q \cdot S + \left(-r_S\right) \cdot V\left(\text{substrate balance}\right) \tag{4.11}$$

$$r_P \cdot V = Q \cdot P\left(\text{product balance}\right) \tag{4.12}$$

Equations (4.10)–(4.12) can be re-arranged by introducing the dilution rate D, which is the reciprocal of the residence time τ:

$$D\left(\frac{1}{d}\right) = \frac{Q}{V} = \frac{1}{\tau} \tag{4.13}$$

Introducing equation (4.13), equations (4.10) –(4.12) become

$$r_X = D \cdot X \tag{4.14}$$

$$D\left(S_0 - S\right) = \left(-r_S\right) \tag{4.15}$$

$$r_P = D \cdot P \tag{4.16}$$

Equations (4.14)–(4.16) can be used to calculate the dilution rate, or the residence time, that is required to obtain the desired concentrations of substrate, microorganisms and product in the effluent stream. Using the Monod kinetics defined in Chapter 3, equation (4.14) becomes

$$\mu = \frac{\mu_{max} S}{K_S + S} = D \tag{4.17}$$

Equation (4.17) can be re-arranged as

$$S = \frac{DK_S}{\mu_{max} - D} \tag{4.18}$$

Equation (4.18) allows the calculation of the substrate concentration in a continuous fermenter and in the effluent stream as a function of the dilution rate if the kinetic parameters of the Monod equation are known. Equation (4.18) shows that, interestingly, the effluent substrate concentration in a continuous fermenter does not depend on the influent substrate concentration. This behaviour is not observed for conventional nonfermentation reactions, where the effluent concentration of the starting material does depend on its influent concentration, as well as on the residence time.

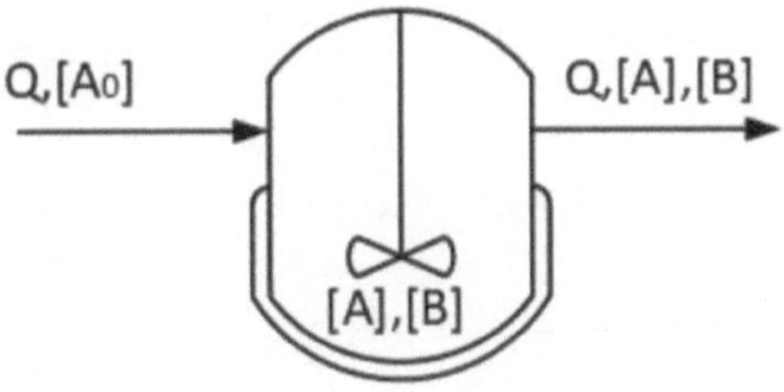

FIGURE 4.4 Scheme of a nonfermentation reaction taking place in a CSTR.

For example, let us consider the generic nonfermentation first-order reaction $A \rightarrow B$, with first-order kinetics $r_A\left(\frac{\text{kmol}}{\text{m}^3\ \text{s}}\right) = -k[A]$, taking place in a CSTR (Figure 4.4). The effluent concentration of the starting material A is given by

$$[A] = \frac{[A_0]}{1 + k\frac{V}{Q}} \tag{4.19}$$

Equation (4.19), which is obtained from the mass balance of species A, can be compared with equation (4.18), which is the analogous equation for fermentation reactions, showing that for nonfermentation reactions the effluent concentration of the starting material depends on its influent concentration, while for fermentation reactions it does not. The reason for this different behaviour is that in fermentation processes the rate of substrate removal is proportional to the concentration of the product, i.e. the microorganisms. If the substrate concentration in the feed increases, so does the concentration of microorganisms in the reactor (see also equation (4.20)), giving higher substrate removal rates and maintaining the same substrate concentration in the reactor.

The substrate and the product balances (4.15) and (4.16) can also be re-arranged to calculate the biomass and the product concentration in the reactor and effluent:

$$X = Y_{X/S}(S_0 - S) \tag{4.20}$$

$$P = X \cdot Y_{P/X} \tag{4.21}$$

In summary, equations (4.18), (4.20) and (4.21) give the concentrations S, X and P as a function of the dilution rate D (or of its reciprocal, the residence time), if the kinetic parameters are known.

The typical profiles of S, X and P as a function of D are shown in Figure 4.5.

Figure 4.5 can be explained as follows. Low values of D correspond to large values of the residence time and the substrate concentration is low because the

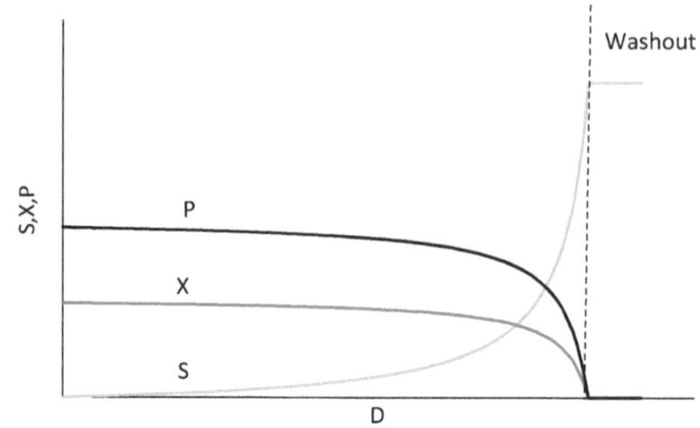

FIGURE 4.5 Typical profiles of X, P and S as a function of D.

microorganisms have a long time to grow on the substrate. Consequently, low values of D also correspond to the highest values of X and P. As D increases, and correspondingly the residence time decreases, S increases and X and P decrease. There is a maximum value of D for which the substrate concentration becomes equal to the influent concentration, with no production of microorganisms or products. This condition ($D = D_{max}$) corresponds to the wash-out of the microorganisms, which cannot grow fast enough to be maintained in the reactor. The condition for the wash-out is obtained by using $S = S_0$ in equation (4.17):

$$D_{max} = \frac{\mu_{max \times S_0}}{K_S + S0} \tag{4.22}$$

A continuous fermenter needs to be operated with a value of D lower than D_{max}, in order to avoid wash-out of the microorganisms. Since in many cases $S_0 >> K_S$, then D_{max} is often very close to μ_{max}.

The mass balances (4.18), (4.20) and (4.21) can also be used for the calculation of the biomass and product productivity as a function of the dilution rate. The productivity is the biomass or product produced per unit of reactor volume per unit of time.

$$\text{Biomass productivity}\left(\frac{\text{kg}}{\text{m}^3\text{d}}\right) = \frac{QX}{V} = DX = DY_{X/S}\left(S_0 - S\right) \tag{4.23}$$

$$\text{Product productivity}\left(\frac{\text{kg}}{\text{m}^3\,\text{d}}\right) = \frac{QP}{V} = DP = DXY_{P/X} \tag{4.24}$$

The productivity of biomass and product depends on the dilution rate. When the dilution rate is low, the productivity will also be low because the rate of biomass and

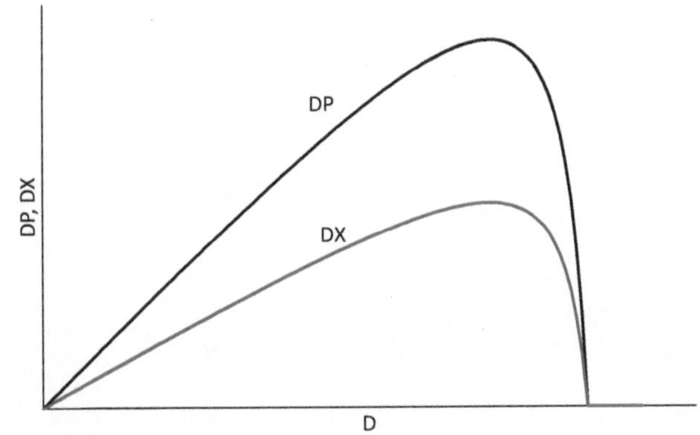

FIGURE 4.6 Typical profiles of biomass and product productivities in a continuous fermenter as a function of the dilution rate.

product generation per unit volume is low. As the dilution rate increases the productivity increases because the biomass and product production per unit volume increases. However, when the dilution rate approaches D_{max}, the productivity falls again because we are approaching wash-out of the microorganisms. It follows that the biomass and product productivity will have a maximum for a certain value of the dilution rate D. Figure 4.6 shows the typical profile of the biomass and product productivities as a function of the dilution rate.

Plots like the one in Figure 4.6 can be obtained by plotting equations (4.23) and (4.24) as a function of D, once S, X and P have been obtained by the solution of equations (4.18), (4.20), (4.21).

Example 4.4

The anaerobic fermentation of glucose to produce ethanol is carried out in a CSTR. The influent glucose concentration is 150 g/L. Calculate the profiles of microorganisms, glucose and ethanol as a function of the dilution rate. Also calculate the value of D, which gives the highest ethanol productivity and the value of the maximum productivity. What is the volume of the reactor that gives the highest ethanol productivity if the feed flow rate is 1000 m^3/d? What is the value of this productivity?

Stoichiometry of the reaction:

$$C_6H_{12}O_6 + 0.31NH_3 \rightarrow 0.31C_5H_7O_2N + 1.64C_2H_6O + 1.17CO_2 + 0.464H_2O$$

Kinetic parameters: $\mu_{max} = 5.3\ d^{-1}$ and $K_S = 3$ g/L.

Solution

The profiles of glucose, microorganisms and ethanol are given by equations (4.18), (4.20) and (4.21), respectively. In this case,

$$Y_{X/S} = \frac{0.31 \cdot 113}{180} = 0.195 \frac{\text{kg}}{\text{kg}}$$

$$Y_{P/X} = \frac{1.64 \cdot 46}{0.31 \cdot 113} = 2.15 \frac{\text{kg}}{\text{kg}}$$

Therefore, equations (4.18), (4.20) and (4.21) become

$$S = \frac{DK_S}{\mu_{max} - D} = \frac{D \cdot 3}{5.3 - D}$$

$$X = 0.195(150 - S)$$

$$P = X \cdot 6.68$$

where S, X and P are in kg/m^3 and D is in d^{-1}.

In this case the maximum dilution rate, corresponding to the wash-out of the microorganisms, is

$$D_{max} = \frac{5.3 \cdot 150}{3 + 150} = 5.196 \text{ d}^{-1}$$

The plot of glucose, microorganisms and ethanol concentrations as a function of the dilution rate is shown in Figure 4.7

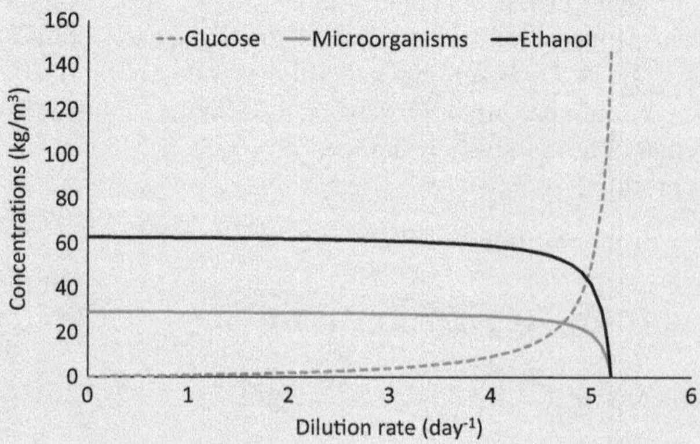

FIGURE 4.7 Results of Example 4.4.

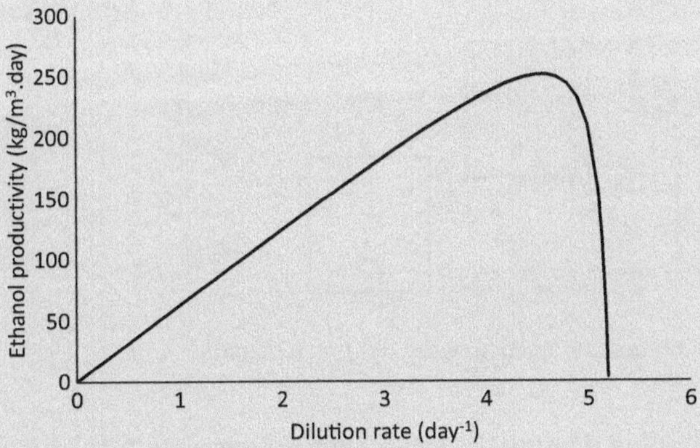

FIGURE 4.8 Ethanol productivity for Example 4.4.

The ethanol productivity (=$D \cdot P$) can be calculated using equation (4.24), and is plotted in Figure 4.8.

From the graph, the highest ethanol volumetric productivity is obtained for $D = 4.6\ \text{d}^{-1}$. For the given feed flow rate, the reactor volume is

$$V = \frac{Q}{D} = \frac{1{,}000 \frac{\text{m}^3}{\text{d}}}{4.6\ \text{d}^{-1}} = 217.39\ \text{m}^3$$

For $D = 4.6\ \text{d}^{-1}$, the ethanol concentration is 54.6 kg/m^3, therefore the ethanol productivity is

$$Q \cdot P = 1000 \frac{\text{m}^3}{\text{d}} \cdot 54.6 \frac{\text{kg}}{\text{m}^3} = 54{,}600 \frac{\text{kg}}{\text{d}}$$

4.4 MASS BALANCES FOR FED-BATCH REACTORS

The fed-batch reactor sits in the middle-ground between the batch and the continuous reactor. In the fed-batch reactor a volume of feed containing the substrate is fed to the reactor over a period of time. When the volume of the liquid (or liquid-solid mixture) reaches the maximum volume, a fraction of the volume is withdrawn bringing the volume back to the initial value, and a new cycle is started. The operational sequence of the fed-batch reactors is shown in Figure 4.9.

Let us consider what happens during the start-up of the fed-batch reactor. Let us assume that the reactor is inoculated with a low concentration of microorganisms when feeding starts. Initially, substrate concentration increases in the reactor, because the substrate feeding rate exceeds the substrate consumption rate, due to the low

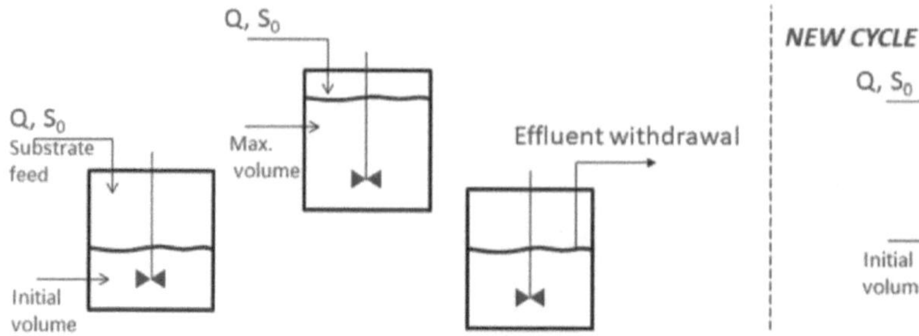

FIGURE 4.9 Operational sequence of the fed-batch reactor.

biomass concentration. During the cycle biomass concentration increases, however, and, after the effluent withdrawal, which marks the end of the first cycle, at the next cycle the substrate concentration will rise more slowly. After a few cycles, a "quasi-steady state" will be reached when the substrate concentration will be very low and the biomass (and products, if present) concentration will be approximately constant during the cycle.

The calculation of the substrate, biomass and product at the quasi-steady state of the fed-batch reactor can be done using mass balances. The mass balance for biomass in the fed-batch reactor is

$$\frac{d(XV)}{dt}=\frac{\mu_{max}S}{K_s+S}XV \tag{4.25}$$

Equation (4.25) can be re-arranged as

$$X\frac{dV}{dt}+V\frac{dX}{dt}=\frac{\mu_{max}S}{K_s+S}XV \tag{4.26}$$

Considering that

$$\frac{dV}{dt}=Q \tag{4.27}$$

Equation (4.26) becomes

$$XQ+V\frac{dX}{dt}=\frac{\mu_{max}S}{K_s+S}XV \tag{4.28}$$

At the "quasi-steady state" it will be that: $\frac{dX}{dt}=0$ and therefore equation (4.28) will become

$$\frac{Q}{V} = D = \frac{\mu_{max} S}{K_S + S} \tag{4.29}$$

Equation (4.29) is the same as equation (4.17) for the continuous fermenter. From equation (4.29), the concentration of the substrate at the quasi-steady state can be calculated using equation (4.18):

$$S = \frac{DK_S}{\mu_{max} - D} \tag{4.18}$$

Equations (4.29) and (4.18) show that the substrate concentration in the fed-batch reactor at the quasi-steady state and in the continuous reactor can be calculated with the same equations. However, note that D is constant for a continuous fermenter, while it varies during the cycle in the fed-batch reactor because the reactor volume V is variable. This means that the substrate concentration will be variable during the quasi steady state. However, in practice in a well-operated fed-batch reactor, the substrate concentration during the cycle, although variable, will always be very low and much lower than the feed concentration S_0.

The biomass concentration can be calculated from the substrate balance:

$$\frac{d(VS)}{dt} = -\frac{\mu_{max} S}{K_S + S} \cdot \frac{X \cdot V}{Y_{X/S}} + QS_0 \tag{4.30}$$

which becomes

$$V\frac{dS}{dt} + S\frac{dV}{dt} = -\frac{\mu_{max} S}{K_S + S}\frac{X \cdot V}{Y_{X/S}} + QS_0 \tag{4.31}$$

Using equation (4.27) and observing that at the quasi-steady state $\frac{dS}{dt} = 0$ we obtain

$$Q(S_0 - S) = \frac{\mu_{max} S}{K_S + S}\frac{X \cdot V}{Y_{X/S}} \tag{4.32}$$

Note that equation (4.32) is the same as equation (4.15) seen for the continuous fermenter. Equation (4.32) can be re-arranged to give

$$X = Y_{X/S}(S_0 - S) \cong Y_{X/S} S_0 \tag{4.33}$$

Equation (4.33), which is essentially the same as equation (4.20) already seen for the continuous fermenter, gives the biomass concentration at the quasi-steady state of the fed-batch reactor.

The product concentration at the quasi-steady state of the fed-batch reactor can be calculated with a similar procedure by starting with the mass balance for the product:

$$\frac{d(PV)}{dt} = \frac{\mu_{maxS}}{K_S + S} \frac{X \cdot V}{Y_{X/S}} Y_{P/S} \tag{4.34}$$

With similar steps and assumptions as done for the biomass and substrate balance, the product balance (4.34) can be re-arranged at the quasi-steady state as

$$P = S_0 \cdot Y_{P/S} \tag{4.35}$$

Equation (4.35) gives the product concentration at the quasi-steady state in the fed-batch reactor.

Note that equations (4.18), (4.33) and (4.35) allow calculatation of the concentrations of substrate, biomass and product in the fed-batch reactor at quasi-steady state only under the hypothesis that the influent flow rate Q is low enough to allow the microorganisms to maintain a very low substrate concentration ($S << S_0$) throughout all the feeding cycle.

The biomass and product productivity can also be calculated for the fed-batch reactor. To calculate the product productivity, we need to consider that the total mass of microorganisms in the reactor, X_{tot}, during a cycle at the quasi-steady state is given by

$$X_{tot}\,(\text{kg}) = X \cdot V = Y_{X/S} S_0 \left(V_0 + Qt\right) = Y_{X/S} S_0 V_0 + Y_{X/S} S_0 Qt \tag{4.36}$$

Obviously, X_{tot} increases during the cycle. The biomass productivity is calculated by taking the derivative of X_{tot} with respect to time, i.e.

$$\frac{dX_{tot}}{dt}\left(\frac{\text{kg biomass}}{\text{d}}\right) = Y_{X/S} S_0 Q \tag{4.37}$$

Similarly, for the product we can define the total mass of product in the reactor, P_{tot}:

$$P_{tot} = P \cdot V = Y_{P/S} S_0 \left(V_0 + Qt\right) = Y_{P/S} S_0 V_0 + Y_{P/S} S_0 Qt \tag{4.38}$$

From equation (4.33), the product productivity can be calculated as

$$\frac{dP_{tot}}{dt}\left(\frac{\text{kg product}}{\text{d}}\right) = Y_{P/S} S_0 Q \tag{4.39}$$

Note that equations (4.37) and (4.39) are valid under the same assumptions used to derive the quasi-steady state concentrations, i.e. that the feed flow rate is low enough to ensure that the substrate concentration in the reactor is always much lower than the feed concentration. The biomass and product productivities can also be expressed per unit of reactor volume, as done in the case of the continuous fermenter.

Example 4.5

It is planned to produce ethanol by fermentation of glucose by the yeast *Saccharomyces cerevisiae* in a fed-batch reactor. The initial volume of the fermentation broth is V_0 and the final volume is V_{final}. The feed time is t_{feed} and the number of cycles that can be carried out in a year is n_{cycles}. The amount of ethanol that is required to be produced per year is $F_{ethanol}$.

(a) Calculate the feed flow rate to the reactor and the required concentration of glucose in the feed.
(b) The nitrogen source for microorganisms' growth is $(NH_4)_2SO_4$, which is purchased externally and added to the feed. Calculate the annual cost of $(NH_4)_2SO_4$ in order to supply the minimum amount of nitrogen necessary for the microorganism growth.

Data:

$V_0 = 10$ m^3 (initial volume in the fed-batch reactor)
$V_{final} = 50$ m^3 (final volume in the fed-batch reactor)
$t_{feed} = 2$ h (feed time in the fed-batch reactor)
$n_{cycles} = 2000$ (cycles per year in the fed-batch reactor)
$F_{ethanol} = 1840$ tonnes/year (required ethanol production rate in the fed-batch reactor)

Cost of $(NH_4)_2SO_4$ = £2/kg

Stoichiometry of microorganism growth and ethanol (C_2H_5OH) production:

$$C_6H_{12}O_6 + 0.12NH_3 \rightarrow 0.12C_5H_7O_2N + 1.8C_2H_5OH + 1.8CO_2 + 0.36H_2O$$

Solution

From the fermentation stoichiometry we get $Y_{P/S} = 0.46$ kg/kg.

The feed flow rate to the reactor is

$$Q = \frac{V_{final} - V_0}{t_{feed}} = \frac{(50-10)\text{m}^3}{2\text{h}} = 20\frac{\text{m}^3}{\text{h}}$$

Since there are 2000 cycles per year, this means that the feed flow rate per year is

$$Q = 20\frac{\text{m}^3}{\text{h}}\cdot\frac{2\ \text{h}}{\text{cycle}}\frac{2000\ \text{cycle}}{\text{year}} = 80,000\frac{\text{m}^3}{\text{year}}$$

The required glucose concentration in the feed is

$$S_0 = \frac{F_{eth}}{Y_{P/S}Q} = \frac{1.84\cdot 10^6\ \dfrac{\text{kg}}{\text{year}}}{0.46\dfrac{\text{kg}}{\text{kg}}\cdot 80,000\dfrac{\text{m}^3}{\text{year}}} = 50\frac{\text{kg}}{\text{m}^3}$$

(b) The amount of nitrogen required by microbial growth is

$$\text{Nitrogen consumed} = 1.84\cdot 10^6\,\frac{0.12\cdot 14}{1.8\cdot 46} = 37,333\frac{\text{kg N}}{\text{year}}$$

The amount of ammonium sulphate required is equal to 37,333·132/28=176·10^3 kg $(NH_4)_2SO_4$/year.

At a cost of £2/kg this corresponds to £352,000/year.

4.5 MODIFICATIONS OF THE MASS BALANCES FOR SECONDARY METABOLITES, MAINTENANCE AND ENDOGENOUS METABOLISM

The equations in previous sections refer to products that are primary metabolites, i.e. produced simultaneously to biomass growth. When the product is a secondary metabolite, there is no production during biomass growth but there is production after growth has stopped. Secondary metabolites are usually produced in batch fermentation or, in some cases, in fed-batch processes, so in this section we will only see the mass balances for batch fermentation. Using equations (3.32), (3.33) the production rate of a secondary metabolite is

$$\frac{dP}{dt} = \alpha X \tag{4.40}$$

or

$$\frac{dP}{dt} = Y_{P/X}r_X + \alpha X \tag{4.41}$$

Equation (4.41) applies when product production occurs both during biomass growth and when growth has stopped.

Another modification of the mass balances is due to the occurrence of maintenance metabolism. If we assume that biomass metabolism follows the maintenance model according to equation (3.34) for the rate of substrate removal, then the substrate balance for batch fermentation will be

$$\frac{dS}{dt} = -\frac{\mu \cdot X}{Y_{X/S}} - m_S X \tag{4.42}$$

The mass balances for biomass and product in batch fermentation will still be given by equations (4.3) and (4.5).

In a continuous reactor, the maintenance metabolism modifies equation (4.15) as follows:

$$D(S_0 - S) = \frac{\mu \cdot X}{Y_{X/S}} + m_S X \tag{4.43}$$

Equation (4.43) can be manipulated to obtain the biomass concentration at steady state:

$$X = \frac{Y_{X/S}(S_0 - S)}{1 + \frac{m_S Y_{X/S}}{D}} \tag{4.44}$$

The comparison of equations (4.20) and (4.44) indicates that, in a model with the maintenance metabolism, the steady-state biomass concentration in a continuous fermenter is lower than in a model without maintenance. The reason for this is that part of the substrate is used for the maintenance metabolism and not for biomass growth. For a continuous fermenter with maintenance metabolism, the steady-state substrate and product concentrations are still given by equations (4.18) and (4.21). The substrate concentration will be the same as in the absence of maintenance, while the product concentration will be lower due to the lower biomass concentration.

If we assume the endogenous metabolism model, in a batch reactor the mass balances for the substrate and product will be given by equations (4.4) and (4.5), while the mass balance (4.3) modifies to

$$\frac{dX}{dt} = \frac{\mu_{max} S}{K_S + S} X - bX \tag{4.45}$$

In a continuous reactor with endogenous metabolism the mass balances for substrate and product will remain the same as (4.15) and (4.16) while the biomass balance (4.14) modifies into

$$\mu \cdot X - bX = DX \tag{4.46}$$

Equation (4.45) can be re-arranged to give

$$\mu = D + b \tag{4.47}$$

and

$$S = \frac{K_S(D+b)}{\mu_{max} - (D+b)} \tag{4.48}$$

Substituting equation (4.47) into (4.15) and (4.16) and re-arranging, we obtain the biomass and product concentrations in a continuous reactor for the model with endogenous metabolism:

$$X = \frac{DY_{X/S}(S_0 - S)}{(D+b)} \tag{4.49}$$

$$P = Y_{P/S}(S_0 - S) \tag{4.50}$$

The comparison of equations (4.18) and (4.48) indicates that, in a model with endogenous metabolism, the steady-state substrate concentration in a continuous fermenter is higher than in a model without endogenous metabolism. The reason is that endogenous metabolism reduces the biomass concentration, therefore increasing the substrate concentration.

Example 4.6

Compare the batch fermentation profiles in the absence of maintenance metabolism of Example 4.1, with the profiles in the same process with maintenance metabolism. Assume a maintenance parameter $m_S = 2.50\ d^{-1}$, and the same parameter values as Example 4.1.

Solution
In this case the mass balances for substrate, biomass and product become

$$\frac{dX}{dt} = \frac{5.2S}{0.05+S}X$$

$$\frac{dS}{dt} = -\frac{5.2S}{0.05+S}\frac{X}{0.15} - 2.5X$$

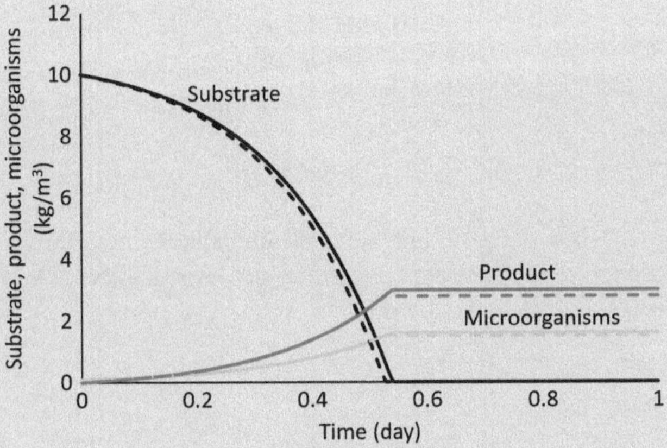

FIGURE 4.10 Solution of Example 4.6, comparison of batch fermentation profiles with (dashed lines) and without (solid lines) maintenance metabolism.

$$\frac{dP}{dt} = \frac{5.2S}{0.05+S} X \cdot 2.70$$

The solution of the mass balances is shown in Figure 4.10. The maintenance metabolism gives a higher substrate removal rate with the same biomass and product concentration. The final concentration of biomass and product will be lower than in the absence of maintenance because part of the substrate is used for maintenance rather than for biomass and product formation. With typical values of the maintenance metabolism parameter, the effect of maintenance is quite modest, as Figure 4.10 shows.

Example 4.7

Compare the glucose, ethanol and biomass profiles vs D obtained for the fermenter in Example 4.4, with the profiles in a model with a maintenance coefficient equal to 2.5 d^{-1}. Assume the same stoichiometry as in Example 4.4.

Solution

In a model with maintenance, the mass balances for glucose, ethanol and biomass become

$$S = \frac{DK_S}{\mu_{max} - D} = \frac{D \cdot 3}{5.3 - D}$$

$$X = \frac{0.195(150 - S)}{1 + \dfrac{2.5 \cdot 0.195}{D}}$$

$$P = X \cdot 6.68$$

The solution is shown in Figure 4.11. Maintenance metabolism causes a decrease in product and substrate concentrations, especially at the lowest dilution rate, where the growth is slowest.

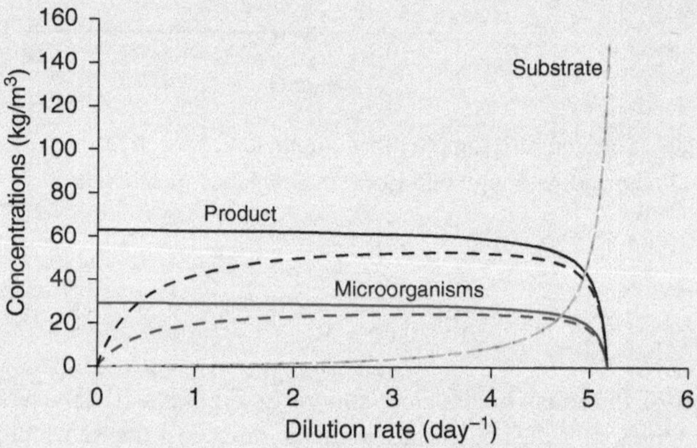

FIGURE 4.11 Solution of Example 4.7, comparison of continuous fermenter profiles with (dashed lines) and without (solid lines) maintenance metabolism.

Example 4.8

Compare the batch fermentation profiles of Example 4.1, with the profiles in the same process with endogenous metabolism. Assume an endogenous metabolism parameter $b = 0.2\ d^{-1}$, and the same parameter values as Example 4.1.

Solution

In this case the mass balances for substrate, biomass and product become

$$\frac{dX}{dt} = \frac{5.2S}{0.05 + S} X - 0.2X$$

$$\frac{dS}{dt} = -\frac{5.2S}{0.05 + S} \frac{X}{0.15}$$

$$\frac{dP}{dt} = \frac{5.2S}{0.05+S} X \cdot 2.70$$

The solution of these differential equations is reported in Figure 4.12.

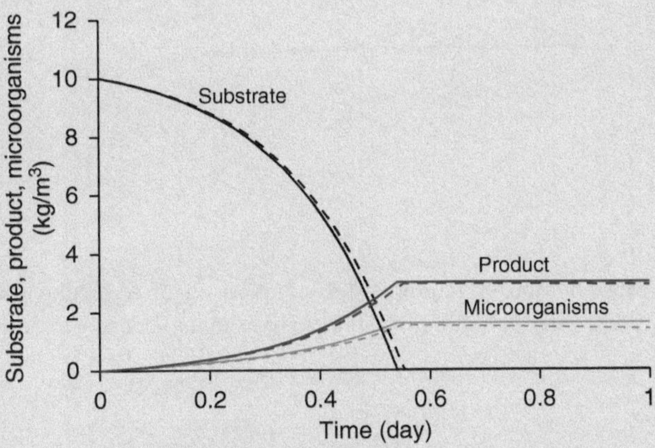

FIGURE 4.12 Solution of Example 4.8, comparison of batch profiles with (dashed lines) and without (solid lines) endogenous metabolism.

Example 4.9

Compare the glucose, ethanol and biomass profiles vs D obtained for the fermenter in Example 4.4, with the profiles in a model with endogenous metabolism coefficient equal to 0.2 d^{-1}. Assume the same stoichiometry as in Example 4.4.

Solution

The mass balances for glucose, ethanol and biomass give in this case

$$S = \frac{3(D+0.2)}{5.3-(D+0.2)}$$

$$X = \frac{D \cdot 0.195(150-S)}{(D+0.2)}$$

$$P = 0.42(150-S)$$

The solution of these equations is shown in Figure 4.13.

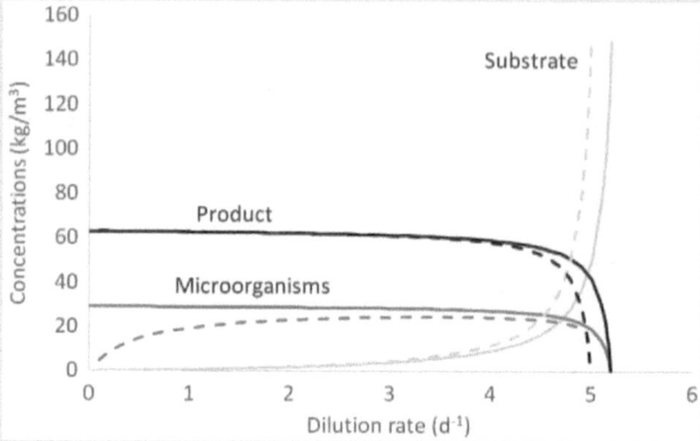

FIGURE 4.13 Solution of Example 4.9, comparison of fermenter profiles with (dashed lines) and without (solid lines) endogenous metabolism.

4.6 USE OF CONTINUOUS REACTORS FOR THE ESTIMATION OF KINETIC PARAMETERS

In Chapter 3 we have seen that kinetic parameters can be estimated from batch tests. Another method is to use continuous cultures. For example, for a microorganism growing in a continuous reactor, the substrate concentration at steady state is obtained by solving equation (4.17):

$$\frac{\mu_{max} S}{K_S + S} = D \tag{4.17}$$

This equation can be re-arranged as follows:

$$\frac{1}{S} = \frac{\mu_{max}}{K_S} \frac{1}{D} - \frac{1}{K_S} \tag{4.51}$$

Equation (4.51) shows that, collecting data of substrate concentration in a fermenter at different dilution rates allows for the calculation of the kinetic parameters μ_{max} and K_S. In particular, a plot of $\frac{1}{S}$ vs $\frac{1}{D}$ should give a straight line with $\frac{\mu_{max}}{K_S}$ as the slope and $-\frac{1}{K_S}$ as the intercept.

The parameter $Y_{X/S}$ can be calculated from equation (4.52) at any dilution rate:

$$Y_{X/S} = \frac{X}{(S_0 - S)} \tag{4.52}$$

Assuming the microbial metabolism is described by the maintenance model, the parameters μ_{max} and K_S can still be determined using equation (4.51), while the parameters $Y_{X/S}$ and m_S can be determined by re-arranging equation (4.44) in a linearised form:

$$\frac{(S_0 - S)}{X} = \frac{1}{Y_{X/S}} + \frac{m_S}{D} \tag{4.53}$$

Equation (4.53) shows that a plot of $\frac{(S_0 - S)}{X}$ vs $\frac{1}{D}$ should give a straight line with the slope equal to m_S and the intercept equal to $1/Y_{X/S}$. Note that if $m_S = 0$, equation (4.53) coincides with equation (4.52).

Example 4.10

The data in Table 4.1 have been obtained in a continuous fermenter at various dilution rates. Calculate the parameters μ_{max}, K_S, $Y_{X/S}$ (ignore maintenance or endogenous metabolism). The substrate concentration in the feed is 50 kg/m^3.

Solution

μ_{max} and K_S can be calculated from the data in Table 4.1 using equation (4.51). The plot is reported in Figure 4.14.

TABLE 4.1
Data for Example 4.10

Dilution Rate (d^{-1})	S (kg/m^3)	X (kg/m^3)
0.2	0.005	12.5
0.4	0.01	12.5
0.6	0.017	12.5
0.8	0.023	12.5
1	0.032	12.5
1.5	0.06	12.5
2	0.101	12.5
2.5	0.20	12.5
3	0.48	12.4
3.2	0.85	12.3

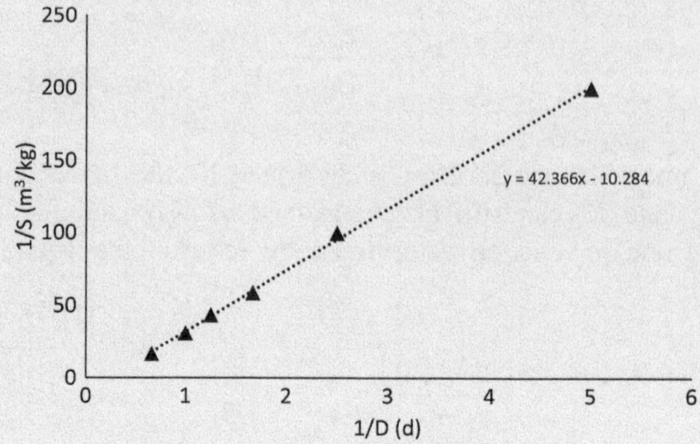

FIGURE 4.14 Plot of Equation (4.51) with the data in Table 4.1.

From the slope and intercept of Figure 4.13 we calculate $K_S = 0.097$ kg/m^3 and $\mu_{max} = 4.12$ d^{-1}. The parameter $Y_{X/S}$ can be calculated from equation (4.52) using values at any of the dilution rates, or the average of these values.

4.7 COMPARISON OF BATCH, CONTINUOUS AND FED-BATCH REACTORS

Fermentation reactions have some distinctive features in terms of reactor design when compared to nonfermentation reactions. For most nonfermentation reactions, the time required for a batch reaction to go to completion is lower than the residence time required for the same reaction in a continuous-flow reactor. This is because for most nonfermentation reactions, the reaction rate is proportional to the concentration of the reagents, and in batch reactors the reagents' concentration is higher than in continuous-flow reactors. However, for fermentation reactions, the reaction rate depends not only on the substrate concentration, but also on the microorganisms' concentration, and the microorganisms' concentration at the start of batch processes is typically low. For a given kinetics of the fermentation reaction, i.e. for given parameters μ_{max} and K_S, the rate of a batch fermentation process depends on the initial concentration of microorganisms. This is not the case for continuous and fed-batch fermentations at steady state (the "quasi-steady state" for bed batch), where the concentration of microorganisms is constant with time and depends only on the substrate concentration in the feed, but not on the microorganisms' concentration in the inoculum.

Figure 4.15 compares the average reaction rate for a batch fermentation (defined as the initial substrate concentration divided by the time required to achieve 99% removal of the initial substrate) at different initial microorganisms' concentrations with the reaction rate for the same fermentation in a continuous reactor. Depending

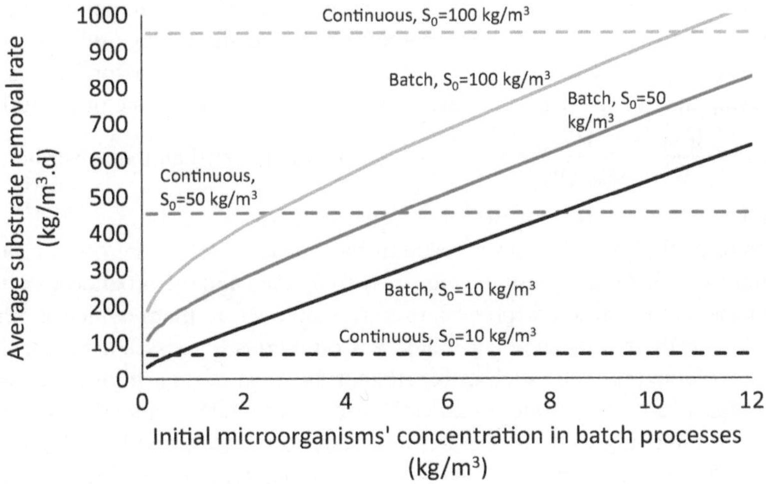

FIGURE 4.15 Comparison of the average substrate removal rate in batch and continuous reactors. The assumed kinetic parameters were $\mu_{max} = 10\ d^{-1}$, $K_S = 0.05\ kg/m^3$, $Y_{X/S} = 0.2\ kg/kg$. Substrate was assumed to be completely removed when the concentration was 1% of the initial value. The rates for the continuous reactor were calculated for the dilution rate that gave an effluent substrate concentration equal to 1% of the feed concentration (these dilution rates were equal to 6.7, 9.1 and 9.5 d^{-1} for S_0 equal to 10, 50 and 100 kg/m^3, respectively).

on the initial microorganisms' concentration in the inoculum the average substrate removal rate in a batch reactor can be higher or lower than in a continuous reactor, where the rate is independent on the initial inoculum concentration. The higher the initial feed substrate concentration, the more kinetically favoured the continuous reactor becomes compared to the batch, as shown by the higher initial inoculum concentration in the batch required to achieve the same rate as in the continuous fermenter. The reason is that in a continuous process the substrate removal rate is approximately linearly proportional to the substrate concentration in the feed (apart from the small effect of the substrate concentration on the optimum dilution rate), while in a batch reactor the initial rate only depends on the inoculum concentration and not on the substrate concentration (assuming that $S_0 >> K_S$). In a batch fermenter, the substrate removal rate increases with time and this increase is higher at higher initial substrate concentration (because of the higher biomass concentration that will be obtained); however, the average substrate removal rate during the batch fermentation is less proportional to the initial substrate concentration than for a continuous reactor.

As far as the fed-batch reactor is concerned, the dilution rate that gives a certain effluent substrate concentration is the same as for the continuous reactor, for a given fermentation reaction. However, the dilution rate varies during the feeding of a fed-batch reactor. In order to guarantee that the substrate concentration in the fed-batch reactor is not larger than a certain desired concentration over the whole length of the feed, the dilution rate needs to be equal to the value given by equation (4.18) at the start of the cycle. However, at the start of the cycle the reactor volume is at the lowest value, so the design flow rate for a fed-batch reactor will be lower, per unit volume of the full reactor, than in a continuous reactor. Therefore, the average substrate removal rate in a fed-batch reactor will be lower than in a continuous reactor. This is evident considering that the average substrate removal rate per unit volume of the full reactor is given by $\frac{QS_0}{V} = \frac{QS_0}{V_0 + Qt_{feed}} = \frac{DS_0}{1 + Dt_{feed}}$, which is lower than the substrate removal rate in the continuous reactor (which is equal to $D \cdot S_0$).

In spite of their usually lower reaction rate, batch reactors are often used in the fermentation industry, because they are easy to operate and the products can be easily traced back to the particular process used to manufacture them, which is important especially for the pharmaceutical industry. Continuous processes are typically used for large volume products, such as bioethanol, because of their higher reaction rates and the absence of downtime. Fed-batch reactors are also frequently used, because of they are relatively easy to operate, like batch processes, and have high productivity, comparable (even though lower) to continuous fermenters. Fed-batch reactors are particularly convenient when the process conditions have to change during the fermentation, for example for secondary metabolites, which are only produced at the end of the growth phase when some nutrients become limiting or when there is a need to add certain chemicals to trigger the excretion of the metabolites. In these cases, fed-batch reactors are suitable because the feed composition can change during the cycle, which is not easily possible in batch and continuous processes.

QUESTIONS AND PROBLEMS

4.1 Calculate the profiles of substrate, microorganisms, and product for the batch fermentation reaction below, where the product is citric acid. The microorganisms have a doubling time of 2.2 h and a K_S of 0.07 kg/m^3. The batch is inoculated with a concentration of microorganisms of 400 mg/L and the initial concentration of glucose is 90 kg/m^3.

$$C_6H_{12}O_6 + 2.53O_2 + 0.1NH_3 \rightarrow 0.1C_5H_7O_2N + 0.66C_6H_8O_7 + 1.54CO_2 + 3.16H_2O$$

4.2 A microorganism grows in a continuous fermenter. At the dilution rate of 1.0 d^{-1}, the effluent substrate concentration is 0.05 kg/m^3. At the dilution rate of 1.3 d^{-1}, the effluent substrate concentration is 0.15 kg/m^3. Calculate the values of μ_{max} and K_S for this microorganism.

4.3 A fermentation process is carried out in a fed-batch reactor, with an initial volume of 1 m^3. The kinetic parameters of the microorganisms are μ_{max} = 9.6 d^{-1}, K_S = 0.08 kg/m^3. Calculate the maximum feed flow rate that allows for the substrate concentration during feeding to remain at any time below 0.1 kg/m^3.

4.4 Consider the fermentation reaction for the production of glutamic acid represented by the stoichiometry below.

$$C_6H_{12}O_6 + 2.215O_2 + 0.83NH_3 \rightarrow 0.10C_5H_7O_2N + 0.73C_5H_9O_4N + 3.61H_2O + 1.85CO_2$$

The fermentation is carried out in a continuous reactor. Plot the biomass, substrate, glutamic acid concentrations and oxygen consumption per unit volume of reactor vs the dilution rate. Assume that the kinetic parameters for the microorganisms are μ_{max} = 8.5 d^{-1}, K_S = 0.06 kg/m$^{3.}$ Assume that the influent feed concentration of the substrate is 100 g/L.

4.5 It is desired to produce 20,000 t/year of ethanol. The stoichiometry of the fermentation is reported below.

$$C_6H_{12}O_6 + 0.12NH_3 \rightarrow 0.12C_5H_7O_2N + 1.8C_2H_5OH + 1.8CO_2 + 0.36H_2O$$

The kinetic parameters of the microorganisms are μ_{max} = 12.1 d^{-1}, K_S = 0.1 kg/m^3. Glucose is available at a concentration of 120 kg/m^3 and it is desired to convert at least 99% of it. Compare the volume of the reactor required in the three cases:

(a) batch fermentation, starting with an inoculum concentration of 500 mg/L. The downtime between two consecutive batches is 2 h;
(b) continuous fermentation;
(c) fed-batch reactor. Assume that each feed is operated for 11 h, with 1 h downtime between batches.

Assume that the reactors operate all year round (365 d) without interruptions.

4.6 Assume the stoichiometry of citric acid production reported in Problem 4.1. The microorganisms have μ_{max} = 10.0 d^{-1} and K_S = 0.15 kg/m$^{3.}$ Glucose is available at a concentration of 100 kg/m^3 and the desired effluent concentration is 0.5 kg/m$^{3.}$ Calculate the annual production rate of citric acid per unit volume of reactor under the following process configurations:

(a) batch reactor inoculated with 200 mg/L of microorganisms and with a downtime of 3 h;
(b) continuous reactor;
(c) fed-batch reactor, with a feed length of 11 h and a downtime of 1 h.

4.7 A microorganism is grown in a continuous reactor. The following data (Table 4.2) are obtained as a function of the dilution rate. Assuming a model with maintenance metabolism, calculate the parameters $Y_{X/S}$ and m_S for this microorganism.

4.8 For a certain microorganism, the effect of temperature on the maximum growth rate is described by equation (2.19), with Ea = 90,000 J/mol. If the doubling time is 2 h at T = 30 °C, what will be the doubling time at T = 35 °C.

4.9 The data in Table 4.3 are obtained in continuous cultures of a microorganism at three different temperatures. Assuming the μ_{max} of the microorganism depends

TABLE 4.2
Data for Problem 4.7

Dilution Rate (d^{-1})	S (kg/m^3)	X (kg/m^3)
0.1	0.001	8.3
0.5	0.006	23.5
1	0.013	31.2
2	0.03	36.5
3	0.054	39.2
4	0.09	40.1
5	0.15	41.3

TABLE 4.3
Data for Problem 4.9

Dilution Rate (d^{-1})	S, T = 20 °C (kg/m^3)	S, T = 25 °C (kg/m^3)	S, T = 30 °C (kg/m^3)
0.2	0.004	0.002	0.001
0.4	0.008	0.004	0.002
1	0.023	0.01	0.005
2	0.064	0.023	0.01
3	0.16	0.04	0.017
4	0.64	0.064	0.024
5		0.1	0.032
6		0.16	0.042
7		0.28	0.053
8		0.64	0.068

on temperature according to equation (2.19), calculate the parameters *Ea* and *A* for this microorganism.

4.10 Consider the anaerobic fermentation of sucrose ($C_{12}H_{22}O_{11}$) to produce lactic acid ($C_3H_6O_3$) by the microorganism *Lactobacillus delbrueckii*:

$$C_{12}H_{22}O_{11} + 0.576NH_3 \rightarrow 0.576C_5H_7O_2N + 3.04C_3H_6O_3 + 0.728H_2O$$

The microorganism grows on sucrose according to the Monod kinetics, with μ_{max} = 1.8 d^{-1} and K_S = 0.4 kg/m^3. The fermentation is carried out in a continuous fermenter. The sucrose concentration in the feed is 100 g/L and it is desired to achieve a sucrose concentration in the effluent of 3 g/L.
Calculate the volume of the reactor required to produce 15 t/d of lactic acid.

4.11 A fermentation process is carried out in two continuous reactors in series. The volumetric flow rate of the feed is 10 m^3/d and the volumes of the reactors are 5 m^3 (first reactor) and 7.5 m^3. The feed to the first reactor is sterile with no product and contains the substrate at a concentration of 100 g/L. The effluent of the first reactor has the following concentrations: biomass = 5 g/L; substrate = 3 g/L; product = 35 g/L. It is known that the microorganisms follow Monod kinetics with μ_{max} = 8.0 d^{-1}.

Calculate

- The composition (biomass, substrate product) at the outlet of the second reactor.
- The volume that would be required to carry out the process in one single reactor, achieving the same final substrate concentration as in the final effluent of the two-stage process.

4.12 A fermentation process is carried out in a fed-batch reactor. The volume at the start of the cycle is 2 m^3 and the volume at the end of the feed is 10 m^3. The feed (substrate concentration 60 g/L) takes place over 5 h. The concentrations in the effluent are as follows: biomass 6 g/L; product 20 g/L. It can be assumed that the substrate concentration is much lower than the feed concentration throughout all the feed time. It is now desired to run the process in a chemostat with a feed of 80 g/L aiming to obtain the same effluent concentrations as obtained for the fed-batch reactor. Calculate the required residence time in the continuous reactor and the effluent concentrations of biomass and product.

4.13 A secondary metabolite is produced in a fed-batch reactor in two phases. In the first phase the reactor is operated as a fed-batch with initial volume 1 m^3 and final volume 4 m^3. The feed takes place over a time of 3 h. The feed has a substrate concentration of 50 g/L and we assume the substrate concentration in the reactor is negligible at any time. The biomass yield is $Y_{X/S}$ = 0.15 g/g. At the end of the feed phase the pH is changed and the microorganisms start producing the desired product with a rate $r_P = 8.9X$, where r_P is in g/L.d and X is in g/L. The length of the product generation phase is 2 h. At the end of the product production, the effluent is discharged, and the cycle is started again after 1 h of downtime. Calculate the product concentration at the end of the cycle and the product productivity (kg/d). Assume that the process runs for 8000 h per year.

4.14 A fermentation process is studied at lab-scale in continuous (working volume 1 L) conditions. The first continuous experiment uses a flow rate of 3 L/d, a feed concentration of 30 g/L. This experiment obtains an effluent substrate concentration of 0.5 g/L, biomass concentration of 4.5 g/L and product concentration of 8.2 g/L. The second experiment uses a flow rate of 5 L/d and a feed concentration of 40 g/L, obtaining an effluent substrate concentration of 1.1 g/L.

Calculate the optimum product productivity (kg m^{-3} d^{-1}) and the product production rate (kg/d) under the conditions of optimum productivity for a full-scale reactor having a volume of 2 m^3 with a feed concentration of 50 g/L. For these conditions, also calculate the biomass concentration.

4.15 Assume that the growth of a microorganism on a substrate is described by the Monod model with maintenance metabolism. Two continuous experiments are carried out, with a substrate concentration in the feed of 5 g/L and with a reactor of 2 L working volume. In the first experiment, with flow rate 0.3 L/h, the biomass and substrate concentration in the reactor are 2.2 and 0.05 g/L, respectively. The second experiment is run with a flow rate of 0.5 L/h, obtaining a substrate concentration of 0.10 g/L and a biomass concentration of 2.4 g/L. Determine the parameters μ_{max}, K_S and m_S for this system.

4.16 A process is carried out in a fed-batch reactor. The initial volume before the feed start is 200 l and the feed concentration is 20 g/L. The kinetic parameters of the microorganisms on the substrate are $\mu_{max} = 7.3$ d^{-1}, $K_S = 0.1$ g/L. Calculate the maximum influent rate that can be used to ensure that, at the quasi-steady state, the substrate concentration is never higher than 1% of the feed concentration.

4.17 Derive the quasi-steady equations for the substrate, biomass and product concentration for a fed-batch reactor assuming the Monod model with (a) maintenance metabolism and (b) endogenous metabolism

4.18 A process is carried out in a fed-batch reactor. The metabolism of the microorganisms can be described by the Monod model with maintenance with the following kinetic parameters: $\mu_{max} = 15.0$ d^{-1}, $K_S = 0.08$ kg/m^3, $m_S = 0.8$ d^{-1}, $Y_{X/S} = 0.2$ g/g and $Y_{P/S} = 0.3$ g/g. It is desired that the substrate concentration never exceeds 1% of the feed concentration, which is 30 g/L. Calculate the maximum product and biomass productivity (in kg/d) that it is possible to achieve in the fed-batch process, assuming that the initial volume is 1 m^3, the final volume 10 m^3, and the downtime between end of the cycle and start of a new cycle is 0.5 h. Assume that the flow rate is constant throughout the feed period and that the process operates for 8000 h/year.

5 Fermentation Processes
Oxygen Transfer

The design of the aeration system is a critical step for aerobic fermentations. As examples of oxygen consumption and related energy consumption, cells of *Escherichia coli* growing on glucose can consume up to 0.5 kg O_2/kg cell dry weight/h and the power requirements for oxygen delivery in fermenters can be higher than 2 kWh/kg O_2. The aeration system must be designed so that it is able to supply the oxygen required by the microorganisms while minimising energy consumption and capital and operating costs.

5.1 THEORY

We have seen that aerobic fermentations consume oxygen and that the oxygen consumption rate by the microorganisms can be expressed by the following equation:

$$r_{O_{2biom}}\left(\frac{\mathrm{kgO_2}}{\mathrm{m^3d}}\right) = -\frac{r_X}{Y_{X/O_2}} \tag{3.13}$$

Oxygen needs to be provided to the microorganisms by the aeration system. Typically, air or pure oxygen are pumped into the vessel using spargers (Figure 5.1).

Oxygen transfers from the bubbles to the liquid phase where it is used by the microorganisms for their metabolism. This section describes a theory to express the rate of mass transfer of a species from the gas to the liquid phase, or vice versa.

The mass transfer theory used in this book is based on the "two-film" model, which is schematised in Figure 5.2. In Figure 5.2 oxygen is used as the substance being transferred, but the model can be applied to any substance.

According to the "two-film" model, all the resistance to mass transfer is located in two boundary layers (films), located in the gas and in the liquid phases (gas and liquid films). Oxygen (or in general the species being transferred) is present in the bulk gas phase at a partial pressure p_{O2} (typically expressed in atm) and in the bulk liquid phase at a concentration C_{O2} (expressed in mass or mol units, e.g. mg/L, kg/m^3, or mol/L). In Figure 5.2, oxygen concentration in the bulk gas and bulk liquid phases is not in equilibrium, i.e. p_{O2} and C_{O2} are not in equilibrium with each other. If they were at equilibrium the rate of mass transfer would be zero. The two-film model assumes that at the interface between the liquid and gas phase oxygen concentration on the two sides of the interface is in equilibrium, i.e. $p_{O2,i}$ and $C_{O2,i}$ are in equilibrium with each other. We assume here that the equilibrium relationship is linear at all concentrations, i.e.

DOI: 10.1201/9781003633822-5

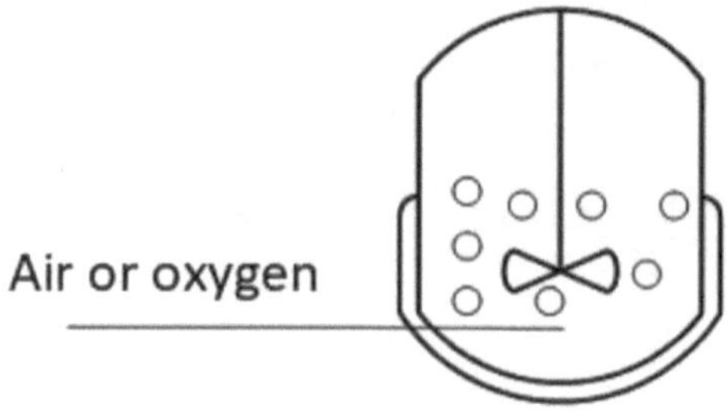

FIGURE 5.1 Conceptual scheme of an aerated fermenter.

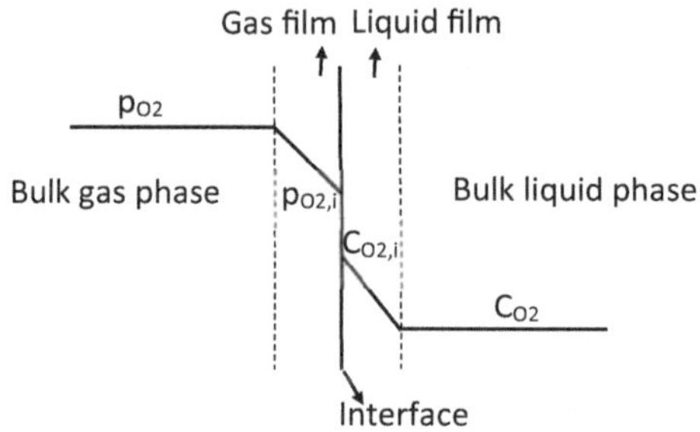

FIGURE 5.2 Scheme of the two-film model (oxygen is used as an example of the substance being transferred).

$$C_{O2,i} = k_{eq} \cdot p_{O2,i} \tag{5.1}$$

Since, according to the model, the mass transfer resistance is only present in the gas and liquid films, the rate of mass transfer per unit of transfer area (or mass transfer flux), J_{O2}, can be expressed as

$$J_{O2}\left(\frac{kg_{O_2}}{m^2 \cdot d}\right) = k_g \cdot \left(p_{O2} - p_{O2,i}\right) = k_l \cdot \left(C_{O2,i} - C_{O2}\right) \tag{5.2}$$

where k_g and k_l are the local mass transfer coefficients for the gas and liquid phases, respectively. Equation (5.2) defines the coefficients k_g and k_l. We assume that in the gas and liquid films the movement of oxygen only occurs via molecular diffusion, i.e. oxygen transfer only occurs because of the concentration gradient. With this assumption, an expression for k_g and k_l can be obtained by comparing equation (5.2) with the corresponding form of Fick's law of diffusion. The general form of Fick's law of diffusion is

$$J_A = -D_{AB} \frac{dC_A}{dx} \tag{5.3}$$

In equation (5.3), J_A is the flux (kg m^{-2} d^{-1}) of species A in species B. D_{AB} is the diffusivity of species A in species B, dC_A/dx is the concentration gradient of species A in the boundary layer. Applying equation (5.3) to the flow of oxygen in the gas and liquid films:

$$J_{O2} = -D_{O2,g}\frac{dC_{O2,g}}{dx_g} = -D_{O2,l}\frac{dC_{O2,l}}{dx_l} \tag{5.4}$$

In equation (5.4), $C_{O2,g}$ and $C_{O2,l}$ are the concentrations of oxygen (kg/m^3) in the gas and liquid phases, respectively, $D_{O2,g}$ and $D_{O2,l}$ are the diffusivities of oxygen in the gas and the liquid phase, respectively. x_l and x_g are the thickness of the gas and liquid films, respectively. It is more usual to represent the oxygen concentration in the gas phase as partial pressure, rather than concentration. Using the ideal gas law, the relationship between partial pressure and concentration is

$$C_{O2,g} = \frac{p_{O2}\cdot MW_{O2}}{RT} \tag{5.5}$$

Using equation (5.5), equation (5.4) becomes

$$J_{O2} = -D_{O2,g}\cdot\frac{MW_{O2}}{RT}\frac{dp_{O2,g}}{dx_g} = -D_{O2,l}\frac{dC_{O2,l}}{dx_l} \tag{5.6}$$

Comparing equation (5.6) with (5.2), and assuming that the profile of oxygen in the gas and liquid films is linear, we obtain equations for k_g and k_l:

$$k_g = \frac{D_{O2,g}}{x_g}\cdot\frac{MW_{O2}}{RT} \tag{5.7}$$

$$k_l = \frac{D_{O2,l}}{x_l} \tag{5.8}$$

Equations (5.7) and (5.8) show that the local mass transfer coefficients are directly proportional to the oxygen diffusivity in the respective phases, and inversely proportional to the thickness of the films.

Equation (5.2) still does not allow for an easy calculation of the mass transfer rate because, even assuming that the mass transfer coefficients are known from the literature or from experiments, the concentrations at the interface, $p_{O2,i}$ and $C_{O2,i}$ are not known. It is desirable to express the mass transfer rate as a function of the known and easily measurable concentrations p_{O2} and C_{O2}. In order to do this, the concentration

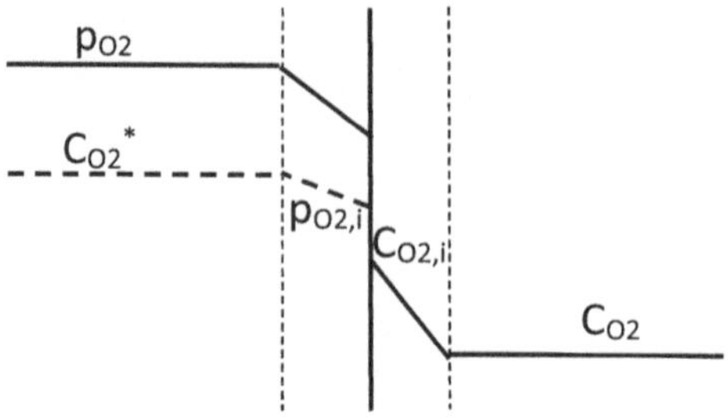

FIGURE 5.3 Two-film model with the introduction of the concentration C^*_{O2}.

C^*_{O2} is introduced (Figure 5.3). C^*_{O2} is the oxygen concentration in the liquid phase that would be in equilibrium with the bulk gas phase, i.e.

$$C^*_{O2} = k_{eq} \cdot p_{O2} \tag{5.9}$$

While the concentrations p_{O2}, $p_{O2,i}$, $C_{O2,i}$ and C_{O2} exist physically in the system, the concentration C^*_{O2} is fictitious and does not exist physically. However, the advantage of introducing C^*_{O2} is that it can be easily calculated from p_{O2}, if the equilibrium relationship is known. Introducing C^*_{O2} the mass transfer flux can also be expressed as

$$J_{O2}\left(\frac{\text{kg}_{O_2}}{\text{m}^2 \cdot \text{day}}\right) = k_L \cdot \left(C^*_{O2} - C_{O2}\right) = k_L \cdot \left(k_{eq} p_{O2} - C_{O2}\right) \tag{5.10}$$

Equation (5.10) represents definition of k_L, the "global" mass transfer coefficient. Global coefficient means that it refers to both phases, while the local coefficients k_g and k_l refer to only one phase. The advantage of using equations (5.10) over equations (5.2) for expressing the mass transfer rates lies in the fact that in equation (5.10) the variables p_{O2} and C_{O2} represent values in the bulk values which are readily measurable, while equation (5.2) includes the interface values $p_{O2,i}$ and $C_{O2,i}$, which are not measurable.

The value of the global coefficient k_L can be related to the values of the local coefficients k_g and k_l. From equations (5.1) and (5.2) we have

$$k_g \cdot \left(p_{O2} - p_{O2,i}\right) = k_l \cdot \left(k_{eq} p_{O2,i} - C_{O2}\right) \tag{5.11}$$

From equation (5.11) the value of $p_{O2,i}$ can be expressed as a function of the other variables:

$$p_{o2,i} = \frac{k_g p_{o2} + k_l C_{O2}}{k_g + k_{eq} k_l} \tag{5.12}$$

Introducing this value of $p_{o2,i}$ into $k_g\left(p_{O2}-p_{O2,i}\right)$ we obtain

$$k_g\cdot\left(p_{O2}-p_{O2,i}\right)=k_g\cdot\left(p_{O2}-\frac{k_g p_{o2}+k_l c_{O2}}{k_g+k_{eq}k_l}\right)=\frac{k_g k_l}{k_g+k_{eq}k_l}\left(k_{eq}p_{o2}-C_{O2}\right) \quad (5.13)$$

By equalling equation (5.13) with equation (5.10) we obtain an expression for k_L as a function of k_g, k_l and k_{eq}:

$$k_L=\frac{k_g k_l}{k_g+k_{eq}k_l} \quad (5.14)$$

Equation (5.14) shows that the global coefficient k_L depends on the local coefficients k_g and k_l and on the equilibrium constant k_{eq}. Equation (5.14) shows that if the gas has a high solubility (high value of k_{eq}), then $k_L \approx k_g/k_{eq}$, i.e. the controlling resistance is on the gas side. On the other hand, if the gas has a low solubility (low value of k_{eq}), then $k_L \approx k_l$, i.e. the controlling resistance is on the liquid side. Oxygen is a gas with low solubility, therefore in this case the controlling resistance is on the liquid side.

So far, we have expressed the mass transfer flux J, i.e. the rate of mass transfer per unit mass transfer area. In bioprocesses we are interested in the volumetric mass transfer rate, i.e. mass transfer rate per unit volume of the biological reactor. We call the volumetric mass transfer rate (again using oxygen as an example) $r_{O_2 transfer}\left(\frac{\text{kgO}_2}{\text{m}^3\cdot\text{d}}\right)$. $r_{O_2 transfer}$ and J_{O2} are related as follows:

$$r_{O_2 transfer}\left(\frac{\text{kg}_{O_2}}{\text{m}^3\cdot\text{d}}\right)=J_{O2}\left(\frac{\text{kg}_{O_2}}{\text{m}^2\cdot\text{d}}\right)\frac{A}{V} \quad (5.15)$$

In equation (5.15) A is the total area of the bubbles present in the reactor and V is the liquid volume in the reactor. We call the ratio $\frac{A}{V}=a$, where a represents the specific area of mass transfer, i.e. the mass transfer area per unit volume. Therefore, the desired volumetric mass transfer rate can be expressed as

$$r_{O2}\left(\frac{\text{kg}_{O_2}}{\text{m}^3\cdot\text{day}}\right)=k_L\cdot a\cdot\left(C^*_{O2}-C_{O2}\right)=k_L a\cdot\left(k_{eq}p_{O2}-C_{O2}\right) \quad (5.16)$$

The two parameters k_L and a have a totally different physical meaning. k_L is related to the resistance to mass transfer, while a accounts for the number and size of the bubbles in the reactor. However, it is difficult with simple experimental means to determine the values of the parameters k_L and a independently. In general, the combined values of $k_L\cdot a$ are determined and therefore the term k_La is usually considered one single parameter, rather than two individual parameters. In equation (5.16), the term k_La is the mass transfer coefficient and the term $\left(C^*_{O2} - C_{O2}\right)$ is the driving force.

Note that equation (5.16) can be used to describe the mass transfer of any substance, not just oxygen, from the gas to the liquid phase (or vice versa). For example, if the substance being transferred is carbon dioxide, equation (5.16) can be written as $r_{CO_2 transfer}\left(\frac{kg_{CO2}}{m^3 \cdot d}\right) = k_L a_{CO2}\left(k_{eqCO2} p_{CO2} - C_{CO2}\right)$ and similar equations describe the mass transfer for any other substances.

Example 5.1

Calculate the rate of mass transfer for oxygen in a fermenter when the concentration of dissolved oxygen is 2.0 mg/L. Repeat the calculation for a concentration of dissolved oxygen of 6.0 mg/L. Assume that the fermented is sparged with air with C^*_{O2}equal to 9.0 mg/L. The mass transfer coefficient is $k_La = 0.3$ min^{-1}.

Solution

When $C_{O2} = 2.0$ mg/L

$$r_{O_2 transfer} = k_L a \cdot \left(C^*_{O2} - C_{O2}\right) = 0.3\,\text{min}^{-1} \cdot (9.0 - 2.0)\frac{\text{mg}}{\text{L}} = 2.1\frac{\text{mg}}{\text{L} \cdot \text{min}}$$

When $C_{O2} = 6.0$ mg/L

$$r_{O_2 transfer} = k_L a \cdot \left(C^*_{O2} - C_{O2}\right) = 0.3\,\text{min}^{-1} \cdot (9.0 - 6.0)\frac{\text{mg}}{\text{L}} = 0.9\frac{\text{mg}}{\text{L} \cdot \text{min}}$$

This example shows that the mass transfer rate is higher when the dissolved oxygen concentration is lower, because the driving force is higher.

Example 5.2

When a fermenter is sparged with air the oxygen mass transfer rate for a dissolved oxygen concentration of 2.0 mg/L, is 5000 kg m^{-3} d^{-1}. Calculate the oxygen transfer rate when the same fermenter is sparged with pure oxygen, rather than air. Assume that in air the partial pressure of oxygen is 0.21 atm. Assume that the k_La is the same with oxygen and air. Assume that the equilibrium relationship between oxygen in the gas phase and in the liquid phase is given by

$C^*_{O2} = 42.9 \cdot p_{O2}$ where C^*_{O2} is in mg/L and p_{O2} is in atm.

Solution

$$r_{O_{2\,transferair}} = k_L a \cdot \left(k_{eq} p_{O2air} - C_{O2}\right)$$

$$r_{O_{2\,transferoxygen}} = k_L a \cdot \left(k_{eq} p_{O2oxygen} - C_{O2}\right)$$

Therefore,

$$r_{O_{2\,transferoxygen}} = r_{O_{2\,tranferair}} \frac{k_{eq} p_{O2oxygen} - C_{O2}}{k_{eq} p_{O2air} - C_{O2}} = 5000 \frac{\text{kg}}{\text{m}^3\text{d}} \cdot \frac{42.9 \frac{\text{mg}}{\text{L} \cdot \text{atm}} \cdot 1\ \text{atm} - 2.0}{42.9 \frac{\text{mg}}{\text{L} \cdot \text{atm}} \cdot 0.21\ \text{atm} - 2.0}$$

$$= 29{,}200 \frac{\text{kg}}{\text{m}^3\text{d}}$$

This example shows that the mass transfer rate is higher when pure oxygen is used, because the driving force is larger.

5.2 MEASUREMENT OF MASS TRANSFER COEFFICIENTS

Many procedures have been reported in the literature to measure $k_L a$. The experiment described here is one of the easiest to measure the $k_L a$ for the oxygen-air system.

The procedure requires an agitated vessel, where air is provided by diffusers. The experimental procedure is very simple:

1. Water is added to the vessel to the desired level and a sensor for dissolved oxygen is inserted into the liquid.
2. Nitrogen is sparged to the vessel in order to bring down the concentration of dissolved oxygen in water. Nitrogen sparging is stopped when oxygen concentration decreases below approximately 2 mg/L.
3. Air (or pure oxygen) at known flow rate is sparged into the vessel. Oxygen concentration starts to rise and oxygen concentration is recorded as a function of time.
5. The experiment is terminated when oxygen concentration reaches a value which is close to the saturation value with the inlet gas.

The $k_L a$ for the system can be calculated by applying the mass balance for oxygen. Indeed:

$$\frac{dC_{O2}}{dt} = k_L a \cdot \left(C^*_{O2} - C_{O2}\right) \tag{5.17}$$

Integrating this equation between time 0, where the oxygen concentration is $C_{O2,in}$ and time 0, with generic oxygen concentration C_{O2}, we obtain

$$\ln\left(C^*_{O2} - C_{O2}\right) = \ln\left(C^*_{O2} - C_{O2,in}\right) - k_L a \cdot t \tag{5.18}$$

Therefore, from a plot of $\ln\left(C^*_{O2} - C_{O2}\right)$ vs t the value of $k_L a$ can be calculated as the slope of the line.

It is important to observe that calculation of $k_L a$ based on this procedure is based on several assumptions:

- Both the gas and the liquid phase are perfectly mixed.
- The dynamics of the oxygen probe are negligible compared to the dynamics of oxygen transfer.
- If air is used as gas, the amount of oxygen that transfers to the liquid phase is negligible compared to the amount of oxygen that is fed to the vessel. This ensures that oxygen partial pressure in the bubbles in the vessel is the same as the in the atmosphere, and so C^*_{O2} is known and constant during the experiment.

This procedure can be modified to measure the $k_L a$ in the presence of microorganisms. In this case, the fermenter is filled with the suspension of water and active microorganisms at the desired concentration. When the dissolved oxygen concentration is high and close to the maximum value, aeration is stopped and the decrease in oxygen concentration due to the microbial metabolism is recorded. When the oxygen concentration has dropped sufficiently, aeration is started again and the increase in oxygen concentration is recorded. From the mathematical analysis of the data, the $k_L a$ can be calculated.

Another method to measure $k_L a$ is the sulphite oxidation method. In this method, a concentrated solution of sodium sulphite (Na_2SO_3) is added into the fermenter. In the presence of dissolved oxygen sulphite is converted into sulphate (Na_2SO_4) with a very fast reaction. The rate at which oxygen transfers into the liquid phase can be determined by measuring the rate of sulphite oxidation into sulphate. From the rate of oxygen transfer, the $k_L a$ can be calculated easily.

In the methods for $k_L a$ measurement based on the measurement of the dissolved oxygen, oxygen sensors of various types (e.g. polarographic, optical) are used. In some methods, oxygen concentration in the gas phase is measured using mass spectroscopic analysis.

Example 5.3

An experiment is carried out to measure the mass transfer coefficient for oxygen in water. The experimental data are reported in Table 5.1. Calculate the mass transfer coefficient. It is known that $C^*_{O2} = 9.10 \frac{\text{mg}}{\text{L}}$.

Solution

The plot of the experimental data in Table 5.1 is shown in Figure 5.4.

To calculate the $k_L a$, the data need to be correlated using Equation (5.18):

$$\ln\left(C^*_{O2} - C_{O2}\right) = \ln\left(C^*_{O2} - C_{O2,in}\right) - k_L a \cdot t$$

The plot of $\ln\left(C^*_{O2} - C_{O2}\right)$ vs time is shown in Figure 5.5. From Figure 5.5, $k_L a$ can be obtained from the slope of the curve. In this case$k_L a = 0.34\,\text{min}^{-1}$

TABLE 5.1
Experimental Data for Example 5.3

Time (min)	Dissolved Oxygen (mg/L)
0	1.69
1	3.36
2	4.67
3	5.90
4	6.68
5	7.26
6	7.72
7	8.09
8	8.36
9	8.57
10	8.73
11	8.90
12	8.96
13	9.01

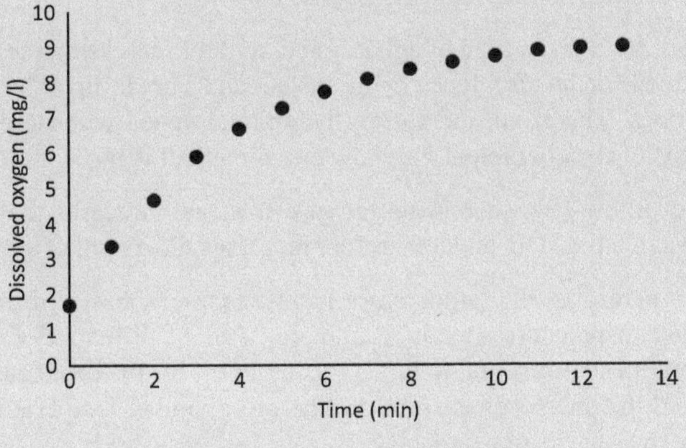

FIGURE 5.4 Plot of the experimental data in Example 5.3.

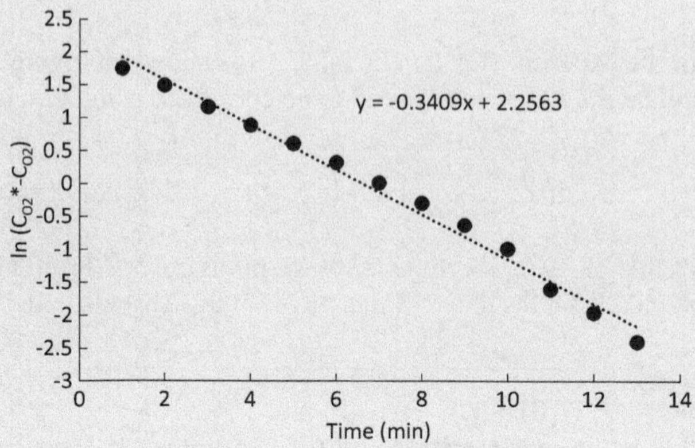

FIGURE 5.5 Plot of $\ln\left(C^*_{O2} - C_{O2}\right)$ vs time for Example 5.3.

5.3 CORRELATIONS FOR MASS TRANSFER COEFFICIENTS

The value of the mass transfer coefficient k_La is affected by the physical properties of the system and by the fluid dynamic conditions in the vessel. The fluid dynamic conditions in the vessel are determined by the geometry of the vessel, by the type of agitator, its size and agitation speed and by the air (or oxygen) flow rate. The physical properties of the system that have the most effect on mass transfer are temperature and viscosity.

Vessel geometry: the geometry of the vessel affects the k_La because it affects the fluid dynamic conditions and the turbulence in the system. The presence of baffles, of probes, etc, affect the fluid dynamic conditions and have an important effect on the k_La.

Agitator type: the type of agitator determines the movement of the fluid in the vessel and therefore affects the mass transfer coefficient.

Agitator speed: for a given type of agitator and agitator increasing the agitation speed causes an increase in the turbulence in the system and a break-up of larger gas bubbles into smaller ones. Therefore, increasing the agitator speed will increase both the k_L and the a terms of the coefficient k_La, however the effect is more important for a.

Air (or oxygen) flow rate: increasing the gas flow rate increases the number and/or size of bubbles in the system and it therefore increases the a term of the k_La coefficient.

Temperature: increasing the temperature increases the mobility of the molecules in the liquid phase, which corresponds to an increase in the diffusivity $D_{O2,l}$. In addition, higher temperatures correspond to larger gas bubbles, for the same gas flow rate on a mass basis, and this increases the a term. Therefore, higher temperature corresponds to an increase in k_La.

Viscosity: higher viscosity increases the size of the liquid and gas films, therefore it decreases the local mass transfer coefficient k_l, causing a decrease in k_La.

The effect of physical properties and agitation conditions on k_La is often described by means of correlations.

For example, one of the correlations used for the interfacial area a is the following:

$$a = 1.44\left[\left(\frac{P_G}{V}\right)^{0.4}\left(\frac{\rho_L}{\sigma^3}\right)^{0.2}\right]\left(\frac{v_g}{v_t}\right)^{0.5} \tag{5.19}$$

In equation (5.19), a is in m²/m³, P_G (W) is the power transferred from the agitator to the liquid phase (the subscript $_G$ indicates that this is the power when the agitator is sparged with the gas, we will use the symbol P to indicate the agitator power draw without gas sparging), V (m³) is the volume of the liquid phase, ρ_L (kg/m³) is the density of the liquid phase, σ (N/m) is the interfacial tension, v_g is the superficial gas velocity (m/s, defined as the gas flow rate Q_G divided by the cross sectional area of the vessel A, $v_g = \frac{Q_G}{A}$), v_t (m/s) is the terminal rising velocity of a single bubble. v_t depends on the bubble diameter, which also in turn depends on the agitator power per unit volume, on the gas flow rate and on the physical properties. Correlations for v_t are available in the literature, however for approximate calculations v_t can be assumed equal to 0.21 m/s.

Equation (5.19) shows that the main parameters that affect the interfacial area a are the agitator power P_G and the gas flow rate (included in the term v_g). As expected, higher values of the power draw per unit volume and of the gas flow rate give higher values of a.

There are also correlations for the mass transfer coefficient k_l. For gas liquid systems in most cases the gas has low solubility in the liquid phase, so $k_l \approx k_L$. A correlation for k_l is given by equation (5.20):

$$Sh_L = 2.0 + 0.31\left(\frac{d^3_{bubbles}\rho_L g}{\mu_L D_{g,L}}\right)^{\frac{1}{3}} \tag{5.20}$$

where Sh_L is the Sherwood number for the liquid phase, defined as $Sh_L = \frac{k_l \cdot d_{bubbles}}{D_{g,L}}$.

In equation (5.20), k_l is obtained in m/s, $d_{bubbles}$ (m) is the diameter of the gas bubbles, g (m/s²) is the gravity acceleration, μ_L (Pa·s) is the viscosity of the liquid phase, $D_{g,L}$ (m²/s) is the diffusivity of the gas in the liquid phase, ρ_L (kg/m³) is the density of the liquid phase.

Equation (5.20) shows that the main physical property that affects k_l is the diffusivity $D_{g,L.}$ k_l increases as the diffusivity increases. This is expected based on the mass transfer theory presented in Section 5.1 (equation (5.8)). k_l is lower for higher values of the liquid viscosity, because higher viscosity corresponds to a larger thickness of the boundary layer (term x_l in equation (5.8)). k_l is essentially independent of the diameter of the bubbles, this means that k_l is independent of the power draw according to this correlation.

For scale-up considerations, it is often useful to measure the mass transfer coefficient in a small vessel, and then use correlations to calculate the coefficient for the large vessel to be used in the commercial process. In these scale-up studies, the physical properties of the gas and liquid phase need to be the same between the small and the large vessels. Furthermore, the small and large vessels need to be geometrically similar. Agitated vessels are geometrically similar if they have the same type of agitator and the ratio between all the geometrical dimensions is the same. For example, the ratio between the diameter of the agitator (D_{ag}) and of the vessel (D) should be the same in the large and small vessels:

$$\left(\frac{D_{ag}}{D}\right)_{small} = \left(\frac{D_{ag}}{D}\right)_{large} \tag{5.21}$$

If the conditions of same physical properties and geometrically similar vessels are verified, a correlation frequently used in scale-up (or scale-down) studies is the following:

$$k_L a = k\left(\frac{P_G}{V}\right)^a v_g^b \tag{5.22}$$

In equation (5.22), the variables have the same meaning as in equation (5.19), i.e. P_G is the agitator power transferred to the liquid phase, V is the volume of the liquid phase, v_g is the superficial gas velocity. Correlations like equation (5.22) can be used to calculate the value of the mass transfer coefficient for a large vessel once the value of $k_L a$ for a smaller vessel is known. The parameter a in equation (5.22) can be determined by measuring $k_L a$ for a range of $\frac{P_G}{V}$ values, keeping v_g constant, and plotting $\ln(k_L a)$ vs $\ln\left(\frac{P_G}{V}\right)$. A similar procedure can be used to determine b, by measuring $k_L a$ for a range of v_g values, keeping $\frac{P_G}{V}$ constant, and plotting $\ln(k_L a)$ vs $\ln(v_g)$. Once a and b are determined, the parameter k in equation (5.22) can also be calculated from the intercepts of the $\ln(k_L a)$ vs $\ln\left(\frac{P_G}{V}\right)$ or $\ln(k_L a)$ vs $\ln(v_g)$ plots. For each experiment the $k_L a$ can be determined using the method described in Section 5.2. The $\frac{P_G}{V}$ value can be varied by changing the agitation speed (see Section 5.3.1) and the v_g can be changed by manipulating the gas flow rate.

Correlation (5.22) is valid in the most common case that the fluid dynamic behaviour in the fermenter is, with good approximation, Newtonian, i.e. with the viscosity independent of the shear rate. However, in some cases, due to the type of microorganisms present, to their high concentration or to the production of extracellular polymeric substances, the fluid dynamic behaviour in the fermenter is better approximated by the pseudoplastic model, where the viscosity decreases as the shear rate increases. Examples of processes with pseudoplastic behaviour are the production of poly-glutamic acid by *Bacillus subtilis*, some fermentations with filamentous fungi for the production of pharmaceutically important products, production of alginate and xanthan by various microorganisms. In some cases, the pseudoplastic behaviour is due to the fermentation product (e.g. xanthan or alginate), in other cases is due to the microorganisms. For example, filamentous fungi at high concentrations tend to become entangled, increasing the viscosity of the medium at low shear rates. When fermentations with filamentous fungi are carried out in batch, initially the rheology of the medium is approximately Newtonian because of the low biomass concentration, however when the fungi grow and their concentration increases, the viscosity increases and the rheology tends to become pseudoplastic. In these cases, generally the mass transfer coefficient is lower than when the fluid is Netwonian, because of the bubble coalescence in zones with lower shear rate (e.g. in zones far from the agitator). When the fluid behaviour is well described by the pseudoplastic model, equation (5.22) can be adapted with an additional term, which includes the average viscosity in the reactor, calculated empirically, and which reduces the k_La.

Correlation (5.22) is no longer valid if the gas flow rate is too high and the agitator becomes unable to disperse the gas in the vessel. This condition is called flooding and is obtained for high gas flow rates combined with low agitator speed. The gas flow rate and the agitator speed in a bioreactor need to be chosen so that flooding is avoided, because flooding leads to a low value of the mass transfer coefficient. There are empirical correlations, which allow estimatation of the gas flow rate that corresponds to the onset of flooding, one example is equation (5.23). The maximum gas flow rate that we can have in the vessel should be lower than the value of Q_{gF} given by equation (5.23):

$$Q_{gF} = K\frac{N^3 D_{ag}^{7.5}}{gD^{3.5}} \tag{5.23}$$

In equation (5.23) Q_{gF} is the gas flow rate at the onset of flooding (m^3/s), N is the agitation speed (s^{-1}), g is the gravitational acceleration (9.8 m/s^2), D_{ag} and D are the diameter (m) of the agitator and of the vessel, respectively. K is a constant that depends on the type of agitator, it can be assumed equal to 30 for Rushton turbines and equal to 170 for Chemineer BT-6 agitators.

So far, we have discussed the effect of these parameters on the mass transfer coefficient k_La. However, the rate of mass transfer is given by the product of the mass transfer coefficient and of the driving force, equation (5.16). The driving force is mainly affected by the temperature and pressure of the system. Higher temperature decreases the solubility of gases in liquids, so C^*_{O2} decreases as the temperature increases. Higher pressure increases the solubility of gases in liquids and therefore increases the driving force. The composition of the inlet gas used for aeration also affects the driving force. The saturation concentration of oxygen in water, C_{O2}^{*}, depends on the gas composition and it is the higher, the higher the oxygen content in the gas. If we use atmospheric air, where the oxygen content is approximately 21% in volume, C_{O2}^{*} is in the approximate range 7–9 mg/L (depending on the process temperature), while if we use pure oxygen, C_{O2}^{*} is in the approximate range 30–45 mg/L. Therefore, clearly using pure oxygen increases the driving force and the overall mass transfer rate compared to the use of atmospheric air.

5.3.1 Correlations for Power Draw in Agitated Vessels

In order to use equations (5.19) and (5.22), we need to know the power that the agitator transfers to the liquid phase. This power depends on the agitator type, size and stirring speed and on the physical properties of the fluids. The physical properties that mostly affect the power draw of agitators are density and viscosity of the liquid phase. The agitator properties that affect the power draw are the type of agitator, its size and its rotational speed. For each type of agitator, the power draw is obtained by experimental correlations, available in the literature that relate the power number (P_0) to the Reynolds number (Re).

$$P_0 = \frac{P}{\rho_L N^3 D_{ag}^5} \tag{5.24}$$

$$\text{Re} = \frac{\rho_L N D_{ag}^2}{\mu_L} \tag{5.25}$$

In equations (5.24) and (5.25), in addition to the symbols already defined, N is the agitator speed. Experimental data indicate that for each type of agitator there is a certain power curve. A power curve (Figure 5.6) is the relationship between P_0 and Re.

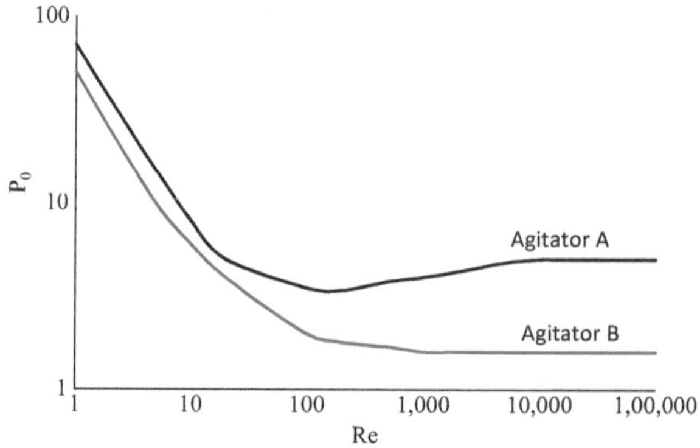

FIGURE 5.6 Typical profiles P_0 vs Re of power curves.

If the power curve for the considered agitator is known, then the power draw of the agitator can be calculated immediately, as a function of the properties of the fluid (density and viscosity) and as a function of the agitator speed and size.

Once the power draw with no gas flowing has been calculated, the power draw with gas flowing can be corrected using correlations, such as equations (5.26):

$$\frac{P_G}{P} = 1 - 12.2\frac{Q_G}{ND_{ag}^3} \quad \left(\frac{Q_G}{ND_{ag}^3} < 0.037\right) \tag{5.26a}$$

$$\frac{P_G}{P} = 0.62 - 1.85\frac{Q_G}{ND_{ag}^3} \quad \left(\frac{Q_G}{ND_{ag}^3} > 0.037\right) \tag{5.26b}$$

Equations (5.26) were obtained experimentally using disk flat-blade turbines. For this type of agitator, if $\frac{Q_G}{ND_{ag}^3} > 0.11$, the ratio P_G/P can be assumed to be constant and equal to 0.43.

The procedure below shows the calculation of the power draw for a given agitator if the agitator type, its geometry and agitation speed are known:

(a) Obtain from the literature the power curve for the agitator you are interested in.
(b) From the geometry of the agitator and its speed and from the physical properties of the liquid medium, calculate the Reynolds number (equation (5.25)).
(c) With the Reynolds number and the power curve, calculate the power number P_0.
(d) From the power number, calculate the power transferred from the agitator to the medium (equation (5.24)).

Equation (5.22) can also be used to calculate the power draw that is required for a certain agitator to achieve the desired value of the k_La. Once the power draw is known, power curves can be used to calculate the required agitation speed. In this case we need an iterative procedure (for given values of the physical properties and of the agitator diameter): we set a value for N, calculate Re, P_G/P, P, then calculate P_0 and calculate N, and repeat the procedure until convergence. In many cases, but not always, the agitation conditions in fermentations fall in the region where P_0 is approx. constant with Re (i.e. independent on Re). In addition, in some cases the ration P_G/P can be assumed constant and independent of N (e.g. when the ratio $\frac{Q_G}{ND_{ag}^3}$ is higher than 0.11 for flat-disk turbines). If P_0 and P_G/P are constant and known values, the procedure to calculate the agitation speed can be greatly simplified: from the required value of P_G, P is calculated. Then, the constant value of P_0 is used to calculate N.

Example 5.4

A fermenter has a diameter of 1.5 m and agitator with a diameter of 0.5 m. The fermenter is sparged with air with a flow rate of 4000 m^3/d. The level of the liquid phase in the reactor is 1.5 m. The power number of the agitator under ungassed conditions is 4.5. The agitator speed is 120 rpm.

Calculate the mass transfer coefficient k_La for oxygen in the fermenter. Assume $v_t = 0.21$ m/s and $d_{bubbles} = 5.0\cdot 10^{-4}$ m

Physical properties:

$$\mu_L = 8.9\cdot 10^{-4}\frac{\text{kg}}{\text{m}\times\text{s}}$$

$$\sigma = 0.072\frac{\text{N}}{\text{m}}$$

$$\rho_L = 1000\frac{\text{kg}}{\text{m}^3}$$

$$D_{O2,L} = 2.1\cdot 10^{-9}\frac{\text{m}^2}{\text{s}}$$

Solution

Equation (5.19) gives the specific surface area of the bubbles. From the power number we calculate the ungassed power draw using equation (5.24):

$$P = P_o\rho N^3 D_{ag}^5 = 4.5\cdot 1000\frac{\text{kg}}{\text{m}^3}\cdot\left(\frac{120\,\text{min}^{-1}}{60\frac{\text{s}}{\text{min}}}\right)^3\cdot(0.5\,\text{m})^5 = 1125\,W$$

The gassed power draw is calculated from equation (5.26). In this case,

$$\frac{Q_G}{ND_{ag}^3} = \frac{\left(\dfrac{1000\frac{\text{m}^3}{\text{d}}}{86{,}400\frac{\text{s}}{\text{d}}}\right)}{2\ \text{s}^{-1}(0.5\ \text{m})^3} = 0.046$$

so equation (5.26b) will be used.

$$\frac{P_G}{P} = 0.62 - 1.85\cdot 0.046 = 0.53$$

so

$$P_G = 0.53 \cdot 1125 \text{ W} = 596 \text{ W}$$

The superficial gas velocity is

$$v_g = \frac{Q_G}{A} = \frac{\left(\dfrac{1000}{86,400}\right)\dfrac{\text{m}^3}{\text{s}}}{\dfrac{\pi 1.5^2}{4}\text{m}^2} = 0.0066\frac{\text{m}}{\text{s}}$$

Therefore, the specific surface area of the bubbles is (equation (5.19)):

$$a = 1.44\left[\left(\frac{596\dfrac{\text{kg}\cdot\text{m}^2}{\text{s}^3}}{2.7\text{ m}^3}\right)^{0.4}\left(\frac{1000\dfrac{\text{kg}}{\text{m}^3}}{0.072\dfrac{\text{kg}}{\text{s}^2}}\right)^{0.2}\right]\left(\frac{0.0066}{0.21}\right)^{0.5} = 14.90\frac{\text{m}^2}{\text{m}^3}$$

The Sherwood number for the liquid phase coefficient k_l is given by equation (5.20):

$$Sh_L = 2.0 + 0.31\left(\frac{(5\cdot 10^{-4}\text{ m})^3\, 1000\dfrac{\text{kg}}{\text{m}^3}\cdot 9.81\dfrac{\text{m}}{\text{s}^2}}{8.9\cdot 10^{-4}\dfrac{\text{kg}}{\text{m}\cdot\text{s}}2.1\cdot 10^{-9}\dfrac{\text{m}^2}{\text{s}}}\right)^{\frac{1}{3}} = 28.94$$

From the definition of the Sherwood number,

$$k_l = \frac{Sh_L D_{O2,L}}{d_{bubbles}} = \frac{28.94\cdot 2.1\cdot 10^{-9}\dfrac{\text{m}^2}{\text{s}}}{5\cdot 10^{-4}\text{ m}} = 1.21\cdot 10^{-3}\frac{\text{m}}{\text{s}}$$

Since $k_l \approx k_L$, the overall mass transfer coefficient $k_L a$ is

$$k_L a = 14.90\frac{\text{m}^2}{\text{m}^3}\cdot 1.21\cdot 10^{-3}\frac{\text{m}}{\text{s}} = 0.018\text{ s}^{-1}$$

Example 5.5

A fermenter has a diameter of 0.9 m and a volume of 1 m^3. The fermenter is aerated with an air flow rate of 500 m^3/d. The diameter of the agitator is 0.3 m and the agitator speed is 150 rpm. What agitation speed is required to have the same k_La in a geometrically similar vessel, having a volume of 10 m^3? The large fermenter is aerated with an air flow rate of 5000 m^3/d. Assume that the power number of the agitator is 4.0 for both vessels.

For these vessels, the mass transfer coefficient is described by the correlation:

$$k_L a\left(\mathrm{s}^{-1}\right) = 0.0015\left(\frac{P_G}{V}\right)^{0.6}\left(\frac{Q_G}{A}\right)^{0.3} \quad \text{with } P_G/V \text{ in W/m}^3 \text{ and } Q_G/A \text{ in m/s.}$$

Solution

The liquid height for the small fermenter is

$$H = \frac{4V}{\pi D^2} = \frac{4 \cdot 1\ \mathrm{m}^3}{\pi 0.9^2\ \mathrm{m}^2} = 1.6\,\mathrm{m}$$

Therefore, the ratio between the liquid height and the diameter of the fermenter is, for the small fermenter:

$$\frac{H}{D} = \frac{1.6\ \mathrm{m}}{0.9\ \mathrm{m}} = 1.8$$

To be geometrically similar, the large fermenter will have the same ratio between the liquid height and the diameter of the fermenter:

$$\left(\frac{H}{D}\right)_{small} = \left(\frac{H}{D}\right)_{large}$$

Therefore, the diameter of the large fermenter will be

$$D = \sqrt[3]{\frac{4V}{\pi 1.8}} = \sqrt[3]{\frac{4 \cdot 10\ \mathrm{m}^3}{\pi 1.8}} = 1.9\ \mathrm{m}$$

The superficial gas velocities for the small and large vessels are

$$v_{g,small} = \left(\frac{Q_G}{A}\right)_{small} = \frac{250\dfrac{\mathrm{m}^3}{\mathrm{d}}}{\dfrac{\pi 0.9^2\ \mathrm{m}^2}{4}} = 393.1\frac{\mathrm{m}}{\mathrm{d}} = 0.0045\ \mathrm{m/s}$$

$$v_{g,large} = \left(\frac{Q_G}{A}\right)_{large} = \frac{2,500\frac{m^3}{d}}{\frac{\pi 1.9^2 m^2}{4}} = 882.2 \text{ m/d} = 0.0102 \text{ m/s}$$

For the small vessel, the ungassed power draw is

$$P = P_0 \rho N^3 D_{ag}^5 = 4.0 \cdot 1000 \frac{kg}{m^3} \cdot \left(\frac{150 \min^{-1}}{60\frac{s}{\min}}\right)^3 (0.3 \text{ m})^5 = 151.9 \text{ W}$$

The factor $\frac{Q_G}{ND_{ag}^3}$ for the small vessel is

$$\frac{Q_G}{ND_{ag}^3} = \frac{\left(\frac{250\frac{m^3}{d}}{86,400\frac{s}{d}}\right)}{2.5 \text{ s}^{-1}(0.3 \text{ m})^3} = 0.043$$

So, according to equation (5.26b), for the small vessel

$$\frac{P_G}{P} = 0.62 - 1.85\frac{Q_G}{ND_{ag}^3} = 0.54$$

For the small vessel, the term gassed power draw is therefore equal to

$$P_G = 0.54 \cdot 151.9 \text{ W} = 82.0 \text{ W}$$

And:

$$\left(\frac{P_G}{V}\right)_{small} = \frac{82.0 \text{ W}}{1 \text{ m}^3} = 82.0\frac{W}{m^3}$$

The power draw per unit volume required for the large vessel is obtained from the mass transfer correlation for the two vessels:

$$\frac{k_L a_{small}}{k_L a_{large}} = \frac{0.0015\left(\frac{P_G}{V}\right)_{small}^{0.6}\left(\frac{Q_G}{A}\right)_{small}^{0.3}}{0.0015\left(\frac{P_G}{V}\right)_{large}^{0.6}\left(\frac{Q_G}{A}\right)_{large}^{0.3}}$$

Therefore, the same $k_L a$ is obtained in the two vessels if

$$\left(\frac{P_G}{V}\right)_{large} = \left(\frac{\left(\frac{P_G}{V}\right)_{small}^{0.6}\left(\frac{Q_G}{A}\right)_{small}^{0.3}}{\left(\frac{Q_G}{A}\right)_{large}^{0.3}}\right)^{\frac{1}{0.6}} = \left(\frac{82.0^{0.6}0.0045^{0.3}}{0.0102^{0.3}}\right)^{\frac{1}{0.6}} = 54.5\frac{\text{W}}{\text{m}^3}$$

Note that the power per unit volume required in the large vessel is lower than in the small vessel, because the superficial gas velocity is higher in the large vessel.

The gassed power draw for the large vessel is

$$P_G = 54.5\frac{\text{W}}{\text{m}^3}\cdot 10\ \text{m}^3 = 545\,\text{W}$$

We need to calculate the ungassed power draw and the corresponding agitation speed. P and N can be obtained by the simultaneous solutions of equations (5.26b) and (5.24):

$$\frac{545}{P} = 0.62 - 1.85\frac{\frac{2500}{86{,}400}\ \frac{\text{m}^3}{\text{s}}}{N\cdot 0.57^3\,\text{m}^3} = 0.62 - \frac{0.29}{N}$$

$$P = P_o\rho N^3 D_{ag}^5 = 4.0\cdot 1000\frac{\text{kg}}{\text{m}^3}\cdot N^3\cdot(0.57\,\text{m})^5 = 241N^3$$

The simultaneous solution of the two equations (either using Solver in Excel, or by iteration) gives

$$P = 1141\,\text{W}$$

$$N = 1.35\ \text{s}^{-1}$$

For the large vessel, an agitator speed of 1.35 s^{-1} is required to obtain the same $k_L a$ as in the small vessel.

We verify that the ratio $\frac{Q_G}{ND_{ag}^3}$ for the large vessel is larger than 0.037 and lower than 0.11, to justify the use of equation (5.26b):

$$\frac{Q_G}{ND_{ag}^3} = \frac{\left(\frac{2,500\frac{m^3}{d}}{86,400\frac{s}{d}}\right)}{1.73\ s^{-1}(0.57\ m)^3} = 0.090$$

The use of equation (5.26b) is justified.

Example 5.6

The power curve of the desired agitator for a fermentation process is shown in Figure 5.7. The fermenter has a volume of 2 m^3 and a diameter of 1.5 m. The diameter of the agitator is 0.5 m and the agitation speed is 210 rpm. The air flow rate is 300 m^3/d. Calculate the k_La for this vessel.

For this vessel, the mass transfer coefficient is described by the correlation:

$$k_La(s^{-1}) = 0.0025\left(\frac{P_G}{V}\right)^{0.7} v_g^{0.2} \text{ (with } P_G \text{ in W, } V \text{ in m}^3, v_g \text{ in m/s)}$$

Solution

First of all, we calculate the ratio between the gassed and ungassed power draw for this vessel.

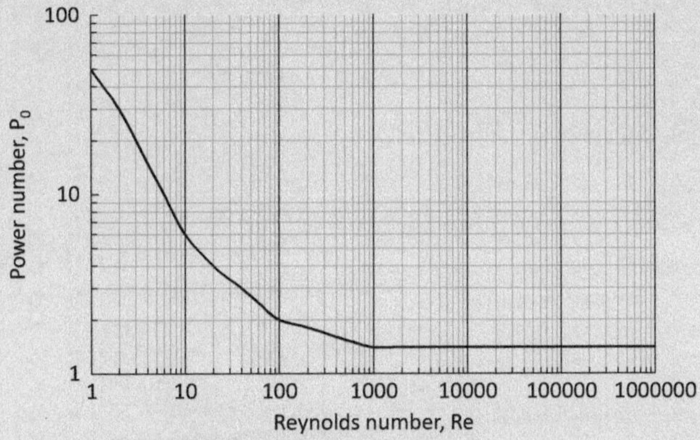

FIGURE 5.7 Power curve for Example 5.6.

$$\frac{Q_G}{ND_{ag}^3} = \frac{\left(\dfrac{300\dfrac{\text{m}^3}{\text{d}}}{86{,}400\dfrac{\text{s}}{\text{d}}}\right)}{3.5\ \text{s}^{-1}\left(0.5\ \text{m}\right)^3} = 0.032$$

Therefore, the ratio between P_G and P will be given by equation (5.26a):

$$\frac{P_G}{P} = 1 - 12.2\frac{Q_G}{ND_{ag}^3} = 1 - 12.2 \cdot 0.032 = 0.61$$

The superficial gas velocity v_g is

$$v_g = \frac{Q_G}{A} = \frac{\dfrac{300}{86{,}400}\dfrac{\text{m}^3}{\text{s}}}{\dfrac{\pi 1.5^2\,\text{m}^2}{4}} = 0.0078\ \text{m/s}$$

To calculate the ungassed power draw P, we need to use the power curve in Figure 5.5. The Reynolds number is

$$\text{Re} = \frac{\rho N D_{ag}^2}{\mu} = \frac{1000\dfrac{\text{kg}}{\text{m}^3} \cdot \dfrac{210}{60}\text{s}^{-1} \cdot 0.5^2\ \text{m}^2}{8.9 \cdot 10^{-4}\dfrac{\text{kg}}{\text{m} \cdot \text{s}}} = 983{,}146$$

In this region of Reynolds numbers, the power number is constant and equal to approximately 1.4. Therefore, the ungassed power draw is given by

$$P = P_0 \rho N^3 D_{ag}^5 = 1.4 \cdot 1000\frac{\text{kg}}{\text{m}^3} \cdot \left(\frac{210\text{min}^{-1}}{60\dfrac{\text{s}}{\text{min}}}\right)^3 \left(0.5\ \text{m}\right)^5 = 1876\,\text{W}$$

The gassed power draw is therefore

$$P_G = 0.61 \cdot 1{,}876 = 1144\ \text{W}$$

The $k_L a$ for this vessel is therefore

$$k_L a = 0.0025\left(\frac{1144}{2}\right)^{0.7} 0.0078^{0.2} = 0.08\ \text{s}^{-1}$$

5.4 OXYGEN BALANCES IN BATCH REACTORS

If the rate of oxygen transfer from the gas to the liquid phase and the rate of oxygen consumption by the microorganisms are known, the mass balance for oxygen in the liquid phase can be written and solved.

For a batch fermentation, the mass balance for oxygen can be written as

$$\frac{d\left(VC_{O2}\right)}{dt}=k_La\left(C^*_{O2}-C_{O2}\right)V+r_{O_{2biom}}V \tag{5.27}$$

In equation (5.27) $r_{O_{2biom}}$ is the rate of oxygen consumption by the microorganisms defined in Chapter 3.

Equation (5.27) can be re-arranged by eliminating the constant volume of the liquid phase and by introducing the definition of r_{O2biom} seen in Chapter 3:

$$\frac{dC_{O2}}{dt}=k_La\left(C^*_{O2}-C_{O2}\right)-\frac{\mu_{max}S}{K_S+S}\frac{X}{Y_{X/O2}} \tag{5.28}$$

Equation (5.26) gives the profile of the oxygen concentration in the liquid phase as a function of time during a batch fermentation. Solution of equation (5.28) requires knowledge of the values of the kinetic parameters μ_{max}, K_S, $Y_{X/O2}$ and of the mass transfer coefficient k_La. It also requires knowledge of the oxygen solubility C^*_{O2} and of an initial condition for the oxygen concentration in the liquid phase. Note that equation (5.28) cannot be solved in isolation, but it needs to be coupled with the mass balances for substrate and biomass in a batch fermentation seen in Chapter 3.

In summary, the profiles of substrate, microorganisms and dissolved oxygen in batch fermentation are given by the solution of the system of equations (5.29).

$$\begin{cases}\dfrac{dX}{dt}\left(\dfrac{\text{kg biomass}}{\text{m}^3\cdot\text{d}}\right)=\dfrac{\mu_{max}S}{K_S+S}X\\ \dfrac{dS}{dt}\left(\dfrac{\text{kg substrate}}{\text{m}^3\cdot\text{d}}\right)=-\dfrac{\mu_{max}S}{K_S+S}\dfrac{X}{Y_{X/S}}\\ \dfrac{dP}{dt}\left(\dfrac{\text{kg product}}{\text{m}^3\cdot\text{d}}\right)=\dfrac{\mu_{max}S}{K_S+S}XY_{P/X}\\ \dfrac{dC_{O2}}{dt}\left(\dfrac{\text{kg oxygen}}{\text{m}^3\cdot\text{d}}\right)=k_La\left(C^*_{O2}-C_{O2}\right)-\dfrac{\mu_{max}S}{K_S+S}\dfrac{X}{Y_{X/O2}}\end{cases} \tag{5.29}$$

Note that, as discussed in Chapter 3, the use of the Monod equation with only the organic substrate as the limiting substrate, assumes that the concentration of other nutrients and oxygen are not limiting. Therefore, the design of the fermentation process requires making sure that the oxygen concentration always stays above a certain minimum level, below which oxygen would become rate limiting. This minimum oxygen concentration is typically in the range 1–2 mg/L. The typical profiles of X, S and C_{O2} obtained from the solution of equations (5.29) are shown in Figure 5.8.

From Figure 5.8 it is evident that the profile of dissolved oxygen follows the profile of the substrate. As microorganisms' concentration increases, the rate of substrate removal increases, and the oxygen concentration starts to decrease more rapidly, because the rate of oxygen consumption is higher than the rate of oxygen transfer. When the substrate is completely removed, oxygen concentration starts to increase because oxygen consumption ceases.

In many aerobic fermentations, the value of the mass transfer coefficient k_La determines the maximum value of the initial substrate concentration that can be used in the process. Indeed, as shown in Figure 5.9, if the initial substrate concentration

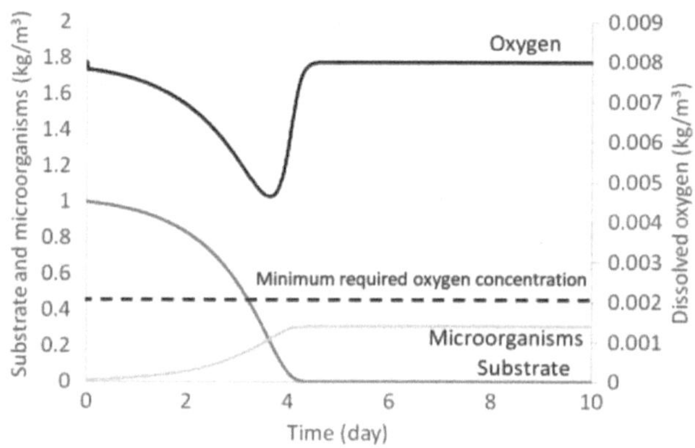

FIGURE 5.8 Typical profiles of oxygen, microorganisms and substrates during a batch fermentation.

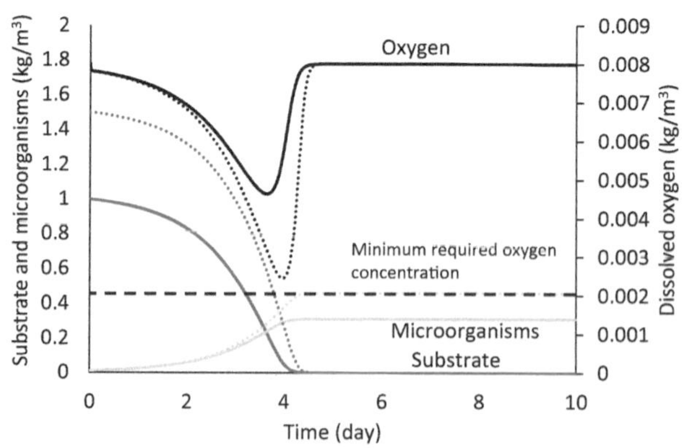

FIGURE 5.9 Effect of the initial substrate concentration on the dissolved oxygen profile in batch fermentation. Solid lines: profiles for initial substrate 1.0 kg/m^3; dotted lines: profiles for initial substrate 1.5 kg/m^3.

increases, the oxygen concentration decreases to a lower level. The maximum initial substrate concentration that we can have in a given fermentation is the one for which the dissolved oxygen concentration reaches the minimum allowable level (i.e. the concentration corresponding to the rate limiting condition).

The maximum initial substrate concentration can be calculated by trial and error by solving the system of equations (5.29) with different values of the initial substrate concentration until the oxygen profile reaches the minimum allowable concentration.

Example 5.7

Acetic acid is produced by the aerobic fermentation of ethanol by the microorganism *Acetobacter*. The stoichiometry of the fermentation can be written as follows:

$$C_2H_5OH + O_2 + 0.10NH_3 \rightarrow 0.10C_5H_7O_2N + 0.75CH_3COOH + 0.3H_2O$$

The fermentation is started with an ethanol concentration of 10 g/L and a concentration of microorganisms of 1 g/L. The initial concentration of dissolved oxygen is 7.0 mg/L. The fermenter is sparged with air with a $k_L a$ of 0.1 s^{-1} and the saturation concentration of oxygen under the fermentation conditions is 9.0 mg/L. The kinetic parameters are

$$\mu_{max} = 7d^{-1}; K_S = 0.02\frac{kg}{m^3}$$

Calculate the time profiles of ethanol, biomass, acetic and oxygen during the fermentation.

Solution

The profiles are obtained from the solution of the system of differential equations (5.29). From the given stoichiometry we can calculate the parameters $Y_{X/S}$, $Y_{X/O2}$ and $Y_{P/X}$:

$$Y_{X/S} = \frac{0.1 \cdot 113}{46} = 0.25\frac{kg}{kg}$$

$$Y_{X/O2} = \frac{0.1 \cdot 113}{32} = 0.35\frac{kg}{kg}$$

$$Y_{P/X} = \frac{0.75 \cdot 60}{0.1 \cdot 113} = 3.98\frac{kg}{kg}$$

The system of equations (5.29) in this case is

$$\begin{cases} \dfrac{dX}{dt}\left(\dfrac{\text{kg biomass}}{\text{m}^3\cdot\text{d}}\right) = \dfrac{7S}{0.02+S}X \\ \dfrac{dS}{dt}\left(\dfrac{\text{kg substrate}}{\text{m}^3\cdot\text{d}}\right) = -\dfrac{7S}{0.02+S}\dfrac{X}{0.35} \\ \dfrac{dP}{dt}\left(\dfrac{\text{kg product}}{\text{m}^3\cdot\text{d}}\right) = \dfrac{7S}{0.02+S}X\cdot 3.98 \\ \dfrac{dC_{O2}}{dt}\left(\dfrac{\text{kg oxygen}}{\text{m}^3\cdot\text{d}}\right) = 0.1\cdot 86{,}400\left(0.009 - C_{O2}\right) - \dfrac{7S}{0.02+S}\dfrac{X}{0.35} \end{cases}$$

The resulting profiles are shown in Figure 5.10:

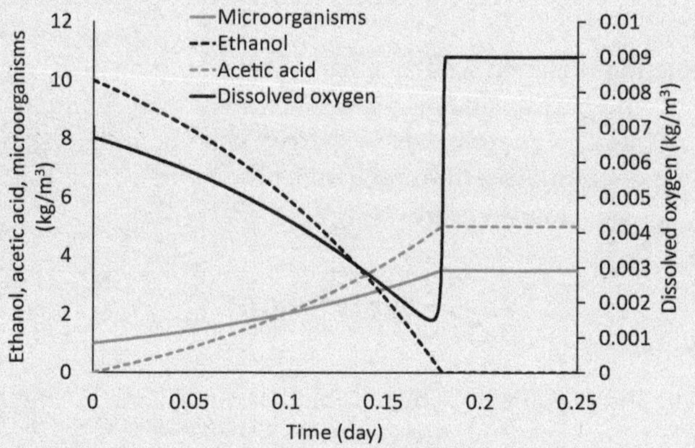

FIGURE 5.10 Calculated profiles for Example 5.7.

Example 5.8

The microorganism *Bacillus subtilis* is used for the production of the enzyme protease ($C_3H_5O_2N$). The fermentation reaction is aerobic and uses glucose as the substrate and ammonia as the nitrogen source. The products are microorganisms, the enzyme, water and carbon dioxide. It is known that 60 g of carbon dioxide and 20 g of microorganisms are produced per 100 g of glucose consumed.

The fermentation is carried out in an agitated vessel having a volume of 20 m³. The vessel has a diameter of 2.1 m. The diameter of the agitator is 0.7 m and the maximum practical agitation speed is 180 rpm. Under the fermentation conditions the power number of the agitator is 5.0. The vessel is sparged with

air with a flow rate of 10,000 m^3/d. Under these conditions we can assume that the ratio between the gassed and ungassed power draw is equal to 0.43.
The fermentation starts with 100 mg/L of microorganisms and with an oxygen concentration of 9.0 mg/L. A minimum dissolved oxygen concentration of 2.0 mg/L is required for the fermentation to proceed. For this vessel the k_La is given by the following correlation:

$$k_La\left(s^{-1}\right) = 0.020\left(\frac{P_G}{V}\right)^{0.5} v_g^{0.3} \text{ (with } P_G \text{ in W, } V \text{ in m}^3\text{, } v_g \text{ in m/s)}$$

The kinetic parameters are

$$\mu_{max} = 5\,d^{-1}; K_S = 0.03\frac{kg}{m^3}$$

(a) Calculate the maximum initial glucose concentration that it is possible to have at the start of the fermentation, assuming that glucose is completely consumed when the fermentation ends.
(b) Assuming that the downtime between consecutive batches is 2 h, and the process is operated for 8000 h/year, calculate the annual protease productivity (in kg) with this process.

Solution
(a) The first step is the calculation of the reaction stoichiometry, which will allow the calculation of the coefficients $Y_{X/S}$, $Y_{P/X}$, $Y_{X/O2}$. The unbalanced stoichiometry for this reaction is

$$C_6H_{12}O_6 + aO_2 + bNH_3 \rightarrow cC_5H_7O_2N + dC_3H_5O_2N + eCO_2 + fH_2O$$

We know that

$$\frac{e \cdot MW_{CO_2}}{MW_{C_6H_{12}O_6}} = 0.60 \quad \Rightarrow \quad e = \frac{0.60 \cdot 180}{44} = 2.45$$

$$\frac{c \cdot MW_{C_5H_7O_2N}}{MW_{C_6H_{12}O_6}} = 0.20 \quad \Rightarrow \quad c = \frac{0.20 \cdot 180}{113} = 0.32$$

The other coefficients a, b, d, f can be obtained from the elemental balances of C, H, N, O:

$$\begin{cases} C-balance: 5c+3d+e=6 \\ O-balance: 6+2a=2c+2d+2e+f \\ H-balance: 12+3b=7c+5d+2f \\ N-balancec: b=c+d \end{cases}$$

From the C-balance we obtain $d = 0.65$; from the N-balance: $b = 0.97$; from the H-balance: $f = 4.71$; from the O-balance: $a = 2.78$.

Hence, the balanced stoichiometry is

$$C_6H_{12}O_6 + 2.78O_2 + 0.97NH_3 \rightarrow 0.32C_5H_7O_2N + 0.65C_3H_5O_2N + 2.45CO_2 + 4.71H_2O$$

The stoichiometric coefficients are therefore

$$Y_{X/S} = \frac{0.32 \cdot 113}{180} = 0.20\frac{\text{kg}}{\text{kg}}$$

$$Y_{X/O2} = \frac{0.32 \cdot 113}{2.78 \cdot 32} = 0.41\frac{\text{kg}}{\text{kg}}$$

$$Y_{P/X} = \frac{0.65 \cdot 87}{0.32 \cdot 113} = 1.56\frac{\text{kg}}{\text{kg}}$$

The superficial gas velocity is

$$v_t = \frac{Q_G}{A} = \frac{\dfrac{10{,}000\ \text{m}^3}{86{,}400\ \text{s}}}{\dfrac{\pi 2.1^2\,\text{m}^2}{4}} = 0.033\ \text{m/s}$$

The ungassed power draw is

$$P = P_o \rho N^3 D_{ag}^5 = 5.0 \cdot 1000\frac{\text{kg}}{\text{m}^3} \cdot \left(\frac{180\text{min}^{-1}}{60\dfrac{\text{s}}{\text{min}}}\right)^3 (0.7\ \text{m})^5 = 22{,}689\ \text{W}$$

And the gassed power draw

$$P_G = 0.43P = 9756\ \text{W}$$

The $k_L a$ for this vessel is

$$k_L a\left(\text{s}^{-1}\right) = 0.020\left(\frac{9{,}756}{20}\right)^{0.5} 0.033^{0.3} = 0.159\ \text{s}^{-1}$$

The mass balances that describe this process are therefore

$$\begin{cases} \dfrac{dX}{dt}\left(\dfrac{\text{kg biomass}}{\text{m}^3\text{d}}\right) = \dfrac{5S}{0.03+S}X \\ \dfrac{dS}{dt}\left(\dfrac{\text{kg substrate}}{\text{m}^3\text{d}}\right) = -\dfrac{5S}{0.03+S}\dfrac{X}{0.20} \\ \dfrac{dP}{dt}\left(\dfrac{\text{kg product}}{\text{m}^3\text{d}}\right) = \dfrac{5S}{0.03+S}X\cdot 1.56 \\ \dfrac{dC_{O2}}{dt}\left(\dfrac{\text{kg oxygen}}{\text{m}^3\text{d}}\right) = 0.033\cdot 86{,}400\left(0.009 - C_{O2}\right) - \dfrac{5S}{0.03+S}\dfrac{X}{0.20} \end{cases}$$

These mass balances can be solved for various values of the initial substrate concentration. The maximum allowable initial substrate concentration is the value for which the oxygen concentration drops to 2.0 mg/L without going below this value. Figure 5.11 shows the oxygen profiles for various values of the initial substrate concentration in the batch. The maximum allowable initial substrate concentration is 40 kg/m^3.

(b) When the substrate concentration is 40 kg/m^3, the product concentration at the end of each batch is 12.57 kg/m^3 (from the reaction stoichiometry). Since the volume of the reactor is 20 m^3, in each batch 251.4 kg of enzyme are produced. The substrate is completely removed in 0.90 day = 21.6 h. Adding the 2 hours of downtime, the total time of a batch is 23.6 h. The process is operated for 8000 h per year, so the number of batches in a year is equal to 339. So, in one year, this fermenter will produce 339·251.4 = 85,225 kg of enzyme.

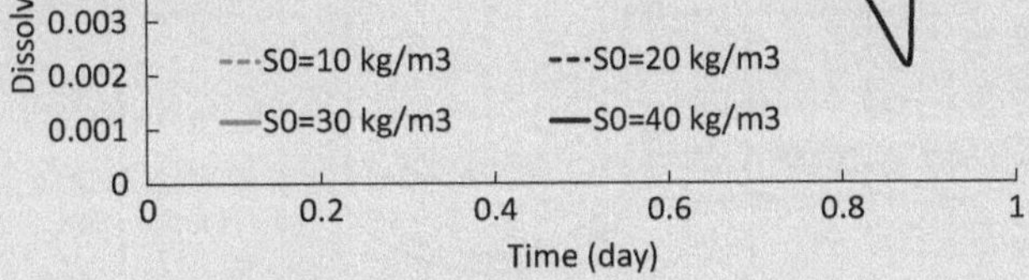

FIGURE 5.11 Solution of Example 5.8.

5.5 OXYGEN BALANCE IN CONTINUOUS REACTORS

In addition to calculating the required residence time, in continuous reactor design for aerobic fermentations it is also important to design the aeration system so that it provides the oxygen required by the microorganisms. Design of the aeration system means essentially selecting an aeration system that is able to transfer oxygen with the required mass transfer coefficient k_La. The k_La that is required for a certain fermentation can be calculated from a mass balance of oxygen on the continuous fermenter (Figure 5.12).

In Figure 5.10, in addition to the symbols already defined, C_{O20} and C_{O2} are the concentrations of dissolved oxygen in the feed and in the reactor, respectively. For a continuous fermenter at steady-state, the mass balance of oxygen can be done using the general equation for mass balances, Equation (3.1). In this case Equation (3.1) becomes:

$$Q \cdot C_{O20} + r_{O_{2transfer}} V = Q \cdot C_{O2} + \left(-r_{O_{2biom}}\right) \cdot V \tag{5.29}$$

In equation (5.29), the terms $r_{O_{2transfer}}$ and $r_{O_{2biom}}$ are expressed as seen earlier:

$$r_{O_{2transfer}} \left(\frac{\mathrm{kg}_{O_2}}{\mathrm{m}^3\ \mathrm{d}}\right) = k_L a \cdot \left(C^*_{O2} - C_{O2}\right)$$

$$r_{O_{2biom}} \left(\frac{\mathrm{kg}_{O2}}{\mathrm{m}^3\ \mathrm{d}}\right) = -\frac{\mu_{max} S}{K_S + S} \frac{X}{Y_{X/O2}}$$

with $C^*_{O2} = k_{eq} p_{O2}$, where p_{O2} is taken as the partial pressure of oxygen in the influent gas, ignoring the decrease in oxygen concentration in the gas bubbles due to oxygen transfer from the gas phase.

Once the equations for $r_{O_{2transfer}}$ and $r_{O_{2biom}}$ are substituted, equation (5.29) becomes

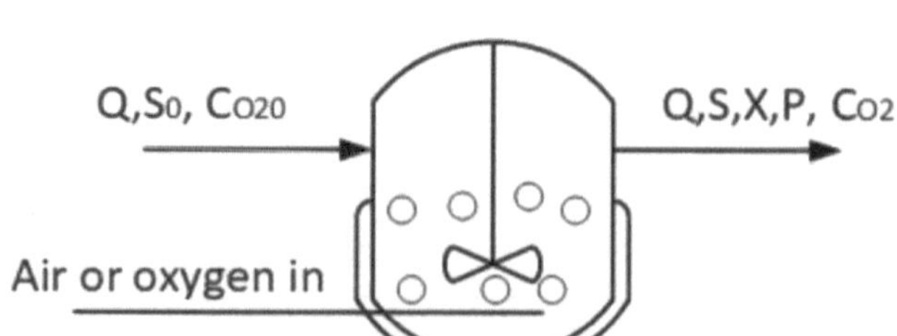

FIGURE 5.12 Scheme of a continuous fermenter including oxygen transfer for oxygen mass balance.

$$Q \cdot C_{O20} + k_L a \cdot \left(C^*_{O2} - C_{O2}\right) V = Q \cdot C_{O2} + \frac{\mu_{max} S}{K_S + S} \frac{X}{Y_{X/O2}} \tag{5.30}$$

Equation (5.30) can be re-arranged into equation (5.31):

$$k_L a = \frac{D \cdot \left[C_{O2} - C_{O20} + \frac{X}{Y_{X/O2}} \right]}{\left(C^*_{O2} - C_{O2}\right)} \tag{5.31}$$

Equation (5.31) can be used to calculate the value of $k_L a$ that the aeration system needs to provide in order to maintain the desired concentration of dissolved oxygen in the reactor. Note that at the numerator of equation (5.31), the term representing the oxygen consumption by the microorganisms ($\frac{X}{Y_{X/O2}}$) is usually much larger than the term representing the transport of oxygen in and out of the reactor (C_{O2}-C_{O20}).

According to equation (5.31), for a given value of the biomass concentration in the reactor, X, the required $k_L a$ is linearly proportional to the dilution rate D. The reason is that the higher the amount of feed processed per unit volume of the reactor, the higher are the oxygen consumption and the required oxygen transfer rate per unit volume.

The value chosen for the dissolved oxygen concentration in the reactor, C_{O2}, also affects the required $k_L a$. The higher the C_{O2}, the higher the required $k_L a$, because the driving force $C_{O2}{}^*$-C_{O2} will be lower. For given values of the dilution rate and of the biomass concentration in the reactor, it is beneficial to maintain the required $k_L a$ at the lowest possible level, because a higher value of the mass transfer coefficient will require higher values of the agitator power draw and/or higher values of the inlet gas flow rate with consequently higher energy costs. Therefore, the oxygen concentration in a bioreactor needs to be maintained at the lowest possible level that ensures oxygen does not become rate limiting for the growth of the microorganisms. Figure 5.13 shows the general profile of the required $k_L a$ vs the dissolved oxygen concentration in the reactor, C_{O2}.

The required $k_L a$ also depends on the substrate concentration in the feed, S_0, because the microorganisms' concentration X depends on S_0. Higher values of S_0 will require higher $k_L a$ values, because the concentration of the microorganisms and their oxygen consumption rate will be higher. To show this effect, equation (5.31) can be re-arranged into equation (5.32) by replacing X as a function of S_0 (equation (3.20)), and by solving for S_0, under the assumption that $S<<S_0$ (i.e. assuming that the dilution rate D is always low enough to guarantee (almost) complete substrate removal):

$$S_0 = \left[\frac{k_L a}{D} \left(C^*_{O2} - C_{O2}\right) - \left(C_{O2} - C_{O20}\right) \right] \frac{Y_{X/O2}}{Y_{X/S}} \tag{5.32}$$

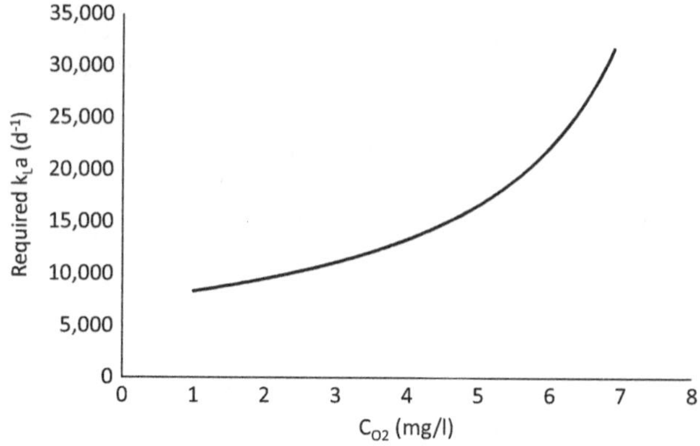

FIGURE 5.13 Plot of equation (5.31) showing the effect of the oxygen concentration in the continuous fermenter on the required k_La. Plot obtained with $D = 5$ d^{-1}, $C_{O20} = 5$ mg/L, $X = 2$ kg/m^3, $Y_{X/O2} = 0.15$ kg/kg, $C_{O2}^* = 9$ mg/L.

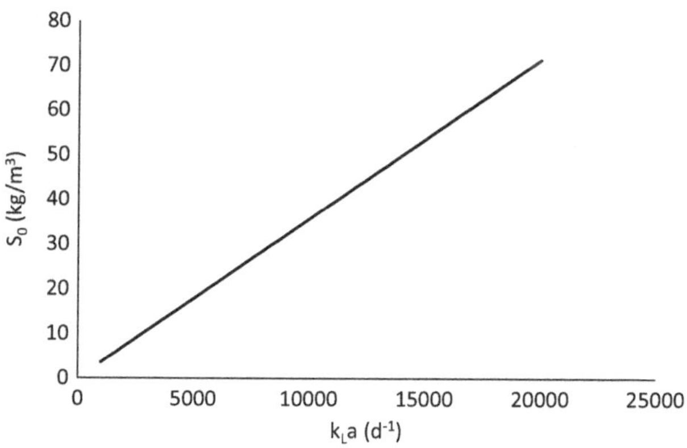

FIGURE 5.14 Plot of equation (5.32) showing the effect of the mass transfer coefficient on the maximum substrate concentration that we can have in the feed of a continuous fermenter to maintain the oxygen concentration at the desired level. Plot obtained with $C_{O2} = 1.5$ mg/L, $C_{O2}^* = 9$ mg/L, $D = 5$ d^{-1}, $C_{O20} = 5$ mg/L, $X = 2$ kg/m^3, $Y_{X/O2} = 0.15$ kg/kg, $Y_{X/S} = 0.063$ kg/kg.

Equation (5.32), plotted in Figure 5.14, gives the maximum substrate concentration that we can have in the feed of a continuous fermenter to maintain the oxygen concentration at the desired value C_{O2}. The higher the k_La in the fermenter, the higher the substrate concentration that we can have in the feed and therefore the higher the productivity of the process. Therefore, if the higher process productivity outweighs the higher energy costs associated with higher k_La values, it may be beneficial to increase the mass transfer coefficient to increase the productivity of the process. Note that in equation (5.32), as in (5.31), the term due to oxygen transfer is usually much higher than the contribution of the dissolved oxygen in and out of the reactor (C_{O2}-C_{O20}).

The efficiency of oxygen transfer from the gas to the liquid phase can also be calculated. The efficiency (η) can be defined as the ratio between the oxygen transferred from the gas to the liquid phase and the total oxygen fed with the gas phase. For a continuous fermenter, the equation for the efficiency is

$$\eta\left(\frac{\text{kgO}_2\text{transferred}}{\text{kgO}_2\text{ provided}}\right)=\frac{k_L aV\left(C^*_{O2}-C_{O2}\right)}{Q_{air}\rho_{air}\cdot 0.23} \tag{5.33}$$

In equation (5.33) the factor 0.23 represents the mass fraction of oxygen in air. The same equation can also be used to calculate the efficiency for a batch reactor, however in a batch process the oxygen concentration C_{O2} will change during the process and so the efficiency will also be variable with time.

Example 5.9

Consider the aerobic fermentation for the production of citric acid:

$$C_6H_{12}O_6+2.35O_2+0.1NH_3\rightarrow 0.1C_5H_7O_2N+0.7C_6H_8O_7+1.3CO_2+3H_2O$$

This reaction is carried out in a continuous fermenter, with a feed flow rate if 1.5 m^3/h. The fermenter has a volume of 10 m^3 and a diameter of 2 m. The concentration of glucose in the fed is 100 g/L and we can assume that glucose is entirely consumed by the microorganisms. Calculate the k_La that is required to maintain the dissolved oxygen concentration to 3.0 mg/L.

Assume that the oxygen flow rate is 10 times the minimum amount required for this reaction. Assume that the correlation for the mass transfer coefficient is in this case:

$$k_La\left(s^{-1}\right)=0.050\left(\frac{P_G}{V}\right)^{0.63}v_g^{0.40}\text{ with }P_G\text{ in W, }V\text{ in m}^3,\ v_g\text{ in m/s}$$

What is the required ungassed power draw? Assume that under these conditions $P_G/P = 0.43$.

Assume that the saturation concentration of oxygen in water under the fermentation conditions is $C^*_{O2} = 9.0$ mg/L.

Solution

We start from the oxygen balance, which is, with the hypotheses of this problem:

$$k_La\left(C^*_{O2}-C_{O2}\right)V=Q\cdot C_{O2}+\left(-r_{O_{2biom}}\right)\cdot V$$

The glucose consumption rate is equal to $100\frac{\text{kg}}{\text{m}^3}\cdot 1.5\frac{\text{m}^3}{\text{h}}=150\frac{\text{kg}}{\text{h}}$.

From the reaction stoichiometry 0.42 kg of oxygen are consumed by the microorganisms per 1 kg of glucose removed. Therefore, the oxygen consumption by the microorganisms is

$$\left(-r_{O_{2biom}}\right) = 150\frac{\text{kg}}{\text{h}} \cdot 0.42\frac{\text{kgO}_2}{\text{kg glucose}} = 63\frac{\text{kgO}_2}{\text{h}}$$

Therefore, the required k_La is

$$k_La = \frac{20\frac{\text{m}^3}{\text{h}} \cdot 0.003\frac{\text{kg}}{\text{m}^3} + 63\frac{\text{kg}}{\text{h}}}{(0.009 - 0.003)\frac{\text{kg}}{\text{m}^3} 10\ \text{m}^3} = 1051\ \text{h}^{-1} = 0.29\ \text{s}^{-1}$$

To calculate the superficial gas velocity v_g we need the gas flow rate. Since the gas flow rate is 10 times the minimum requirement, this will be equal to 630 kg_{O2}/hr. Using the ideal gas law

$$V_{O2} = \frac{nRT}{P} = \frac{\frac{630{,}000\text{g}}{32\frac{\text{g}}{\text{mol}}} \cdot 0.0821\frac{\text{L} \cdot \text{atm}}{\text{mol} \cdot \text{K}} \cdot 298.15\ \text{K}}{1\ \text{atm}} = 481{,}912\frac{\text{L}}{\text{h}}$$

Since atmospheric air is made of 21% in volume by oxygen the air flow rate will be

$$V_{air} = \frac{481{,}912}{0.21} = 2{,}294{,}823\frac{\text{L}}{\text{h}} = 0.637\frac{\text{m}^3}{\text{s}}$$

The superficial gas velocity is

$$v_g = \frac{Q_G}{A} = \frac{0.637\frac{\text{m}^3}{\text{s}}}{\frac{\pi 2^2\ \text{m}^2}{4}} = 0.203\ \text{m/s}$$

Therefore, the gassed power draw required:

$$P_G = V\left(\frac{k_La}{0.050 \cdot v_t^{0.40}}\right)^{\frac{1}{0.63}} = 10\left(\frac{0.29}{0.050 \cdot 0.203^{0.40}}\right)^{\frac{1}{0.63}} = 448\ \text{W}$$

Therefore, the ungassed power draw will be $P_G/0.43 = 1042$ W.

5.6 CONSIDERATIONS FOR SCALE-UP AND SCALE-DOWN

It is often desirable to reproduce the results of a lab-scale fermentation at larger (commercial) scale (scale-up), or, on the other hand, to replicate a large-scale fermentation at lab-scale (scale-down). Scale-up is made with the obvious aims to obtain more product or more microorganisms or to process more wastewater, the aims of scale-down are often to investigate at lab-scale how altering the process conditions (e.g. the residence time, temperature or pH) affects the fermentation. We have already seen an example of a scale-up calculation in Example 5.5. In this section we will cover the general theory of scale-up/scale-down for oxygen transfer in fermentation processes, assuming they are carried out in agitated vessels.

If we want to maintain the same dissolved oxygen concentration in the scaled-up or scaled-down vessel as in the original fermentation, the aeration and agitation system needs to be designed appropriately. We assume that the temperature, pH, residence time, feed composition and all the other fermentation parameters are kept constant on scale-up or scale-down, so here we will only focus on oxygen transfer. We will see in the next chapter how to deal with scale-up/scale-down for temperature control.

The oxygen balance in a continuous fermenter is given by equation (5.34). This equation shows, that, in order to maintain the same oxygen concentration in the scaled-up/scale-down vessel as in the original vessel, the k_La must be maintained to the same value. Indeed, with the same feed composition and residence time, the substrate and biomass concentration in the vessels will be the same. In order to maintain the same k_La in the scaled-up/scaled-down vessels, assuming that, as it is good practice, the scale-up/scale-down is made using geometrically similar vessels, we need to look at the correlation for k_La (5.22). According to this correlation, maintaining the same k_La means maintaining the same value of the term $\left(\frac{P_G}{V}\right)^a v_g^b = \left(\frac{P_G}{V}\right)^a \left(\frac{Q_G}{A}\right)^b$. In this equation the terms V and A are fixed by the required volumes and scale-up/scale-down ratio, whilst the values of P_G and Q_G in the scale-up/scale-down vessel need to be calculated or assigned. If we assign a value for Q_G for the scale-up/scale-down vessel, then the value of P_G for this vessel will be calculated, vice versa if the value of P_G for the scale-up/scale-down vessel is assigned, the value of Q_G for this vessel will be calculated. In our example procedure explained below, we assume that the scale-up/scale-down is made maintaining the same gas flow rate per unit volume of reactor, i.e. $\left(\frac{Q_G}{V}\right)_{large} = \left(\frac{Q_G}{V}\right)_{small}$. This is not an absolute requirement, but it is often a convenient way to scale-up/scale-down the gas flow rate. We will also assume, for simplicity's sake, that the ratio $\frac{P_G}{P}$ is the same in the large and small vessels and that the power number P_0 is independent on the Reynolds number for both vessels (flat region of the P_0 vs Re curve). These assumptions simplify the analysis but do not alter the significance of the results.

With the assumptions in the paragraph above, the scale-up/scale-down problem means finding the value of the power draw P_G required in the scale-down/

scaled-down vessel once the geometry of the vessels and the agitation conditions in the original vessel are known. We assume we know the scale-up/scale-down ratio, i.e. the ratio between the vessel volumes, $\frac{V_{large}}{V_{small}}$. As we have seen the condition of same $k_L a$ for the large and small vessels is

$$\left(\frac{P_G}{V}\right)^a_{large}\left(\frac{Q_G}{A}\right)^b_{large}=\left(\frac{P_G}{V}\right)^a_{small}\left(\frac{Q_G}{A}\right)^b_{small} \tag{5.34}$$

Equation (5.34) can be re-arranged as

$$\left(\frac{P_{G,large}}{P_{G,small}}\right)^a=\left(\frac{V_{large}}{V_{small}}\right)^a\left(\frac{A_{large}}{A_{small}}\right)^b\left(\frac{Q_{G,small}}{Q_{G,large}}\right)^b \tag{5.35}$$

With the conditions that the gas flow rate per unit volume remains constant on scale-up/scale-down, equation (5.35) can be re-arranged as

$$\left(\frac{P_{G,large}}{P_{G,small}}\right)^a=\left(\frac{V_{large}}{V_{small}}\right)^{a-b}\left(\frac{A_{large}}{A_{small}}\right)^b \tag{5.36}$$

For cylindrical vessels the volume and the cross-sectional area are given by

$$V=\frac{\pi D^2}{4}H \tag{5.37}$$

$$A=\frac{\pi D^2}{4} \tag{5.38}$$

In equations (5.37) and (5.38) D is the vessel diameter and H the vessel height. We define α as the ratio between the height and diameter of the vessel:

$$\alpha=\frac{H}{D} \tag{5.39}$$

For geometrically similar vessels the parameter α will stay constant on scale-up/scale-down. With the introduction of α we can write the vessel volume as

$$V=\frac{\alpha\pi D^3}{4} \tag{5.40}$$

From equations (5.38) and (5.40) we obtain

$$\frac{D_{large}}{D_{small}}=\left(\frac{V_{large}}{V_{small}}\right)^{\frac{1}{3}} \tag{5.41}$$

$$\frac{A_{large}}{A_{small}} = \left(\frac{V_{large}}{V_{small}}\right)^{\frac{2}{3}} \tag{5.42}$$

Introducing equation (5.42) in equation (5.36) we obtain

$$\left(\frac{P_{G,large}}{P_{G,small}}\right)^{a} = \left(\frac{V_{large}}{V_{small}}\right)^{a-\frac{b}{3}} \Rightarrow \frac{P_{G,large}}{P_{G,small}} = \left(\frac{V_{large}}{V_{small}}\right)^{1-\frac{b}{3a}} \tag{5.43}$$

Equation (5.43) can be rewritten in terms of the power per unit volume:

$$\frac{\dfrac{P_{G,large}}{V_{large}}}{\dfrac{P_{G,small}}{V_{small}}} = \left(\frac{V_{large}}{V_{small}}\right)^{-\frac{b}{3a}} \tag{5.44}$$

Equations (5.43) and (5.44) show that in the scale-up/scale-down the ratio between the power draw to be provided in the vessels is proportional to the scale-up/scale-down ratio, however the proportionality is less than linear and the power draw per unit volume decreases on scale-up. The reason for the lower than linear increase of $\frac{P_{G,large}}{P_{G,small}}$ with $\frac{V_{large}}{V_{small}}$ is that, with the assumption of $\left(\frac{Q_G}{V}\right)_{large} = \left(\frac{Q_G}{V}\right)_{small}$, the superficial gas velocity increases as $\frac{V_{large}}{V_{small}}$ increases, giving a higher contribution to the k_La. This is positive since it allows the achievement of the same k_La in large reactors with lower power-to-volume ratios than in small reactors, limiting the increase in agitator energy consumption on scale-up. The relative role of the superficial gas velocity and of the power draw depends of course on the values of the coefficients a and b of equation (5.22). Figures 5.15 and 5.16 show, for arbitrary but realistic values of the parameters a and b, the shape of the curves defined by equations (5.43) and (5.44).

Once the required power in the scaled-up/scale-down vessel has been calculated with equations (5.43) or (5.44), the next step is to calculate the agitation speed that is required in the scaled-up/scaled-down vessel if the agitation speed in the original vessel is known.

The relationship between the agitator power draw and the power number is given by equation (5.24). Since we are assuming that $\frac{P_G}{P}$ and P_0 stay constant with scale-up/

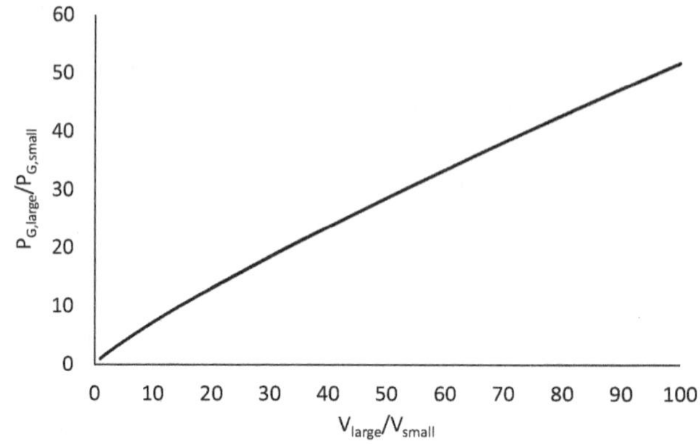

FIGURE 5.15 Plot of equation (5.43) with $a = 0.7$, $b = 0.3$.

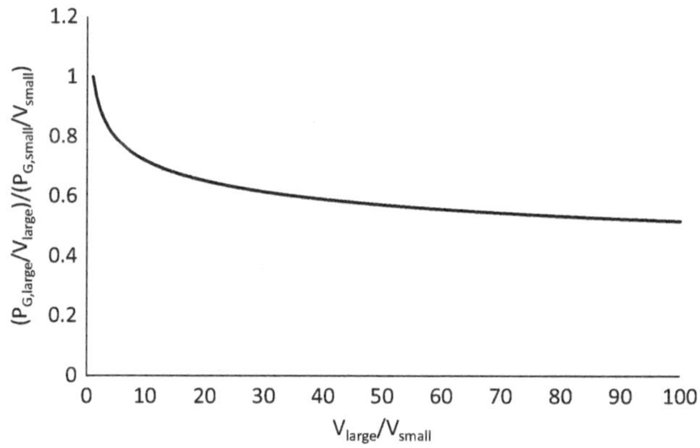

FIGURE 5.16 Plot of equation (5.44) with $a = 0.7$, $b = 0.3$.

scale-down, and the density of the medium is also constant, equation (5.24) leads to equation (5.45):

$$\left(\frac{P_{G,large}}{P_{G,small}}\right) = \left(\frac{N_{large}}{N_{small}}\right)^3 \left(\frac{D_{ag,large}}{D_{ag,small}}\right)^5 \tag{5.45}$$

For geometrically similar vessels we have

$$\frac{D_{ag,large}}{D_{ag,small}} = \frac{D_{large}}{D_{small}} = \left(\frac{V_{large}}{V_{small}}\right)^{\frac{1}{3}} \tag{5.46}$$

Inserting equation (5.46) into (5.44), combining with equation (5.43) and re-arranging we obtain the equation for the required agitation speed in the scaled-up/scaled-down vessel:

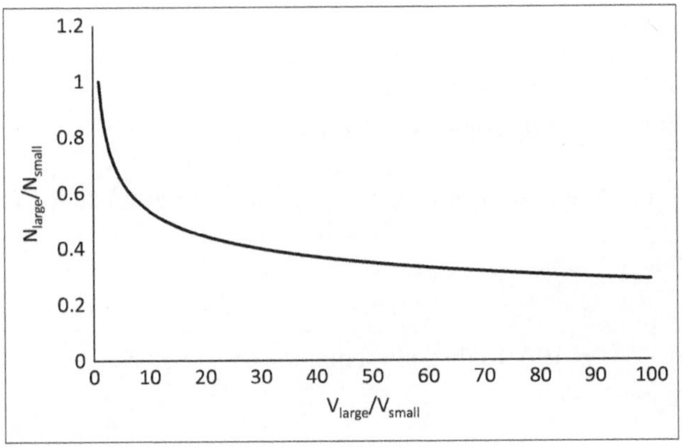

FIGURE 5.17 Plot of equation (5.47) with $a = 0.7$, $b = 0.3$.

$$\frac{N_{large}}{N_{small}} = \left(\frac{V_{large}}{V_{small}}\right)^{-\frac{1}{9}\left(2+\frac{b}{a}\right)} \tag{5.47}$$

Equation (5.47) shows that the agitation speed required in the scaled-up/scaled-down vessel is inversely proportional to the ratio $\frac{V_{large}}{V_{small}}$. This is because of the large effect of the agitator diameter (fifth power) on the power draw. Figure 5.17 shows the plot of $\frac{N_{large}}{N_{small}}$ vs $\frac{V_{large}}{V_{small}}$ for the same values of the parameters a and b used in Figure 5.15.

Equations (5.43) and (5.47) allow for the calculation of the power draw and agitation speed in scaled-up/scaled-down vessels. The same methodology shown in this section can be used for the scale-up/scale-down of fermentation vessels under different assumptions, e.g. $\left(\frac{Q_G}{V}\right)_{large} \neq \left(\frac{Q_G}{V}\right)_{small}$, $\left(\frac{P_G}{P}\right)_{large} \neq \left(\frac{P_G}{P}\right)_{small}$ or $P_{0,large} \neq P_{0,small}$.

Example 5.10

A fermentation process is carried out at commercial scale in a 10 m^3 vessel. The vessel is operated as a continuous fermenter and stirred with a 0.80 m agitator rotating at 90 rpm, giving a gassed power draw of 5.8 kW. In order to identify strategies to optimise the fermentation, it is desired to scale down the process to a 100 L pilot scale fermenter, geometrically similar to the one at commercial scale. It is desired to maintain the same dissolved oxygen concentration in the small reactor as in the large reactor. Assume that all the fermentation parameters (temperature, pH, residence time) remain constant with the scale down. Assume that the gas flow rate per unit volume, the ratio between the gassed and ungassed power draw and the power number remain constant on scale down. For this vessel the $k_L a$ is described by equation (5.22) with $a = 0.65$ and $b = 0.28$.

Calculate the required gassed power draw and agitation speed in the small reactor.

Solution

The ratio $\frac{V_{large}}{V_{small}}$ is 100. In this case equation (5.41) gives

$$\frac{P_{G,large}}{P_{G,small}} = (100)^{1-\frac{0.28}{3 \cdot 0.65}} = 51.5$$

The required gassed power draw in the small reactor is therefore

$$P_{G,small} = \frac{5.8 \text{ kW}}{51.5} = 0.11 \text{ kW}$$

The required agitation speed is given by equation (5.44), which in this case is

$$\frac{N_{large}}{N_{small}} = (100)^{-\frac{1}{9}\left(2+\frac{0.28}{0.65}\right)} = 0.29$$

The required agitation speed in the small reactor is therefore

$$N_{small} = \frac{N_{large}}{0.29} = 310 \text{ rpm}$$

QUESTIONS AND PROBLEMS

5.1 Explain which factors affect the oxygen transfer rate in agitated vessels.

5.2 An experiment has been carried out in an agitated vessel to measure the mass transfer coefficient $k_L a$, obtaining the graph in Figure 5.18. Calculate the rate of oxygen transfer in this vessel (kg m^{-3} min^{-1}), under the conditions of the experiment, if the equilibrium concentration of oxygen in water is 9.2 mg/L and the oxygen concentration in the fermentation medium is 2.1 mg/L.

5.3 A fermenter has a diameter of 2 m, an agitator diameter of 0.6 m and a liquid height of 2 m. Which are the diameter, agitator diameter and liquid height for a geometrically similar vessel having a working volume of 15 m^3?

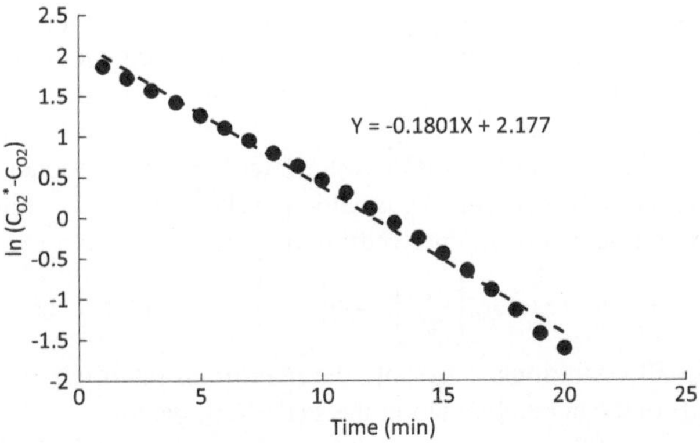

FIGURE 5.18 Plot for Problem 5.2. The dashed line represents the linear regression of the experimental data.

5.4 A fermenter has a diameter of 0.9 m and a volume of 1 m^3. The fermenter is aerated with an air flow rate of 250 m^3/d. The diameter of the agitator is 0.3 m and the agitator speed is 150 rpm. What agitation speed is required to have the same k_La in a geometrically similar vessel, having a volume of 10 m^3? The large fermenter is aerated with an air flow rate of 2500 m^3/d. Assume that the power number of the agitator is 4.0 for both vessels. Assume that the ratio P_G/P is the same for both vessels in all conditions.

For these vessels, the mass transfer coefficient is described by the correlation:

$$k_La = \alpha\left(\frac{P_G}{V}\right)^{0.62}\left(\frac{Q_G}{A}\right)^{0.34} \quad \text{with } P_G/V \text{ in W/m}^3 \text{ and } Q_G/A \text{ in m/s.}$$

5.5 The stoichiometry of a microorganism growing on glucose as carbon source can be described by the following equation:

$$C_6H_{12}O_6 + 3O_2 + 0.48NH_3 \rightarrow 0.48C_6H_{10}NO_3 + 4.32H_2O + 3.12CO_2$$

The microorganism grows in a continuous fermenter operated at a dilution rate of 0.5 d^{-1}. The glucose concentration is 40 g/L in the influent and 4 g/L in the effluent. The fermenter is aerated with air, and the oxygen concentration in equilibrium with air can be assumed to be 8 mg/L. The oxygen concentration in the fermentation broth is measured and it is equal to 2 mg/L.

Calculate the oxygen mass transfer coefficient (k_La) for this system. Ignore the oxygen concentration in the feed.

5.6 It is desired to grow a certain microorganism aerobically in a continuous fermenter (the microorganism will then be used to produce an enzyme in a subsequent process step). The substrate used for growing the microorganism is glucose ($C_6H_{12}O_6$) and the growth yield is known ($Y_{X/S}$). So far the microorganism has been grown in a continuous fermenter of volume V_1 and experience has shown that the maximum oxygen transfer rate that can be obtained in this vessel is $Q_{O2,max1}$. $Q_{O2,max1}$ is obtained when the solubility of oxygen is C_{O2sat} and with an oxygen concentration in the liquid equal to C_{O2min} (the minimum oxygen concentration for microorganisms growth).

It is now decided to scale-up the process to a volume V_2, maintaining geometric similarity, in order to increase the microorganism productivity. In scaling-up the process the superficial gas velocity is kept constant, while the power per unit volume is changed, due to limitations in the maximum power available in the scaled-up reactor: $\left(\frac{P_G}{V}\right)_2 = \gamma\left(\frac{P_G}{V}\right)_1$. Assume that the ratio P_G/P is the same in all conditions. Calculate the maximum microorganism productivity (in kg/d) that can be obtained in the scaled-up reactor.

Assumptions

- assume that the products of glucose fermentation are only microorganisms ($C_5H_7O_2N$), water and carbon dioxide. Assume that nitrogen is supplied in the feed as ammonia (NH_3).
- assume that for the considered vessels the mass transfer coefficient for oxygen in water (k_La) is given by $k_La = \alpha\left(\frac{P_G}{V}\right)^{0.7}\left(\frac{Q_G}{A}\right)^{0.2}$.
- in carrying out the mass balances for oxygen in the fermenter, ignore the contribution of the dissolved oxygen in the inlet and outlet streams.

Data

- $Q_{O2,max1}$ = 50 g/min (maximum oxygen transfer rate in reactor of volume V1)
- C_{O2sat} = 8 mg/L (oxygen concentration in water/ fermentation broth at equilibrium with air)
- C_{O2min} = 1 mg/L (minimum oxygen concentration to be maintained in the reactor)
- $Y_{X/S}$ = 0.3 g/g (biomass produced per unit of glucose consumed)
- γ = 0.5 (ratio of the power per unit volume in the two reactors)
- V_1 = 1 m^3 (volume of the small reactor)
- V_2 = 10 m^3 (volume of the scaled-up reactor)

5.7 It is desired to produce a certain microorganism at a high rate in a continuous process. The microorganism grows aerobically using acetic acid (CH_3COOH) as substrate and ammonia as nitrogen source and producing new microorganisms (assume an empirical formula $C_5H_7O_2N$), carbon dioxide and water. The fermenter will be operated with a substrate concentration in the feed equal to 20 g/L and at a dilution rate $D = 0.9\,\mu_{max}$ where μ_{max} is the maximum growth rate of the microorganism.

A batch test on the same microorganisms have produced the data in Figure 5.19 below. The batch test was carried out with an initial acetic acid

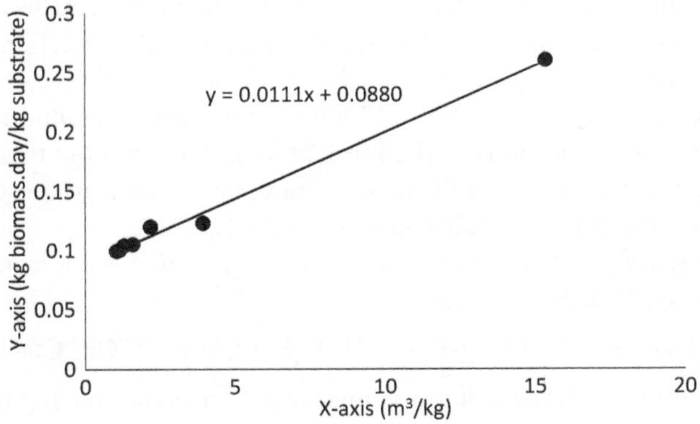

FIGURE 5.19 Plot for Problem 5.7.

concentration equal to 1 kg/m^3 and, at the end of the test when all the substrate was removed, the observed production of microorganisms was 0.3 kg/m^3.

In this graph the variable on the Y-axis is $\frac{X}{(-r_S)}$, where X is the biomass concentration and $(-r_S)$ is the substrate removal rate, and the variable on the X-axis is $\frac{1}{S}$ where S is the substrate concentration.

Which minimum value of k_La needs to be provided in the continuous fermenter, in order to maintain the oxygen concentration at least to 2 mg/L? Assume that air is used to provide oxygen in the fermenter. Assume that the solubility of oxygen in water under the conditions of the fermenter is 8.5 mg/L. Consider negligible the oxygen concentration in the feed.

5.8 Consider the fermentation reaction for the production of citric acid below:

$$C_6H_{12}O_6 + 2.35O_2 + 0.1NH_3 \rightarrow 0.1C_5H_7O_2N + 0.7C_6H_8O_7 + 1.3CO_2 + 3H_2O$$

Assume that the fermentation is carried out in batch and that each batch is inoculated with 100 mg/L of the microorganism, already acclimated to the substrate. Aeration is carried out by diffusers with a constant air flow rate. It is desired to maintain a constant oxygen concentration in the liquid phase of 4 mg/L throughout all the fermentation and this will be achieved by adjusting the stirring speed of the agitator.

Calculate the stirring speed that will be required after 4.5 h from the start of the fermentation.

Assumptions and data

- For the particular reactor where the fermentation takes place and for the applied air flow rate, the mass transfer coefficient (k_La) is related to the agitation power (P) by the equation $k_La = 0.020\ P^{0.8}$ and the agitation power relates to the stirring speed (N) according to the equation $P = 2.40 \cdot N^3$. In these equations k_La is in s^{-1}, P is in W and N is in s^{-1}.

- The microorganism follows Monod growth kinetics with $S >> K_S$.
- Under the conditions of the fermentation, the doubling time of the microorganism is 58 min and the oxygen concentration in the liquid phase in equilibrium with the air phase is 8.6 mg/L.

5.9 The reaction by which acetic acid bacteria produce acetic acid (CH_3COOH) from ethanol is the following:

$$C_2H_5OH + O_2 + 0.10NH_3 \rightarrow 0.10C_5H_7O_2N + 0.75CH_3COOH + 0.3H_2O$$

The reaction is carried out in a continuous fermenter having a volume of 10 m^3. Under the initial conditions, the process is operated with an ethanol concentration in the feed equal to 60 g/L and with a feed flow rate of 20 m^3/d. Under these conditions, the acetic acid and dissolved oxygen concentration in the reactor are 48 g/L and 4.2 mg/L, respectively. It is now desired to increase the acetic acid production rate to 1500 kg/d by increasing the ethanol concentration in the feed, leaving any other operating parameters of the reactor unchanged.

(a) Calculate the required ethanol concentration in the feed under the new conditions.

(b) The minimum oxygen concentration, which needs to be maintained in the reactor is 1.5 mg/L, otherwise growth of the microorganisms will stop. Determine whether it is possible to carry out the process under the new desired conditions, without making any changes to the aeration and agitation system.

Note: ignore any changes in volume due to the water produced by the reaction and any effects of the microorganisms' concentration on the mass transfer coefficient. Assume that the saturation concentration for oxygen in water is in all conditions 8.0 mg/L.

5.10 An aerobic fermentation process is carried out as a continuous process. The reactor has a volume of 5 m^3 and a cylindrical shape. The agitator rotates at 180 rpm. In this reactor the oxygen concentration in the fermentation medium is 4.0 mg/L.

It is desired to scale-up the process in a 25 m^3 vessel, geometrically similar to the smaller vessel. In the scale-up, the concentration of the feed and the residence time are kept constant. Furthermore, the air flow rate per unit volume of reactor is also kept constant. To save energy, in the large vessel it is desired to maintain the oxygen concentration at 2.0 mg/L. It is expected that the metabolism of the microorganisms will not be affected by the change in oxygen concentration.

Calculate the agitator speed which is required in the larger vessel.

The power number of the agitator ($P_0 = \dfrac{P}{\rho_L N^3 D_{ag}^5}$) can be assumed to be the same in all fermentation conditions. Assume that the density of the liquid phase is the same in all conditions. For these vessels, the mass transfer coefficient is described by the following correlation:

$$k_L a = k\left(\frac{P_G}{V}\right)^{0.75}\left(\frac{Q_G}{A}\right)^{0.45}$$

Assume that the ratio P_G/P is the same in all conditions and that the equilibrium concentration of oxygen in water is $C^*_{O2} = 8.0$ mg/L in all conditions. In the oxygen balance ignore the contribution of the oxygen in the feed and in the outlet stream.

5.11 A fermentation is carried out in a lab-scale continuous process having a volume of 5 L, with an agitator running at 120 rpm. For vessels of this geometry, the mass transfer coefficient for oxygen in water is given by the following correlation:

$$k_L a = 0.03\left(\frac{P_G}{V}\right)^{0.69}\left(\frac{Q_G}{A}\right)^{0.39}$$

In this correlation, $k_L a$ is in s^{-1}, P_G in W, V in m^3, Q_G in m^3/s and A in m^2. Assume that in all conditions $P_G/P = 0.43$ and $P_0 = 3.6$.

It is now desired to scale-up the process using a 30 L reactor, to be operated as a continuous fermenter with the same feed and residence time as the lab-scale reactor. In the scale-up process, it is desired to maintain the superficial gas velocity at the same value as for the lab-scale reactor. Calculate the agitation rate that is required in the large reactor to maintain the same dissolved oxygen as in the lab-scale reactor. Assume that the oxygen saturation concentration will stay the same in the large and small reactors.

5.12 Under certain agitation and gas flow conditions, the mass transfer rate for oxygen in a fermenter at 25 °C is equal to 0.14 kg/m^3.s, with a dissolved oxygen concentration of 2.5 mg/L. Calculate the mass transfer rate for oxygen in the same vessel when the temperature is 35 °C (physical properties in Table 5.2), for the same agitation conditions, mass flow of the inlet air and dissolved oxygen concentration. Assume that the parameters a and k_l are described by equations (5.19) and (5.20), respectively. Assume that the interfacial tension, the liquid density (1000 kg/m^3), the superficial gas velocity and the terminal gas velocity do not depend on temperature. Assume that the diameter of gas bubbles is in all conditions equal to $3.0 \cdot 10^{-3}$ m.

TABLE 5.2
Physical Properties for Problem 5.12

Property	25 °C	35 °C
C^*_{O2} (mg/L)	9.3	7.5
Diffusivity of oxygen in water (cm²/s)	$2.2 \cdot 10^{-5}$	$3.0 \cdot 10^{-5}$
Viscosity of water (Pa·s)	$8.90 \cdot 10^{-4}$	$7.20 \cdot 10^{-4}$

5.13 A fermentation reaction is characterised by the following values of the parameters: $Y_{X/S}$ = 0.10 kg/kg, $Y_{X/O2}$ = 0.18 kg/kg, $Y_{P/X}$ = 5.20 kg/kg. It is desired to carry out this fermentation in a 2 m³ continuous fermenter with a residence time of 8 h, maintaining the oxygen concentration at at least 2.0 mg/L. The maximum k_La that is possible to achieve in this vessel is 0.087 s^{-1} and the saturation concentration for oxygen is 9.0 mg/L. Calculate the maximum product productivity (in kg/d) that it is possible to achieve in this process and the maximum substrate concentration that we can have in the feed, assuming that the effluent substrate concentration is negligible compared to the one in the feed.

5.14 A continuous fermentation process is operated with a residence time of 12 h. Under these conditions, the substrate in the feed is converted practically completely. The dissolved oxygen concentration is 3.5 mg/L and the biomass concentration is 4.0 kg/m³. In order to increase the productivity of the process, it is proposed to decrease the residence time to 10 h, still expecting total conversion of the substrate. Determine whether it is possible to decrease the residence time to 10 h, whilst maintaining a dissolved oxygen concentration of at least 1.5, without any changes to the agitation rate and to the gas flow rate. Assume for the parameter $Y_{X/O2}$ a value of 0.25 kg/kg and for the oxygen saturation in water a value of 9.2 mg/L.

5.15 Assuming that we only have one available fermenter of a given volume, describe a series of experiments aimed at measuring the parameters of k, a and b of equation (5.22).

5.16 Consider equation (5.36) for the required k_La in a continuous process. Derive modified equations for a microbial metabolism based on the Monod model with (a) maintenance and (b) endogenous metabolism.

5.17 A fermentation process is carried out in batch in a 3 m³ reactor. The kinetic parameters of the fermentation are μ_{max} = 4.8 d^{-1}, $Y_{X/S}$ = 0.20 kg/kg; $Y_{X/O2}$ = 0.25 kg/kg. The batch is started with a substrate concentration of 20 kg/m³ and inoculated with biomass concentration of 0.1 kg/m³. The fermentation is stopped when the substrate concentration reaches 0.2 kg/m³ and it can be assumed that, during all the fermentation, $S >> K_S$. It is desired to maintain the oxygen concentration to 3.0 mg/L throughout all the length of the batch by manipulating the air flow rate while keeping the agitator power draw constant. The oxygen saturation is 9.0 mg/L.

Derive and plot an equation for the time profile of the air flow rate during this batch process.

The k_La for this reactor is given by the following correlation:

$$k_La = 0.045\left(\frac{P_G}{V}\right)^{0.60}\left(\frac{Q_G}{A}\right)^{0.42}$$

In this correlation, k_La is in s^{-1}, P_G in W, V in m^3, Q_G in m^3/s and A in m^2. Assume that in all conditions $P_G = 1.0$ kW. Assume $A = 2.8\ m^2$.

5.18 A fermentation process is carried out in a continuous process having an agitator of 1.3 m diameter. Under the standard process conditions the agitation speed is 180 rpm and the dissolved oxygen concentration in the reactor is 4.0 mg/L. To save energy, it is desired to reduce the oxygen concentration to 2.0 mg/L by reducing the agitation speed, without changing the gas flow rate or any other fermentation parameters. It is believed that the performance of the microorganisms will not be affected by the lower oxygen concentration.

Calculate the new required agitation rate and the energy saved over one year (assume the reactor operates for 8000 h/year). The mass transfer coefficient is described by the following correlation:

$$k_La = 0.025\left(\frac{P_G}{V}\right)^{0.70}\left(\frac{Q_G}{A}\right)^{0.26}$$

In this correlation, k_La is in s^{-1}, P_G in W, V in m^3, Q_G in m^3/s and A in m^2. Assume that the ratio $P_G/P = 0.43$ and $P_0 = 4.2$ in all conditions. In the oxygen balance, ignore the contributions of the mass flow of dissolved oxygen with the influent and outlet streams. Assume the saturation concentration for oxygen in water is 8.5 mg/L and the parameter $Y_{X/O2}$ is 0.15 kg/kg.

5.19 A continuous fermenter with a working volume of 30 m^3 is aerated with an air flow rate of 3000 m^3/h. The inlet air (composed of 21% oxygen by volume) is at 25 °C and 1 atm. A fermentation process is carried out characterised by a dilution rate of 1.9 d^{-1} and a biomass concentration of 3.5 kg/m^3. For this process, $Y_{X/O2} = 0.29$ kg/kg. Calculate the efficiency of oxygen transfer in this process. In the oxygen balance, ignore the contributions of the mass flow of dissolved oxygen with the influent and outlet streams. Assume ideal gas law for the inlet stream with $R = 0.0821$ L atm mol^{-1} K^{-1}.

5.20 A fermentation process is characterised by the following kinetic parameters: $\mu_{max} = 5.8\ d^{-1}$; $K_S = 0.08\ kg/m^3$; $Y_{X/S} = 0.15$ kg/kg; $Y_{X/O2} = 0.22$ kg/kg; $Y_{P/S} = 0.40$ kg/kg. It is desired to run the process in a continuous fermenter having a working volume of 10 m^3, with an influent substrate concentration of 45 kg/m^3 and an effluent substrate concentration not higher than 0.1 kg/m^3. Calculate the product productivity (kg/d) and the k_La required for this process, ignoring the contributions of the oxygen mass flow in the inlet and outlet streams. Assume that the oxygen saturation is 9 mg/L and the desired oxygen concentration in the reactor is 2 mg/L.

6 Fermentation Processes
Heat Generation and Heat Balances

Microorganisms only work in a certain temperature range and have an optimum temperature for their metabolism. Most *Saccharomyces cerevisiae* strains used for bioethanol production have an optimum temperature in the range 30–35 °C, while the optimum temperature for citric acid production with *Aspergillus niger* and other fungi is in the range 25–30 °C. Anaerobic digestion processes, which use mixed microbial culture to convert organic waste and biomass into methane, are usually operated in the temperature range 35–40 °C.

If the fermentation temperature goes outside the optimum range, i.e. below the minimum or above the maximum allowable temperature, the performance of the process will be severely compromised. It is therefore important to make sure that the temperature of the fermentation remains in the right range.

6.1 ENTHALPY OF REACTION

Like any other chemical reactions, fermentation reactions have heat effects. Fermentation reactions may generate or absorb heat. For a generic chemical reaction:

$$A + \beta B \rightarrow \gamma C + \delta D \tag{6.1}$$

The heat of reaction is defined as

$$\Delta H_r \left(\frac{\text{J}}{\text{mol}}\right) = \gamma H_C + \delta H_D - H_A - \beta H_B \tag{6.2}$$

In equation (6.2) the term H_i represents the enthalpies of the various chemical species, calculated according to the general equation (6.3):

$$H_i(T)\left(\frac{\text{J}}{\text{mol}}\right) = \Delta H_{fi}^0 + \left[\lambda_{phase\ change\left(\frac{and}{or\ mixing}\right)}(25°C)\right] + \int_{25}^{T} c_{Pi}\, dT \tag{6.3}$$

DOI: 10.1201/9781003633822-6

In equation (6.3), the term ΔH^0_{fi} is the enthalpy of formation of species i in its reference state at 25 °C, λ is the enthalpy associated to the phase change from the reference state to the final state and/or to the dissolution of the substance, and c_p is the specific heat of the substance in its final state.

If at the temperature T the substance is still in its reference state, if we ignore any heat of mixing and if the specific heat is constant in the temperature range 25 °C-T, then equation (6.3) can be simplified into

$$H_i(T)\left(\frac{\text{J}}{\text{mol}}\right) = \Delta H^0_{fi} + c_{Pi}(T-25) \tag{6.4}$$

The standard enthalpy of reaction, ΔH^0_r, is the enthalpy of the reaction when all the species are in their reference state at 25 °C, equation (6.5):

$$\Delta H^0_r\left(\frac{\text{J}}{\text{mol}}\right) = \gamma\Delta H^0_{fC} + \delta\Delta H^0_{fD} - \Delta H^0_{fA} - \beta\Delta H^0_{fB} \tag{6.5}$$

Note that for the generic reaction (6.1) the enthalpy of reaction ΔH_r (J/mol) is expressed per mol of reagent A converted. From the enthalpy of reaction, the rate of heat generation per unit volume of reactor can be calculated:

$$r_H\left(\frac{\text{J}}{\text{m}^3\text{d}}\right) = (-r_A)(-\Delta H_r) \tag{6.6}$$

where r_A is the rate of reaction referred to component A, as mol m^{-3} s^{-1}. The rate of heat generation can be expressed as a function of the production/consumption rate of other reactants if the stoichiometry of the reaction is known. For example if the rate of production of component C is known (r_C, mol m^{-3} s^{-1}) and the reaction stoichiometry is equation (6.1), then the rate of heat generation is

$$r_H\left(\frac{\text{J}}{\text{m}^3\,\text{d}}\right) = \frac{r_C}{\gamma}(-\Delta H_r) \tag{6.7}$$

Example 6.1

Consider the fermentation stoichiometry for citric acid production (properties in Table 6.1):

$$C_6H_{12}O_6 + 2.35O_2 + 0.1NH_3 \rightarrow 0.1C_5H_7O_2N + 0.7C_6H_8O_7 + 1.3CO_2 + 3H_2O$$

(a) Calculate the enthalpy of the reaction at 25 °C and with each substance in its reference state.

TABLE 6.1
Physical Properties and Enthalpies for Example 6.1

Substance	Reference State	ΔH_f^0 (J/kg)	Enthalpy of Phase Change (any T)	Specific Heat (J kg^{-1} K^{-1})
Glucose	solid	$-7.074 \cdot 10^6$	(solid to dissolved in water) $6.11 \cdot 10^4$	(solid) $1.22 \cdot 10^3$
Oxygen	gas	0	(gas to dissolved in water) $-3.656 \cdot 10^5$	(gas) $9.10 \cdot 10^2$
Ammonia	gas	$-2.717 \cdot 10^6$	(gas to dissolved in water) $-1.794 \cdot 10^6$	(gas) $1.95 \cdot 10^2$
Biomass	Solid	$-8.000 \cdot 10^6$	–	(solid) $3.00 \cdot 10^3$
Citric acid	solid	$-1.021 \cdot 10^7$	(solid to dissolved in water) $9.484 \cdot 10^4$	(solid) $1.18 \cdot 10^3$
Carbon dioxide	gas	$-9.000 \cdot 10^6$	–	(gas) $8.71 \cdot 10^2$
Water	liquid	$-1.600 \cdot 10^7$	–	(liquid) $4.19 \cdot 10^3$

(b) Calculate the actual enthalpy of reaction under the fermentation conditions (35 °C), where glucose, oxygen, ammonia, citric acid and water are in the liquid phase, microorganisms are solid and carbon dioxide is a gas.

Solution
(a) Standard enthalpy of formation:

$$\begin{aligned}
\Delta H_r^0 &= 0.1\Delta H_{fC_5H_7O_2N}(solid, 25\,°\mathrm{C}) + 0.7\Delta H_{fC_6H_8O_7}(solid, 25\,°\mathrm{C}) \\
&\quad + 1.3\Delta H_{fCO_2}(gas, 25\ °\mathrm{C}) + 3\Delta H_{fH_2O}(water, 25\ °\mathrm{C}) \\
&\quad - \Delta H_{fC_6H_{12}O_6}(solid, 25\ °\mathrm{C}) \\
&\quad - 2.35\Delta H_{fO_2}(gas, 25\,°\mathrm{C})0.1\Delta H_{fNH_3}(gas, 25\ °\mathrm{C}) \\
&= 0.1\left(-8.000 \cdot 10^6 \frac{\mathrm{J}}{\mathrm{kg}}\right)113\frac{\mathrm{kg}}{\mathrm{kmol}} \\
&\quad + 0.7\left(-1.021 \cdot 10^7 \frac{\mathrm{J}}{\mathrm{kg}}\right)192\frac{\mathrm{kg}}{\mathrm{kmol}} \\
&\quad + 1.3\left(-9.000 \cdot 10^6 \frac{\mathrm{J}}{\mathrm{kg}}\right)44\frac{\mathrm{kg}}{\mathrm{kmol}} \\
&\quad + 3\left(-1.6 \cdot 10^7 \frac{\mathrm{J}}{\mathrm{kg}}\right)18\frac{\mathrm{kg}}{\mathrm{kmol}} \\
&\quad - \left(-7.074 \cdot 10^6 \frac{\mathrm{J}}{\mathrm{kg}}\right)180\frac{\mathrm{kg}}{\mathrm{kmol}} - 0 - 0.1\left(-2.717 \cdot 10^6 \frac{\mathrm{J}}{\mathrm{kg}}\right)17\frac{\mathrm{kg}}{\mathrm{kmol}} \\
&= -1.563 \cdot 10^9 \frac{\mathrm{J}}{\mathrm{kmol}}
\end{aligned}$$

(b) Under the actual fermentation conditions:

$$\begin{aligned}\Delta H_r = &0.1H_{C_5H_7O_2N}(solid, 35°C) + 0.7H_{C_6H_8O_7}(liquid, 35°C)\\ &+ H_{CO_2}(gas, 35°C) + 3H_{H_2O}(liquid, 35°C)\\ &- H_{C_6H_{12}O_6}(liquid, 35°C) - 2.35H_{O_2}(gas, 35°C)\\ &- 0.1H_{NH3}(liquid, 35°C)\end{aligned}$$

$$\begin{aligned}H_{C_5H_7O_2N}(solid, 35°C) &= -8.000 \cdot 10^6 \frac{J}{kg} \cdot 113 \frac{kg}{kmol}\\ &+ 3.00 \cdot 10^3 \frac{J}{kg\ °C} \cdot 113 \frac{kg}{kmol}(35-25)^o C\\ &= -9.006 \cdot 10^8 \frac{J}{kmol}\end{aligned}$$

$$\begin{aligned}H_{C_6H_8O_7}(liquid, 35°C) &= -1.021 \cdot 10^7 \frac{J}{kg} \cdot 192 \frac{kg}{kmol} + 9.484 \cdot 10^4 \frac{J}{kg} \cdot 192 \frac{kg}{kmol}\\ &+ 1.18 \cdot 10^3 \frac{J}{kg^oC} \cdot 192 \frac{kg}{kmol}(35-25)^o C\\ &= -1.938 \cdot 10^9 \frac{J}{kmol}\end{aligned}$$

$$\begin{aligned}H_{CO_2}(gas, 35°C) &= -9.000 \cdot 10^6 \frac{J}{kg} \cdot 44 \frac{kg}{kmol}\\ &+ 8.71 \cdot 10^2 \frac{J}{kg°C} \cdot 44 \frac{kg}{kmol}(35-25)°C\\ &= -3.956 \cdot 10^8 \frac{J}{kmol}\end{aligned}$$

$$\begin{aligned}H_{H_2O}(liquid, 35°C) &= -1.600 \cdot 10^7 \frac{J}{kg} \cdot 18 \frac{kg}{kmol}\\ &+ 4.19 \cdot 10^3 \frac{J}{kg°C} \cdot 18 \frac{kg}{kmol}(35-25)°C\\ &= -2.872 \cdot 10^8 \frac{J}{kmol}\end{aligned}$$

$$\begin{aligned}H_{C_6H_{12}O_6}(dissolved, 35°C) &= -7.074 \cdot 10^6 \frac{J}{kg} \cdot 180 \frac{kg}{kmol} + 6.11 \cdot 10^4 \frac{J}{kg} \cdot 180 \frac{kg}{kmol}\\ &+ 1.22 \cdot 10^3 \frac{J}{kg°C} \cdot 180 \frac{kg}{kmol}(35-25)°C\\ &= -1.240 \cdot 10^9 \frac{J}{kmol}\end{aligned}$$

$$H_{O_2}\left(dissolved, 35°C\right) = 0 + 3.656 \cdot 10^5 \frac{J}{kg} \cdot 32 \frac{kg}{kmol}$$
$$+ 9.10 \cdot 10^2 \frac{J}{kg°C} \cdot 32 \frac{kg}{kmol}(35-25)°C$$
$$= -1.199 \cdot 10^7 \frac{J}{kmol}$$

$$H_{NH_3}\left(dissolved, 35°C\right) = -2.717 \cdot 10^6 \frac{J}{kg} \cdot 17 \frac{kg}{kmol} - 1.794 \cdot 10^6 \frac{J}{kg} \cdot 17 \frac{kg}{kmol}$$
$$+ 1.95 \cdot 10^2 \frac{J}{kg°C} \cdot 17 \frac{kg}{kmol}(35-25)°C$$
$$= -7.665 \cdot 10^7 \frac{J}{kmol}$$

Therefore,

$$\Delta H_r = -1.428 \cdot 10^9 \frac{J}{kmol}$$

Example 6.2

Consider the fermentation reaction of Example 6.1, with $\Delta H_r^0 = -1.563 \cdot 10^6 \frac{J}{mol}$. Assuming that all the reagents and products are in their reference state and at the reference temperature, what is the rate of heat generation when the rate of citric acid ($C_6H_8O_7$) production is 10 kg m^{-3} d^{-1}?

Solution

From the reaction stoichiometry we produce 0.75 kg of citric acid per kg of glucose converted. To produce 10 kg m^{-3} d^{-1} of citric acid we need to convert 10/0.75 = 13.33 kg of glucose m^{-3} d^{-1}. In mol terms, this is 74.07 mol of glucose m^{-3} d^{-1} and the rate of heat generation is

$$r_H\left(\frac{J}{m^3\,d}\right) = 74.07 \frac{\text{mol glucose}}{m^3\,d}\left(1.563 \cdot 10^6 \frac{J}{\text{mol glucose}}\right) = 1.158 \cdot 10^8 \frac{J}{m^3\,d}$$

6.2 RATE OF HEAT GENERATION OF FERMENTATION REACTIONS

In fermentation reactions, it is often useful to express the rate of heat generation as a function of the rate of biomass growth. To link the rate of heat generation with the rate of biomass growth, we introduce the coefficient Y_H, defined as

$$Y_H\left(\frac{\text{kg}}{\text{J}}\right)=\frac{\text{Biomass produced}}{\text{Heat generated}} \tag{6.8}$$

Using the coefficient Y_H, the rate of heat generation r_H can be expressed as

$$r_H\left(\frac{\text{J}}{\text{m}^3\,\text{d}}\right)=\frac{r_X}{Y_H}\frac{\left(\dfrac{\text{kg biomass}}{\text{m}^3\,\text{d}}\right)}{\left(\dfrac{\text{kg biomass}}{\text{J}}\right)}=\frac{\mu\cdot X}{Y_H} \tag{6.9}$$

Note that according to these definitions, Y_H and r_H are positive for exothermic reactions, which generate heat and have a negative enthalpy of reaction ΔH_r. Most fermentations are exothermic. The coefficient Y_H can be calculated from the enthalpy of the fermentation reaction. If the enthalpy of the fermentation reaction is not known, Y_H can still be calculated if the enthalpy of combustion of the substrate and of the microorganisms are known, and if the growth yield $Y_{X/S}$ is also known. The calculation of the coefficient Y_H from the enthalpies of combustion of the substrate and biomass is shown with reference to the fermentation stoichiometry (6.10), as an example, but the equations obtained have general validity:

$$C_6H_{12}O_6+2.55O_2+0.60NH_3\rightarrow 0.60C_5H_7O_2N+3CO_2+3.9H_2O \tag{6.10}$$

Using equation (6.10), the enthalpy of the reactants in their reference state at 25 °C is (the enthalpy of formation of molecular oxygen is zero, as it is an element in its reference state):

$$H_{reactants}=0.60H_{NH3}+H_{glucose} \tag{6.11}$$

And the enthalpy of the products is

$$H_{products}=0.60H_{biom}+3H_{CO2}+3.9H_{H2O} \tag{6.12}$$

Since we are calculating the enthalpy of reaction with the substances in their reference state at 25 °C, the enthalpy of each species will be equal to their enthalpy of formation. The enthalpy of the reaction (6.10) can be written therefore as:

$$\Delta H_R = 0.60\Delta H_{f,biom} + 3\Delta H_{f,CO2} + 3.9\Delta H_{f,H2O} - 0.60\Delta H_{f,NH3} - \Delta H_{f,glucose} \quad (6.13)$$

Equation (6.13) can be re-written by introducing the combustion reaction for the substrate (glucose) and for the biomass.

For glucose, the combustion reaction is

$$C_6H_{12}O_6 + 6O_2 \rightarrow 6CO_2 + 6H_2O \quad (6.14)$$

The enthalpy of reaction (6.14) at 25 °C is

$$\Delta H_{comb,glucose}\left(\frac{\text{J}}{\text{mol glucose}}\right) = 6\Delta H_{f,CO_2} + 6\Delta H_{f,H_2O} - \Delta H_{f,glucose} \quad (6.15)$$

Similarly for the biomass, its combustion reaction is

$$C_5H_7O_2N + 5O_2 \rightarrow 5CO_2 + 2H_2O + NH_3 \quad (6.16)$$

The enthalpy of reaction (6.16) at 25 °C is

$$\Delta H_{comb,biom}\left(\frac{\text{J}}{\text{mol biom}}\right) = 5\Delta H_{f,CO_2} + 2\Delta H_{f,H_2O} + \Delta H_{f,NH_3} - \Delta H_{f,biom} \quad (6.17)$$

We can now eliminate the enthalpies of formation of glucose and biomass from equation (6.13) using equations (6.15) and (6.17). We obtain

$$\Delta H_R\left(\frac{\text{J}}{\text{mol}}\right) = \Delta H_{comb,glucose} - 0.60\Delta H_{comb,biom} \quad (6.18)$$

Equation (6.18) shows that the enthalpy of a fermentation reaction can be calculated from the enthalpy of combustion of the substrate and of the biomass.

For the fermentation reaction (6.10), the coefficient Y_H can be calculated as

$$Y_H\left(\frac{\text{kg}}{\text{J}}\right) = \frac{\text{Biomass produced}}{\text{Heat generated}} = \frac{\dfrac{\text{Biomass produced}}{\text{mol glucose}}}{\dfrac{\text{Heat generated}}{\text{mol glucose}}} = \frac{0.60MW_{biomass}}{\left(-\Delta H_R\right)} \quad (6.19)$$

The coefficient Y_H can be expressed as a function of the enthalpies of combustion of the substrate and biomass by introducing equation (6.18) into equation (6.19):

$$Y_H = \frac{0.60 MW_{biom}}{\left(-\Delta H_{comb,glucose}\right) - 0.60\left(-\Delta H_{comb,biom}\right)} \tag{6.20}$$

In equation (6.20) the minus sign has been used before the terms $\Delta H_{comb,glucose}$ and $\Delta H_{comb,biom}$ because these enthalpies are negative but Y_H needs to be positive by definition. Equation (6.20) shows the expression for the calculation of the Y_H for the reaction stoichiometry (6.10). However, equation (6.20) can be made more general by substituting the factor 0.60 with the yield coefficient $Y_{X/S}$. Indeed, $Y_{X/S}$ can be expressed as

$$Y_{X/S}\left(\frac{kg\ biomass}{kg\ substrate}\right) = \frac{0.60 MW_{biom}}{MW_{glucose}} \tag{6.21}$$

Combining equation (6.21) with equation (6.20) we obtain

$$Y_H = \frac{Y_{X/S}}{\frac{\left(-\Delta H_{comb,glucose}\right)}{MW_{glucose}} - \frac{Y_{X/S}\cdot\left(-\Delta H_{comb,biom}\right)}{MW_{biom}}} \tag{6.22}$$

Equation (6.22) gives the general expression for the calculation of Y_H once $Y_{X/S}$ and the enthalpies of combustion of biomass and substrate are known. Note that so far the enthalpies of all the reaction have been expressed in J/mol. However, the terms $\frac{\left(-\Delta H_{comb,biom}\right)}{MW_{glucose}}$ and $\frac{\left(-\Delta H_{comb,biom}\right)}{MW_{biom}}$ correspond to the enthalpies of combustion of the substrate and biomass expressed in J/g. Equation (6.22) is therefore equivalent to equation (6.23):

$$Y_H = \frac{Y_{X/S}}{\left(-\Delta H_{comb,glucose}\right)\left(\frac{\mathrm{J}}{\mathrm{g}}\right) - Y_{X/S}\cdot\left(-\Delta H_{comb,biom}\right)\left(\frac{\mathrm{J}}{\mathrm{g}}\right)} \tag{6.23}$$

In summary, the coefficient Y_H, which allows the calculation of the rate of heat generation per unit of biomass produced, can be calculated from the growth yield and from the enthalpies of combustion of the substrate and biomass. As stated above, Y_H can also be calculated from the enthalpy of the fermentation reaction.

Equation (6.23) can be extended to anaerobic reactions with other products in addition to microorganisms. For example, let us consider the anaerobic fermentation of glucose for the production of ethanol:

$$C_6H_{12}O_6 + 0.31NH_3 \rightarrow 0.31C_5H_7O_2N + 1.64C_2H_6O + 1.17CO_2 + 0.464H_2O \quad (6.24)$$

The enthalpy of this reaction is

$$\begin{aligned}\Delta H_R = 0.31\Delta H_{f,biom} + 1.64\Delta H_{f,C_2H_6O} + 1.17\Delta H_{f,CO_2} \\ + 0.464\Delta H_{f,H_2O} - 0.31\Delta H_{f,NH_3} - \Delta H_{f,glucose}\end{aligned} \quad (6.25)$$

The combustion of ethanol is described by the following reaction:

$$C_2H_6O + 3O_2 \rightarrow 2CO_2 + 3H_2O \quad (6.26)$$

With enthalpy of combustion of

$$\Delta H_{comb,C_2H_6O}\left(\frac{J}{mol\ ethanol}\right) = 2\Delta H_{f,CO_2} + 3\Delta H_{f,H_2O} - \Delta H_{f,C_2H_6O} \quad (6.27)$$

Using equations (6.27), (6.17) and (6.15), the enthalpy of the reaction (6.24) can be written as

$$\Delta H_R\left(\frac{\text{J}}{\text{mol}}\right) = \Delta H_{comb,glucose} - 0.31\Delta H_{comb,biom} - 1.64\Delta H_{comb,ethanol} \quad (6.28)$$

The coefficient Y_H in this case is given by

$$Y_H = \frac{0.31MW_{biom}}{\left(-\Delta H_{comb,,glucose}\right) - 0.31\left(-\Delta H_{comb,biom}\right) - 1.64\left(-\Delta H_{comb,ethanol}\right)} \quad (6.29)$$

The definitions of $Y_{X/S}$ and $Y_{P/S}$ are, in this case

$$Y_{X/S}\left(\frac{\text{kg biomass}}{\text{kg substrate}}\right) = \frac{0.31MW_{biom}}{MW_{glucose}} \quad (6.30)$$

$$Y_{P/S}\left(\frac{\text{kg biomass}}{\text{kg substrate}}\right) = \frac{1.64MW_{ethanol}}{MW_{glucose}} \quad (6.31)$$

Using equations (6.30) and (6.31) the coefficient Y_H can be written as

$$Y_H = \frac{Y_{X/S}}{\left(-\Delta H_{comb,glucose}\right)\left(\frac{\text{J}}{\text{g}}\right) - Y_{X/S}\cdot\left(-\Delta H_{comb,biom}\right)\left(\frac{\text{J}}{\text{g}}\right) - Y_{P/S}\cdot\left(-\Delta H_{comb,ethanol}\right)\left(\frac{\text{J}}{\text{g}}\right)} \tag{6.32}$$

Equation (6.32) can be written in a more general form:

$$Y_H = \frac{Y_{X/S}}{\left(-\Delta H_{comb,substrate}\right)\left(\frac{\text{J}}{\text{g}}\right) - Y_{X/S}\cdot\left(-\Delta H_{comb,biom}\right)\left(\frac{\text{J}}{\text{g}}\right) - \Sigma_i Y_{Pi/S}\cdot\left(-\Delta H_{comb,product\ i}\right)\left(\frac{\text{J}}{\text{g}}\right)} \tag{6.33}$$

In equation (6.33), $\Delta H_{comb,product\ i}$ accounts for the enthalpy of combustion of all the products of the fermentation reaction except biomass (which is included separately in equation (6.33)), carbon dioxide and water (which cannot be combusted). Using the general equation (6.33), the rate of heat generation in a fermentation reaction can be written as

$$\begin{aligned} r_H\left(\frac{\text{J}}{\text{m}^3\text{d}}\right) &= \frac{r_X}{Y_H} \\ &= \frac{r_X}{Y_{X/S}}\left[\left(-\Delta H_{comb,substrate}\right)\left(\frac{\text{J}}{\text{g}}\right) - Y_{X/S}\cdot\left(-\Delta H_{comb,biom}\right)\left(\frac{\text{J}}{\text{g}}\right) - \Sigma_i Y_{Pi/S}\cdot\left(-\Delta H_{comb,product\ i}\right)\left(\frac{\text{J}}{\text{g}}\right)\right] \\ &= \left(-r_S\right)\left[\left(-\Delta H_{comb,substrate}\right)\left(\frac{\text{J}}{\text{g}}\right) - Y_{X/S}\cdot\left(-\Delta H_{comb,biom}\right)\left(\frac{\text{J}}{\text{g}}\right) - \Sigma_i Y_{Pi/S}\cdot\left(-\Delta H_{comb,product\ i}\right)\left(\frac{\text{J}}{\text{g}}\right)\right] \end{aligned} \tag{6.34}$$

Equation (6.34) shows that fermentation reactions, which mainly oxidise the substrate with low generation of microorganisms and products (other than carbon dioxide and water), generate higher heat than reactions with high generation of microorganisms and products. Typically, even though not exclusively, reactions with product formation are anaerobic reactions, and for these reactions the heat generated per unit of substrate removed is much lower than for aerobic reactions.

Example 6.3

Calculate the coefficient Y_H (kg/J), the enthalpy of the reaction (J/mol) and the rate of heat generation (J m^{-3} h^{-1}) when the growth rate is 1 kg/m^3.h for the following reactions:

(a) aerobic conversion of glucose to microorganisms, water and carbon dioxide, with the production of 0.38 kg of microorganisms per 1 kg of glucose removed;

(b) anaerobic conversion of glucose to ethanol, microorganisms, water and carbon dioxide, with the production of 0.19 kg of microorganisms and 0.42 kg of ethanol per 1 kg of glucose removed.

Standard enthalpies of combustion: glucose = $-1.556\cdot10^7$ J/kg; biomass=-$1.503\cdot10^7$ J/kg; ethanol = $-2.957\cdot10^7$ J/kg

(a) From equation (6.23),

$$Y_H = \frac{0.38\dfrac{kg}{kg}}{1.556\cdot10^7\dfrac{J}{kg} - 0.38\cdot1.503\cdot10^7\dfrac{J}{kg}} = 0.386\cdot10^{-7}\frac{kg}{J}$$

If $Y_{X/S} = 0.38$ kg/kg, the stoichiometric coefficient for biomass in the fermentation reaction is 0.60, so, using equation (6.18):

$$\begin{aligned}\Delta H_R\left(\frac{J}{mol}\right) &= \Delta H_{comb,glucose} - 0.60\Delta H_{comb,biom}\\ &= -1.556\cdot10^7\frac{J}{kg}\cdot0.180\frac{kg}{mol} - 0.60\left(-1.503\cdot10^7\frac{J}{kg}\cdot0.113\frac{kg}{mol}\right)\\ &= -1.78\cdot10^6\frac{J}{mol}\end{aligned}$$

The rate of heat generation is given by equation (6.9):

$$r_H\left(\frac{J}{m^3\cdot day}\right) = \frac{1\dfrac{kg\ biomass}{m^3\cdot d}}{0.386\cdot10^{-7}\dfrac{kg}{J}} = 2.59\cdot10^7\frac{J}{m^3\ d}$$

(b) In this case $Y_{X/S} = 0.19$ kg/kg and $Y_{P/S} = 0.42$ kg/kg.
From equation (6.32),

$$\begin{aligned}Y_H &= \frac{0.19\dfrac{kg}{kg}}{1.556\cdot10^7\dfrac{J}{kg} - 0.19\cdot1.503\cdot10^7\dfrac{J}{kg} - 0.42\cdot2.957\cdot10^7\dfrac{J}{kg}}\\ &= 6.67\cdot10^{-7}\frac{kg}{J}\end{aligned}$$

From equation (6.28),

$$\Delta H_R\left(\frac{\mathrm{J}}{\mathrm{mol}}\right)=\Delta H_{comb,glucose}-0.31\Delta H_{comb,biom}-1.64\Delta H_{comb,ethanol}$$
$$=-1.556\cdot 10^7\,\frac{\mathrm{J}}{\mathrm{kg}}\cdot 0.180\frac{\mathrm{kg}}{\mathrm{mol}}-0.31\left(-1.503\cdot 10^7\,\frac{\mathrm{J}}{\mathrm{kg}}\cdot 0.113\frac{\mathrm{kg}}{\mathrm{mol}}\right)$$
$$-1.64\left(-2.957\cdot 10^7\,\frac{\mathrm{J}}{\mathrm{kg}}\cdot 0.046\frac{\mathrm{kg}}{\mathrm{mol}}\right)=-4.35\cdot 10^4\,\frac{\mathrm{J}}{\mathrm{mol}}$$

The rate of heat generation is given by equation (6.9):

$$r_H\left(\frac{\mathrm{J}}{\mathrm{m^3\cdot d}}\right)=\frac{1\,\dfrac{\mathrm{kg\ biomass}}{\mathrm{m^3\cdot d}}}{6.67\cdot 10^{-7}\,\dfrac{\mathrm{kg}}{\mathrm{J}}}=1.50\cdot 10^6\,\frac{\mathrm{J}}{\mathrm{m^3\,d}}$$

Note that the rate of heat generation is much higher for the aerobic reaction than for the anaerobic one.

6.3 HEAT BALANCES

The general form of heat balances is

$$\text{Enthalpy accumulated} = \text{Enthalpy in} - \text{Enthalpy out} + \text{Enthalpy added / removed} \tag{6.35}$$

6.3.1 Heat Balances for Continuous Fermentation

Let us consider the continuous fermenter in Figure 6.1 at steady state.

The enthalpy balance (6.35) for the fluid in the reactor becomes:

$$\text{Enthalpy in} - \text{Enthalpy out} + \text{Enthalpy added / removed} = 0 \tag{6.36}$$

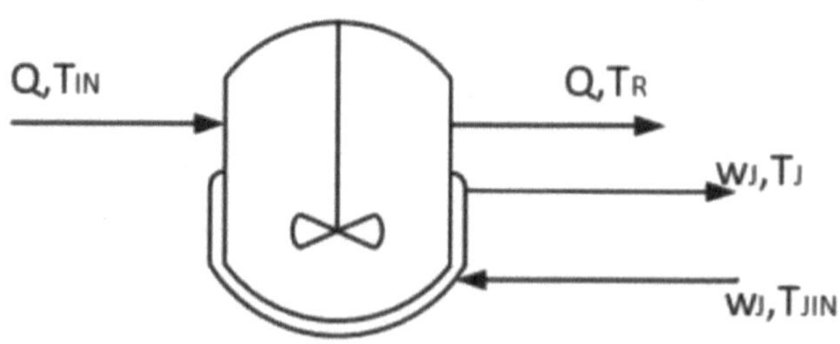

FIGURE 6.1 Generic scheme of a continuous fermenter for heat balances.

The term Enthalpy in is given by

$$\text{Enthalpy in} = \sum_i \dot{m}_i H_i\left(T_{IN}\right) \tag{6.37}$$

where m_i are the mass flow rates and H_i are the enthalpies (per unit of mass) of the components in the feed.

The term enthalpy out is given by

$$\text{Enthalpy out} = \sum_i \dot{m}_i H_i\left(T_R\right) \tag{6.38}$$

where the enthalpy terms are calculated at the outlet temperature $T_{R.}$

The term "Enthalpy added/removed" corresponds to the rate of enthalpy provided/removed to/from the fluid in the reactor by the external system and it is positive if enthalpy is added and negative if enthalpy is removed. For an adiabatic reactor the Enthalpy added/removed is equal to zero. If heat transfer occurs via a completely-mixed jacket, with the fluid in the jacket at a uniform temperature T_J, the term Enthalpy added/removed is given by:

$$\text{Enthalpy added/removed} = UA\left(T_J - T_R\right) \tag{6.39}$$

In equation (6.39) if $T_J > T_R$ (positive term) enthalpy is added to the fluid in the reactor, if $T_J < T_R$ enthalpy is removed. Therefore, in this case the enthalpy balance becomes

$$\sum_i \dot{m}_i H_i\left(T_{IN}\right) - \sum_i \dot{m}_i H_i\left(T_R\right) + UA\left(T_J - T_R\right) = 0 \tag{6.40}$$

We want now to show that the enthalpy balance (6.40) can be written, if we ignore the enthalpy in with any inlet gas stream, as

$$w_{feed} C_{Pfeed}\left(T_{IN} - T_R\right) + r_H V = UA\left(T_R - T_J\right) \tag{6.41}$$

In equation (6.41), w_{feed} represents the mass flow rate of the feed and C_{Pfeed} its specific heat.

The equivalence between equations (6.40) and (6.41) will be shown using, as an example, a typical stoichiometry for the anaerobic fermentation to ethanol:

$$C_6H_{12}O_6 + 0.12NH_3 \rightarrow 0.12C_5H_7O_2N + 1.8C_2H_5OH + 1.8CO_2 + 0.36H_2O \tag{6.42}$$

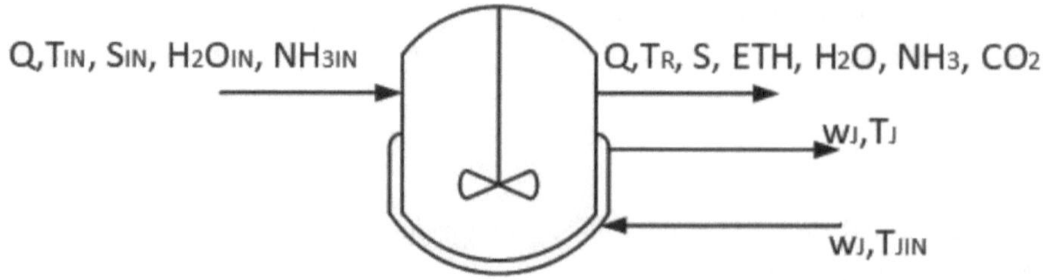

FIGURE 6.2 Scheme of a jacketed fermenter for hat balances using reaction (6.42). Q represents the volumetric flow rate (units, e.g., m³/d) and the terms S_{IN}, S, H_2O_{IN}, H_2O, etc represent the concentrations of the components (units, e.g. kg/m³).

We will develop the heat balances referring to the scheme in Figure 6.2:

For the reaction stoichiometry (6.42), the various terms in the enthalpy balance (6.36) are shown below.

Enthalpy in:

$$w_{NH_3,IN} H_{NH_3}(T_{IN}) + w_{S,IN} H_S(T_{IN}) + w_{H_2O,IN} H_{H_2O}(T_{IN}) \tag{6.43}$$

Enthalpy out:

$$\begin{aligned} &w_{X,OUT} H_X(T_R) + w_{ETH,OUT} H_{ETH}(T_R) + w_{CO_2,OUT} H_{CO_2}(T_R) \\ &+ w_{H_2O,OUT} H_{H_2O}(T_R) + w_{S,OUT} H_S(T_R) + w_{NH_3,OUT} H_{NH_3}(T_R) \end{aligned} \tag{6.44}$$

Enthalpy removed:

$$UA(T_R - T_J) \tag{6.39}$$

In equations (6.43) and (6.44) and in the rest of this section the terms w_i represent the mass flow rate (in units of, e.g., kg/d) of component i at the inlet or outlet of the reactor. Each term w_i can be calculated by the product of the volumetric flow rate Q and the component concentration, e.g. $w_{S,IN} = Q \cdot S_{IN}$, $w_{ETH,OUT} = Q \cdot ETH$.

The overall enthalpy balance is shown by Equation (6.45):

$$\begin{aligned} &w_{NH_3,IN} H_{NH_3}(T_{IN}) + w_{S,IN} H_S(T_{IN}) + w_{H_2O,IN} H_{H_2O}(T_{IN}) \\ &= w_{X,OUT} H_X(T_R) + w_{ETH,OUT} H_{ETH}(T_R) + w_{CO_2,OUT} H_{CO_2}(T_R) \\ &+ w_{H_2O,OUT} H_{H_2O}(T_R) + w_{S,OUT} H_S(T_R) + w_{NH_3,OUT} H_{NH_3}(T_R) + UA(T_R - T_J) \end{aligned} \tag{6.45}$$

Equation (6.45) can be shown to correspond to equation (6.41) if we do several rearrangements. We start by writing the enthalpy out term (6.44) as a function of the reaction rate. We express the outlet concentration of each species as a function of the biomass concentration using equations (6.46)

$$\begin{cases} ETH\left(\frac{\text{kg}}{\text{m}^3}\right) = \frac{1.8MW_{ETH}}{0.12MW_{BIOM}}X \\ CO_2\left(\frac{\text{kg}}{\text{m}^3}\right) = \frac{1.8MW_{CO2}}{0.12MW_{BIOM}}X \\ H_2O\left(\frac{\text{kg}}{\text{m}^3}\right) = H_2O_{IN} + \frac{0.36MW_{H2O}}{0.12MW_{BIOM}}X \\ S\left(\frac{\text{kg}}{\text{m}^3}\right) = S_{IN} - \frac{MW_S}{0.12MW_{BIOM}}X \\ NH_3\left(\frac{\text{kg}}{\text{m}^3}\right) = NH_{3IN} - \frac{0.12MW_{NH3}}{0.12MW_{BIOM}}X \end{cases} \tag{6.46}$$

Substituting equations (6.46) into equation (6.44), the enthalpy out term becomes

$$\begin{aligned} & w_{X,OUT}H_X\left(T_R\right)+w_{X,OUT}\frac{1.8MW_{ETH}}{0.12MW_{BIOM}}H_{ETH}\left(T_R\right) \\ & + w_{X,OUT}\frac{1.8MW_{CO_2}}{0.12MW_{BIOM}}H_{CO_2}\left(T_R\right)+w_{H_2O,IN}H_{H_2O}\left(T_R\right) \\ & + w_{X,OUT}\frac{0.36MW_{H_2O}}{0.12MW_{BIOM}}H_{H_2O}\left(T_R\right)+w_{S,IN}H_S\left(T_R\right)+ \\ & w_{X,OUT}\left(-\frac{MW_S}{0.12MW_{BIOM}}\right)H_S\left(T_R\right)+w_{NH_3,IN}H_{NH_3}\left(T_R\right)+w_{X,OUT}\left(-\frac{MW_{NH_3}}{MW_{BIOM}}\right)H_{NH_3}\left(T_R\right) \end{aligned} \tag{6.47}$$

Equation (6.47) can be re-written by noting the equivalence between enthalpies expressed on a mass and on a molar basis:

$$H_i\left(\frac{\text{J}}{\text{kg}}\right) = \frac{H_{i,mol}\left(\frac{\text{J}}{\text{kmol}}\right)}{MW_i} \tag{6.48}$$

By substituting equations (6.48) into (6.47) the enthalpy out can be re-written as

$$w_{X,OUT}\frac{H_{X,mol}(T_R)}{MW_{BIOM}}+w_{X,OUT}\frac{1.8}{0.12MW_{BIOM}}H_{ETH,mol}(T_R)$$
$$+w_{X,OUT}\frac{1.8}{0.12MW_{BIOM}}H_{CO_2,mol}(T_R)+w_{H_2O,IN}H_{H_2O}(T_R)+$$

$$w_{X,OUT}\frac{0.36}{0.12MW_{BIOM}}H_{H_2O,mol}(T_R)+w_{S,IN}H_S(T_R)-w_{X,OUT}\frac{1}{0.12MW_{BIOM}}H_{S,mol}(T_R)$$
$$+w_{NH_3,IN}H_{NH_3}(T_R)-w_{X,OUT}\frac{1}{MW_{BIOM}}H_{NH_3,mol}(T_R) \tag{6.49}$$

The enthalpy of reaction (6.42) can be written as

$$\Delta H_{R,mol}=0.12H_{X,mol}+1.8H_{ETH,mol}+1.8H_{CO_2,mol}+0.36H_{H_2O,mol}-H_{S,mol}-0.12H_{NH_3,mol} \tag{6.50}$$

By substituting equation (6.50) into equation (6.49) the enthalpy out term becomes

$$w_{H_2O,IN}H_{H_2O}(T_R)+w_{S,IN}H_S(T_R)+w_{NH_3,IN}H_{NH_3}(T_R)+w_{X,OUT}\frac{\Delta H_R(T_R)}{0.12MW_{BIOM}} \tag{6.51}$$

For reaction (6.42) the coefficient Y_H is defined as

$$Y_H=\frac{0.12MW_{substrate}}{(-\Delta H_{R,mol})} \tag{6.52}$$

Equation (6.52) becomes

$$(-\Delta H_{R,mol})=\frac{0.12MW_{biomass}}{Y_H} \tag{6.53}$$

In addition, from the mass balance for the biomass in the fermenter we have

$$r_X V=QX$$

Introducing these equations in Equation (6.51) we have

$$w_{H_2O,IN}H_{H_2O}(T_R)+w_{S,IN}H_S(T_R)+w_{NH_3,IN}H_{NH_3}(T_R)-\frac{r_X V}{Y_H} \tag{6.54}$$

Equation (6.54) is equal to

$$w_{H_2O,IN}H_{H_2O}\left(T_R\right)+w_{S,IN}H_S\left(T_R\right)+w_{NH_3,IN}H_{NH_3}\left(T_R\right)-r_HV \tag{6.55}$$

Equation (6.55) represents the enthalpy out as a function of the rate of heat generation. Using equation (6.55), the overall heat balance (6.45) can be re-written as

$$\begin{aligned}&w_{NH3,IN}H_{NH3}\left(T_{IN}\right)+w_{S,IN}H_S\left(T_{IN}\right)+w_{H2O,IN}H_{H2O}\left(T_{IN}\right)=w_{NH3,IN}H_{NH3}\left(T_R\right)\\&+w_{S,IN}H_S\left(T_R\right)+w_{H2O,IN}H_{H2O}\left(T_R\right)-r_HV+UA\left(T_R-T_J\right)\end{aligned} \tag{6.56}$$

The term on the left-hand side of equation (6.56) is the rate at which enthalpy enters the reactor with the inlet feed:

$$w_{NH3,IN}H_{NH3}\left(T_{IN}\right)+w_{S,IN}H_S\left(T_{IN}\right)+w_{H2O,IN}H_{H2O}\left(T_{IN}\right)=w_{feed}H_{feed}\left(T_{IN}\right) \tag{6.57}$$

In the right-hand side of equation (6.56), the first term corresponds to the enthalpy flow of a stream having the flow rate and composition of the feed but the temperature of the reactor:

$$w_{NH_3,IN}H_{NH_3}\left(T_R\right)+w_{S,IN}H_S\left(T_R\right)+w_{H_2O,IN}H_{H_2O}\left(T_R\right)=w_{feed}H_{feed}\left(T_R\right) \tag{6.58}$$

Using equations (6.57) and (6.58), the enthalpy balance for the fluid in the reactor, equation (6.56), becomes

$$w_{feed}H_{feed}\left(T_{IN}\right)-w_{feed}H_{feed}\left(T_R\right)=-r_HV+UA\left(T_R-T_J\right) \tag{6.59}$$

Equation (6.59) can be re-arranged into equations (6.60) and (6.41). Feeds to fermenters are usually diluted water-based mixtures and in many cases the enthalpy of the feed can be approximated by the enthalpy of water:

$$w_{feed}\left[H_{feed}\left(T_{IN}\right)-H_{feed}\left(T_R\right)\right]+r_HV=UA\left(T_R-T_J\right) \tag{6.60}$$

$$w_{feed}C_{Pfeed}\left(T_{IN}-T_R\right)+r_HV=UA\left(T_R-T_J\right) \tag{6.41}$$

In summary, the enthalpy balance for the fluid in the reactor (equation (6.40)) can also be expressed by equation (6.41). Note that $w_{feed}=Q\rho_{feed}$ and that the density and specific heat of the feed to fermenters are often similar to the values of water, because fermentation feeds are often diluted water solutions. Equation (6.41) is a convenient form to use if the reaction rate and the enthalpy of reaction are known. On the other hand, equation (6.40) is the equation to use to calculate the heat effects if the inlet and outlet compositions are known, but the reaction stoichiometry and enthalpy are not known.

Equation (6.41) can be solved for T_R (expressing r_H according to equation (6.9) and using the steady-state mass balances for the continuous fermentation process):

$$T_R = \frac{\dfrac{w_{feed} C_{Pfeed} T_{IN}}{UA} + \dfrac{w_{feed} X}{\rho_{feed} UAY_H} + T_J}{1 + \dfrac{w_{feed} C_{Pfeed}}{UA}} \tag{6.61}$$

Equation (6.61) shows that, if the jacket is used for cooling (T_J<T_R), higher value of the UA term corresponds to lower value of the temperature in the reactor. This shows that higher values of the heat transfer coefficient and/or of the heat transfer area are a benefit for temperature control in fermenters. Assuming that the rate of reaction and the heat of reaction are known (these are included in the terms X and Y_H in equation (6.61)), equations (6.61) allow the calculation of the reactor temperature T_R if the jacket temperature T_J is known. To determine T_J, we use the heat balance for the fluid in the jacket, which can be derived using the same general form, equation (6.35), which in this case becomes

$$W_J C_{PJ} \left(T_{JIN} - 25\right) + UA\left(T_R - T_J\right) = W_J C_{PJ} \left(T_J - 25\right) \tag{6.62}$$

Equation (6.62) immediately becomes

$$UA\left(T_R - T_J\right) = W_J C_{PJ} \left(T_J - T_{JIN}\right) \tag{6.63}$$

So, in summary the temperatures of the reactor and of the fluid in the jacket (T_R and T_J) can be calculated by the simultaneous solution of equations (6.61) (or of its equivalent (6.41)) and (6.63). Instead of equations (6.61) or (6.41), depending on the data available, the general form of the heat balance for the fluid in the reactor (6.40) can be used.

Equations (6.61) and (6.63) can be used to understand the effect of the fermentation parameters and fermenter design on the process temperature T_R and how to control this temperature within the desired range. By solving equation (6.63) for T_J we obtain equation (6.64) and by substituting the value of T_J in equation (6.61) we obtain equation (6.65), which shows T_R as a function of the fermentation and fermenter design parameters:

$$T_J = \frac{UAT_R + w_J C_{PJ} T_{JIN} T_{IN}}{UA + w_J C_{PJ}} \tag{6.64}$$

$$T_R = \frac{\dfrac{w_{feed} C_{Pfeed} T_{IN}}{UA} + \dfrac{w_{feed} X}{\rho_{feed} UAY_H} + \dfrac{w_J C_{PJ} T_{JIN}}{UA + w_J C_{PJ}}}{1 + \dfrac{w_{feed} C_{Pfeed}}{UA} - \dfrac{UA}{UA + w_J C_{PJ}}} \tag{6.65}$$

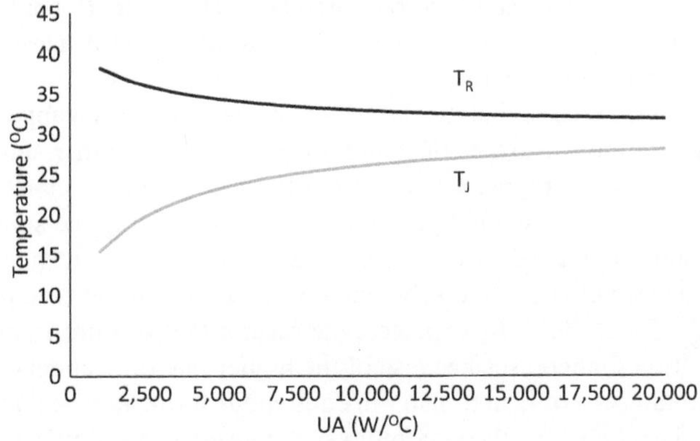

FIGURE 6.3 Effect of *UA* on the reactor temperature T_R (equation (6.65)) and on the jacket temperature T_J (equation (6.64)). Parameter values used: w_{feed} = 2 kg/s, $C_{Pfeed} = C_{PJ}$ = 4,186 J kg^{-1} °C^{-1}, T_{IN} = 25 °C, ρ_{feed} = 1000 kg/m^3, X = 2 kg/m^3, $Y_H = 3.0 \cdot 10^{-8}$ kg/J, w_J = 1 kg/s, T_{JIN} = 10 °C.

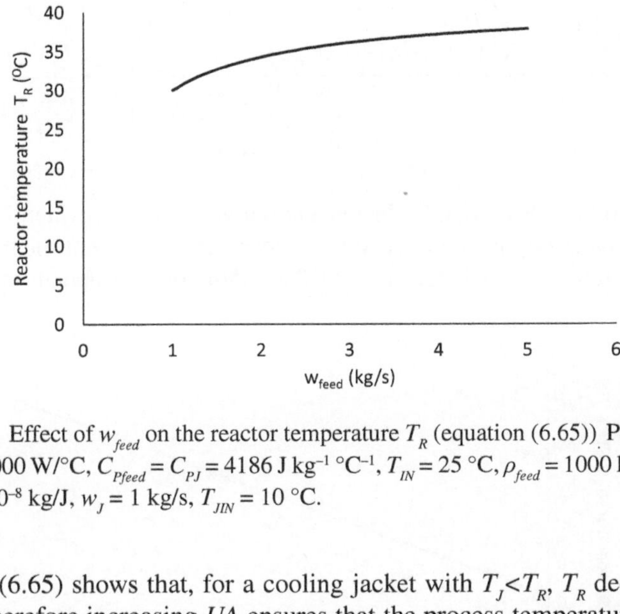

FIGURE 6.4 Effect of w_{feed} on the reactor temperature T_R (equation (6.65)). Parameter values used: UA = 5,000 W/°C, $C_{Pfeed} = C_{PJ}$ = 4186 J kg^{-1} °C^{-1}, T_{IN} = 25 °C, ρ_{feed} = 1000 kg/m^3, X = 2 kg/m^3, $Y_H = 3.0 \cdot 10^{-8}$ kg/J, w_J = 1 kg/s, T_{JIN} = 10 °C.

Equation (6.65) shows that, for a cooling jacket with $T_J<T_R$, T_R decreases as *UA* increases. Therefore increasing *UA* ensures that the process temperature T_R does not exceed the desired value. The effect of *UA* on T_R and T_J is shown in Figure 6.3. With the data used for Figure 6.3, assuming that the maximum allowable temperature in the reactor is 35 °C, a *UA* value of at least 4000 W/°C is required.

Equation (6.64) also shows the effect of the feed flow rate on the reactor temperature T_R (Figure 6.4). Assuming that X remains unaffected by w_{feed}, i.e. that w_{feed} is always low enough to allow the (almost) complete removal of the fed substrate, increasing w_{feed} increases the rate of heat generation because more substrate is reacted in the same reactor volume. Therefore, everything else being constant, the temperature

T_R tends to increase as the feed flow rate increases. However, T_R tends to an asymptotic value for high values of w_{feed}, because increasing w_{feed} also increases the rate of heat removal with the outlet stream.

It is important to observe that the biomass concentration X that appears in equation (6.65) depends on the substrate concentration in the feed. Higher substrate concentration in the feed gives higher values of the biomass concentration, assuming that the residence time is large enough to guarantee (almost) complete substrate removal (under this assumption $X = Y_{X/S}S_0$). If we replace X with $Y_{X/S}{\cdot}S_0$ in equation (6.65) we are able to show the effect of the substrate concentration in the feed on the reaction temperature T_R (Figure 6.5). As expected, the reactor temperature T_R increases as the feed concentration S_0 increases because of the higher rate of heat generation.

As a further analysis of heat transfer in continuous fermenters, equation (6.65) can be manipulated to show how the maximum substrate concentration in the feed that we can have without exceeding a certain maximum temperature for the microorganisms depends on the heat removal properties of the system. If we rearrange equation (6.65) solving for S_0 we obtain equation (6.66):

$$S_0 = \frac{T_R\left(1+\frac{w_{feed}C_{Pfeed}}{UA}-\frac{UA}{UA+w_JC_{PJ}}\right)-\frac{w_{feed}C_{Pfeed}T_{IN}}{UA}-\frac{w_JC_{PJ}T_{JIN}}{UA+w_JC_{PJ}}}{\frac{w_{feed}Y_{X/S}}{\rho_{feed}UAY_H}} \tag{6.66}$$

Equation (6.66) shows that by increasing the value of UA, which is proportional to the heat transfer rate from the vessel, we obtain an increase of the maximum substrate concentration that we can have in the continuous fermentation feed without

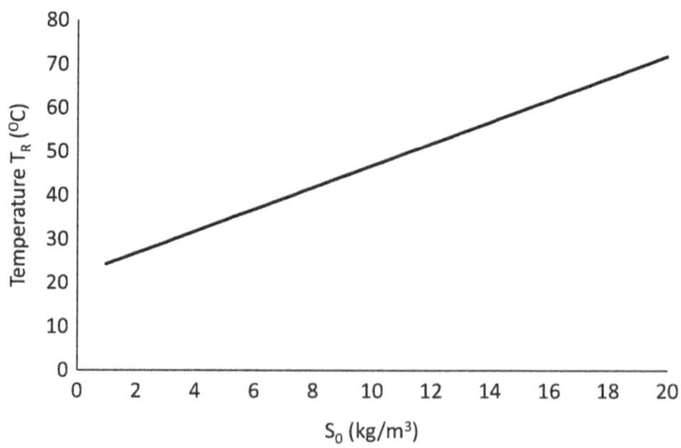

FIGURE 6.5 Effect of substrate concentration in the feed (under the assumption that the effluent substrate is negligible compared to the influent) on the reactor temperature T_R. Parameter values used: $w_{feed} = 2$ kg/s, $UA = 5000$ W/°C, $C_{Pfeed} = C_{PJ} = 4186$ J/kg.°C, $T_{IN} = 25$ °C, $\rho_{feed} = 1000$ kg/m³, $Y_{X/S} = 0.4$ kg/kg, $Y_H = 3.0{\cdot}10^{-8}$ kg/J, $w_J = 1$ kg/s, $T_{JIN} = 10$ °C.

exceeding a maximum desired reactor temperature. Therefore, increasing the heat removal capacity of the vessel can lead to an increase of the fermentation productivity, i.e. of the amount of substrate that we can process in a given reactor volume. This effect is shown in Figure 6.6.

Overall, the analysis in this section shows that heat transfer, analogously to oxygen mass transfer (Chapter 5), can limit the performance of fermentation process. Therefore, it is important to optimise the heat transfer system according to the requirements of the fermentation.

Note that equation (6.41) does not consider explicitly the enthalpy of the inlet gas phase, if there is any. In aerobic fermentations, there is an inlet gas composed by oxygen and, if air is used, nitrogen. In many cases, the enthalpy of the inlet gas gives only a minor contribution to the enthalpy balance and can be ignored, however if needed it can be accounted for easily with equation (6.67), which is a modification of equation (6.41):

$$w_{feed}C_{Pfeed}\left(T_{IN}-T_R\right)+w_{gas}C_{Pgas}\left(T_{IN}-T_R\right)+r_HV=UA\left(T_R-T_J\right) \qquad (6.67)$$

In equation (6.67) w_{gas} is the total mass flow rate of the inlet gas and C_{Pgas} the specific heat of this stream. In equation (6.67) it is assumed that the gas phase enters and leaves the reactor at the same temperatures T_{IN} and T_R as the liquid phase. For an exothermic reaction, the effect of the enthalpy of the gas phase, accounted for in equation (6.67), is to reduce the temperature increase (T_{IN}-T_R) of the fermentation mixture caused by the heat of reaction.

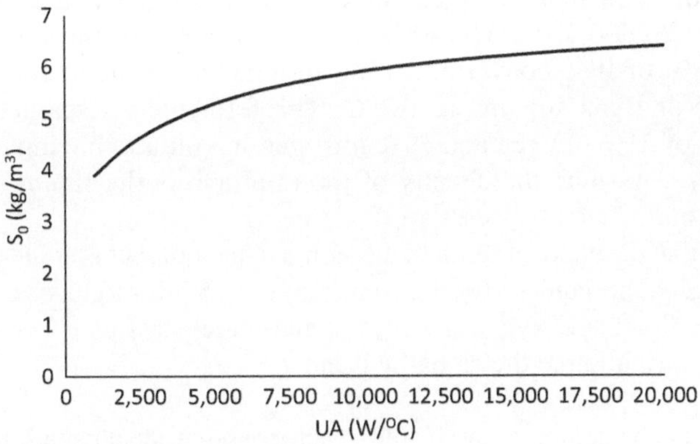

FIGURE 6.6 Effect of *UA* on the maximum substrate concentration that we can have in the feed of a continuous fermentation process (under the assumption that the effluent substrate is negligible compared to the influent) without exceeding the maximum desired reactor temperature T_R. Parameter values used: T_R = 35 °C, UA = 5000 W/°C, $C_{Pfeed} = C_{PJ}$ = 4186 J kg^{-1} °C^{-1}, T_{IN} = 25 °C, ρ_{feed} = 1000 kg/m^3, $Y_{X/S}$ = 0.4 kg/kg, Y_H = 3.0·10^{-8} kg/J, w_J = 1 kg/s, T_{JIN} = 10 °C.

Example 6.4

Consider the fermentation reaction (properties in Table 6.2):

TABLE 6.2
Physical Properties and Enthalpies for Example 6.4

Substance	Reference State	ΔH_f^0(J/kg)	Enthalpy of Phase Change (any T)	Specific Heat (J kg^{-1} K^{-1})
Glucose	solid	$-7.074 \cdot 10^6$	(solid to dissolved in water) $6.11 \cdot 10^4$	(solid) $1.22 \cdot 10^3$
Oxygen	gas	0	(gas to dissolved in water) $-3.656 \cdot 10^5$	$9.10 \cdot 10^2$
Nitrogen	gas	0	(gas to dissolved in water) $- 3.929 \cdot 10^5$	$1.04 \cdot 10^3$
Ammonia	gas	$-2.717 \cdot 10^6$	(gas to dissolved in water) $-1.794 \cdot 10^6$	(gas) $1.95 \cdot 10^2$
Biomass	Solid	$-8.000 \cdot 10^6$	–	(solid) $3.00 \cdot 10^3$
Citric acid	solid	$-1.021 \cdot 10^7$	(solid to dissolved in water) $9.484 \cdot 10^4$	(solid) $1.18 \cdot 10^3$
Carbon dioxide	gas	$-9.000 \cdot 10^6$	–	$8.71 \cdot 10^2$
Water	liquid	$-1.600 \cdot 10^7$	–	$4.19 \cdot 10^3$

$$C_6H_{12}O_6 + 2.35O_2 + 0.1NH_3 \rightarrow 0.1C_5H_7O_2N + 0.7C_6H_8O_7 + 1.3CO_2 + 3H_2O$$

This reaction is carried out in a continuous fermenter. The total inlet mass flow rate is 20,000 kg/d. The glucose concentration in the influent is 15 g/L and 95% of it is converted in the fermenter. The temperature of the inlet liquid and gas streams is 18 °C. The fermenter is sparged with air (composed of 21% oxygen and 79% nitrogen in volume) having a flow rate of 7500 m^3/d. Assume the density of the inlet and outlet liquid streams to be 1000 kg/m^3.

Assume that the concentration of ammonia in the influent is equal to the stoichiometric amount required for the conversion of 15 g/L of glucose.
Assume the dissolved oxygen in the inlet and outlet streams is negligible and ignore its contribution to the enthalpy balance.

(a) Calculate the temperature in the reactor, assuming adiabatic operation. Compare the use of equation (6.40) and (6.41), assuming, for equation (6.41) that the physical properties of the feed are the ones of water. In the use of equation (6.41), assume that the enthalpy of reaction is equal at all temperatures to its standard value.

(b) Calculate the flow rate of the cooling fluid that is required to maintain the fermentation temperature to 38 °C. Assume the cooling fluid is available at 10 °C and assume UA = 800 W/°C. The heat capacity of the cooling fluid is $4.19 \cdot 10^3$ J/kg.°C.

Solution

(a) The problem can be solved with either equation (6.40) or (6.41), with $UA = 0$.
equation (6.40) in this case is

$$\begin{aligned}
&\dot{m}_{glucose,in} H_{glucose}\left(T_{IN}\right) + \dot{m}_{O2,ingas} H_{O2}\left(T_{IN}\right) \\
&+ \dot{m}_{N2,ingas} H_{N2}\left(T_{IN}\right) + \dot{m}_{water,in} H_{water}\left(T_{IN}\right) + \dot{m}_{NH3,in} H_{NH3}\left(T_{IN}\right) \\
&= \dot{m}_{glucose,out} H_{glucose}\left(T_R\right) + \dot{m}_{O2,outgas} H_{O2}\left(T_R\right) + \dot{m}_{N2,outgas} H_{N2}\left(T_R\right) \\
&+ \dot{m}_{NH3,out} H_{NH3}\left(T_R\right) + \dot{m}_{biom,out} H_{Biom}\left(T_R\right) + \dot{m}_{citric,out} H_{citric}\left(T_R\right) \\
&+ \dot{m}_{CO2,outgas} H_{CO2}\left(T_R\right) + \dot{m}_{water,out} H_{water}\left(T_R\right)
\end{aligned}$$

We need to calculate the mass flow rates of all the species in and out of the process.

$$\dot{m}_{Glucose,in} = \frac{20,000 \dfrac{\text{kg}}{\text{day}} \cdot 15 \dfrac{\text{kg}}{\text{m}^3}}{1,000 \dfrac{\text{kg}}{\text{m}^3}} = 300 \frac{\text{kg}}{\text{d}}$$

$$\dot{m}_{O2,ingas} = 7.5 \cdot 10^6 \frac{1}{\text{day}} \cdot 0.21 \frac{1}{1} \cdot \frac{1\text{atm} \cdot \text{mol} \cdot \text{K}}{0.08211 \cdot \text{atm} \cdot 291.15\text{K}} \cdot 0.032 \frac{\text{kg}}{\text{mol}} = 2,108 \frac{\text{kg}}{\text{d}}$$

$$\dot{m}_{N2,ingas} = 7.5 \cdot 10^6 \frac{1}{\text{day}} \cdot 0.79 \frac{1}{1} \cdot \frac{1\text{atm} \cdot \text{mol} \cdot \text{K}}{0.08211 \cdot \text{atm} \cdot 291.15\text{K}} \cdot 0.028 \frac{\text{kg}}{\text{mol}} = 6940 \frac{\text{kg}}{\text{d}}$$

$$\dot{m}_{NH3,in} = \frac{300 \dfrac{\text{kg}}{\text{day}}}{180 \dfrac{\text{kg}}{\text{kmol}}} \cdot 0.1 \frac{\text{mol NH}_3}{\text{mol glucose}} \cdot 17 \frac{\text{kg}}{\text{kmol}} = 2.8 \frac{\text{kg}}{\text{d}}$$

$$\dot{m}_{water,in} = \left(20,000 - 300 - 2.8\right) \frac{\text{kg}}{\text{d}} = 19,697.2 \frac{\text{kg}}{\text{d}}$$

$$\dot{m}_{glucose,out} = 300\left(1 - 0.95\right) \frac{\text{kg}}{\text{day}} = 15 \frac{\text{kg}}{\text{d}}$$

$$\dot{m}_{O2,out} = 2108\frac{\text{kg}}{\text{d}} - 300\cdot 0.95\frac{\text{kg}}{\text{d}}\cdot 0.42\frac{\text{kgO2}}{\text{kg glucose}} = 1988\frac{\text{kg}}{\text{d}}$$

$$\dot{m}_{N2,outgas} = 6940\frac{kg}{d}$$

$$\dot{m}_{NH3,out} = 2.8\frac{kg}{d} - 300(0.95)\frac{\text{kg}}{\text{d}}\cdot 0.0094\frac{\text{kg NH3}}{\text{kg glucose}} = 0.12\frac{\text{kg}}{\text{d}}$$

$$\dot{m}_{biomass,out} = 300\cdot 0.95\frac{\text{kg}}{\text{d}}\cdot 0.063\frac{\text{kgbiom}}{\text{kgglucose}} = 17.9\frac{\text{kg}}{\text{d}}$$

$$\dot{m}_{citric,out} = 300\cdot 0.95\frac{\text{kg}}{\text{d}}\cdot 0.75\frac{\text{kgcitric}}{\text{kgglucose}} = 212.8\frac{\text{kg}}{\text{d}}$$

$$\dot{m}_{CO2,outgas} = 300\cdot 0.95\frac{\text{kg}}{\text{d}}\cdot 0.32\frac{\text{kgCO2}}{\text{kgglucose}} = 91\frac{\text{kg}}{\text{d}}$$

$$\dot{m}_{water,out} = 19{,}697.2\frac{\text{kg}}{\text{d}} + 300\cdot 0.95\frac{\text{kg}}{\text{d}}\cdot 0.30\frac{\text{kgwater}}{\text{kgglucose}} = 19{,}782.7\frac{\text{kg}}{\text{d}}$$

The enthalpies of the components per unit mass are:

$$H_{glucose}(T_{IN}) = -7.074\cdot 10^6\frac{\text{J}}{\text{kg}} + 1.22\cdot 10^3\frac{\text{J}}{\text{kg}\cdot{}^\circ\text{C}}(18-25)\,{}^\circ\text{C} + 6.11\cdot 10^4 = -7.021\cdot 10^6\frac{\text{J}}{\text{kg}}$$

$$H_{O2gas}(T_{IN}) = 9.10\cdot 10^2\frac{\text{J}}{\text{kg}\cdot{}^\circ\text{C}}(18-25)^\circ C = -6.37\cdot 10^3\frac{\text{J}}{\text{kg}}$$

$$H_{N2gas}(T_{IN}) = 1.04\cdot 10^3\frac{\text{J}}{\text{kg}\cdot{}^\circ\text{C}}(18-25)^\circ C = -7.28\cdot 10^3\frac{\text{J}}{\text{kg}}$$

$$H_{NH3}(T_{IN}) = -2.717\cdot 10^6\frac{\text{J}}{\text{kg}} + 1.95\cdot 10^2\frac{\text{J}}{\text{kg}\cdot{}^\circ\text{C}}(18-25)^\circ\text{C} + 1.794\cdot 10^6 = -9.243\cdot 10^5\frac{\text{J}}{\text{kg}}$$

$$H_{water}(T_{IN}) = -1.600\cdot 10^7\frac{\text{J}}{\text{kg}} + 4.19\cdot 10^3\frac{\text{J}}{\text{kg}\cdot{}^\circ\text{C}}(18-25)^\circ\text{C} = -1.602\cdot 10^7\frac{\text{J}}{\text{kg}}$$

$$H_{glucose}\left(T_R\right) = -7.074\cdot 10^6\,\frac{\text{J}}{\text{kg}} + 1.22\cdot 10^3\,\frac{\text{J}}{\text{kg}\cdot{}^\circ\text{C}}\left(T_R - 25\right)^\circ C + 6.11\cdot 10^4$$

$$= -7.043\cdot 10^6\,\frac{\text{J}}{\text{kg}} + 1.22\cdot 10^3 T_R H_{O2gas}\left(T_R\right)$$

$$= 9.10\cdot 10^2\,\frac{\text{J}}{\text{kg}\cdot{}^\circ\text{C}}\left(T_R - 25\right)^\circ\text{C}$$

$$= -2.28\cdot 10^4\,\frac{\text{J}}{\text{kg}} + 9.10\cdot 10^2 T_R$$

$$H_{N2gas}\left(T_R\right) = 1.04\cdot 10^3\,\frac{\text{J}}{\text{kg}\cdot{}^\circ\text{C}}\left(T_R - 25\right)^\circ\text{C} = -2.60\cdot 10^4\,\frac{\text{J}}{\text{kg}} + 1.04\cdot 10^3 T_R$$

$$H_{NH3}\left(T_R\right) = -2.717\cdot 10^6\,\frac{\text{J}}{\text{kg}} + 1.95\cdot 10^2\,\frac{\text{J}}{\text{kg}\cdot{}^\circ\text{C}}\left(T_R - 25\right)^\circ C + 1.794\cdot 10^6$$

$$= -9.279\cdot 10^5\,\frac{\text{J}}{\text{kg}} + 1.95\cdot 10^2 T_R H_{water}\left(T_R\right)$$

$$= -1.600\cdot 10^7\,\frac{\text{J}}{\text{kg}} + 4.19\cdot 10^3\,\frac{\text{J}}{\text{kg}\cdot{}^\circ\text{C}}\left(T_R - 25\right)^\circ\text{C}$$

$$= -1.610\cdot 10^7\,\frac{\text{J}}{\text{kg}} + 4.19\cdot 10^3 T_R$$

$$H_{biomass}\left(T_R\right) = -8.000\cdot 10^6\,\frac{\text{J}}{\text{kg}} + 3.00\cdot 10^3\,\frac{\text{J}}{\text{kg}\cdot{}^\circ\text{C}}\left(T_R - 25\right)^\circ\text{C}$$

$$= -8.075\cdot 10^6\,\frac{\text{J}}{\text{kg}} + 3.00\cdot 10^3 T_R$$

$$H_{citric}\left(T_R\right) = -1.021\cdot 10^7\,\frac{\text{J}}{\text{kg}} + 1.18\cdot 10^3\,\frac{\text{J}}{\text{kg}\cdot{}^\circ\text{C}}\left(T_R - 25\right)^\circ C + 9.484\cdot 10^4$$

$$= -1.014\cdot 10^7\,\frac{\text{J}}{\text{kg}} + 1.18\cdot 10^3 T_R H_{CO2,outgas}\left(T_R\right)$$

$$= -9.000\cdot 10^6\,\frac{\text{J}}{\text{kg}} + 8.71\cdot 10^2\,\frac{\text{J}}{\text{kg}\cdot{}^\circ\text{C}}\left(T_R - 25\right)^\circ C$$

$$= -9.022\cdot 10^6\,\frac{\text{J}}{\text{kg}} + 8.71\cdot 10^2 T_R$$

The enthalpy of the inlet stream is

$$\dot{m}_{glucose,in} H_{glucose}\left(T_{IN}\right) + \dot{m}_{O_2,ingas} H_{O_2}\left(T_{IN}\right)$$

$$+ \dot{m}_{N_2,ingas} H_{N_2}\left(T_{IN}\right) + \dot{m}_{water,in} H_{water}\left(T_{IN}\right) + \dot{m}_{NH_3,in} H_{NH_3}\left(T_{IN}\right)$$

$$= 300\,\frac{\text{kg}}{\text{d}}\cdot\left(-7.021\cdot 10^6\,\frac{\text{J}}{\text{kg}}\right)$$

$$+2{,}108\frac{\text{kg}}{\text{d}}\cdot\left(-6.37\cdot 10^{3}\frac{\text{J}}{\text{kg}}\right)+6940\frac{\text{kg}}{\text{d}}\cdot\left(-7.28\cdot 10^{3}\frac{\text{J}}{\text{kg}}\right)$$
$$+2.8\frac{\text{kg}}{\text{d}}\cdot\left(-9.243\cdot 10^{5}\frac{\text{J}}{\text{kg}}\right)+19{,}697.2\frac{\text{kg}}{\text{d}}\cdot\left(-1.602\cdot 10^{7}\frac{\text{J}}{\text{kg}}\right)$$
$$=-3.177\cdot 10^{11}\frac{\text{J}}{\text{d}}$$

The enthalpy of the outlet stream is

$$\dot{m}_{glucose,out}H_{Glucose}\left(T_R\right)+\dot{m}_{O_2,outgas}H_{O_2}\left(T_R\right)+\dot{m}_{N_2,outgas}H_{N_2}\left(T_R\right)$$
$$+\dot{m}_{NH_3,out}H_{NH_3}\left(T_R\right)+\dot{m}_{biom,out}H_{biom}\left(T_R\right)+\dot{m}_{citric,out}H_{citric}\left(T_R\right)$$
$$+\dot{m}_{CO2,outgas}H_{CO2}\left(T_R\right)+\dot{m}_{water,out}H_{water}\left(T_R\right)$$
$$=15\frac{\text{kg}}{\text{day}}\cdot\left(-7.043\cdot 10^{6}+1.22\cdot 10^{3}T_R\right)+1988\frac{\text{kg}}{\text{day}}\cdot\left(-2.28\cdot 10^{4}+9.10\cdot 10^{2}T_R\right)$$
$$+6940\frac{\text{kg}}{\text{day}}\cdot\left(-2.60\cdot 10^{4}+1.04\cdot 10^{3}T_R\right)+0.12\frac{\text{kg}}{\text{day}}\cdot\left(-9.279\cdot 10^{5}+1.95\cdot 10^{2}T_R\right)$$
$$+19.9\frac{\text{kg}}{\text{day}}\cdot\left(-8.075\cdot 10^{6}+3.00\cdot 10^{3}T_R\right)+212.8\frac{\text{kg}}{\text{day}}\cdot\left(-1.014\cdot 10^{7}+1.18\cdot 10^{3}T_R\right)$$
$$+91\frac{\text{kg}}{\text{day}}\cdot\left(-9.022\cdot 10^{6}+8.71\cdot 10^{2}T_R\right)+19{,}782.7\frac{\text{kg}}{\text{day}}\cdot\left(-1.610\cdot 10^{7}+4.19\cdot 10^{3}T_R\right)$$
$$=-3.220\cdot 10^{11}+9.232\cdot 10^{7}T_R$$

T_R can be calculated from

$$-3.177\cdot 10^{11}=-3.220\cdot 10^{11}+9.232\cdot 10^{7}T_R$$

which gives $T_R = 46.6$ °C

Using equation (6.41) with the physical properties of water:

$$Q\rho_{H2O}C_{PH2O}\left(T_{IN}-T_R\right)=20\frac{\text{m}^3}{\text{d}}.1000\frac{\text{kg}}{\text{m}^3}\cdot 4.19\cdot 10^{3}\frac{\text{J}}{\text{kg}\cdot{}^{\circ}\text{C}}\left(18-T_R\right)$$
$$=8.38\cdot 10^{7}\frac{\text{J}}{\text{kg}\cdot{}^{\circ}\text{C}}\left(18-T_R\right)$$

$$r_H V\left(\frac{\text{J}}{\text{d}}\right)=\left[\left(-r_S\right)V,\frac{\text{kg}}{\text{d}}\right]\left(-\Delta H_R,\frac{\text{J}}{\text{kg}}\right)$$

The rate of substrate removal is

$$(-r_S)V = (300-15)\frac{\text{kg}}{\text{d}} = 285\frac{\text{kg}}{\text{d}}$$

The enthalpy of this reaction under standard conditions was calculated in Example 6.1:

$$\Delta H_R = -1.563\cdot 10^9 \frac{\text{J}}{\text{kmol}} = -1.563\cdot 10^9 \frac{\text{J}}{180\text{kg}} = -8.683\cdot 10^6 \frac{\text{J}}{\text{kg}}$$

So:

$$r_H V\left(\frac{\text{J}}{\text{d}}\right) = 285\frac{\text{kg}}{\text{d}}\cdot 8.683\cdot 10^6 \frac{\text{J}}{\text{kg}} = 2.475\cdot 10^9 \frac{\text{J}}{\text{d}}$$

So the reactor temperature T_R is calculated by

$$8.38\cdot 10^7 \frac{\text{J}}{\text{kg}\cdot {}^\circ\text{C}}(18-T_R) = -2.475\cdot 10^9 \frac{\text{J}}{\text{d}}$$

which gives T_R = 47.5 °C, in good agreement with the temperature calculate with equation (6.41).

(b) In this case the enthalpy balance is

$$\begin{aligned} &\dot{m}_{glucose,in} H_{Glucose}(T_{IN}) + \dot{m}_{O2,ingas} H_{O2}(T_{IN}) \\ &+ \dot{m}_{N2,ingas} H_{N2}(T_{IN}) + \dot{m}_{water,in} H_{water}(T_{IN}) + \dot{m}_{NH3,in} H_{NH3}(T_{IN}) \end{aligned}$$

$$\begin{aligned} &= \dot{m}_{glucose,out} H_{Glucose}(T_R) + \dot{m}_{O2,outgas} H_{O2}(T_R) + \dot{m}_{N2,outgas} H_{N2}(T_R) \\ &+ \dot{m}_{NH3,out} H_{NH3}(T_R) + \dot{m}_{biom,out} H_{biom}(T_R) + \dot{m}_{citric,out} H_{citric}(T_R) \\ &+ \dot{m}_{CO2,outgas} H_{CO2}(T_R) + \dot{m}_{water,out} H_{water}(T_R) + UA(T_R - T_J) \end{aligned}$$

And the enthalpy balance for the fluid in the jacket is

$$UA(T_R - T_J) = W_J \cdot C_{PJ}(T_J - T_{JIN})$$

The enthalpy of the inlet stream is the same as calculated at the previous point = -3.177·10[11]J/d. The enthalpy of components the outlet stream at 38 °C is

$$H_{glucose}(T_R) = -7.074\cdot10^6\,\frac{\text{J}}{\text{kg}} + 1.22\cdot10^3\,\frac{\text{J}}{\text{kg}\cdot{}^\circ\text{C}}(38-25)^\circ\text{C} + 6.11\cdot10^4 = -6.997\cdot10^6\,\frac{\text{J}}{\text{kg}}$$

$$H_{O2gas}(T_R) = 9.10\cdot10^2\,\frac{\text{J}}{\text{kg}\cdot{}^\circ\text{C}}(38-25)^\circ\text{C} = 1.183\cdot10^4\,\frac{\text{J}}{\text{kg}}$$

$$H_{N2gas}(T_R) = 1.04\cdot10^3\,\frac{\text{J}}{\text{kg}\cdot{}^\circ\text{C}}(38-25)^\circ\text{C} = 1.352\cdot10^4\,\frac{\text{J}}{\text{kg}}$$

$$H_{NH3}(T_R) = -2.717\cdot10^6\,\frac{\text{J}}{\text{kg}} + 1.95\cdot10^2\,\frac{\text{J}}{\text{kg}\cdot{}^\circ\text{C}}(38-25)^\circ\text{C} + 1.794\cdot10^6 = -9.205\cdot10^5\,\frac{\text{J}}{\text{kg}}$$

$$H_{water}(T_R) = -1.600\cdot10^7\,\frac{\text{J}}{\text{kg}} + 4.19\cdot10^3\,\frac{\text{J}}{\text{kg}\cdot{}^\circ\text{C}}(38-25)^\circ\text{C} = -1.594\cdot10^7\,\frac{\text{J}}{\text{kg}}$$

$$H_{biomass}(T_R) = -8.000\cdot10^6\,\frac{\text{J}}{\text{kg}} + 3.00\cdot10^3\,\frac{\text{J}}{\text{kg}\cdot{}^\circ\text{C}}(38-25)^\circ\text{C} = -7.961\cdot10^6\,\frac{\text{J}}{\text{kg}}$$

$$H_{citric}(T_R) = -1.021\cdot10^7\,\frac{\text{J}}{\text{kg}} + 1.18\cdot10^3\,\frac{\text{J}}{\text{kg}\cdot{}^\circ\text{C}}(38-25)^\circ\text{C} + 9.484\cdot10^4 = -1.010\cdot10^7\,\frac{\text{J}}{\text{kg}}$$

$$H_{CO2,outgas}(T_R) = -9.000\cdot10^6\,\frac{\text{J}}{\text{kg}} + 8.71\cdot10^2\,\frac{\text{J}}{\text{kg}\cdot{}^\circ\text{C}}(38-25)^\circ\text{C} = -8.989\cdot10^6\,\frac{\text{J}}{\text{kg}}$$

Therefore, the total enthalpy of the components in the outlet streams is

$$\begin{aligned}
&\dot{m}_{glucose,out}H_{Glucose}(T_R) + \dot{m}_{O_2,outgas}H_{O_2}(T_R) + \dot{m}_{N_2,outgas}H_{N_2}(T_R)\\
&+\dot{m}_{NH_3,out}H_{NH_3}(T_R) + \dot{m}_{biom,out}H_{Biom}(T_R) + \dot{m}_{citric,out}H_{citric}(T_R)\\
&+\dot{m}_{CO2,outgas}H_{CO2}(T_R) + \dot{m}_{water,out}H_{water}(T_R)
\end{aligned}$$

$$\begin{aligned}
&= 15\,\frac{\text{kg}}{\text{d}}\cdot(-6.997\cdot10^6)\,\frac{\text{J}}{\text{kg}} + 1{,}988\,\frac{\text{kg}}{\text{d}}\cdot(1.183\cdot10^4)\,\frac{\text{J}}{\text{kg}}\\
&\quad + 6940\,\frac{\text{kg}}{\text{d}}\cdot(1.352\cdot10^4)\,\frac{\text{J}}{\text{kg}} + 0.12\,\frac{\text{kg}}{\text{d}}\cdot(-9.205\cdot10^5)\,\frac{\text{J}}{\text{kg}}
\end{aligned}$$

$$+19.9\frac{\text{kg}}{\text{d}}\cdot\left(-7.961\cdot10^{6}\right)\frac{\text{J}}{\text{kg}}+212.8\frac{\text{kg}}{\text{d}}\cdot\left(-1.010\cdot10^{7}\right)\frac{\text{J}}{\text{kg}}$$
$$+91\frac{\text{kg}}{\text{d}}\cdot\left(-8.989\cdot10^{6}\right)\frac{\text{J}}{\text{kg}}+19{,}782.7\frac{\text{kg}}{\text{d}}\cdot\left(-1.594\cdot10^{7}\right)\frac{\text{J}}{\text{kg}}$$
$$=-3.184\cdot10^{11}\ \frac{\text{J}}{\text{d}}$$

Therefore, from the heat balance for the fluid in the reactor we can calculate T_J:

$$-3.177\cdot10^{11}\frac{J}{d}=-3.184\cdot10^{11}\frac{J}{d}+\frac{800J}{s°C}86{,}400\frac{s}{d}\left(38-T_J\right)$$

which gives $T_J = 27.9$ °C.

The required flow rate of the cooling fluid is obtained from the heat balance on the cooling fluid:

$$W_J=\frac{UA\left(T_R-T_J\right)}{C_{PJ}\left(T_J-T_{JIN}\right)}=\frac{800\frac{\text{J}}{°\text{C}}\cdot\frac{86{,}400\text{s}}{d}(38-27.9)°\text{C}}{4.19\cdot10^{3}\frac{\text{J}}{\text{kg}°\text{C}}(27.9-10)°\text{C}}=9308\frac{\text{kg}}{\text{d}}$$

6.3.2 Heat Balances for Batch Fermentation

We will refer to the general scheme of a batch reactor shown in Figure 6.7.

Using the general form of heat balance (6.35), the heat balance for the fluid in a batch reactor is

$$M_R C_{PR}\frac{dT_R}{dt}=r_H V-UA\left(T_R-T_J\right)\tag{6.68}$$

In equation (6.68), M_R is the mass of fluid inside the reactor and C_{PR} is its specific heat. Equation (6.68) corresponds to the analogous heat balance for a continuous

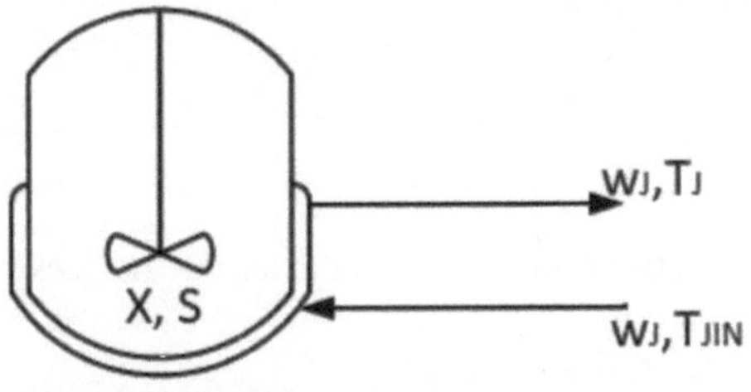

FIGURE 6.7 Scheme of a jacketed batch reactor.

system, equation (6.41), and can be derived with the same methodology. In order to solve equation (6.68) for T_R, we need an equation for T_J. This equation is the heat balance for the fluid in the jacket:

$$M_J C_{PJ} \frac{dT_J}{dt} = W_J C_{PJ} \left(T_{JIN} - 25\right) + UA\left(T_R - T_J\right) - W_J C_{PJ} \left(T_J - 25\right) \tag{6.69}$$

Equation (6.69) becomes

$$M_J C_{PJ} \frac{dT_J}{dt} = W_J C_{PJ} \left(T_{JIN} - T_J\right) + UA\left(T_R - T_J\right) \tag{6.70}$$

Equation (6.68) and (6.70) can be solved simultaneously to calculate the profiles of T_R and T_J as a function of time in a batch reactor. Note that, similarly as for the continuous fermentation, the heat balances (6.68) and (6.70) need to be coupled with the mass balances because the biomass concentration, and therefore the rate of heat generation, changes with time.

Typical profiles of T_R and T_J in a batch reactor are shown in Figure 6.8. Typically, for exothermic fermentations, T_R and T_J both increase until the substrate has been completely removed, at which point the heat generation stops and the temperature starts to drop. Note that, when the substrate is present, the rate of temperature rise increases with time, because the biomass concentration increases with time and so the rate of heat generation also increases with time. Figures like Figure 6.8 and equations (6.68), (6.70) can be used to design the batch fermentation process, to ensure that the process temperature stays within the desired range, without exceeding the maximum allowable temperature for the microorganisms. Similarly to what was

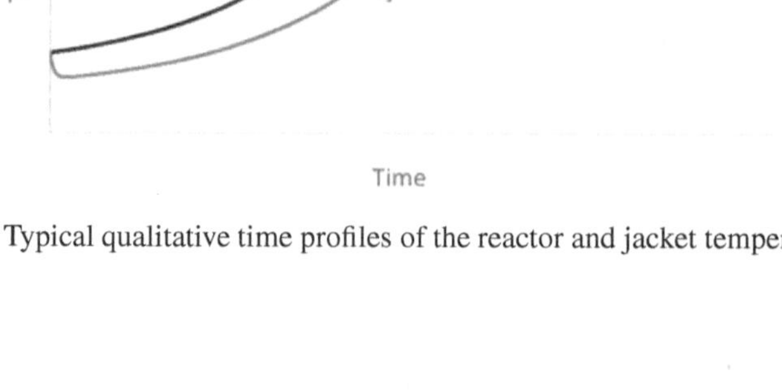

FIGURE 6.8 Typical qualitative time profiles of the reactor and jacket temperatures in batch fermentation.

discussed for the heat balances in continuous fermenters, for batch reactors the maximum initial substrate concentration can be limited by the heat transfer ability of the vessel.

Note that equation (6.68) does not account for the enthalpy of the inlet gas, if there is any. In case of aerobic fermentations, the enthalpy of the inlet gas is accounted for in equation (6.71):

$$M_{feed} C_{Pfeed} \frac{dT_R}{dt} + w_{gas} C_{Pgas} \left(T_R - T_{IN}\right) = r_H \cdot V - UA\left(T_R - T_J\right) \tag{6.71}$$

In equation (6.71) w_{gas} is the mass flow rate of the inlet gas and C_{Pgas} its specific heat. Equation (6.71) assumes that the gas enters the reactor at temperature T_{IN} and leaves at the same temperature of the reaction mixture, T_R.

Example 6.5

Consider the fermentation:

$$C_6H_{12}O_6 + 0.31NH_3 \rightarrow 0.31C_5H_7O_2N + 1.64C_2H_6O + 1.17CO_2 + 0.464H_2O$$

This fermentation is carried out in a batch reactor. The batch is started with a concentration of microorganisms of 1 kg/m^3 and with a glucose concentration of 100 kg/m^3. The initial temperature is 25 °C. Assume that the density of the fermentation mixture is 1,000 kg/m^3, the heat capacity s 4.19 J kg^{-1} °C^{-1} and the volume is 10 m^3. The kinetic parameters for the growth of the microorganism are

$$\mu_{max} = 5.5 \text{ d}^{-1}; K_S = 0.1 \text{ kg/m}^3$$

Calculate the temperature profile of the fermentation fluid vs time,

(a) for an adiabatic reaction;
(b) when a cooling fluid is used in the jacket. The flow rate of the cooling fluid is 1000 kg/d its inlet temperature is 10 °C. The heat capacity of the cooling fluid is 4.19 J kg^{-1} °C^{-1} and UA = 500 W/°C. Assume that the initial temperature in the jacket is 25 °C. The volume of the jacket is 50 L.

Solution

(a) For the adiabatic case, the heat balance is

$$\rho C_{P_R} \frac{dT_R}{dt} = r_H$$

The standard enthalpy of the reaction at 25 °C is –4.35·10^4 J/mol of glucose. We need to express this per unit of mass of microorganisms produced. According to the reaction stoichiometry, the removal of 1 mol of glucose generates 35.03 g of microorganisms, so the heat of reaction per unit mass of microorganisms formed is

$$\Delta H_R = -\frac{-4.35\cdot 10^4 \dfrac{\text{J}}{\text{mol}}}{35.03\dfrac{\text{g}}{\text{mol}}} = -1.242\cdot 10^3 \frac{\text{J}}{\text{g}} = -1.242\cdot 10^6 \frac{\text{J}}{\text{kg}}$$

The heat balance is:

$$1000\frac{\text{kg}}{\text{m}^3}4.19\cdot 10^3 \frac{\text{J}}{\text{kg°C}}\frac{dT_R}{dt} = \frac{5.5\text{ day}^{-1}S}{0.1\dfrac{\text{kg}}{\text{m}^3}+S}X\cdot 1.242\cdot 10^6 \frac{\text{J}}{\text{kg}}$$

Which simplifies to:

$$\frac{dT_R}{dt}\left(\frac{°C}{d}\right) = 0.296\frac{5.5S}{0.1+S}X$$

The heat balance has to be solved together with the mass balances for the substrate and biomass:

$$\frac{dX}{dt} = \frac{5.5S}{0.1+S}X$$

$$\frac{dS}{dt} = -\frac{5.5S}{0.1+S}\frac{X}{0.195}$$

The profiles of S, X, P and T are shown in Figure 6.9.

(b) For the case with a jacket, the heat balances for the fluid in the reactor and in the jacket are

$$\begin{cases} \rho C_{P_R}\dfrac{dT_R}{dt} = r_H - \dfrac{UA}{V}\left(T_R - T_J\right) \\ V_J C_{P_J}\dfrac{dT_J}{dt} = W_J\cdot C_{PJ}\left(T_{JIN} - T_J\right) + UA\left(T_R - T_J\right) \end{cases}$$

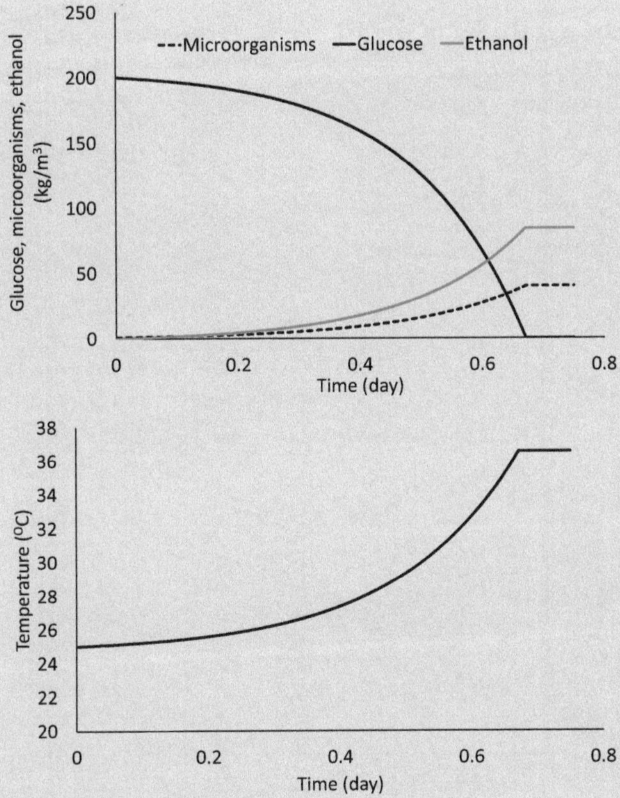

FIGURE 6.9 Profiles for Example 6.5a.

In this case the equations can be written as (using kg, m³, J, °C and d as units):

$$\begin{cases} 1000\cdot 4.19\cdot 10^3\dfrac{dT_R}{dt}=\dfrac{5.5S}{0.1+S}X-\dfrac{500\cdot 86,400}{10}\left(T_R-T_J\right) \\ 0.050\cdot 1000\cdot 4.19\cdot 10^3\dfrac{dT_J}{dt}=W_J\cdot 4.19\cdot 10^3\cdot\left(10-T_J\right)+500\cdot 86,400\left(T_R-T_J\right) \end{cases}$$

which can be simplified to

$$\begin{cases} \dfrac{dT_R}{dt}=0.296\dfrac{5.5S}{0.1+S}X-1.03\left(T_R-T_J\right) \\ \dfrac{dT_J}{dt}=0.02W_J\left(10-T_J\right)+206.21\left(T_R-T_J\right) \end{cases}$$

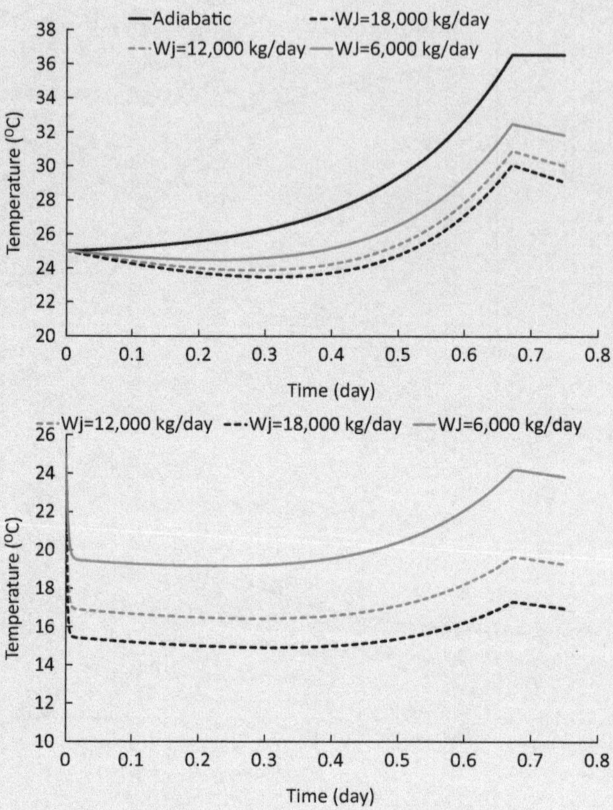

FIGURE 6.10 Profiles for Example 6.5b.

The profiles of the reactor and jacket temperature in this case are shown in the Figure 6.10.

6.4 EFFECT OF VARIABLE TEMPERATURE ON REACTION RATES

So far, we have not considered explicitly the effect of temperature on reaction rates. This effect can be considered, as seen in Chapter 3, by writing an equation for the dependency of μ_{max} on temperature, e.g. equations (3.19) and (3.20). Accounting for the effect of temperature on μ_{max} does not change the heat balance equations for continuous and batch reactors, but it changes the mass balance equations, which give the substrate, product and biomass profiles. Therefore, the temperatures in continuous and batch processes will be affected by the temperature dependency of kinetic parameters.

For a continuous process, the reactor temperature will still be given by the solution of equations (6.65), (4.19) and (4.18), but if we take the effect of temperature on

μ_{max} into account (e.g. by using equations (3.19) or (3.20)), the three equations will need to be solved simultaneously rather than independently. For a batch fermentation, the temperature profiles are in any cases obtained from the simultaneous solution of equations (6.68), (6.70), (4.3), (4.4) and, if we need to account for the effect of temperature on reaction rates, equations (3.19) or (3.20) will be used for μ_{max} in equations (4.3) and (4.4).

6.5 HEAT TRANSFER COEFFICIENTS IN AGITATED VESSELS

The heat transfer coefficient U is the combination of the heat transfer coefficients of the process fluid, of the jacket fluid and of the material that separates the reactor from the jacket (the construction material of the vessel). Due to the thickness of the construction material, the vessel will have an external diameter D_{ext} and an internal diameter D_{int}, the thickness of the material being expressed as $x = \left(D_{ext} - D_{int}\right)/2$. If referred to the external area of the vessel, the heat transfer coefficient will be expressed as

$$\frac{1}{U} = \frac{1}{h_{jacket}} + \frac{D_{ext}}{D_{int} h_{reactor}} + \frac{D_{ext} \ln\left(\frac{D_{ext}}{D_{int}}\right)}{2k} \tag{6.72}$$

In equation (6.72), h_{jacket} and $h_{reactor}$ are the heat transfer coefficients of the reactor and of the jacket, respectively, while k is the conductivity of the material used for the vessel walls.

The coefficients h_{jacket} and $h_{reactor}$ are available in the literature as correlations. $h_{reactor}$ is given by correlations such as equation (6.73):

$$Nu = \frac{h_{reactor} D_{vessel}}{k_L} = a_1 \, \mathrm{Re}^{a2} \, \mathrm{Pr}^{0.33} \left(\frac{\mu_L}{\mu_{WL}}\right)^{a_3} \tag{6.73}$$

In equation (6.73) Nu is the Nusselt number for the fluid in the reactor, Re and Pr are the Reynolds and Prandtl numbers for the fluid in the reactor, defined as

$$\mathrm{Re} = \frac{N D_{ag}^2 \rho_L}{\mu_L} \tag{6.74}$$

$$\mathrm{Pr} = \frac{c_{PL} \mu_L}{k_L} \tag{6.75}$$

In equations (6.74) and (6.75) N is the agitator stirring speed (revs/s), D_{ag} the diameter of the agitator (m), ρ_L, μ_L, c_{PL} and k_L the density, viscosity, specific heat and conductivity of the fluid in the vessel, respectively. μ_{WL} is the viscosity of the temperature of the vessel wall. Note that the factor $\left(\frac{\mu_L}{\mu_{WL}}\right)^{a_3}$ is only important for very viscous fluids and usually negligible for most fermentation processes. The Reynolds number for the agitator, defined by equation (6.74) is analogous to the Reynolds number for fluids in pipes, with the fluid velocity taken in this case as the velocity at the tip of the agitator (which is ND_{ag}) and the characteristic dimension as the diameter of the agitator. Equations (6.73) and (6.74) indicate that increasing the agitation speeds increases the heat transfer coefficient for the fluid in the reactor.

The coefficients a_1, a_2, a_3 in equation (6.73) depend on the type of agitator, on the presence or absence of baffles and other geometrical characteristics of the vessels and, in some cases, on the Reynolds number. For example, for flat-blade disc turbines and baffled vessels, an arrangement often used in fermentation vessels, for Re>400, $a_1 = 0.74$, $a_2 = 0.67$, $a_3 = 0.14$.

For the fluid in the jacket, assuming it is a nonviscous liquid, the heat transfer coefficient h_{jacket} is often given by equation (6.76)

$$Nu_{jacket} = \frac{h_{jacket} D_e}{k_J} = 0.023\,\mathrm{Re}_J^{0.8}\, Pr_J^{0.33} \tag{6.76}$$

In equation (6.76) Nu_{jacket} is the Nusselt number for the fluid in the jacket and k_J is the thermal conductivity of the fluid in the jacket. De is the equivalent diameter defined as:

$$De = \frac{4 \times \text{Cross sectional area for fluid flow}}{\text{Wetted perimeter}} \tag{6.77}$$

Re_J is the Reynolds number for the fluid in the jacket, defined as

$$\mathrm{Re}_J = \frac{\rho_J v D_e}{\mu_J} \tag{6.78}$$

In equation (6.78) v is the fluid velocity in the jacket and ρ_L and μ_J are the density and viscosity, respectively, of the fluid in the jacket. Pr_J is the Prandtl number for the fluid in the jacket and is defined in the same way as the Prandtl number for the fluid in the agitator (6.75), using the physical properties for the fluid in the jacket.

Equations (6.76)–(6.78) show that the heat transfer coefficient for the jacket fluid mainly depends on the fluid velocity in the jacket. In order to increase the fluid velocity, jackets are often equipped with baffles (spirally baffled jackets) or nozzles, or alternatively, half-pipe jackets are also common. Note that the jacket can cover the full area of the vessel (including the base area) or just the cylindrical portion of this area. Figure 6.11 shows a scheme of jacketed vessels with various types of jackets.

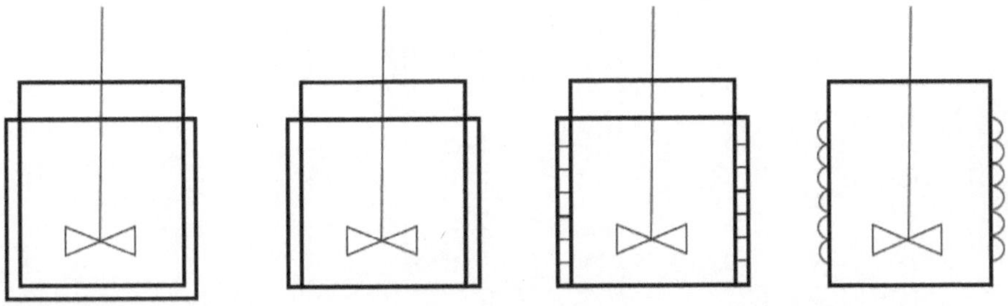

FIGURE 6.11 Scheme of jacketed vessels, from left to right: – jacket that covers the whole external area of the vessel; – jacket that covers only the cylindrical part; – spirally baffled jacket; half-pipe jacket.

Example 6.6

(a) Calculate the heat transfer coefficient for a fermentation vessel. The vessel is cylindrical with internal diameter 2.1 m and thickness 10 mm. The vessel is made of stainless steel. The agitator is a flat disc turbine with diameter 0.70 m, rotating at 150 rpm. The jacket covers the cylindrical section only and extends to a height of 1.5 m. The spacing between the jacket and the external walls of the vessel is 50 mm. The jacket fluid has a flow rate of 800 m^3/d, enters the jacket from the bottom and leaves at the top. In using equation (6.73) assume $a_1 = 0.74$, $a_2 = 0.67$, $a_3 = 0.0$ (this corresponds to ignoring the viscosity correction term, as the fluid in the reactor is not very viscous).

(b) Repeat the calculations of part (a), this time assuming that the jacket is made of a spiral baffle, with pitch between the spirals of 0.10 m.

Physical properties:

$\rho_L = 1000$ kg/m^3 (for both the reactor and the jacket fluid)
$\mu_L = 9 \cdot 10^{-3}$ Pa·s (for the fluid in the reactor)
$\mu_J = 9 \cdot 10^{-4}$ Pa·s (for the jacket fluid)
$k = 16$ W m^{-1} °C^{-1} (for stainless steel)
$k_J = 0.6$ W m^{-1} °C^{-1} (for both the reactor and the jacket fluid)
$C_P = 4186$ J kg^{-1} °C^{-1} (for both the reactor and the jacket fluid)

Solution

(a) The overall heat transfer coefficient U is given by equation (6.72), with $D_{int} = 2.100$ m and $D_{ext} = 2.120$ m. The heat transfer coefficient for the fluid in the reactor is given by equation (6.73) with the Reynolds and Prandtl numbers given by equations (6.74) and (6.75).

$$\mathrm{Re}_{ag} = \frac{N D_{ag}^2 \rho_L}{\mu_L} = \frac{\frac{150}{60} \cdot 0.7^2 \cdot 1000}{0.009} = 1.36 \cdot 10^5$$

$$\Pr_{ag} = \frac{c_{PL}\mu_L}{k_L} = \frac{4,186 \cdot 0.009}{0.6} = 62.8$$

Therefore, $h_{reactor}$ is given by

$$h_{reactor} = \frac{0.6 \cdot 0.74\left(1.36 \cdot 10^5\right)^{0.67} 62.8^{0.33}}{2.1} = 2281 \frac{\text{W}}{\text{m}^2\,{}^{\text{o}}\text{C}}$$

In the jacket, the fluid flows upwards through a rectangular section having sides of $\pi \cdot D_{ext}$ and 0.05 m. Therefore, the cross sectional area is $\pi \cdot 2.120 \cdot 0.05 = 0.33\,\text{m}^2$ and the wetted perimeter is $2 \cdot (\pi \cdot 2.120 + 0.05) = 13.41\,\text{m}$. The equivalent diameter is

$$De = \frac{4 \times 0.33}{13.41} = 0.098 \text{ m}$$

The fluid velocity in the jacket is

$$v = \frac{\frac{800}{86,400}}{0.33} = 0.028 \frac{\text{m}}{\text{s}}$$

The Reynolds number for the jacket is, with equation (6.78):

$$Re_J = \frac{1,000 \cdot 0.028 \cdot 0.098}{0.0009} = 3049$$

The Prandtl umber in the jacket is

$$\Pr_J = \frac{c_{PJ}\mu_J}{k_J} = \frac{4,186 \cdot 0.0009}{0.6} = 6.28$$

h_{jacket} is given by re-arranging equation (6.76):

$$h_{jacket} = \frac{0.6 \cdot 0.023 \cdot 3,049^{0.8} 6.28^{0.33}}{0.098} = 158.6 \frac{\text{W}}{\text{m}^2\,{}^{\circ}\text{C}}$$

According to equation (6.72):

$$\frac{1}{U} = \frac{1}{158.6} + \frac{2.12}{2.10 \cdot 2,281} + \frac{2.12 \ln\left(\frac{2.12}{2.10}\right)}{2 \cdot 16} = 0.0063 + 0.00044 + 0.000627 = 0.0073$$

$$U = 135.7 \frac{\mathrm{W}}{\mathrm{m}^2 \cdot {}^\circ\mathrm{C}}$$

(b) In this case $h_{reactor}$ will be the same of part (a), while h_{jacket} will change. The number of spirals will be equal to the height of the jacket divided by the pitch = 1.5/0.10 = 15 spirals. Each spiral has a rectangular section with sides of 0.10 and 0.05 m, therefore the cross sectional area is $0.10 \cdot 0.05 = 0.005\ \mathrm{m}^2$ and the wetted perimeter is $2 \cdot (0.10 + 0.05) = 0.30\ \mathrm{m}$. The equivalent diameter is:

$$De = \frac{4 \times 0.005}{0.30} = 0.067\ \mathrm{m}$$

The fluid velocity in the jacket is

$$v = \frac{\frac{800}{86{,}400}}{0.005} = 1.85 \frac{\mathrm{m}}{\mathrm{s}}$$

The Reynolds number for the jacket is, with equation (6.78):

$$\mathrm{Re}_J = \frac{1{,}000 \cdot 1.85 \cdot 0.067}{0.0009} = 1.38 \cdot 10^5$$

h_{jacket} is

$$h_{jacket} = \frac{0.6 \cdot 0.023 \cdot \left(1.37 \cdot 10^5\right)^{0.8} 6.28^{0.33}}{0.067} = 4{,}859 \frac{W}{m^2 {}^\circ C}$$

And the overall heat transfer coefficient U:

$$\frac{1}{U} = \frac{1}{4{,}859} + \frac{2.12}{2.10 \cdot 2{,}281} + \frac{2.12 \ln\left(\frac{2.12}{2.10}\right)}{2 \cdot 16}$$
$$= 0.000205 + 0.00044 + 0.000627 = 0.00127 \frac{\mathrm{m}^2\ {}^\circ\mathrm{C}}{\mathrm{W}}$$

$$U = 786 \frac{\mathrm{W}}{\mathrm{m}^2 \cdot {}^\circ\mathrm{C}}$$

6.6 CONSIDERATIONS FOR SCALE-UP AND SCALE-DOWN

As we have seen in Chapter 5, scale-up and scale-down of fermentation processes are often carried out in the industry. In dealing with a scale-up/scale-down problem it is important to ensure that the scaled-up/scaled-down process reproduces the performance of the original reactor. We have seen in Chapter 5 how the scale-up/scale-down affects oxygen transfer. In this section, we will see the effects of scale-up/scale-down on the fermentation temperature and how to deal with these effects.

We will use the same assumptions that we used in Chapter 5, i.e. that the scale-up/scale-down is made maintaining the same residence time, feed concentration and physical properties of the medium. We assume that it is aimed to maintain the temperature of the fermentation constant on scale-up/scale-down. The first equation to consider is the enthalpy balance (6.41). From this equation we see that, if both the original and the scaled-up/scaled-down vessels are adiabatic ($U = 0$) the fermentation temperature T_R will be constant on scale-up/scale-down, as long as the residence time and feed composition are kept constant. Indeed, for constant residence time and feed composition, the terms r_H and $\frac{W_{feed}}{V}$ are constant on scale-up/scale-down. So, the scale-up/scale-down of adiabatic reactors with the aim of maintaining the same temperature is, in principle, straightforward.

The situation is more complicated if the reactor is not adiabatic, but temperature is controlled with an external jacket. In this case the enthalpy balances are equations (6.41) and (6.63) (enthalpy balance for the fluid in the jacket). Equation (6.41) indicates, that, in order for the temperature T_R to be constant on a scale-up/scale-down done with the same feed composition and residence time (same r_H and $\frac{W_{feed}}{V}$), the term $\frac{UA(T_R - T_J)}{V}$ needs to be kept constant between the scaled-up/scaled-down and the original reactors. The term $\frac{A}{V}$ is a purely geometrical term and term and is given, for a cylindrical vessel where all the external area (including the base) is used for heat transfer, by

$$\frac{A}{V} = \left(\frac{1}{\alpha} + 4\right)\sqrt[3]{\frac{\alpha\pi}{4V}} \tag{6.79}$$

In equation (6.79) the parameter α indicates the ratio between the liquid height and the vessel diameter. Equation (6.79) shows that the ratio $\frac{A}{V}$ decreases on scale-up and increases on scale-down. This is an inevitable geometrical effect which needs to be compensated by the term $U(T_R - T_J)$. The enthalpy balance for the jacket fluid (6.63) indicates that, in order to keep the fermentation temperature T_R and therefore the term $\frac{UA(T_R - T_J)}{V}$ constant on scale-up/scale-down, the term $\frac{W_J C_{PJ}(T_J - T_{JIN})}{V}$ needs also to be kept constant. If the term $\frac{W_J}{V}$ is kept constant on scale-up/scale-down (i.e. the scale-up/scale-down is made maintaining the flow rate of the jacket fluid proportional to the reactor volume), then the condition of constant $\frac{UA(T_R - T_J)}{V}$ gives that T_J will also be constant on scale-up/scale-down. Ultimately, therefore, the condition of constant temperature on scale-up/scale-down implies that the term $\frac{UA}{V}$ needs to be kept constant and therefore, the heat transfer coefficient U needs to compensate for the change of $\frac{A}{V}$. In other words, the condition of constant $\frac{UA}{V}$ on scale-up/scale-down translates, using equation (6.79) into

$$\frac{U_{large}}{U_{small}} = \sqrt[3]{\frac{V_{large}}{V_{small}}} \tag{6.80}$$

The heat transfer coefficient U, depends, as we have seen in section 6.5, on the heat transfer coefficients on the reactor side and on the jacket side. Generally, both these heat transfer coefficients are important, however a simplified analysis can be made assuming that the controlling heat transfer resistance is the one on the reactor side (this means that the heat transfer coefficient is much lower for the reactor than for the jacket). This generally corresponds to very viscous fermentations (high microorganisms' concentration). In this case $U \equiv h_{reactor}$ and, using equation (6.73) with constant physical properties on scale-up/scale-down:

$$\frac{U_{large}}{U_{small}} = \left(\frac{N_{large} D^2_{ag,large}}{N_{small} D^2_{ag,small}}\right)^{a_2} \frac{D_{vessel,small}}{D_{vessel,large}} \tag{6.81}$$

Equation (6.81) can be re-written taking geometrical similarity and the relationship between vessel diameter and volume, obtaining

$$\frac{U_{large}}{U_{small}} = \left(\frac{N_{large}}{N_{small}}\right)^{a_2} \left(\frac{V_{large}}{V_{small}}\right)^{\frac{2a_2 - 1}{3}} \tag{6.82}$$

The comparison of equations (6.82) and (6.80) gives how the agitator speed needs to change on scale-up/scale-down to maintain the same fermentation temperature:

$$\left(\frac{N_{large}}{N_{small}}\right)=\left(\frac{V_{large}}{V_{small}}\right)^{\frac{2}{3}\left(\frac{1}{a_2}-1\right)} \tag{6.83}$$

Since the coefficient a_2 is usually <1 equation (6.83) indicates that, in order to maintain the fermentation temperature constant, the agitation rate needs to increase in scale-up and decrease in scale-down. This can be a serious problem for scale-up. Indeed, the required increase in agitation speed would cause a significant increase in the required power draw per unit volume (note that the agitator power draw depends on the agitation speed to the third power and to the agitator diameter to the fifth power and the agitator diameter increases on scale-up), which would probably make the scale-up at constant temperature impossible, especially for large scale-up ratios. Figure 6.12 shows a plot of equation (6.83) for $a_2 = 0.67$.

The change in the agitator power required to maintain the same temperature on scale-up and scale-down is given by

$$\left(\frac{P_{large}}{P_{small}}\right)=\left(\frac{V_{large}}{V_{small}}\right)^{\left(\frac{2}{a_2}-\frac{1}{3}\right)} \tag{6.84}$$

Equation (6.84) can be derived by combining equation (6.83) with the definition of power number (5.24), assuming that the power number remains constant on scale-up/scale-down (this corresponds to being in the region of the power curve where the power number is independent on the Reynolds number), and by assuming geometric similarity (equation (5.21)). Equation (6.84) can also be rewritten to show the change in the required power draw per unit volume

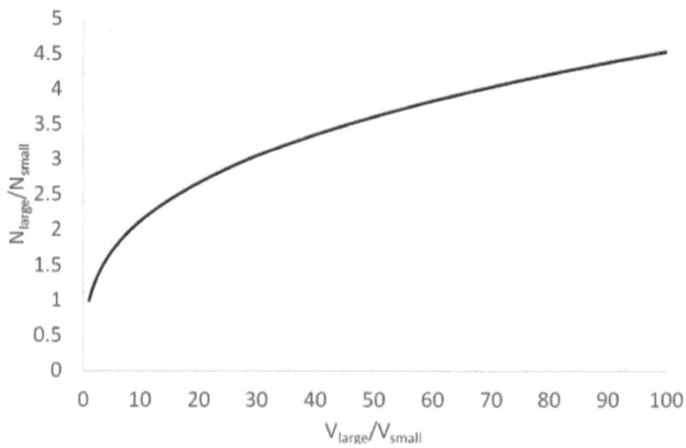

FIGURE 6.12 Plot of equation (6.83) for $a_2 = 0.67$.

$$\frac{\left(\frac{P_{large}}{V_{large}}\right)}{\left(\frac{P_{small}}{V_{small}}\right)} = \left(\frac{V_{large}}{V_{small}}\right)^{\left(\frac{2}{a_2}-\frac{4}{3}\right)} \tag{6.85}$$

Equation (6.85) shows that, since a_2 is generally <1, the required power draw per unit volume increases on scale-up.

To try and limit this problem, other strategies are possible: increasing the jacket fluid flow rate more than linearly with the reactor volume, so that T_J decreases and (T_R-T_J) increases; using a jacket fluid with a lower inlet temperature; increasing the heat transfer area in the large vessel, for example using coils. If the controlling heat transfer resistance is not the one on the reaction side but both the reaction and jacket side resistances are important, the effect of the change in the velocity of the jacket fluid on scale-up/scale-down also becomes important (equation (6.76)). The jacket fluid velocity tends to increase on scale-up, making the scale-up somewhat easier than predicted by equation (6.83). In general, however, temperature control is easier in smaller vessels because of the higher ratio between the heat transfer area and the reactor volume.

Comparing the findings in this section with the section on scale-up/scale-down of oxygen transfer in Chapter 5, it is evident that the design of the scaled-up/scale-down reactor is different. For example, scaling up to maintain the same oxygen concentration requires a decrease in both the agitation speed and the power draw per unit volume, while scaling-up to maintain the same temperature requires an increase in both variables. In other words, scaling-up/down based on one variable causes changes in other variables. This is typical of all scale-up/scale-down problems. Not all the variables can be kept constant on scale-up/scale-down, it has to be decided which variable is more important to control and, based on this decision, the appropriate scale-up/scale-down rules need to be used.

Example 6.7

Assume that a fermentation carried out at 0.1 m^3 needs to be scaled-up to 0.5 m^3. It is desired to maintain the same fermentation temperature. The agitation speed in the small vessel is 100 rpm and the agitator power is 100 W. Assume that the heat transfer is controlled by the resistance of the fluid in the reactor and that the coefficient a_2 is equal to 0.67. Assume that the power number is the same for the large and small vessel. Calculate the stirring speed and agitator power required in the large vessel.

Solution

With the hypothesis of this problem equation (6.83) can be used. The agitator speed in the large reactor is

$$N_{large} = 100(5)^{\frac{2}{3}\left(\frac{1}{0.67}-1\right)} = 170 \text{ rpm}$$

To calculate the required power draw in the large vessel we use equation (6.84):

$$P_{large} = 100 \text{ W} \cdot 5^{\left(\frac{2}{0.67}-\frac{1}{3}\right)} = 7116 \text{ W} = 7.1 \text{ kW}$$

This is a very large power. The power per unit volume increases from 1.0 kW/m^3 in the small vessel to 14.2 kW/m^3 in the large vessel. This is very probably impossible to achieve. In this case, the scale-up at constant temperature would require lowering the jacket temperature T_J, using coils to increase the heat transfer area or using multiple vessels to decrease the scale-up ratio.

QUESTIONS AND PROBLEMS

6.1 Describe the main issues in the scale-up of jacketed agitated vessels for fermentation reactions if the reaction is limited by heat transfer.

6.2 With reference to the stoichiometry of glucose fermentation reported below, calculate the heat generated by the reaction (in kJ) when 1 kg of ethanol is produced, if the reaction is carried out at 40 °C. Assume that at 40 °C all the species are in their reference state and ignore any heat of mixing. Assume that the specific heat is constant in the considered temperature interval. The thermodynamic properties are reported in Table 6.3.

$$C_6H_{12}O_6 + 0.12NH_3 \rightarrow 0.12C_5H_7O_2N + 1.8C_2H_5OH + 1.8CO_2 + 0.36H_2O$$

TABLE 6.3
Thermodynamic Properties for Problem 6.2

Substance	Reference State	Heat of Formation at 25 °C (kJ/mol)	Specific Heat (kJ mol^{-1} °C^{-1})
Glucose	Dissolved in water	−1270	0.021
Ammonia	Dissolved in water	−80.8	0.078
Biomass	Solid	−770	0.333
Ethanol	Liquid	−278	0.113
Carbon dioxide	Gas	−393	0.037
Water	Liquid	−286	0.075

6.3 A microorganism grows anaerobically on a substrate in a continuous fermenter at steady state. Under the initial operating conditions the substrate concentration in the feed is S_{0a}, the substrate and biomass concentration in the reactor are S_a and X_a, respectively. The temperature in the reactor is T_{Ra}. The feed flow rate is Q and the reactor volume used for the fermentation is V. Heat is removed via an external jacket where water flows at a flow rate w_J. The influent temperature of the cooling water is T_{JIN} and the effluent temperature is T_{Ja}.

It is proposed to increase the substrate concentration in the feed to S_{0b}, in order to increase the microorganisms' productivity. Determine whether it is possible to operate the reactor with the influent substrate concentration S_{0b} without exceeding the maximum allowable temperature for the microorganisms, T_{Rmax}.

Assumptions: assume that the enthalpy of reaction and the heat transfer coefficient U remain unchanged under the new conditions and that the cooling fluid flow rate and inlet temperature w_J and T_{Jin} cannot be changed. In the heat balances, neglect the contribution of the enthalpy of the reactor feed and outlet streams. Assume that the jacket is completely mixed. Assume that the microorganism growth rate follows Monod kinetics and that the values of the kinetic parameters are independent of temperature.

Data

V= 10 m^3; Q= 0.001 m^3/s; S_{0a}= 10 kg/m^3; S_a= 2 kg/m^3; X_a= 4 kg/m^3; T_{Ra}= 35°C; S_{0b}=15 kg/m^3; w_J = 1.5 kg/s; C_{PJ} = 4186 J/kg.°C; T_{Jin} = 5°C; T_{Ja} = 15°C; A = 30 m^2; T_{Rmax} = 50°C.

6.4 A microorganism grows in a continuous fermenter, with a substrate concentration in the feed equal to S_0. The dilution rate is D. The fermenter is cooled via a jacket, where cooling water flows at flow rate w_J. The inlet temperature of the cooling water is T_{JIN} and the heat transfer coefficient is U.

Calculate the temperature in the reactor.

Assumptions: the microorganism follows Monod kinetics. The jacket can be assumed perfectly mixed with a uniform temperature equal to $T_{cool,out}$. Neglect the enthalpies of the inlet and outlet streams.

Data: μ_{max} = 6 d^{-1}; K_S = 0.1 g/L; S_0 = 20 g/L; D = 5 d^{-1}; $Y_{X/S}$ = 0.5 g/g; Y_H = 3·10^{-8} kg/J; w_J = 2.0 kg/s; C_{PJ} = 4186 J/kg.K; V = 10 m^3; A = 30 m^2; T_{Jin} = 5 °C; U = 300 W/m^2/K

6.5 The reaction of citric acid ($C_6H_8O_7$) production from glucose by the microorganism *Aspergillus Niger* is

$$C_6H_{12}O_6 + 2.35O_2 + 0.1NH_3 \rightarrow 0.1C_5H_7O_2N + 0.7C_6H_8O_7 + 1.3CO_2 + 3H_2O$$

It is desired to produce 15 t of citric acid per day with a continuous process. Calculate the maximum concentration of glucose (in kg/m^3) that it is possible to have in the feed of the reactor without exceeding the maximum operating temperature of 35 °C.

The feed of the reactor has a temperature of 25 °C. Assume the reactor is adiabatic.

The enthalpy of the reaction is $-1.56 \cdot 10^6$ J/mol glucose. Assume that this enthalpy refers to the reactants in their physical state under the reaction conditions. Assume that the enthalpy of reaction is constant in the temperature range considered here. Assume that the density of all the streams is equal to 1000 kg/m^3 and their heat capacity is 4.19 kJ kg^{-1} °C^{-1}.

Ignore the enthalpy absorbed by the inlet gas stream.

6.6 The reaction of ethanol production from glucose using the bacterium *Zymomonas mobilis* under anaerobic conditions can be schematised as follows:

$$C_6H_{12}O_6 + 0.30NH_3 \rightarrow 0.30C_5H_7O_2N + 1.5C_2H_5OH + 1.5CO_2 + 0.90H_2O$$

The reaction is carried out in a continuous reactor where the ethanol production rate is 150 kg/h. The volume of the reactor is 56 m^3, the dilution rate is 0.8 d^{-1} (assume that the inlet and outlet flow rates are the same) and the temperature of the feed is 20 °C. The reactor is cooled with a jacket, where a fluid having an inlet temperature of 10 °C and a flow rate of 10,000 kg/h circulates. The heat transfer area is 40 m^2 and the heat transfer coefficient is 200 W m^{-1} °C^{-1}

Calculate the temperature in the reactor.

Density of the feed and of the reaction medium: 1000 kg/m^3. Specific heat of the feed, reaction medium and jacket fluid: 4.19 kJ kg^{-1} °C^{-1}. The enthalpy of the reaction, assumed constant in the temperature interval considered here, is $" H_R = -118$ kJ/mol glucose.

6.7 A fermentation process is carried out at pilot-scale (50 L) in a cylindrical continuous fermenter. Cooling is achieved with a jacket that covers all the external area (including the bottom base) of the vessel. The fermenter is mixed with an agitator which rotates at 90 rpm.

It is now desired to scale-up the process to a geometrically similar 0.5 m^3 vessel, maintaining the same residence time and substrate concentration than in the pilot scale process. It is also desired to maintain all the inlet and outlet temperatures for the reactor and the jacket at the same values as in the pilot-scale process. Assuming that in all conditions the heat transfer resistance is controlled by the reaction fluid (i.e. heat transfer coefficients for the vessel walls and for the jacket are much larger than for the fluid in the reactor), calculate the required agitation rate and flow rate of the jacket fluid in the large vessel. Assume that the heat transfer coefficient for the fluid in the reactor is given by the following correlation:

$$\frac{h_{reactor} D_{vessel}}{k_L} = 0.74\,\mathrm{Re}_{ag}^{0.67}\,Pr_{ag}^{0.33}$$

6.8 A fermentation is carried out in a continuous process, with the inlet feed at a temperature of 15 °C and the reactor temperature of 28 °C. The cooling fluid has a flow rate of 100,000 kg/d, enters the jacket at a temperature of 10 °C and leaves at 17 °C. The fermenter has a working volume of 10 m^3 and the feed flow rate is 20 m^3/d (density 1000 kg/m^3). The substrate concentration in the feed is 80 kg/m^3 and it can be assumed that the substrate is entirely removed by the microorganisms. Calculate the enthalpy of the fermentation (J/kg substrate).

6.9 It is known from experimental data that the fermentation of 1 kg of substrate with the addition of 0.02 kg of ammonia will yield the following: 0.12 kg of microorganisms, 0.40 kg of the desired product, 0.35 kg of carbon dioxide, the balance being produced water. The enthalpies of the species are summarised in Table 6.4 (assume that all the species in this reaction are in their reference state). Calculate the maximum substrate concentration (assuming the substrate is completely removed) that it is possible to have in an adiabatic continuous fermenter carrying out this reaction without exceeding the maximum allowable temperature of 35 °C. Assume that the feed enters the reactor at 22 °C. Assume that the feed has a density of 1000 kg/m^3 and a specific heat of 4.19 kJ kg^{-1} K^{-1}.

6.10 A vessel which is being considered for a fermentation process has the following dimensions: volume 25 m^3, diameter 3.2 m, liquid height 3.1 m, agitator diameter 1.0 m. The agitator has a power number (to be assumed constant in all the considered conditions) of 3.5. It is desired not to exceed an agitator power of 15 kW. The jacket is a spiral jacket which covers the cylindrical section only, to the full liquid height. The spacing between the jacket and the external wall is 40 mm and the spiral pitch is 50 mm. The jacket fluid is water entering at 5 °C with a maximum flow rate of 50 m^3/d.

TABLE 6.4
Enthalpy Properties for Problem 6.9. The Specific Heats are Given for the Substances in Their Reference State and Can be Assumed Constant Within the Considered Temperature Range

Substance	Reference State	ΔH_f^0 (J/kg)	Specific Heat (J kg^{-1} K^{-1})
Substrate	Solid	$-7.500 \cdot 10^6$	$1.50 \cdot 10^3$
Ammonia	Dissolved in water	$-4.511 \cdot 10^6$	$4.80 \cdot 10^3$
Biomass	Solid	$-8.000 \cdot 10^6$	$3.00 \cdot 10^3$
Product	Dissolved in water	$-1.080 \cdot 10^7$	$1.18 \cdot 10^3$
Carbon dioxide	gas	$-9.000 \cdot 10^6$	$8.71 \cdot 10^2$
Water	liquid	$-1.600 \cdot 10^7$	$4.19 \cdot 10^3$

It is desired to use this vessel for a fermentation process having a feed of 75 m^3/d at 20 °C and a substrate concentration of 70 kg/m^3. The enthalpy of reaction at the fermentation conditions is –9.50·10^5 J/kg substrate. Determine whether this vessel is suitable for this fermentation, considering that the maximum reaction temperature must not exceed 30 °C. Assume that the substrate is completely removed in the fermentation and ignore the resistance to heat transfer of the vessel walls. In the calculation of the overall heat transfer coefficient U, assume that the ratio between the external and internal diameter of the vessel is equal to 1.

Assume for all the streams, the following physical properties: conductivity 0.6 W m^{-1} °C^{-1}, density 1000 kg/m^3 and specific heat 4.19 kJ kg^{-1} °C^{-1}. Assume a viscosity of 1.5·10^{-1} Pa·s for the fluid in the reactor and viscosity 8.9·10^{-4} Pa·s for the fluid in the jacket.

6.11 A fermentation process is carried out in batch. The kinetic parameters for the process are the following: μ_{max} = 3.8 d^{-1}, K_S = 0.06 kg/m^3, $Y_{X/S}$ = 0.14 kg/kg. The enthalpy of the reaction is —9.67·10^5 J/kg substrate. The process is started at 30 °C with biomass and substrate concentration of 0.5 and 50 kg/m^3, respectively. It is desired to maintain the fermentation temperature to 30 °C throughout the length of the fermentation by manipulating the cooling water flow rate. The cooling water enters the jacket at 15 °C. The overall heat transfer coefficient is U = 100 W m^{-2} °C^{-1}.

Calculate and plot the required time profile for the cooling water in this process.

Assume that all the kinetic parameters and physical properties are constant throughout the length of the fermentation. The reactor has a diameter of 1.5 m, a height of 2 m, and the jacket width is 50 mm. Assume that the jacket covers only the sides of the vessel and not the bottom. Assume that the density of the reactor and jacket fluid is 1000 kg/m^3. The specific heat for all streams is 4.19 kJ kg^{-1} °C^{-1}.

6.12 Calculate the maximum initial substrate concentration and the maximum product productivity (t of product per year) for an adiabatic batch fermentation process characterised by the following parameters: μ_{max} = 4.5 d^{-1}, K_S = 0.07 kg/m^3, $Y_{X/S}$ = 0.12 kg/kg; $Y_{P/S}$ = 0.35 kg/kg. The enthalpy of reaction is –5.50·10^5 J/kg. Each batch is started at a temperature of 20 °C with 200 mg/L of microorganisms and is stopped when the substrate concentration is below 1 kg/m^3 and the temperature reaches 35 °C. Assume that the process runs for 8000 h per year and the downtime between consecutive batches is 2 h. The volume of the reactor is 10 m^3, the fluid density 1000 kg/m^3 and the specific heat 4.19 kJ kg^{-1} °C^{-1}.

6.13 A fermentation is carried out in a 10 m^3 continuous reactor. Under the standard fermentation conditions, the feed flow rate is 25 m^3/d and has a substrate concentration of 60 kg/m^3. The effluent substrate and product concentrations are 2 kg/m^3 and 25 kg/m^3 respectively, the jacket flow rate is 200 m^3/d, the jacket inlet and outlet temperatures are 10 and 20 °C, respectively. The feed of the reactor has a temperature of 22 °C and the reactor temperature is 30 °C. The reactor is mixed with an agitator rotating at a speed of 80 rpm.

It is now desired to increase the substrate concentration in the feed to 70 kg/m^3, to increase the product productivity, maintaining the same temperature in the reactor. Assuming that the heat transfer rate is controlled by the resistance of the fluid in the reactor, calculate the new required value of the agitation rate. Also calculate the new product productivity and the increase in agitator power draw due to the increased agitation rate. Assume that the power number of the agitator is the same in all cases.

Assume that the heat transfer coefficient for the fluid in the reactor is given by the following correlation:

$$\frac{h_{reactor} D_{vessel}}{k_L} = 0.74\,\mathrm{Re}_{ag}^{0.67}\, Pr_{ag}^{0.33}$$

7 Downstream Processes

The term "downstream processes" generally indicates all processes that follow the enzymatic reaction or the fermentation stages. The aim of downstream processes is to separate the microorganisms and enzymes and to purify the product(s) to the required specification. For example, in the production of penicillin by the microorganism *Penicillium chrysogenum*, at the end of the fermentation penicillin is present in the reaction medium at concentrations lower than 5 g/L (less than 0.5% of the total mass of the medium) and, to be used for medical applications, needs to be purified to >98% w/w by removing water, microorganisms, salts and any other medium components. The chemical and physical nature of microorganisms, enzymes, components of the reaction medium (e.g. salts) and products is very different, as well as different products may have very different physical and chemical properties. Therefore, depending on the type of substances to separate and purify, very different downstream processes may be required. Many downstream processes used in biochemical engineering, for example distillation, drying, crystallisation, are also used, often at much larger scale, in other sectors of chemical engineering and are covered in other chemical engineering courses, therefore they will not be covered here. In this section, we will cover those separation and purification processes that are mainly, although certainly not exclusively, used as downstream processes after enzymatic and fermentation reactors: separation of microorganisms and crossflow filtration; ion exchange; adsorption; chromatography.

7.1 SEPARATION OF MICROORGANISMS AND CROSSFLOW FILTRATION

7.1.1 Separation of Microorganisms

After fermentation processes, often the first step is the removal of microorganisms from the rest of the reaction mixture, which contains the desired product. Microorganisms are usually removed by pressure filtration or centrifugal filtration (Figure 7.1), which remove from the solution any particles larger than approximately 1–5 μm (depending on the pore size of the filter medium). Pressure and centrifugal filtration are also called cake filtration because suspended solids form a cake on the filter medium.

DOI: 10.1201/9781003633822-7

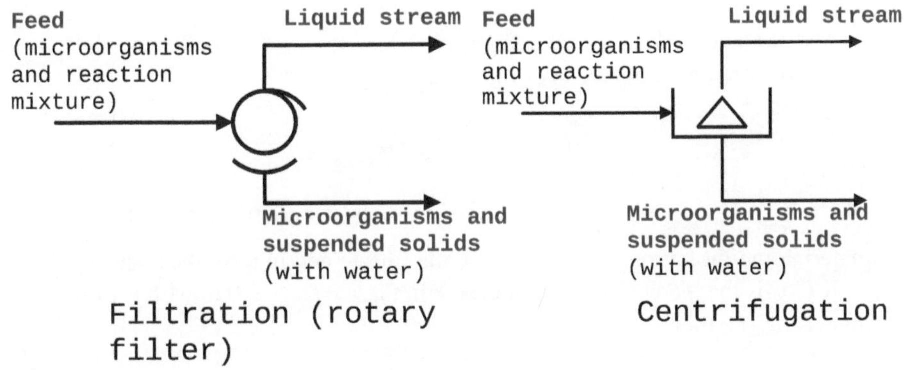

FIGURE 7.1 Conceptual scheme of filtration and centrifugation to remove microorganisms after fermentation.

The filter medium can be made of various materials, e.g. canvas cloth, metal cloth, synthetic fabrics. For separation of microorganisms, filter media with porosity in the range 0.2–5 μm are typically used. After the first layer of microorganisms (or more in general suspended solids) has deposited on the filter, the filtration action is mainly done by the solids on the filter (cake) rather than by the filter itself. Over the course of the filtration the deposition of the solids increases the resistance to filtration therefore the energy applied must increase. In pressure filtration, the energy required for the fluid to pass through the filter medium is obtained by applying a differential pressure, in centrifugal filtration the energy is obtained by applying centrifugal force.

Note that the cake containing microorganisms and suspended solids also includes the water that fills the pores in the cake, therefore the volume of filtrate (liquid stream) is slightly lower than the volume of liquid in the feed.

Example 7.1

A fermentation medium consisting of a total flowrate of 100 kg/h with microorganisms at concentration of 5 g/L is to be filtered by cake filtration to remove the microorganisms. Assuming that the cake retains all microorganisms and that it has a water content of 60%, what is the flowrate of the filtrate? Assume that microorganisms are the only suspended solids in the fermentation medium.

Solution

The mass of microorganisms to be removed is (note that 5 g/L corresponds to 0.005 kg/kg if we assume a density of the fermentation medium of 1000 kg/m^3, i.e. the density of water): $100\frac{kg}{h}\cdot 0.005\frac{kg}{kg} = 0.5\frac{kg}{h}$. The mass flowrate of water and soluble materials in the fermentation medium is $100 - 0.5 = 99.5\frac{kg}{h}$.

The mass flow rate of filtrate, made of water and soluble materials, is obtained by a mass balance over the filtration system:

$$m_{win} = m_{wfiltrate} + m_{wcake}$$

where $m_{win}, m_{wfiltrate}, m_{wcake}$ are the mass flow rates (in kg/h) of water (plus soluble materials) in the fermentation broth, in the filtrate and in the cake respectively.

If f_{wcake} is the water content in the cake in mass ($f_{wcake} = 0.6$ with the data of this example), then

$$m_{wcake} = m_{cake} f_{wcake}$$

where m_{cake} is the total mass flowrate of the cake leaving the system. Note that the cake is typically removed discontinuously from the filter, once enough cake has accumulated, therefore m_{cake} and m_{wcake} are to be intended as average mass flow rate over the length of the whole filtration process.

Calling m_{scake} the mass flowrate of microorganisms (suspended solids) in the cake (with the same comment as for m_{cake} and m_{wcake}) which in this case is equal to 0.5 kg/h, the total mass flowrate of the cake is given by

$$m_{cake} = m_{scake} + m_{wcake} = m_{scake} + m_{cake} f_{wcake}$$

from which we obtain the expression for the total mass flowrate of the cake:

$$m_{cake} = \frac{m_{scake}}{1 - f_{wcake}}$$

Therefore, the mass flowrate of the water (plus soluble substances) in the cake is

$$m_{wcake} = \frac{m_{scake}}{1 - f_{wcake}} f_{wcake} = \frac{0.5 \frac{\text{kg}}{\text{h}}}{1 - 0.6} 0.6 = 0.75 \frac{\text{kg}}{\text{h}}$$

Therefore, the mass flowrate of the filtrate is

$$m_{wfiltrate} = m_{win} - m_{wcake} = (99.5 - 0.75) \frac{\text{kg}}{\text{h}} = 98.75 \frac{\text{kg}}{\text{h}}$$

7.1.2 Crossflow Filtration

After microorganisms and suspended solids have been removed, fine particles or soluble species can be removed by crossflow filtration. In crossflow filtration, the feed

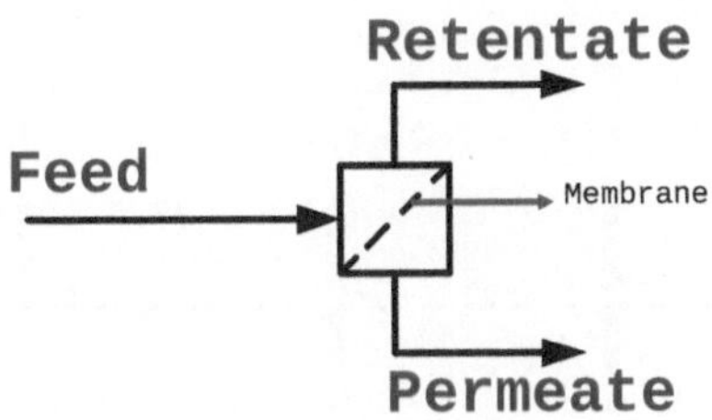

FIGURE 7.2 Conceptual scheme of crossflow filtration.

flows on the surface of the separation membranes, which are more permeable to some components of the feed mixture than to others. The stream that passes through the membrane is called permeate, the stream that does not pass is called retentate (or concentrate) (Figure 7.2).

Crossflow filtration is classified depending on the type of membrane and on the size of the substances retained by the membrane. Often the type of membrane is characterised by its cutoff, i.e. the smallest size or lowest molecular weight of the substances retained. However, in general separation and retention is never complete and there is usually some permeation of substances even above the cutoff size.

Microfiltration retains small suspended solids and fine particles of size typically 0.5–5 μm. Some of these particles could also be separated by cake filtration, however with small particles the cake resistance would often be too high giving very low filtration rate. Ultrafiltration retains colloidal materials (size smaller than 0.5 μm) or large soluble molecules (e.g. some proteins). Nanofiltration retains soluble organics of different molecular weights depending on the type of membrane. Reverse osmosis retains essentially all organic and inorganic species (salts present in fermentation mixtures) allowing permeation of only pure water.

Figure 7.3 shows a scheme of the separation operated by the different types of crossflow membranes.

Common materials for crossflow membranes are polysulfone, polyimide, polycarbonate, cellulose acetate (Figure 7.4). Many other materials are also used. The pore size and cutoff can be varied by changing the membrane preparation method and conditions.

Figure 7.5 shows the flow of feed, retentate and permeate inside a membrane module for crossflow filtration.

Multiple membrane modules can be arranged in series on in parallel (Figure 7.6) or in other ways.

Table 7.1 shows some examples of the application of crossflow filtration in the enzyme or fermentation industries

The area required for a certain crossflow filtration process is usually determined experimentally by measuring the permeate flux as a function of the pressure applied:

$$Flux = Q_m \cdot \Delta P \tag{7.1}$$

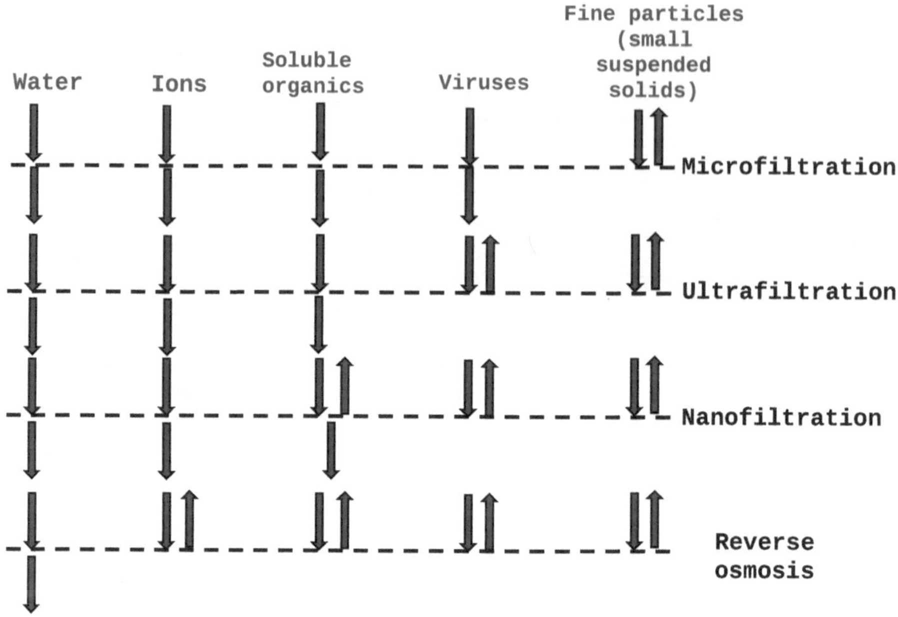

FIGURE 7.3 Conceptual scheme of crossflow filtration. For each type of filtration, the extent of separation depends on the size of the molecules and on the type of membranes. Large soluble organics (e.g. enzymes) may be retained by ultrafiltration membranes.

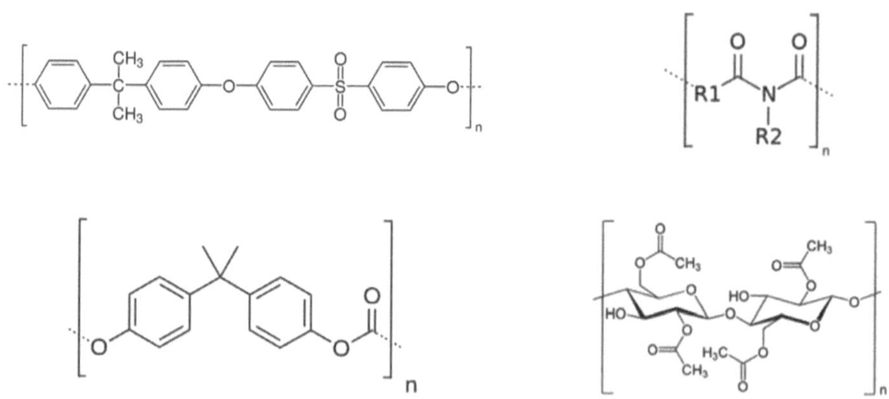

FIGURE 7.4 Chemical structures of common materials for crossflow membranes: polysulfone (top left, https://en.wikipedia.org/wiki/Polysulfone#/media/File:Polysulfone.svg); polyimide (top right); polycarbonate from bisphenol A (bottom left, https://en.wikipedia.org/wiki/Polycarbonate#/media/File:Lexan.svg); cellulose acetate (bottom right, https://en.wikipedia.org/wiki/Cellulose_acetate#/media/File:Cellulose_acetate.svg).

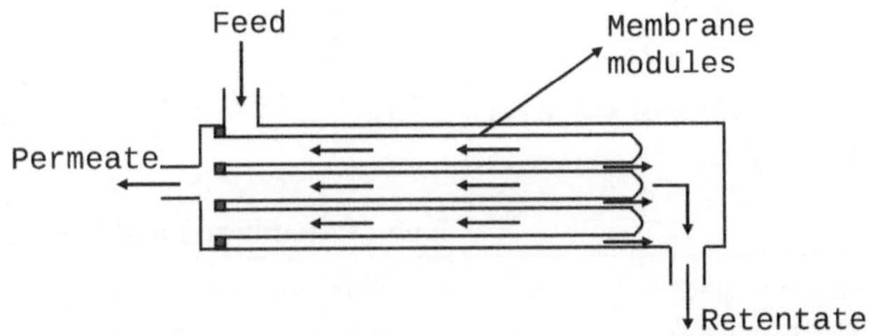

FIGURE 7.5 Conceptual scheme of fluid flow inside a crossflow membrane module.

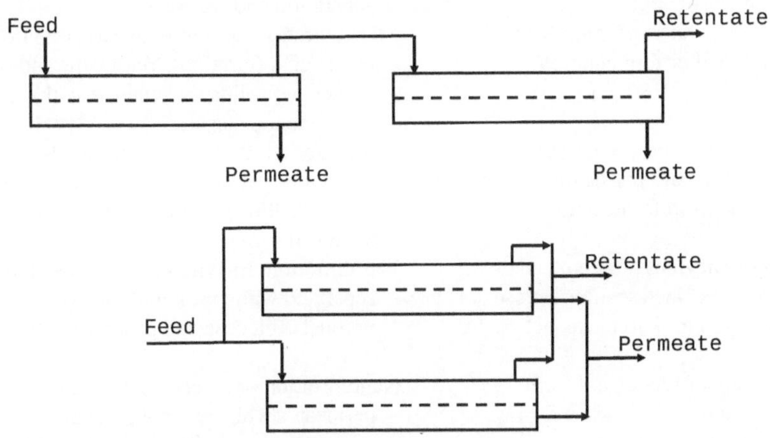

FIGURE 7.6 Scheme of two crossflow membrane modules arranged in series (top) and in parallel (bottom).

Typical units for flux are L m^{-2} h^{-1}. ΔP is the pressure drop across the membrane (transmembrane pressure, typical units are atm). Q_m (typical units L m^{-2} h^{-1} atm^{-1}) is the membrane permeability, which depends on the nature of the membrane (chemical composition, thickness, porosity) and of the separation. From the experimental measurement of the flux and permeability in a pilot unit, the membrane area required for a commercial-scale unit can be calculated.

The total and component mass balances can be calculated for a crossflow filtration at steady-state using the notation in Figure 7.7. No reactions occur in crossflow filtration, so the mass balance is simply input=output for the total mass and for individual components.

Using the notation in Figure 7.7, mass balances can be written for the total mass and for the mass of each component with equations (7.2) and (7.3), respectively:

$$F = R + P \tag{7.2}$$

$$Fx_i = Rr_i + Pp_i \tag{7.3}$$

TABLE 7.1
Examples of Applications of Crossflow Filtration in the Biochemical Engineering Industry

Application	Type of Membrane and Description
Concentration of organic acids (e.g. citric and lactic) from fermentation.	Nanofiltration. Water and unwanted salts permeate, while organic acids are retained and concentrated.
Amino acid and peptide recovery after enzymatic protein hydrolysis	Ultrafiltration. Amino acids permeate while large peptides are retained, allowing their separation and recovery.
Yeast cell removal in beer and wine production after fermentation	Microfiltration (as alternative to pressure or centrifugal filtration). Yeast cells and other suspended solids are retained so that beer and wine are clarified.
Enzyme (e.g. lipases, proteases) concentration and purification after their production from fermentation	Ultrafiltration. Water, salts and small molecules permeate while enzymes are retained, allowing for their concentration and purification
Removal of enzymes after enzymatic reactions (e.g. lactase after lactose hydrolysis in milk to make lactose-free milk)	Ultrafiltration. Enzymes are retained and separated while milk with the hydrolysis products (glucose and galactose) permeates
Concentration of bioethanol from fermentation.	Nanofiltration or reverse osmosis. Water permeates while ethanol is retained, allowing to concentrate the ethanol before distillation.

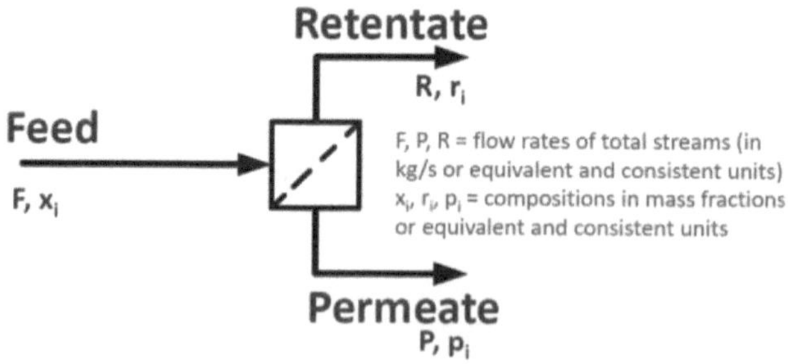

FIGURE 7.7 Scheme of crossflow filtration for mass balances.

In equation (7.3), the subscript i refers to each component in the feed, retentate and permeate feeds. Equations (7.2) and (7.3) allow calculation of unknown flowrates and compositions if other terms are known.

Example 7.2

For a certain fermentation effluent, laboratory scale ultrafiltration experiments indicated that a permeate flux of 50 L m^{-2} h^{-1} was obtained with a transmembrane pressure of 3 bar. Calculate the membrane area required to produce 150 L/h of permeate with a transmembrane pressure of 2.5 bar.

Solution

We can calculate the permeability of the membrane by re-arranging equation (7.1):

$$Q_m = \frac{Flux}{\Delta P} = \frac{50\,\text{L m}^{-2}\,\text{h}^{-1}}{3\,\text{bar}} = 16.7\,\text{L m}^{-2}\ \text{h}^{-1}\text{bar}^{-1}$$

When the transmembrane pressure is 2.5 bar the flux is

$$Flux = Q_m \cdot \Delta P = 16.7\ \text{L m}^{-2}\,\text{h}^{-1}\,\text{bar}^{-1} \cdot 2.5\ \text{bar} = 41.7\ \text{L m}^{-2}\ \text{h}^{-1}$$

Therefore, to produce 150 L/h of permeate, the required membrane area is

$$\text{Membrane area} = \frac{150\ \text{L h}^{-1}}{41.7\ \text{L m}^{-2}\,\text{h}^{-1}} = 3.6\ \text{m}^2$$

Example 7.3

In a nanofiltration process for the concentration of lactic acid from fermentation, the concentration of lactic acid is 10% w/w in the feed and 2% w/w in the permeate. The total flow rate of the feed is 500 kg/h, the total flow rate of the permeate is 400 kg/h. Calculate the mass flow rate of lactic acid in the retentate, the concentration of lactic acid in the retentate and the mass flowrate of water in the retentate and in the permeate. Assume that lactic acid and water are the only components.

Solution

The mass flowrate of lactic acid in the feed is 500·0.10 = 50 kg/h (this is the term Fx_i in equation (7.3)). The mass flowrate of lactic acid in the permeate is 400·0.02=8 kg/h (this is the term Pp_i in equation (7.3)). Therefore, according to equation (7.3), the mass flowrate of lactic acid in the retentate is

$$Rr_i = Fx_i - Pp_i = (50-8)\frac{\text{kg}}{\text{h}} = 42\frac{\text{kg}}{\text{h}}$$

The mass flowrate of water in the feed is 500 – 50 = 450 kg/h. The mass flow rate of water in the permeate is 400 – 8=392 kg/h. Therefore, the mass flowrate of water in the retentate is 450 – 392=58 kg/h. The total mass flow rate of the retentate is 58+42 = 100 kg/h. The concentration of lactic acid in the retentate is 42 kg/h/100 kg/h = 42% w/w.

7.2 ION EXCHANGE

7.2.1 Introduction

Ion exchange processes remove charged species from a liquid medium by contact with a solid phase (resin) exchanging ions from the liquid with ions from the solid. Usually, the process is carried out in columns packed with the resin with the liquid phase flowing through the column. After the fermentation or enzymatic reactions, ion exchange processes are sometimes used to remove cations and anions (e.g. salts in the mineral medium used to grow the microorganisms, and which are undesired in the final product), charged organic species (e.g. charged organic acids), nucleic acids, etc. A conceptual scheme of ion exchange is shown in Figure 7.8

If the removal of only cations or only anions is required, then the stream to be treated needs only to be passed through one column. If removal of both cations and anions is required, then the stream needs to flow through two columns in series (Figure 7.9).

Table 7.2 reports some examples of the use of ion exchange in the biochemical engineering industry.

7.2.2 Ion Exchange Resins and Column Operation

In most cases, industrial ion exchange processes use synthetic resins, prepared from polymers (e.g. cross-linked polystyrene) to which ion-exchange groups have

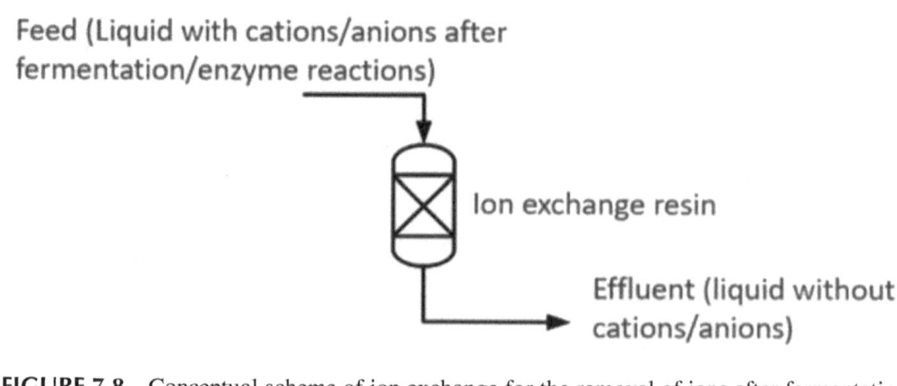

FIGURE 7.8 Conceptual scheme of ion exchange for the removal of ions after fermentation or enzymatic reactions.

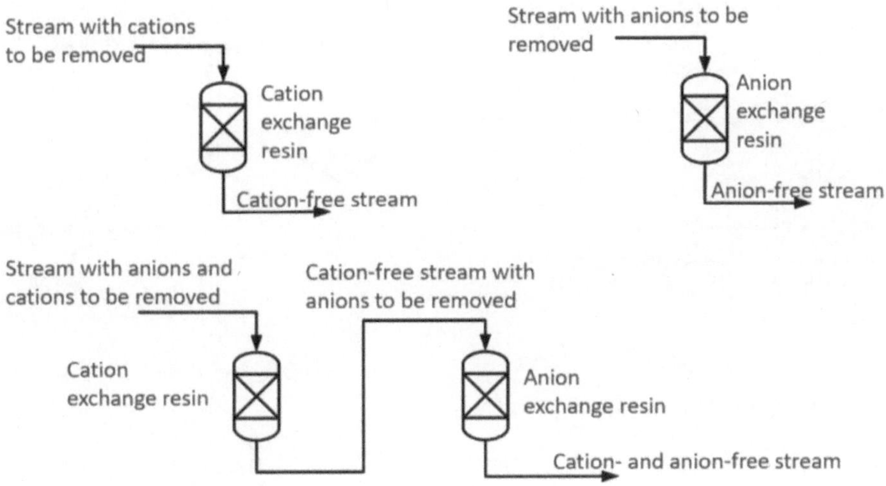

FIGURE 7.9 Conceptual scheme of ion exchange for the removal of cations and anions.

TABLE 7.2
Examples of Applications of Ion Exchange After Fermentation and Enzymatic Reactions

Application	Type of Resin and Description
Removal of metal ions (e.g. Ca^{2+}, Mg^{2+}) after enzymatic hydrolysis of proteins	Cation exchange. Charged metal ions are removed while uncharged hydrolysis products pass through the column
Removal of salts after fermentation to produce hyaluronic acid	Cation and anion exchange. Neutral hyaluronic acid is not retained by the resin while anions and cations from the fermentation broth are retained
Removal of nucleotides and nucleic acids from fermentation broth after production of biopharmaceuticals (e.g. recombinant proteins, monoclonal antibodies)	Anion exchange. Negatively charged nucleotides and DNA/RNA fragments are retained on the resin, allowing for the purification of biopharmaceuticals
Purification of fermentation-derived enzymes from salts that may affect their stability	Cation and anion exchange. Salts from the fermentation medium (e.g. Na^+, K^+, Cl^-, SO_4^{2-}) are removed allowing for the purification and improvement of stability of the enzymes

been added (Figure 7.10). For cation exchangers, the ion-exchange groups include, e.g. sulfonic acid (SO_3H, strong acid), carboxylic groups (COOH, weak acid) and others. Sulfonic acid and carboxylic groups remove cations of the general form A^+ from solutions by exchanging their H^+ ions (e.g. $-SO_3H + A^+ \rightarrow -SO_3A + H^+$).

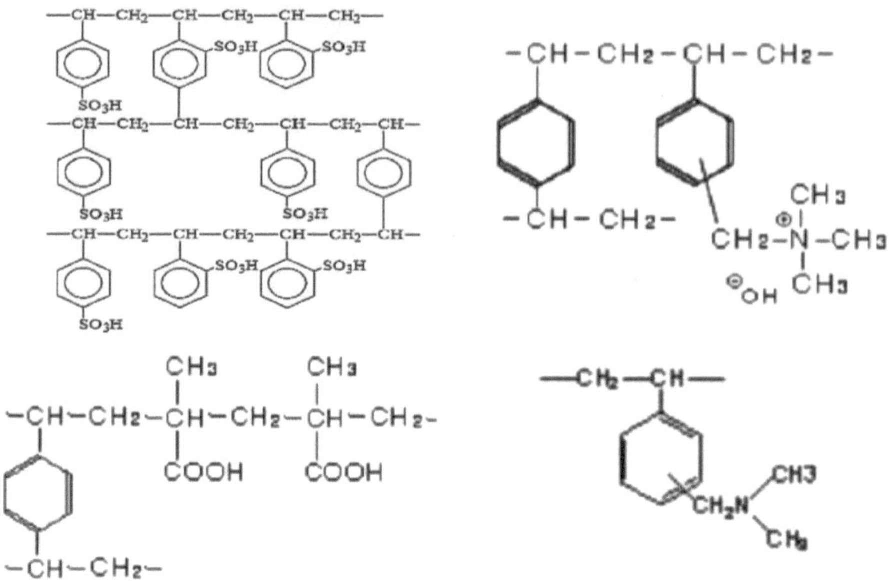

FIGURE 7.10 Examples of ion exchange polystyrene resins with functional groups. Top left: sulfonic acid groups (strong acid for cation exchange) (www.researchgate.net/figure/Chemical-structure-of-the-sulfonated-polystyrene-resin_fig1_351993927). Top right: quaternary ammonium groups (strong base for anion exchange) (www.m-chemical.co.jp/en/products/departments/mcc/ion/product/1201066_8072.html). Bottom left: carboxylic acid groups (weak acid for cation exchange) (www.m-chemical.co.jp/en/products/departments/mcc/ion/product/1201065_8072.html). Bottom right: amine groups (weak base for anion exchange) (www.m-chemical.co.jp/en/products/departments/mcc/ion/product/1201067_8072.html).

Note that in many cases the resin is made of the sodium salts rather than of the acidic form of the functional group (e.g. $NaSO_3$ rather than HSO_3). For anion exchangers, ion-exchange groups include quaternary ammonium groups ($N(CH_3)_3^+$ OH^-, strong base) or weak-base amine groups (NR_2, where R is a generic alkyl group, often CH_3). Anion exchangers groups remove anions (generic formula B^-) from solutions by exchanging their OH- groups or by reacting with the anion (e.g. $N\left(CH_3\right)_3^+ OH^- + B^- \rightarrow N\left(CH_3\right)_3^+ B^- + OH^-$ or $N\left(CH_3\right)_2 + HB \rightarrow NH\left(CH_3\right)_2 B$). Strong acid (e.g. SO_3H) and strong base (e.g. $N(CH_3)_3^+OH^-$) groups dissociate easily and can remove any cations or anions from solutions. On the other hand, weak acid (e.g. COOH) and weak base (e.g. $N(CH_3)_2$) groups only dissociate if the solution is not too acidic (weak acid groups) or too alkaline (weak base groups). For example, the group COOH is a weak acid and does not dissociate and does not work in strongly acidic solutions. Similarly, the group $CH_2N(CH_3)_2$ is a weak base and does not work in strong alkaline solutions.

In the usual application of a packed column, ion exchange is a discontinuous process where the column gradually fills with the ion to be exchanged until the column is full and operation must stop. Let us assume we use ion exchange to remove cations A^+ from a feed at concentration C_{A+0} using a resin with Na^+ counterions. Before the start of the ion exchange, the column is filled with resin in the NaR form. As the feed with A^+ passes through the column, A^+ is exchanged with Na^+ and the effluent contains no A^+. As the ion exchange progresses, the column gradually fills (saturates) with A^+. The sections saturated with A^+ will no longer be active in the ion exchange. When all the column is saturated with A^+, the column is no longer able to remove any A^+ ions, which will start being released into the effluent (breakthrough). At this point, the flow through the column must be stopped and the column must be regenerated in order for the process to start again. Figure 7.11 shows this process. As time progresses from t_1 to t_5, the column is gradually filled with the ion A^+. Until time t_3, there is no A^+ in the effluent. From t_3 onwards, the ion starts being released in the effluent until the effluent concentration becomes equal to the feed (column completely filled with A^+).

The shape of the curves in Figure 7.10 is affected by the mass transfer rate of the ions from the solution to the resin (Figure 7.12). For fast mass transfer, the ion to be removed transfers rapidly to the resin and the effluent concentration shows a rapid breakthrough. This is because for fast mass transfer breakthrough only happens when the column is (almost) completely saturated. On the other hand, for slow mass transfer the ion starts being released in the effluent when some of the column is still unsaturated. This shows the important of having fast mass transfer in ion exchange (and in all packed bed operations). For slow mass transfer, breakthrough happens early, which is a disadvantage as the column can be used for a shorter time than in the case of fast mass transfer. These concepts will be covered more quantitatively later in Section 7.2.5.

7.2.3 Regeneration

After the column has been saturated with the ion to be removed (breakthrough), the column has no ion exchange capacity left and becomes unusable. At this point, we must regenerate the resin by replacing the ion removed from solution with the original counterion (often this is Na^+ or H^+ for cation exchangers, OH^- or Cl^- for anion exchangers). Regeneration is usually done with concentrated solutions containing the ions to be counter exchanged into the resins (Na^+, H^+, OH^-, Cl^-). In practice, ion exchange is typically done with two columns in parallel, so that, while one column is regenerated, the other column is in active use (Figure 7.13). Regeneration is easier for weak-acid or weak-base resins. For these resins, any strong acid or strong base will be an effective regenerant. Instead, strong-acid or strong-base resins are difficult to regenerate because the resin is highly stable in ionised form. For these resins, high concentration acids or bases are needed, and often regeneration is only partial.

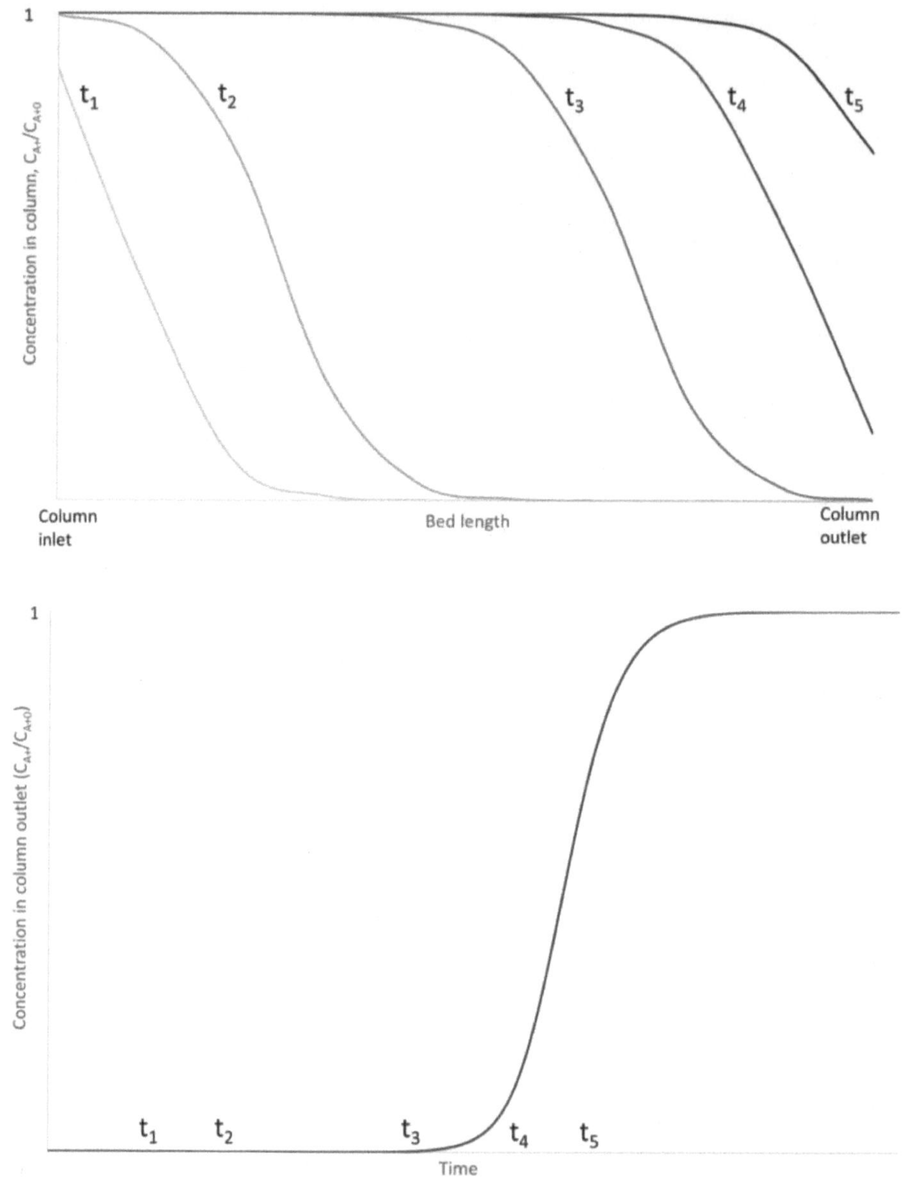

FIGURE 7.11 Scheme of the profiles of a generic exchangeable ion A^+ in solution inside the column bed (top) and at the column outlet (bottom). t_1-t_5 refer to generic times, in progression, during column operation.

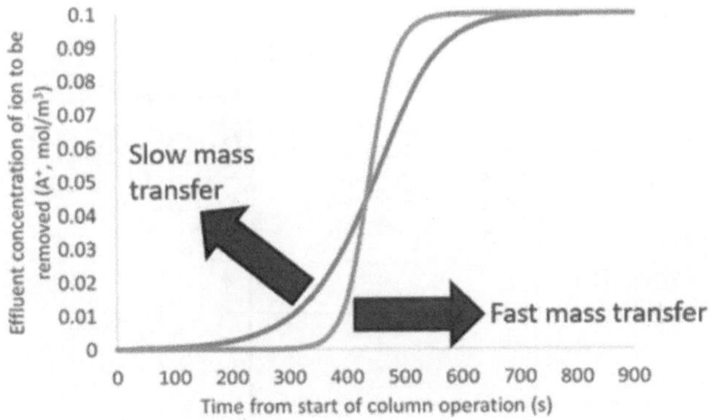

FIGURE 7.12 Qualitative effect of mass transfer on ion exchange.

7.2.4 Ion Exchange Equilibria

Ion exchange reactions can be characterised by the equilibrium reactions between the ion in solution and the ion on the resin. For example, for the exchange of a generic cation A^+ with groups NaR on the resin (Na^+ is in this example the resin cation, R is the resin support), the equilibrium is (note that H^+/HR can also be used instead of Na^+/NaR):

$$A^+ + \text{NaR} \leftrightarrow \text{AR} + \text{Na}^+ \tag{7.4}$$

where AR is the cation removed from solution and attached to the resin. Reaction (7.4) can be characterised by the equilibrium constant K_{eq}:

$$K_{eq} = \frac{C_{\text{Na+}} C_{\text{AR}}}{C_{\text{A+}} C_{\text{NaR}}} \cdot \frac{\gamma_{\text{Na+}} \gamma_{\text{AR}}}{\gamma_{\text{A+}} \gamma_{\text{NaR}}} \tag{7.5}$$

where the C terms represent the concentrations referred to the liquid phase and the γ terms the activities in solution. The activity coefficients γ can be assumed to be approximately constant and independent of concentrations (in the usual concentration ranges used in ion exchange), therefore we can use a concentration-based activity coefficient K':

$$K' = \frac{C_{\text{Na+}} C_{\text{AR}}}{C_{\text{A+}} C_{\text{NaR}}} \tag{7.6}$$

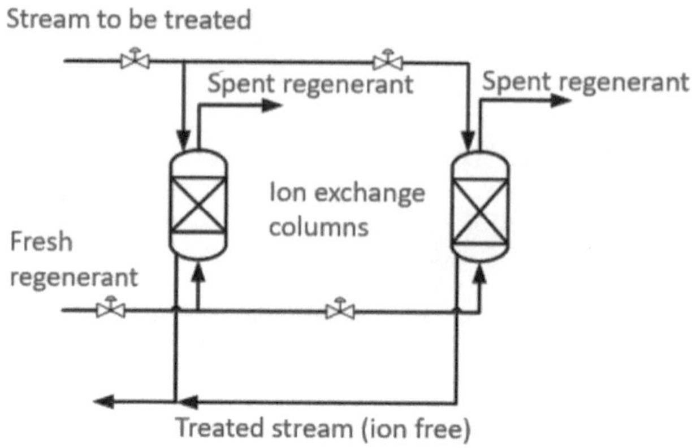

FIGURE 7.13 Scheme of regeneration process for ion exchange columns.

Any resin is characterised by its own total concentration (C_{tot}) of ion exchange groups per unit of liquid phase volume:

$$C_{tot} = C_{\mathrm{NaR}} + C_{\mathrm{AR}} \tag{7.7}$$

For the removal of a divalent ion:

$$\mathrm{B}^{2+} + 2\mathrm{NaR} \leftrightarrow \mathrm{BR}_2 + 2\mathrm{Na}^+ \tag{7.8}$$

The notation BR_2 indicates that the divalent ion B is attached to two ion exchange groups on the resin. In this case the equilibrium constant is:

$$K' = \frac{\left(C_{\mathrm{Na+}}\right)^2 C_{\mathrm{BR2}}}{C_{\mathrm{B2+}}\left(C_{\mathrm{NaR}}\right)^2} \tag{7.9}$$

Tables 7.3 and 7.4 report experimental values of K' obtained with cation and anion ion exchange resins. Values referred to Li^+ as reference cation and Cl^- as reference anion (with polystyrene resin). High values of K' indicate preference of the exchanger for the given ion (equilibria shifts to the right and ions more easily removed from solution), however the extent of ion exchange depends on the concentration of the ion in solution. From the values in the table (or in similar tables for other types of resins/active groups), the K' for any ion exchange reaction can be calculated, e.g. if we want to remove K^+ with a NaR resin, then $K' = 2.90/1.98 = 1.46$

TABLE 7.3
Examples of Values of the Equilibrium Constant for a Cation Exchange Resin with Various Cations. The Equilibrium Constants are Referred to the Resin in the LiR Form

Exchangeable Ion	K′
Li^+	1.00
H^+	1.27
Na^+	1.98
NH_4^+	2.55
K^+	2.90
Mg^{2+}	3.29
Cu^{2+}	3.47
Ni^{2+}	3.93
Ca^{2+}	5.16
Ba^{2+}	11.5

Source: Adapted from W.L. McCabe, J.C. Smith, P. Harriott. Unit Operations of Chemical Engineering. McGraw-Hill Chemical Engineering Series, New York NY USA.

TABLE 7.4
Examples of Values of the Equilibrium Constant for an Anion Exchange Resin with Various Anions. The Equilibrium Constants are Referred to the Resin in the RCl Form

Exchangeable Ion	K′
NO_3^-	3.8
Br^-	2.8
CN^-	1.6
Cl^-	1.0
HCO_3^-	0.3
CH_3COO^-	0.2
SO_4^{2-}	0.15
F^-	0.09
OH^-	0.05-0.07

Source: Adapted from W.L. McCabe, J.C. Smith, P. Harriott. Unit Operations of Chemical Engineering. McGraw-Hill Chemical Engineering Series, New York NY USA.

Example 7.4

Considering a cation exchange resin characterised by the equilibria in Table 7.3, calculate the equilibrium constant K' for the exchange of potassium, assuming that the functional groups on the resin are in their Na form (NaR).

Solution

For the exchange of potassium on a NaR resin the equilibrium reaction is:

$$K^+ + \mathrm{NaR} \leftrightarrow KR + \mathrm{Na}^+$$

with the equilibrium constant K' defined as

$$K'_{\mathrm{K+\to Na+}} = \frac{C_{\mathrm{Na+}} C_{\mathrm{KR}}}{C_{\mathrm{K+}} C_{\mathrm{NaR}}}$$

From Table 7.3 we know the equilibrium constants of ion exchange of Na^+ and K^+ when the resin is in the LiR form:

$$\mathrm{Na}^+ + \mathrm{LiR} \leftrightarrow \mathrm{NaR} + \mathrm{Li}^+ \; \mathrm{K}'_{\mathrm{Na+\to Li+}} = \frac{C_{\mathrm{Li+}} C_{\mathrm{NaR}}}{C_{\mathrm{Na+}} C_{\mathrm{LiR}}} = 1.98$$

$$\mathrm{K}^+ + \mathrm{LiR} \leftrightarrow \mathrm{KR} + \mathrm{Li}^+ \mathrm{K}'_{\mathrm{K+\to Li+}} = \frac{C_{\mathrm{Li+}} C_{\mathrm{KR}}}{C_{\mathrm{K+}} C_{\mathrm{LiR}}} = 2.90$$

It follows that

$$\frac{\mathrm{K}'_{\mathrm{K+\to Li+}}}{\mathrm{K}'_{\mathrm{Na+\to Li+}}} = \frac{\dfrac{C_{\mathrm{Li+}} C_{\mathrm{KR}}}{C_{\mathrm{K+}} C_{\mathrm{LiR}}}}{\dfrac{C_{\mathrm{Li+}} C_{\mathrm{NaR}}}{C_{\mathrm{Na+}} C_{\mathrm{LiR}}}} = \frac{C_{\mathrm{Na+}} C_{\mathrm{KR}}}{C_{\mathrm{K+}} C_{\mathrm{NaR}}} = \mathrm{K}'_{\mathrm{K+\to Na+}} = \frac{2.90}{1.98} = 1.46$$

7.2.5 Mass Transfer and Mass Balances in Ion Exchange

So far, we have considered the transfer of ions from the solution to the resin by means of reaction equilibria. However, the transfer of ions from the solution to the resin generally does not proceed at infinitely fast rates so equilibrium conditions between the solution and the resin are not reached instantaneously. The rate of mass transfer for an ion from the solution to the resin can be expressed as

$$r = k_C a\left(C - C^*\right) \tag{7.10}$$

where

r = mass transfer rate of the ion per unit volume of the dry resin, in mol m^{-3} s^{-1} or equivalent units
k_c = mass transfer coefficient of the ion between the solution and the resin, in m s^{-1} or equivalent units
a = contact area between the resin and the solution per unit volume of dry resin, in m^2 m^{-3} or equivalent units
C = concentration of the ion in solution, in mol m^{-3} or equivalent units
C^* = concentration of the ion in solution that would be in equilibrium with the ion concentration on the resin (same units as C)

For example, considering a cation exchange resin and a generic cation A^+ to be removed from solution, the mass transfer rate is expressed as

$$r = k_C a\left(C_{A+} - C^*_{A+}\right) \tag{7.11}$$

Since C^*_{A+} is assumed to be at equilibrium with the ion concentration on the resin, assuming the resin is in the NaR form:

$$K' = \frac{C_{\mathrm{Na+}} C_{\mathrm{AR}}}{C^*_{\mathrm{A+}} C_{\mathrm{NaR}}} \rightarrow C^*_{\mathrm{A+}} = \frac{C_{\mathrm{Na+}} C_{\mathrm{AR}}}{K' C_{\mathrm{NaR}}} \tag{7.12}$$

Equations (7.10) and (7.11) show that reaching the equilibrium between C_{A+} and C^*_{A+} is not instantaneous but is limited by the mass transfer coefficient $k_c a$ (which has the same meaning as the mass transfer coefficient for oxygen in bioreactors that we have seen in earlier chapters). The rate of mass transfer r is the highest at the start of the ion exchange when the concentration of ion A^+ on the resin is 0 ($C_{AR} = 0$ in equation

(7.12)) and becomes equal to 0 when the ion in solution and on the resin reach equilibrium ($C^*_{A+} = C_{A+}$).

The mass transfer coefficient $k_c a$ depends on the fluid dynamic conditions in the ion exchange column and on the physical properties of the ion and of the resin. $k_c a$ increases with increasing fluid velocity in the bed and with increasing values of the effective diffusivity of the ion in the resin.

Using the mass transfer rate in equation (7.11), we can calculate the mass balance for the ion to be exchanged A^+ over the length of the column. The mass balance allows to calculate the profile of A^+ in solution as a function of time and of length inside the column. To derive the mass balance, we will use the conceptual scheme in Figure 7.14. The mass balance assumes piston flow through the column (no axial mixing between different sections) and that any section of volume $dV = AdL$ is perfectly mixed.

Considering the small (ideally infinitesimally small) element of column between lengths L and $L+dL$, the mass balance for ion A^+ in solution is

$$\left.\frac{\partial C_{A+}}{\partial t}\right|_L \varepsilon dV = Q\left(C_{A+}\big|_L - C_{A+}\big|_{L+dL}\right) - r(1-\varepsilon)dV \tag{7.13}$$

In equation (7.13) the terms have the following meaning:

- $\frac{\partial C_{A+}}{\partial t}$ (units of mol m^{-3} s^{-1} or equivalent) represents the change in concentration of the ion A^+ in solution over time.
- εdV is the liquid volume between L and $L+dL$, where ε is the fraction of volume occupied by the liquid ($\varepsilon = \frac{V_{liquid\ in\ column}}{V_{total\ in\ column}}$). dV is the total volume in the column between L and $L+dL$ and is equal to AdL with A being the cross sectional area of the column (in m^2 or equivalent units).
- $Q\left(C_{A+}\big|_L - C_{A+}\big|_{L+dL}\right)$ (units of mol s^{-1} or equivalent) represents the difference in the mass flowrates of A^+ entering and leaving the volume dV. Q (m^3 s^{-1} or

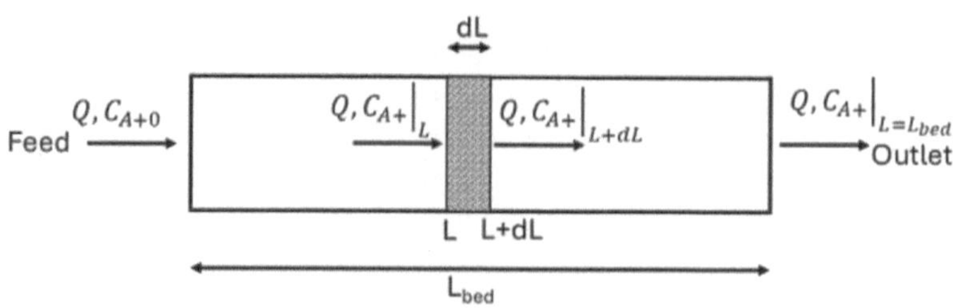

FIGURE 7.14 Conceptual scheme for the mass balance of ion A^+ inside an ion exchange column.

equivalent units) is the volumetric flow rate of solution entering the column and is equal to $u\varepsilon A$ where u is the fluid velocity in the column (m s^{-1} or equivalent units).

- $r(1-\varepsilon)dV$ (units of mol s^{-1} or equivalent) represents the rate of mass transfer of the ion A^+ from the solution to the resin, where r is given by Equation (7.10) and $(1-\varepsilon)dV$ represents the volume occupied by the dry resin ($1-\varepsilon = \frac{V_{dry\ resin}}{V_{total\ in\ column}}$).

By making substitutions as appropriate, equation (7.13) becomes

$$\left.\frac{\partial C_{A+}}{\partial t}\right|_L \varepsilon A dL = u\varepsilon A\left(C_{A+}\big|_L - C_{A+}\big|_{L+dL}\right) - r(1-\varepsilon)AdL \tag{7.14}$$

which can be rearranged as (noting that for infinitesimally small dL, $\frac{\left(C_{A+}\big|_L - C_{A+}\big|_{L+dL}\right)}{dL} = -\frac{\partial C_{A+}}{\partial L}$)

$$\frac{\partial C_{A+}}{\partial t} = -u\frac{\partial C_{A+}}{\partial L} - \frac{(1-\varepsilon)r}{\varepsilon} \tag{7.15}$$

Equation (7.15) represents the mass balance of the ion A^+ in the column and shows that, as expected, the concentration of A^+ in the liquid phase C_{A+} depends both on the time and on the position inside the column.

In solving equation (7.15) we note that the mass transfer rate depends on the concentration of A^+ on the resin C_{AR} as shown by equation (7.12). Considering equation (7.12) we note that the concentrations on the resin C_{AR} and C_{NaR} are linked by the total concentration of exchangeable ions on the resin C_{tot}, which we assume is a known property of the resin:

$$C_{tot} = C_{AR} + C_{NaR} \tag{7.16}$$

Therefore, in equation (7.12) C_{NaR} can be replaced by

$$C_{NaR} = C_{tot} - C_{AR} \tag{7.17}$$

which gives

$$C^*_{A+} = \frac{C_{Na+}C_{AR}}{K'\left(C_{tot} - C_{AR}\right)} \tag{7.18}$$

Therefore, the rate of mass transfer equation (7.10) can be re-written as

$$r = k_C a\left(C_{A+} - \frac{C_{Na+}C_{AR}}{K'\left(C_{tot} - C_{AR}\right)}\right) \tag{7.19}$$

In order to calculate the profile of C_{A+} inside the column from equations (7.15) and (7.19), we need equations for C_{Na+} and C_{AR} that appear in equation (7.19). These equations are the mass balances for Na+ and AR in the column, which are respectively

$$\frac{\partial C_{Na+}}{\partial t} = -u\frac{\partial C_{Na+}}{\partial L} + \frac{(1-\varepsilon)r}{\varepsilon} \tag{7.20}$$

$$\frac{\partial C_{AR}}{\partial t}\varepsilon dV = r(1-\varepsilon)dV \rightarrow \frac{\partial C_{AR}}{\partial t} = \frac{r(1-\varepsilon)}{\varepsilon} \tag{7.21}$$

Equation (7.20) for Na^+ is analogous to equation (7.15) for A^+ with the difference that Na^+ increases rather than decreasing when A^+ is transferred. The mass balance of AR, equation (7.21), is similar to equations (7.15) and (7.21) but accounts for the fact that there is no flow of AR in the column.

Equations (7.15), (7.20) and (7.21), combined with the mass transfer rate (7.19) represent a system of three partial differential equations in the three unknowns C_{A+}, C_{Na+}, C_{AR}, which can be solved to give the profiles of these variables inside the column as a function of time. To solve these equations, it is necessary to set the initial conditions for the concentrations of C_{A+}, C_{Na+}, C_{AR} at the start of the process. Usually, ion exchange is started with the column filled with a fresh solution without the ion to be exchanged and without counterions in solution (i.e. $C_{A+}(t=0) = C_{Na+}(t=0) = 0$ for any length L inside the column) and with a fresh resin entirely in the NaR form (i.e. $C_{AR}(t=0) = 0$ for any length L). In addition, we need boundary conditions at the column inlet for C_{A+}, C_{Na+}. These boundary conditions are the concentrations of A^+ and Na^+ in the feed (i.e. $C_{A+}(L=0) = C_{A+0}$ and $C_{Na+}(L=0) = 0$ at any time). Note that we are assuming that the feed does not include the counterion. The integration of equations (7.15), (7.20) and (7.20) must be carried out up to the maximum time required (which may coincide with the saturation of the resin with A^+) and up to the length of the column (L_{bed}).

Equations (7.15), (7.20) and (7.21) with the mass transfer rate (7.19) can be solved in many ways and using various pieces of software. In Table 7.5, the solution by finite differences and discretisation of the time and length domain is shown, which can be implemented in Microsoft Excel. The time and length domains are discretised in small uniform intervals of size Δt and ΔL respectively. Note that $\Delta L = \frac{L_{bed}}{number\,of\,sections}$. The numerical integration is only accurate enough if the size of the intervals Δt and ΔL is small enough compared with the rate of the processes.

In addition to defining the discretisation intervals Δt and ΔL, the solution of equations (7.15), (7.20) and (7.21) requires the definition of the constant values u, ε, $k_c a$, C_{tot}, K', C_{A+0}. u is the velocity of the liquid inside the column and can be specified or calculated from the feed volumetric flow rate, cross sectional area of the column and liquid fraction in the column. ε is the liquid fraction in the column, which coincides with the empty volume in the resin and is a property of the resin. $k_c a$ is the mass transfer coefficient for the ion from the solution to the resin, which depends on the physical properties of the ion and solution and on the fluid dynamics in the column. C_{tot} is the total concentration of exchangeable ions on the resin, referred to the liquid volume, and is a property of the resin. Note that at the start of the column operation, when there are no A^+ ions in the column, $C_{tot} = C_{NaR}$. The equilibrium constant K' is a property of the resin/ion A^+ system. The concentration of A^+ in the feed, C_{A+0}, is also assumed to be known and constant.

When solving equations (7.15), (7.20) and (7.21), it is important to check the accuracy of the model and of the obtained numerical solution. This can be done, for example, by checking that the material balances of A^+ and of the counterion Na^+ are conserved throughout the column at any given time. For A^+, this means that the total moles that entered the column up to any given time must coincide with the sum of the moles that have left the column, that are in the column in solution or are attached to the resin up to the same time. For Na^+, the material balance means that the total moles present in the resin at the start of the process must be the same as the sum, at any given time, of the moles in solution inside the column, of the moles still attached to the resin and of the moles that have left the column. Translated into formulas, the said materials balances for A^+ and Na^+ become equations (7.22) and (7.23) respectively, where all the terms have the units of mol/m^2 and n is a generic positive number variable from 0 to the number of time steps used in the calculations. Equations (7.22) and (7.23) have to be verified for any values of n.

$$u\varepsilon C_{A+0} n\Delta t = \sum_{t=0}^{n\Delta t} C_{A+}\left(t, L = L_{bed}\right) u\varepsilon\Delta t + \sum_{L=0}^{L_{bed}} C_{A+}\left(n\Delta t, L\right)\varepsilon\Delta L + \sum_{L=0}^{L_{bed}} C_{AR}\left(n\Delta t, L\right)\varepsilon\Delta L \tag{7.22}$$

$$C_{tot}\varepsilon L_{bed} = \sum_{L=0}^{L_{bed}} C_{Na+}\left(n\Delta t, L\right)\varepsilon\Delta L + \sum_{t=0}^{n\Delta t} C_{Na+}\left(t, L = L_{bed}\right) u\varepsilon\Delta t + \sum_{L=0}^{L_{bed}} C_{NaR}\left(n\Delta t, L\right)\varepsilon\Delta L \tag{7.23}$$

In equation (7.22), the terms have the following meanings:

$u\varepsilon C_{A+0} n\Delta t$: moles of A^+, which entered the column up to the time $n\Delta t$ from the start of the process.

TABLE 7.5
Integration of the Mass Balance Equations for Ion Exchange (equations (7.15), (7.20), (7.21)). Only the First Three Time Intervals and the First Two Column Sections are Shown. Generally, the Integration Requires Many Time Intervals and Many Sections. In Each Section, the Concentrations Refer to the Same Section Unless Specified Otherwise

	Section 1	
Time	C_{A+}	C_{AR}
0	0	0
Δt	$C_{A+}(t=0)+\left.\frac{\partial C_{A+}}{\partial t}\right\|_{t=0}\Delta t$	$C_{AR}(t=0)+\left.\frac{\partial C_{AR}}{\partial t}\right\|_{t=0}\Delta t$
2Δt	$C_{A+}(t=\Delta t)+\left.\frac{\partial C_{A+}}{\partial t}\right\|_{t=\Delta t}\Delta t$	$C_{AR}(t=\Delta t)+\left.\frac{\partial C_{AR}}{\partial t}\right\|_{t=\Delta t}\Delta t$
3Δt	$C_{A+}(t=\Delta t)+\left.\frac{\partial C_{A+}}{\partial t}\right\|_{t=2\Delta t}\Delta t$	$C_{AR}(t=\Delta t)+\left.\frac{\partial C_{AR}}{\partial t}\right\|_{t=2\Delta t}\Delta t$
	Section 2	
0	0	0
Δt	$C_{A+}(t=0)+\left.\frac{\partial C_{A+}}{\partial t}\right\|_{t=0}\Delta t$	$C_{AR}(t=0)+\left.\frac{\partial C_{AR}}{\partial t}\right\|_{t=0}\Delta t$
2Δt	$C_{A+}(t=\Delta t)+\left.\frac{\partial C_{A+}}{\partial t}\right\|_{t=\Delta t}\Delta t$	$C_{AR}(t=\Delta t)+\left.\frac{\partial C_{AR}}{\partial t}\right\|_{t=\Delta t}\Delta t$
3Δt	$C_{A+}(t=\Delta t)+\left.\frac{\partial C_{A+}}{\partial t}\right\|_{t=2\Delta t}\Delta t$	$C_{AR}(t=\Delta t)+\left.\frac{\partial C_{AR}}{\partial t}\right\|_{t=2\Delta t}\Delta t$

$\sum_{t=0}^{n\Delta t} C_{A+}(t, L = L_{bed})u\varepsilon\Delta t$: moles of A^+, which left the column up to the time $n\Delta t$.

$\sum_{L=0}^{L_{bed}} C_{A+}(n\Delta t, L)\varepsilon\Delta L$: moles of A^+ in solution inside the column at the time $n\Delta t$.

$\sum_{L=0}^{L_{bed}} C_{AR}(n\Delta t, L)\varepsilon\Delta L$: moles of A^+ on the resin at the time $n\Delta t$.

In equation (7.23), the terms have the following meanings:

C_{Na+}	$\frac{\partial C_{A+}}{\partial t}$	$\frac{\partial C_{AR}}{\partial t}$
0	$\left(-u\frac{\partial C_{A+}}{\partial L}-\frac{(1-\varepsilon)r}{\varepsilon}\right)\Big\vert_{t=0}$	$\frac{r(1-\varepsilon)}{\varepsilon}\Big\vert_{t=0}$
$C_{Na+}(t=0)+\frac{\partial C_{Na+}}{\partial t}\Big\vert_{t=0}\Delta t$	$\left(-u\frac{\partial C_{A+}}{\partial L}-\frac{(1-\varepsilon)r}{\varepsilon}\right)\Big\vert_{t=\Delta t}$	$\frac{r(1-\varepsilon)}{\varepsilon}\Big\vert_{t=\text{"t}}$
$C_{Na+}(t=\Delta t)+\frac{\partial C_{Na+}}{\partial t}\Big\vert_{t=\Delta t}\Delta t$	$\left(-u\frac{\partial C_{A+}}{\partial L}-\frac{(1-\varepsilon)r}{\varepsilon}\right)\Big\vert_{t=2\Delta t}$	$\frac{r(1-\varepsilon)}{\varepsilon}\Big\vert_{t=2\text{"t}}$
$C_{Na+}(t=\Delta t)+\frac{\partial C_{Na+}}{\partial t}\Big\vert_{t=2\Delta t}\Delta t$	$\left(-u\frac{\partial C_{A+}}{\partial L}-\frac{(1-\varepsilon)r}{\varepsilon}\right)\Big\vert_{t=3\Delta t}$	$\frac{r(1-\varepsilon)}{\varepsilon}\Big\vert_{t=3\text{"t}}$

0	$\left(-u\frac{\partial C_{A+}}{\partial L}-\frac{(1-\varepsilon)r}{\varepsilon}\right)\Big\vert_{t=0}$	$\frac{r(1-\varepsilon)}{\varepsilon}\Big\vert_{t=0}$
$C_{Na+}(t=0)+\frac{\partial C_{Na+}}{\partial t}\Big\vert_{t=0}\Delta t$	$\left(-u\frac{\partial C_{A+}}{\partial L}-\frac{(1-\varepsilon)r}{\varepsilon}\right)\Big\vert_{t=\Delta t}$	$\frac{r(1-\varepsilon)}{\varepsilon}\Big\vert_{t=\text{"t}}$
$C_{Na+}(t=\Delta t)+\frac{\partial C_{Na+}}{\partial t}\Big\vert_{t=\Delta t}\Delta t$	$\left(-u\frac{\partial C_{A+}}{\partial L}-\frac{(1-\varepsilon)r}{\varepsilon}\right)\Big\vert_{t=2\Delta t}$	$\frac{r(1-\varepsilon)}{\varepsilon}\Big\vert_{t=2\text{"t}}$
$C_{Na+}(t=\Delta t)+\frac{\partial C_{Na+}}{\partial t}\Big\vert_{t=2\Delta t}\Delta t$	$\left(-u\frac{\partial C_{A+}}{\partial L}-\frac{(1-\varepsilon)r}{\varepsilon}\right)\Big\vert_{t=3\Delta t}$	$\frac{r(1-\varepsilon)}{\varepsilon}\Big\vert_{t=3\text{"t}}$

$C_{tot}\varepsilon L_{bed}$: total moles of Na^+ on the resin at the start of the process (fresh resin)

$\sum_{L=0}^{L_{bed}} C_{Na+}(n\Delta t, L)\,\varepsilon\Delta L$: moles of Na^+ in solution inside the column at the time $n\Delta t$.

$\sum_{t=0}^{n\Delta t} C_{Na+}(t, L=L_{bed})\,u\varepsilon\Delta t$: moles of Na^+, which left the column up to the time $n\Delta t$.

$\sum_{L=0}^{L_{bed}} C_{NaR}(n\Delta t, L)\,\varepsilon\Delta L$: moles of Na^+ on the resin at the time $n\Delta t$.

TABLE 7.5 (Continued)
Integration of the Mass Balance Equations for Ion Exchange (equations (7.15), (7.20), (7.21)). Only the First Three Time Intervals and the First Two Column Sections Are Shown. Generally, the Integration Requires Many Time Intervals and Many Sections. In Each Section, the Concentrations Refer to the Same Section Unless Specified Otherwise

	Section 1	
Time	$\frac{\partial C_{Na+}}{\partial t}$	$\frac{\partial C_{A+}}{\partial L}$
0	$\left.\left(-u\frac{\partial C_{Na+}}{\partial L}-\frac{(1-\varepsilon)r}{\varepsilon}\right)\right\|_{t=0}$	$\frac{\left(C_{A+}(t=0)-C_{A+0}\right)}{\Delta L}$
Δt	$\left.\left(-u\frac{\partial C_{Na+}}{\partial L}-\frac{(1-\varepsilon)r}{\varepsilon}\right)\right\|_{t=\Delta t}$	$\frac{\left(C_{A+}(t=\Delta t)-C_{A+0}\right)}{\Delta L}$
2Δt	$\left.\left(-u\frac{\partial C_{Na+}}{\partial L}-\frac{(1-\varepsilon)r}{\varepsilon}\right)\right\|_{t=2\Delta t}$	$\frac{\left(C_{A+}(t=2\Delta t)-C_{A+0}\right)}{\Delta L}$
3Δt	$\left.\left(-u\frac{\partial C_{Na+}}{\partial L}-\frac{(1-\varepsilon)r}{\varepsilon}\right)\right\|_{t=3\Delta t}$	$\frac{\left(C_{A+}(t=3\Delta t)-C_{A+0}\right)}{\Delta L}$
	Section 2	
0	$\left.\left(-u\frac{\partial C_{Na+}}{\partial L}-\frac{(1-\varepsilon)r}{\varepsilon}\right)\right\|_{t=0}$	$\frac{\left(C_{A+}(t=0)-C_{A+}(t=0)_{Section\,1}\right)}{\Delta L}$
Δt	$\left.\left(-u\frac{\partial C_{Na+}}{\partial L}-\frac{(1-\varepsilon)r}{\varepsilon}\right)\right\|_{t=\Delta t}$	$\frac{\left(C_{A+}(t=\Delta t)-C_{A+}(t=\Delta t)_{Section\,1}\right)}{\Delta L}$
2Δt	$\left.\left(-u\frac{\partial C_{Na+}}{\partial L}-\frac{(1-\varepsilon)r}{\varepsilon}\right)\right\|_{t=2\Delta t}$	$\frac{\left(C_{A+}(t=2\Delta t)-C_{A+}(t=2\Delta t)_{Section\,1}\right)}{\Delta L}$
3Δt	$\left.\left(-u\frac{\partial C_{Na+}}{\partial L}-\frac{(1-\varepsilon)r}{\varepsilon}\right)\right\|_{t=3\Delta t}$	$\frac{\left(C_{A+}(t=3\Delta t)-C_{A+}(t=3\Delta t)_{Section\,1}\right)}{\Delta L}$

$\frac{\partial C_{Na+}}{\partial L}$	C^*_{A+}	r
$\frac{\left(C_{Na+}(t=0)-0\right)}{\Delta L}$	$\left.\frac{C_{Na+}C_{AR}}{K'\left(C_{tot}-C_{AR}\right)}\right\rvert_{t=0}$	$\left.k_C a\left(C_{A+}-C^*_{A+}\right)\right\rvert_{t=0}$
$\frac{\left(C_{Na+}(t=\Delta t)-0\right)}{\Delta L}$	$\left.\frac{C_{Na+}C_{AR}}{K'\left(C_{tot}-C_{AR}\right)}\right\rvert_{t=\Delta t}$	$\left.k_C a\left(C_{A+}-C^*_{A+}\right)\right\rvert_{t=\Delta t}$
$\frac{\left(C_{Na+}(t=2\Delta t)-0\right)}{\Delta L}$	$\left.\frac{C_{Na+}C_{AR}}{K'\left(C_{tot}-C_{AR}\right)}\right\rvert_{t=2\Delta t}$	$\left.k_C a\left(C_{A+}-C^*_{A+}\right)\right\rvert_{t=2\Delta t}$
$\frac{\left(C_{Na+}(t=3\Delta t)-0\right)}{\Delta L}$	$\left.\frac{C_{Na+}C_{AR}}{K'\left(C_{tot}-C_{AR}\right)}\right\rvert_{t=3\Delta t}$	$\left.k_C a\left(C_{A+}-C^*_{A+}\right)\right\rvert_{t=3\Delta t}$

$\frac{\left(C_{Na+}(t=0)-0\right)}{\Delta L}$	$\left.\frac{C_{Na+}C_{AR}}{K'\left(C_{tot}-C_{AR}\right)}\right\rvert_{t=0}$	$\left.k_C a\left(C_{A+}-C^*_{A+}\right)\right\rvert_{t=0}$
$\frac{\left(C_{Na+}(t=\Delta t)-0\right)}{\Delta L}$	$\left.\frac{C_{Na+}C_{AR}}{K'\left(C_{tot}-C_{AR}\right)}\right\rvert_{t=\Delta t}$	$\left.k_C a\left(C_{A+}-C^*_{A+}\right)\right\rvert_{t=\Delta t}$
$\frac{\left(C_{Na+}(t=2\Delta t)-0\right)}{\Delta L}$	$\left.\frac{C_{Na+}C_{AR}}{K'\left(C_{tot}-C_{AR}\right)}\right\rvert_{t=2\Delta t}$	$\left.k_C a\left(C_{A+}-C^*_{A+}\right)\right\rvert_{t=2\Delta t}$
$\frac{\left(C_{Na+}(t=3\Delta t)-0\right)}{\Delta L}$	$\left.\frac{C_{Na+}C_{AR}}{K'\left(C_{tot}-C_{AR}\right)}\right\rvert_{t=3\Delta t}$	$\left.k_C a\left(C_{A+}-C^*_{A+}\right)\right\rvert_{t=3\Delta t}$

Example 7.5

Consider an ion exchange column characterised by the following parameters: A (cross sectional area) = 0.50 m^2, L (column length) = 2 m, ε (liquid fraction) = 0.40, C_{tot} (total concentration of exchangeable ions) = 6 mM. Calculate the maximum theoretical removal of a generic $A+$ ion and for how long a stream of 0.2 L/s containing A^+ at concentration 2 mM can be fed to the column before the column is saturated (assume the ion exchange is instantaneous). Assume that the fresh resin is in the NaR form.

Solution

The maximum number of moles of A^+ that can be removed by the column corresponds to the exchange of all counterions Na^+ on the resin. This is equal to (note the conversion from L to m^3 for unit consistency)

$$C_{tot}\varepsilon AL = 6\frac{\text{mol}}{\text{m}^3_{\text{liquid}}}\cdot 0.40\frac{\text{m}^3_{\text{liquid}}}{\text{m}^3_{\text{tot}}}\cdot 0.50\ \text{m}^2\cdot 2\ \text{m} = 2.4\ \text{mol}$$

A stream at 2 L/s with A^+ at 2 mM carries $0.2\frac{\text{L}}{\text{s}}\cdot 0.002\frac{\text{mol}}{\text{L}} = 0.0004\frac{\text{molA}^+}{\text{s}}$

and can therefore be operated for $\dfrac{2.4\ \text{mol}}{0.0004\ \dfrac{\text{molA}^+}{\text{s}}} = 6000$ s before the column reaches saturation.

Example 7.6

Considering the ion exchange process of Example 7.5, assume now that the process is not instantaneous but is limited by mass transfer according to equation (7.19) with $k_c a = 0.10\ \text{s}^{-1}$. Assume the equilibrium constant is $K' = 2.5$. Calculate the time required for the effluent A^+ concentration to reach 1% of C_{A+0} (breakthrough) and which fraction of the maximum theoretical ion exchange capacity calculated in Example 7.5 has been used up to breakthrough. Also, plot the time profile of A^+, of the counterion Na^+ and of the ion in the resin AR (all concentrations referred to the liquid phase) in three sections inside the column: initial part of the column, middle of the column, column outlet.

In the numerical solution it is advised to use: $\Delta t = 0.1$ s; $\Delta L = 20$ cm.

Solution

The solution is obtained by solving equations (7.15), (7.20), 7.21), with the given parameters and with

$$u = \frac{Q}{\varepsilon A} = \frac{0.0002\ \text{m}^3\text{s}^{-1}}{0.40\cdot 0.50\ \text{m}^2} = 0.001\ \text{m s}^{-1}$$

Using in all cases the units of mol/m^3 for the concentrations and of s for the time, the equation for mass transfer rate, equation (7.19), and the mass balances, equations (7.15), (7.20), (7.21), are in this case is in this case:

$$r = 0.10\left(C_{A+} - \frac{C_{Na+}C_{AR}}{2.5(6 - C_{AR})}\right)$$

$$\frac{\partial C_{A+}}{\partial t} = -0.001\frac{\partial C_{A+}}{\partial L} - \frac{(1-0.4)r}{0.4}$$

$$\frac{\partial C_{Na+}}{\partial t} = -0.001\frac{\partial C_{Na+}}{\partial L} + \frac{(1-0.4)r}{0.4}$$

$$\frac{\partial C_{AR}}{\partial t} = \frac{r(1-0.4)}{0.4}$$

Using for the numerical integration Microsoft Excel with the recommended values of Δt = 0.1 s and ΔL = 20 cm, Table 7.6 reports, as an example, the results of the first time steps and sections in the column. Note that for a column length of 2 m, ΔL = 20 cm corresponds to dividing the column in 10 sections. The breakthrough time is defined in this case as the time required for the outlet concentration to reach 1% of the feed concentration, i.e. 0.02 mol/m^3. Looking at the results for section 10 (last section in the column which corresponds to the column outlet), the concentration of 0.02 mol/m^3 is reached after 5150 s. Up to this time, the total number of moles that have passed through the column are $0.2\frac{\text{L}}{\text{s}}\cdot 0.002\frac{\text{mol}}{\text{L}}\cdot 5{,}150\text{s} = 2.06\text{mol}$, which represents 85.8% of the maximum number of moles of A^+ (2.4 mol) that can be exchanged by the column (from Example 7.5). It is confirmed that the effectiveness of the column is reduced when ion exchange proceeds at finite rate rather than at the instantaneous rate assumed in Example 7.5.

The plots of A^+, Na^+ and AR in the column are shown in Figure 7.15. The considered sections are section 1 (initial part of the column), section 5 (middle part of the column) and section 10 (final part of the column and outlet). The plot shows that over the length of the process A^+ gradually fills all the column. On the other hand, the counterion Na^+ in solution initially accumulates inside the column as it is replaced on the resin by A^+, then is removed from the column by the flow of the feed which is assumed to contain no Na^+. The AR on the resin is initially 0 then it increases to reach C_{tot}, which indicates that the resin has replaced all its exchangeable ions and must be regenerated.

TABLE 7.6
Integration of the Mass Balance Equations for Ion Exchange (equations (7.15), (7.20), (7.21)) for Example 7.6. Only the First Six Time Steps and Three Sections in the Column are Shown. The Whole Simulation for Example 7.6 includes 110,000 Time Steps (11,000 s with $\Delta T = 0.1$ s) and 10 Sections (2 m bed length with $\Delta L = 0.2$ m). Units are m^3 for Volume, mol/m^3 for Concentrations and s for Time

	Section 1									
Time	C_{A+}	C_{AR}	C_{Na+}	$\frac{\partial C_{A+}}{\partial t}$	$\frac{\partial C_{AR}}{\partial t}$	$\frac{\partial C_{Na+}}{\partial t}$	$\frac{\partial C_{A+}}{\partial L}$	$\frac{\partial C_{Na+}}{\partial L}$	C^{*}_{A+}	r
0	0	0	0	0.01	0	0	-10	0	0	0
0.1	0.001	0	0	0.009928	6.67E-05	6.67E-05	-9.995	0	0	0.0001
0.2	0.001993	6.67E-06	6.67E-06	0.009857	0.000133	0.000133	-9.99004	3.33E-05	2.96E-12	0.000199
0.3	0.002979	2E-05	1.99E-05	0.009787	0.000199	0.000198	-9.98511	9.97E-05	2.65E-11	0.000298
0.4	0.003908	8.91E-05	8.9E-05	0.009394	0.000586	0.000586	-9.98046	0.000445	5.29E-10	0.000391
0.5	0.004847	0.000148	0.000148	0.009249	0.000727	0.000726	-9.97576	0.000738	1.45E-09	0.000485
0.6	0.005772	0.00022	0.00022	0.009105	0.000866	0.000865	-9.97114	0.001101	3.24E-09	0.000577

Section 2										
0	0	0	0	0	0	0	0	0	0	0
0.1	0	0	0	0.000005	0	0	-0.005	0	0	0
0.2	5E-07	0	0	9.93E-06	3.33E-08	6.67E-08	-0.00996	-3.3E-05	0	5E-08
0.3	1.49E-06	3.33E-09	6.67E-09	1.48E-05	9.95E-08	1.99E-07	-0.01489	-1E-04	1.48E-18	1.49E-07
0.4	2.94E-06	2.98E-08	5.96E-08	1.91E-05	4.41E-07	8.86E-07	-0.01953	-0.00044	1.18E-16	2.94E-07
0.5	4.85E-06	7.38E-08	1.48E-07	2.35E-05	7.27E-07	1.46E-06	-0.02421	-0.00074	7.3E-16	4.85E-07
0.6	7.2E-06	1.47E-07	2.95E-07	2.77E-05	1.08E-06	2.18E-06	-0.02883	-0.0011	2.88E-15	7.2E-07
Section 3										
0	0	0	0	0	0	0	0	0	0	0
0.1	0	0	0	0	0	0	0	0	0	0
0.2	0	0	0	2.5E-09	0	0	-2.5E-06	0	0	0
0.3	2.5E-10	0	0	7.38E-09	3.75E-11	1.13E-10	-7.4E-06	-7.5E-08	0	2.5E-11
0.4	9.88E-10	3.75E-12	1.13E-11	1.45E-08	1.48E-10	4.46E-10	-1.5E-05	-3E-07	2.81E-24	9.88E-11
0.5	2.44E-09	1.86E-11	5.59E-11	2.39E-08	3.66E-10	1.11E-09	-2.4E-05	-7.4E-07	6.92E-23	2.44E-10
0.6	4.83E-09	5.52E-11	1.67E-10	3.52E-08	7.24E-10	2.2E-09	-3.6E-05	-1.5E-06	6.13E-22	4.83E-10

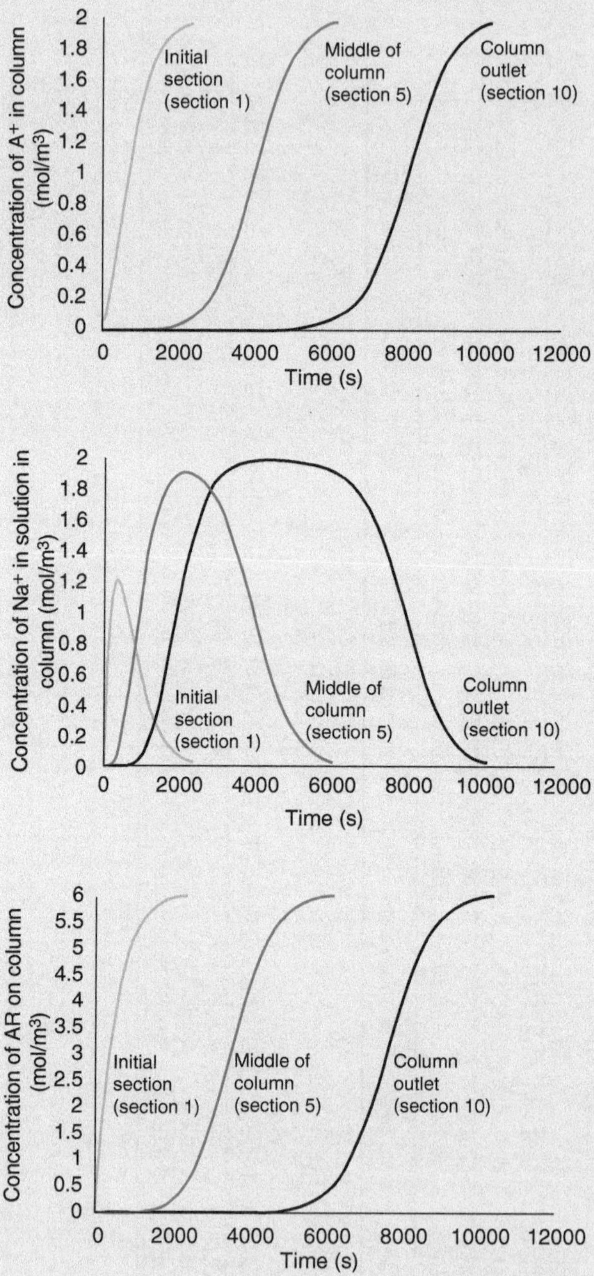

FIGURE 7.15 Plots for Example 7.6. Top: A^+; centre: Na^+; bottom: AR.

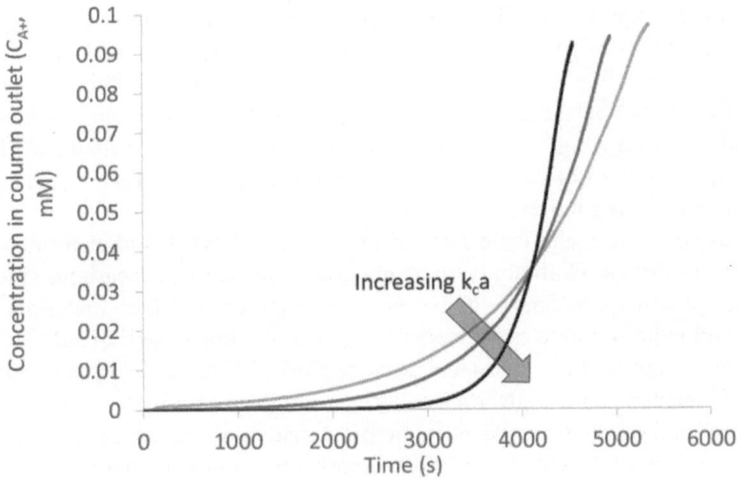

FIGURE 7.16 Effect of the mass transfer coefficient $k_c a$ on the concentration of the exchangeable ion A^+ at the column outlet. Simulations done by solving equations (7.15), (7.20), (7.21).

7.2.5.1 Analysis of Mass Balance Equations and Effect of Process Parameters

The mass balances for ion exchange, equations (7.15), (7.20), (7.21), and the mass transfer rate (7.19), allow for a better understanding of how the process is affected by the operating parameters. One interesting observation is that the cross-sectional area of the column does not appear in the material balances. This means that, for a given ion exchange process, the performance of different columns will be the same as long as the fluid velocity in the column u is maintained the same. This is an important consideration for scale-up and scale-down. Ion exchange can be scaled up or down by maintaining the same fluid velocity in the column. Under these conditions of the same u, the ion exchange performance is expected to be the same on scale up or scale down, as long as the other process conditions (e.g. feed concentration) are also maintained the same.

The material balances also allow for a quantitative analysis of the effect of the mass transfer coefficient $k_c a$ on process performance (Figure 7.16). By increasing $k_c a$, the outlet concentration increases more steeply with time, which is beneficial as the breakthrough happens later and the process can be carried out for longer before starting column regeneration. The mass transfer coefficient $k_c a$ can be increased by improving the turbulence in the column, e.g. by improved design of the resin.

7.3 ADSORPTION

7.3.1 INTRODUCTION

Adsorption is the removal of substances from a gas or liquid phase by contact with a solid surface (adsorbent), which retains the substances (adsorbates) while allowing the bulk of the liquid and solid phase to pass through. Adsorption is widely used in the chemical industry for the treatment of liquid and gas streams. As a downstream process after fermentation or enzymatic reactions, adsorption is often used on liquid streams for product purification, e.g. to remove unwanted byproducts or impurities, or for product recovery, e.g. of proteins from complex fermentation mixtures. Table 7.7 shows some examples of application of adsorption as downstream process in the biochemical engineering industry.

Various substances and materials can be used as adsorbents after fermentation or enzymatic reactions. Commonly used materials are: activated carbon, silica-based materials, polymeric resins (e.g. based on polystyrene). Other materials such as zeolites and metal-organic frameworks are also sometimes used. Ideally, adsorbents should have high surface area (which corresponds to high adsorption capacity as adsorption happens on the adsorbent's surface), high selectivity for the substances to be adsorbed and should be resistant to chemicals and other substances that may be present in fermentation broths. In addition, adsorbents should be able to be regenerated effectively (high degree of regeneration even after many cycles) and rapidly. Indeed, similarly as for ion exchange, during adsorption the resin adsorbs the species to be removed from solution and gradually over time becomes saturated, losing its adsorption capacity. At this point, the resin must be regenerated. Regeneration of adsorbents can be done with various processes, e.g. with steam (at temperature typically in the range 100–250 °C); thermally (with inert gas at temperature typically in the range 600–900 °C), chemically (with solvents, acids or alkaline substances). The scheme

TABLE 7.7
Examples of Applications of Adsorption After Fermentation and Enzymatic Reactions

Application	Process Description
Antibiotic purification	After fermentation and removal of the cells, adsorption on activated carbon retains the antibiotic allowing separation from water and impurities which are not adsorbed
Purification of organic acids (e.g. lactic acid) from fermentation	Pigments and other impurities in the fermentation broth are retained by adsorption on activated carbon, while organic acids are not retained and are purified
Enzyme recovery	After enzyme production by fermentation, impurities (pigments, metabolites, organic contaminants) adsorb onto the resin while enzymes are not retained and are purified
Amino acid purification	After production of the amino acids by fermentation, residual proteins, organic metabolites and other impurities are retained on the adsorbent while amino acid are not retained

of operation and regeneration for adsorption is conceptually the same as seen for ion exchange in Figure 7.13.

7.3.2 Adsorption Equilibrium

Adsorption equilibrium is a mathematical relationship between the concentration of the substance to be adsorbed in the liquid or gas phase in contact with the adsorbent and the concentration of the same substance on the resin. Equilibrium is established after a long enough contact time between the substance to the adsorbed and the resin.

Many mathematical relationships have been developed to describe adsorption equilibrium. Frequently used forms of adsorption equilibrium are the Langmuir (7.24) and Freundlich (7.25) equations (Figure 7.17, they are also called "isotherms" as the values of the parameters is temperature dependent):

$$W = W_{max} \frac{\alpha \cdot c}{1 + \alpha \cdot c} \tag{7.24}$$

$$W = \beta \cdot c^m \tag{7.25}$$

where W is the concentration of adsorbate on the adsorbent (e.g. in kg/kg), c is the concentration of adsorbate in the fluid phase (e.g. in g/m^3), W_{max} (same units as W), α (units reciprocal of c), β, m (units consistent with the units for W and c) are parameters, dependent on the adsorbate, adsorbent and temperature, usually determined from experimental data. Note from Figure 7.17 that the Langmuir model predicts a maximum concentration (W_{max}) on the adsorbent for high concentrations in

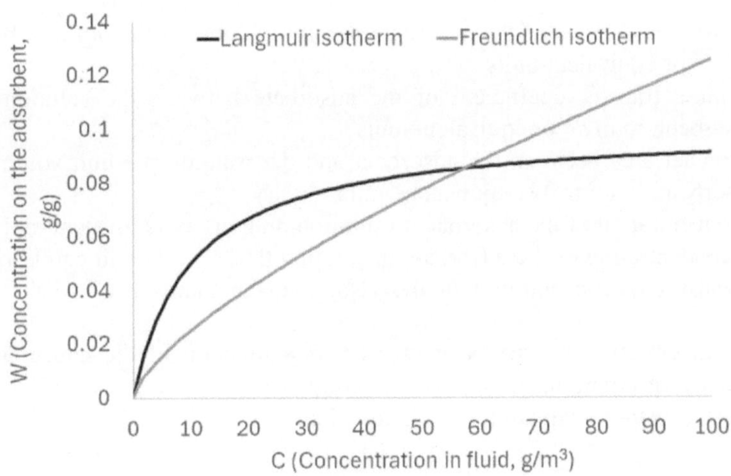

FIGURE 7.17 Comparison of Langmuir and Freundlich isotherms for adsorption.

the fluid, while according to the Freundlich model the concentration on the adsorbent increases indefinitely.

Example 7.7

Assuming the Langmuir isotherm is valid for the system under consideration, calculate the concentration on the solid in equilibrium with a concentration in the liquid of 10 mg/L. Assume the following Langmuir parameters: $W_{max} = 0.45$ g/g, $\alpha = 0.25$ L/mg.

Solution

The concentration on the solid in equilibrium with the liquid is called W in Equation (7.24) representing the Langmuir isotherm. W is obtained from

$$W = W_{max}\frac{\alpha \cdot c}{1+\alpha \cdot c} = 0.45\frac{g}{g}\frac{0.25\dfrac{\text{L}}{\text{mg}}\cdot 10\dfrac{\text{mg}}{\text{L}}}{1+0.25\dfrac{\text{L}}{\text{mg}}\cdot 10\dfrac{\text{mg}}{\text{L}}} = 0.32\frac{g}{g}$$

7.3.3 Mass Transfer and Mass Balances in Adsorption

The rate of mass transfer in an adsorption process can be written in the same form as seen for ion exchange and equation (7.10) also applies for adsorption. When using equation (7.10) to describe the mass transfer rate of the adsorbate from the liquid phase to the solid (adsorbent) phase, the meaning of the symbols is as follows:

r = mass transfer rate of the adsorbate per unit volume of the dry adsorbent, in g m^{-3} s^{-1} or equivalent units

k_c = mass transfer coefficient of the adsorbate between the solution and the adsorbent, in m s^{-1} or equivalent units

a = contact area between the adsorbent and the solution per unit volume of dry adsorbent, in m^2 m^{-3} or equivalent units

C = concentration of the adsorbate in solution, in g m^{-3} or equivalent units

C^* = concentration of the adsorbate in solution that would be in equilibrium with the adsorbate concentration on the adsorbent (same units as C)

The concentration C^* can be expressed as a function of the concentration on the solid by re-arranging either the Langmuir (obtaining equation (7.26)) or the Freundlich (obtaining equation (7.27)) equations:

$$C^* = \frac{W}{\alpha\left(W_{max} - W\right)} \tag{7.26}$$

$$C^* = \left(\frac{W}{\beta}\right)^{\frac{1}{m}} \tag{7.27}$$

Note that with the notation in this section the concentration on the solid W is in equilibrium with C^* and not with C.

In order to calculate the profile of the adsorbate in solution and inside the column, mass balances for the adsorbate in the liquid and in the solid phase must be written. The conceptual scheme of an adsorption column for mass balances is the same already used for ion exchange, with the same assumptions of piston flow and perfect mixing in each section. Figure 7.14, with the only change in notation from C_{A+} to C. For the adsorbate in solution, the mass balance is the same already seen for the ion A^+ in ion exchange, equation (7.15), the only difference being the notation (equation 7.28):

$$\frac{\partial C}{\partial t} = -u\frac{\partial C}{\partial L} - \frac{(1-\varepsilon)r}{\varepsilon} \tag{7.28}$$

For the adsorbate on the solid, the mass balance is similar to the one for the ion on the resin (AR) seen with equation (7.21), however for adsorption the concentration on the solid (W) is referred to the mass of solid rather than to the volume of liquid therefore the mass balance becomes (with ρ_{solid} being the density of the adsorbent not including the void space in g m^{-3} or equivalent units):

$$\frac{\partial W}{\partial t}(1-\varepsilon)dV\rho_{solid} = r(1-\varepsilon)dV \rightarrow \frac{\partial W}{\partial t} = \frac{r}{\rho_{solid}} \tag{7.29}$$

In the derivation of equation (7.29), the term $(1-\varepsilon)dV\rho_{solid}$ represents the mass of solid adsorbent in the volume dV. The terms u, ε in equations (7.28) and (7.29) have the same meaning as the same terms in the mass balances for ion exchange. Note that all terms in equations (7.28) have the units of g of adsorbate s^{-1} (or equivalent) while in equation (7.29) the terms have the units of g g^{-1} s^{-1} or equivalent.

Equations (7.28) and (7.29) give the time profiles of the adsorbate in solution and on the solid over time and over the length of the column. The equations can be solved with the same method used for the solution of the ion exchange mass balances (7.15), (7.20), (7.21), i.e. discretisation of the time domain and of the column length in small Δt intervals and ΔL sections. Note that for adsorption there is no equivalent to the mass balance for the counterion in ion exchange, equation (7.20).

Similarly as for ion exchange, after the numerical solution is obtained, it is important to check the correctness of the model and of the solution by checking the mass balance of the adsorbate across the system. For an adsorption system, the mass balance of the adsorbate states that, at any given time since the start of the process, the adsorbate that entered the column must be equal to the sum of the adsorbate that left the column, of the adsorbed dissolved in solution inside the column and of the adsorbate adsorbed on the solid inside the column. This is translated into equation (7.30), which is the same as equation (7.22) with the difference that for adsorption the concentration of the adsorbate on the adsorbent is referred to the mass of adsorbent hence the presence of $(1-\varepsilon)\rho_{solid}$ in the last term of the equation (note that in equation (7.30) all the terms have the units of g/m^2):

$$u\varepsilon Cn\Delta t = \sum_{t=0}^{n\Delta t} C\left(t, L = L_{bed}\right) u\varepsilon\Delta t + \sum_{L=0}^{L_{bed}} C\left(n\Delta t, L\right)\varepsilon\Delta L + \sum_{L=0}^{L_{bed}} W\left(n\Delta t, L\right)\left(1-\varepsilon\right)\rho_{solid}\Delta L \tag{7.30}$$

Regarding the effect of the mass transfer coefficient $k_c a$ in adsorption, the same considerations already done for ion exchange are valid, i.e. fast mass transfer gives a steeper adsorption zone and a higher utilisation of the column before breakthrough. Similarly, the same considerations done for the scale-up/scale-down of ion exchange columns are also valid for adsorption processes.

Example 7.8

Consider an adsorption process to remove a contaminant from a fermentation broth. The adsorption column is 1 m long and is loaded with an adsorbent of density 1800 kg/m^3 (this is the density of the solid excluding void fractions). The column has a void fraction of 0.7. The feed has a concentration of the contaminant of 100 g/m^3 and enters the column with velocity $u = 0.008$ m/s. Assume that the equilibrium of the adsorbate with the adsorbent is described by the Langmuir model with $W_{max} = 0.005$ g/g and $\alpha = 0.1$ m^3/g. Assume the mass transfer coefficient is 0.2 s^{-1}.

Plot the profile over time of the contaminant in the initial section of the column, in the middle of the column and at the column outlet.

Repeat the calculations and plot the outlet concentration for the value of the mass transfer coefficient of 0.5 s^{-1}. Compare the breakthrough time (defined as the time required for the outlet concentration to reach 1% of the feed concentration) and the ratio between the mass of contaminant fed until breakthrough and the maximum mass of contaminant adsorbable by the column for the $k_c a$ values of 0.2 and 0.5 s^{-1}.

For all simulations, check the accuracy of the model and of the numerical solution.

It is advised to use in all simulations: $\Delta t = 1$ s, $\Delta L = 0.1$ m.

Solution

The solution is obtained by solving differential equations (7.28) and (7.29) by discretising the time domain and the length of the column. With the numerical values of this example, equations (7.28) and (7.29) become (with units of m, s and g):

$$\frac{\partial C}{\partial t} = -0.008\frac{\partial C}{\partial L} - \frac{(1-0.7)r}{0.7}$$

$$\frac{\partial W}{\partial t} = \frac{r}{1.8 \cdot 10^6}$$

with the rate of adsorption given by

$$r = 0.2\left(C - \frac{W}{\alpha\left(W_{max} - W\right)}\right)$$

The solution of the mass balances will give the profiles of C and W over the column length as a function of time. With the suggested value of ΔL, the column is divided into ten sections.

As an example of the calculation results, Table 7.8 reports the initial results for some sections of the column.

Figure 7.18 reports the plot of the concentration of the adsorbate in solution and on the adsorbent vs time for section 1 (initial part of the column), section 5 (middle of the column) and section 10 (end of the column and outlet). The plots show that the column gradually saturates with the contaminant, which eventually leaves the column with the outlet stream. Note that, although the maximum capacity of the adsorbent for the adsorbate is $W_{max} = 0.005\frac{\text{g}}{\text{g}}$, in the column the concentration of the adsorbate on the adsorbent only reaches 0.0045 g/g. This is because, the concentration of the feed C = 100 mg/L is not high enough to reach W_{max}, indeed from the Langmuir equation:

$$W = W_{max}\frac{\alpha \cdot C}{1+\alpha \cdot C} = 0.005\frac{\text{g}}{\text{g}}\frac{0.1\frac{\text{m}^3}{\text{g}} \cdot 100\frac{\text{g}}{\text{m}^3}}{1+0.1\frac{\text{m}^3}{\text{g}} \cdot 100\frac{\text{g}}{\text{m}^3}} = 0.0045\frac{\text{g}}{\text{g}}$$

Figure 7.19 compares the adsorbate outlet concentration with the two $k_c a$ values of 0.2 and 0.5 s^{-1}. As expected from the theory, a higher mass transfer coefficient gives a sharper concentration profile and a longer breakthrough time. From the data of Figure 7.19, the breakthrough time, i.e. the time required for the outlet concentration to reach 1 g/m^3 (1% of the feed concentration of 100 g/m^3) is 2013 s for $k_c a = 0.2$ s^{-1} and 3085 s for $k_c a = 0.5$ s^{-1}. The maximum mass of

TABLE 7.8
Integration of the Mass Balance Equations for Adsorption (equations (7.28), (7.29)) for Example 7.8. Only the First Six time steps and three sections in the column are shown. The whole Simulation for Example 7.7 includes 7200 Time Steps (7200 s with Δt=1 s) and 10 Sections (1 m Bed Length with ΔL = 0.1 m). Units are m³ for Volume, g/m³ for Concentrations in the Liquid, g/g for Concentration on the Solid and s for Time

Time	C	C^*	W	r	$\frac{\partial C}{\partial t}$	$\frac{\partial W}{\partial t}$	$\frac{\partial C}{\partial L}$
				Section 1			
0	0	0	0	0	8	0	-1000
1	8	0	0	1.6	6.674286	8.89E-07	-920
2	14.67429	0.001778	8.89E-07	2.934502	5.568414	1.63E-06	-853.257
3	20.2427	0.005041	2.52E-06	4.047532	4.645928	2.25E-06	-797.573
4	24.88863	0.009545	4.77E-06	4.975816	3.876417	2.76E-06	-751.114
5	28.76504	0.015087	7.53E-06	5.749991	3.234514	3.19E-06	-712.35
6	31.99956	0.021499	1.07E-05	6.395612	2.699059	3.55E-06	-680.004
				Section 2			
0	0	0	0	0	0	0	0
1	0	0	0	0	0.64	0	-80
2	0.64	0	0	0.128	1.067886	7.11E-08	-140.343
3	1.707886	0.000142	7.11E-08	0.341549	1.336407	1.90E-07	-185.348
4	3.044293	0.000522	2.61E-07	0.608754	1.486652	3.38E-07	-218.443
5	4.530945	0.001198	5.99E-07	0.905949	1.550464	5.03E-07	-242.341
6	6.081409	0.002205	1.10E-06	1.215841	1.552377	6.75E-07	-259.181
				Section 3			
0	0	0	0	0	0	0	0
1	0	0	0	0	0	0	0
2	0	0	0	0	0.0512	0	-6.4
3	0.0512	0	0	0.01024	0.128146	5.69E-09	-16.5669
4	0.179346	1.14E-05	5.69E-09	0.035867	0.213824	1.99E-08	-28.6495
5	0.39317	5.12E-05	2.56E-08	0.078624	0.297326	4.37E-08	-41.3777
6	0.690496	0.000139	6.93E-08	0.138072	0.372099	7.67E-08	-53.9091

adsorbate adsorbable by the column with the given feed concentration is given by (in g/m^2, i.e. referred to the unit cross section of the column):

$$W_{max}\frac{\alpha\cdot c}{1+\alpha\cdot c}\cdot(1-\varepsilon)\rho_{solid}\Delta L\cdot 10=2450\frac{\mathrm{g}}{\mathrm{m}^2}$$

The mass of adsorbate that has been fed until breakthrough is given by

$$u\varepsilon C_0 t_{breakthrough}$$

where C_0 is the feed concentration and $t_{breakthrough}$ is the breakthrough time. For $k_c a = 0.2$ s^{-1} and $k_c a = 0.5$ s^{-1} we obtain 1127 and 1728 g respectively, which account for 46% and 70% respectively of the maximum adsorbable mass. This comparison confirms that higher mass transfer coefficients give a better utilisation of the column.

The check of model accuracy and of the numerical solution can be done by verifying that equation (7.30) is valid throughout the bed for any calculated time. With the calculated results, equation (7.30) is verified with less than 0.2% discrepancy between the left- and right-hand sides. For example, for $k_c a = 0.2$ s^{-1} and $t = 3000$ s ($n = 3000$) the terms of equation (7.30) are as follows (all in g/m^2):

$u\varepsilon Cn\Delta t = 1680$ (contaminant that entered the column)

$\sum_{t=0}^{n\Delta t} C(t, L = L_{bed})u\varepsilon\Delta t = 16.05$ (contaminant that left the column)

$\sum_{L=0}^{L_{bed}} C(n\Delta t, L)\varepsilon\Delta L = 42.94$ (contaminant in solution inside the column)

$\sum_{L=0}^{L_{bed}} W(n\Delta t, L)(1-\varepsilon)\rho_{solid}\Delta L = 1{,}621$ (contaminant on the solid in the column)

and $1{,}680 \cong 16.05 + 42.94 + 1{,}621$

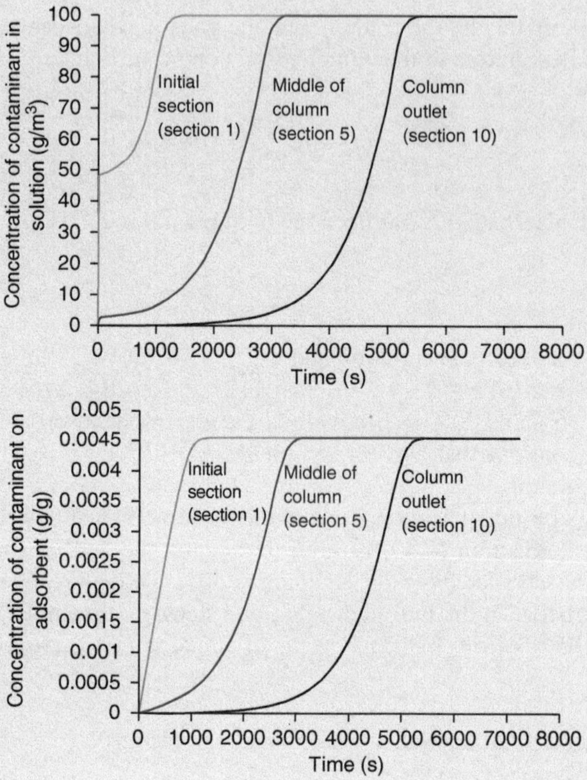

FIGURE 7.18 Solution of Example 7.8. Concentration of the adsorbate in solution (top) and on the adsorbent (bottom).

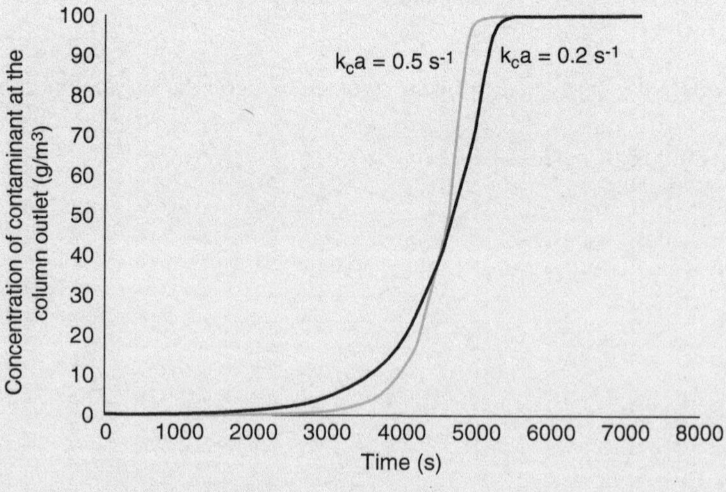

FIGURE 7.19 Solution of Example 7.8. Concentration of the adsorbate in the column outlet for two different values of the mass transfer coefficient.

7.4 CHROMATOGRAPHY

7.4.1 Introduction

Chromatography is a separation technique based, similarly as for ion exchange and adsorption, on the contact between a fluid (gas or liquid) phase (in this case called mobile phase) and a solid phase. In this section we will consider the use of chromatography for separation and purification of substances produced from fermentation or enzymatic reactions. This is also called "preparative chromatography" to distinguish from the frequently used "analytical chromatography". Similarly as for ion exchange and adsorption, in chromatography the substances in the fluid phase are separated by retention on the solid phase (which in this case is called stationary phase). However, differently from ion exchange and adsorption, in chromatography the substances to be separated are not flown continuously through the solid phase but instead are injected pulse wise into the column and carried through the column by the continuous flow of the liquid phase (called in this case eluent). Because of the batchwise feeding of the substances to be separated into a continuous-flow liquid phase that does not contain them, these substances are retained and released by the stationary phase and appear at the outlet of the column as peaks. The term chromatography, from the Greek "chroma" (colour), derives from the initial applications of this technique, which was used to separate plant pigments which, after chromatographic separation, gave bands of different colours. Chromatography is often used as an analytical, to identify and quantify the substances in a mixture, rather than as a separation technique. Indeed, the peaks of the different substances leaving the column are detected using detectors (these can be of different types depending on the type of substance, e.g. flame ionisation, thermal conductivity) and concentration of each substance is quantified by comparison with samples at known concentration (calibration standards). When chromatography is used as analytical technique, the separation and/or recovery of the substances is obtained with other techniques, e.g. ion exchange and adsorption.

Chromatography is classified in many different types, e.g. gas chromatography (GC), high-performance liquid chromatography (HPLC), gel permeation chromatography (GPC), based on the nature of the mobile and of the stationary phase and on the mechanism of the interaction between phases. Various mechanisms of interaction between the substances in the fluid phase and the stationary phase are possible in chromatography, e.g. ion exchange, adsorption, affinity, size exclusion.

The conceptual scheme of chromatography is shown in Figure 7.20.

Table 7.9 shows some examples of the use of chromatography for separation and recovery of products after fermentation and enzymatic reactions.

7.4.2 Mass Transfer and Mass Balances in Chromatography

The rate of mass transfer of a substance from the mobile phase to the stationary phase in chromatography can be expressed conceptually with the same rate equation already used for ion exchange and adsorption (equation (7.10)), i.e. as a product of the mass transfer coefficient and of the driving force. The driving force is in all cases the difference between the concentration of the substance in the mobile phase and

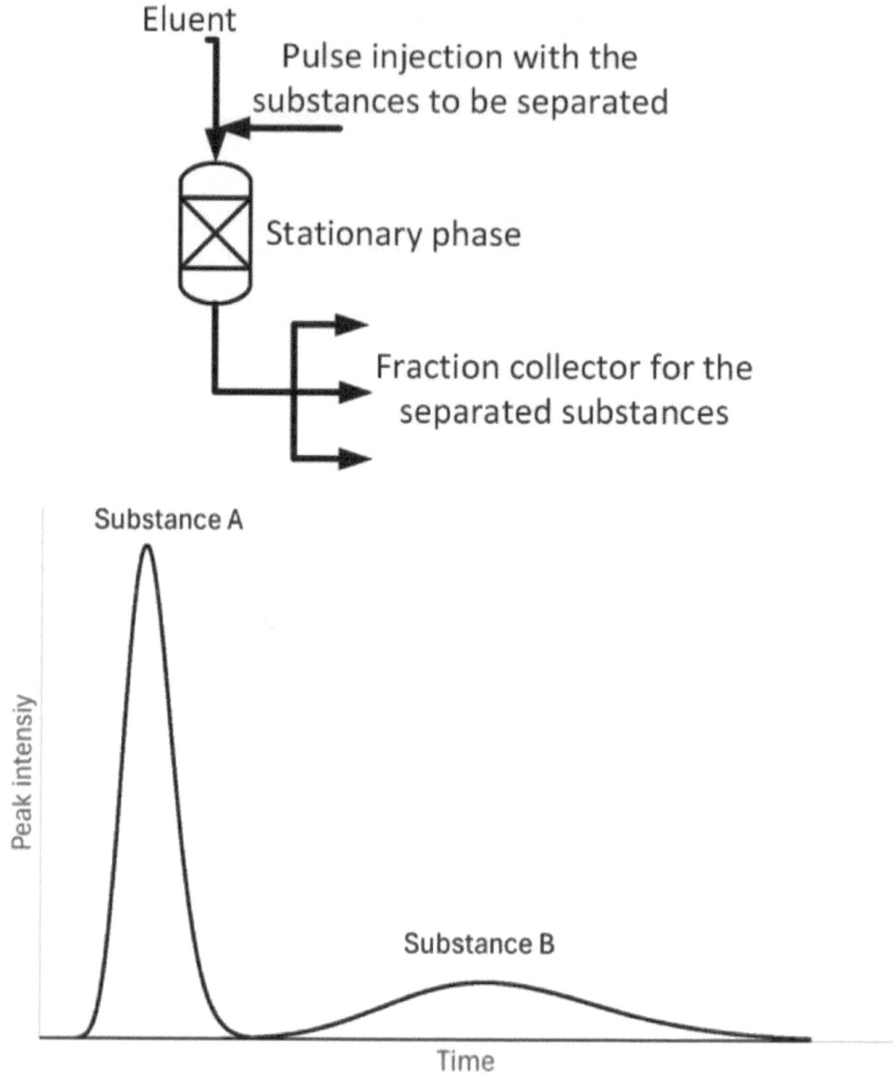

FIGURE 7.20 Conceptual schemes of chromatography for the separation and recovery of substances after fermentation or enzymatic reactions (top) and of peaks obtained for the separation of two substances injected together in the same pulse injection (bottom).

the concentration that would be in equilibrium with the concentration of the same substance on the stationary phase. The form of the equilibrium relationship, and therefore the formula for C^* in equation (7.10) can however be very different for different types of chromatography depending on the type of interaction between the substances and the stationary phase. If we assume, for example, that the mechanism of the interaction between solutes and stationary phase is adsorption, then equation

TABLE 7.9
Examples of Applications of Adsorption After Fermentation and Enzymatic Reactions

Application	Type of Chromatography and Process Description
Influenza vaccine purification	Size-exclusion and anion exchange chromatography. Proteins and DNA are retained by stationary phases and are separated from inactivated viruses (not retained) which are therefore purified.
Purification of enzymes	Ion exchange chromatography. The target charged enzymes are retained on the column while uncharged proteins are washed out leading to enzymes' purification.
Purification of vitamin B12	Affinity or ion exchange chromatography. After synthesis of the vitamin by fermentation and extraction of the vitamin from the microorganisms, the vitamin is retained on the chromatography column while undesired components or impurities are not retained, allowing for vitamin purification.
Purification of NAD^+	Cation exchange, size exclusion, reverse-phase chromatography. After production of NAD^+ from fermentation and other downstream processes, NAD^+ is retained on chromatography columns while impurities are washed out, allowing the separation and purification of the cofactor for various applications (e.g., biocatalysis, nutraceuticals and anti-aging supplements).
Purification of amino acids	Cation exchange chromatography. Amino acids produced from fermentation are, after other downstream processes, retained by the stationary phase while other substances and impurities are not retained, allowing for purification of the amino acids.

(7.10) can be used with the same meaning of the symbols, with easy adaptation of the terminology, as used for the mass transfer rate for adsorption in section 7.3.3. The different substances fed to the chromatography column will have different adsorption parameters, e.g. different W_{max} and α in the Langmuir isotherm, equation (7.24), and therefore different parameters in the equation for C^*, equation (7.26), therefore they will be retained differently and will be separated at the column outlet.

To write the mass balances in a chromatography column, we can use the conceptual scheme in Figure 7.21, which is very similar to the scheme used for ion exchange and adsorption (Figure 7.14) and based on the same assumptions of piston flow and perfect mixing in each section. Differently from ion exchange and adsorption, the eluent does not contain the substance to be separated, however the pulse injection can be represented by an initial concentration C_0 in the first section of the column. With this conceptual model, the mass balances are the same as seen for adsorption, equations (7.28) and (7.29), the only difference being the boundary condition for the

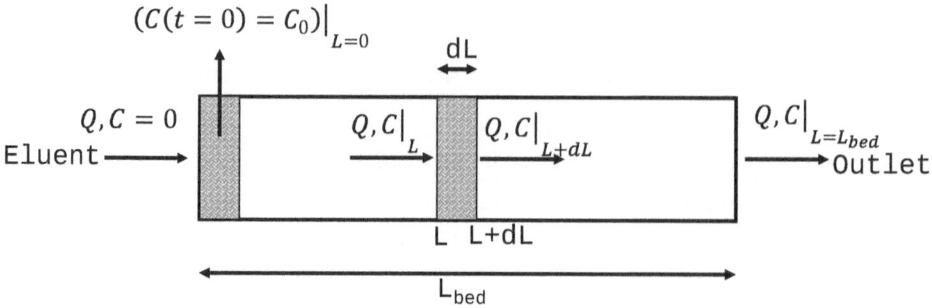

FIGURE 7.21 Conceptual scheme for mass balances in chromatography.

differential equation $\frac{\partial C}{\partial L}$ which is in this case $C = 0$ and the initial condition for $\frac{\partial C}{\partial t}$, which is in this case $C = C_0$ for the first section (and $C = 0$) for all the other sections. The value of C_0 can be calculated based on the chosen discretisation ΔL, as the concentration in the section ΔL that corresponds to the mass of substance injected into the column:

$$C_0 = \frac{\text{Mass of substance injected into the column}}{\varepsilon A \cdot \Delta L} \tag{7.31}$$

where A is the cross sectional area of the column and $\varepsilon A \cdot \Delta L$ is the volume occupied by the liquid in the first section.

The solution of the mass balances will give the profile of the substance inside the column and at the column outlet. Mass balances (7.28) and (7.29) can be written for multiple substances injected into the column, which will have different retention parameters, e.g. different parameters for the Langmuir isotherm, so that the solution of the mass balances will give the extent of separation.

As done for ion exchange and adsorption columns, it is important to check the correctness of the model and of the numerical solution by performing mass balance for the substances injected into the column. For a generic substance injected at the concentration C_0, the mass balance states that the mass of substance injected must be equal to the sum of the substance dissolved in the column, plus the substance retained on the resin, plus the substance that left the column. This is translated into equation (7.32) where all the terms have the units of g/m^2 or equivalent:

$$C_0 \Delta L = \sum_{t=0}^{n\Delta t} C\left(t, L = L_{bed}\right) u\varepsilon\Delta t + \sum_{L=0}^{L_{bed}} C\left(n\Delta t, L\right)\varepsilon\Delta L + \sum_{L=0}^{L_{bed}} W\left(n\Delta t, L\right)(1-\varepsilon)\rho_{solid}\Delta L \tag{7.32}$$

In equation (7.32) the term on the left-hand side represents the substance injected into the column and the terms on the right-hand side have the same meaning as in equation (7.30).

Example 7.9

A 50 mL sample containing substances A (concentration 1 g/L) and B (concentration 2 g/L) is injected into a chromatography column having cross sectional area 0.12 m² and length 0.8 m in order to separate substances A and B. The retention mechanism on the column is adsorption, described by the Langmuir model with $W_{maxA} = 9 \cdot 10^{-6} \frac{g}{g}$, $\alpha_A = 0.1 \frac{m^3}{g}$, $k_c a_A = 2.0\ s^{-1}$, $W_{maxB} = 7 \cdot 10^{-5} \frac{g}{g}$,

$\alpha_B = 0.09 \frac{m^3}{g}$, $k_c a_B = 5.0\ s^{-1}$

The mobile phase has a velocity of 0.03 m/s (referred to the total cross-sectional area). The void fraction inside the column is 0.6 and the density of the stationary phase (not including the void spaces) is 1500 kg/m³.
Calculate the time profiles of both substances inside the column in the middle section of the column and in the column outlet. It is suggested to use $\Delta t = 0.05$ s and $\Delta L = 4$ cm and to run the simulation for a total of 360 s.

Verify the mass balances of both substances in the system.

Solution

The solution is obtained by solving differential equations (7.28) and (7.29) for substances A and B. As for the examples on ion exchange and adsorption, the equations are solved by discretising the time domain and the length of the column. With the numerical values of this example, equations (7.28) and (7.29) become (with units of m, s and g):

$$\frac{\partial C_A}{\partial t} = -0.03 \frac{\partial C}{\partial L} - \frac{(1-0.6) r_A}{0.6}$$

$$\frac{\partial W_A}{\partial t} = \frac{r_A}{1.5 \cdot 10^6}$$

$$r_A = 2.0 \left(C_A - \frac{W_A}{0.1 \left(9 \cdot 10^{-6} - W_A \right)} \right)$$

$$\frac{\partial C_B}{\partial t} = -0.03 \frac{\partial C}{\partial L} - \frac{(1-0.6) r_B}{0.6}$$

$$\frac{\partial W_B}{\partial t} = \frac{r_B}{1.5 \cdot 10^6}$$

$$r_B = 5.0\left(C_B - \frac{W_B}{0.09\left(7\cdot10^{-5} - W_B\right)}\right)$$

The mass for substances A and B can be solved by discretising the time domain and the column length with the suggested values of Δt and ΔL, as done for Example 7.9. For the column length, the boundary condition for substances A and B is their concentration in the feed, which is zero. The initial condition at time $t = 0$ for the concentration of A and B in the column is zero in all sections except the first one where is, according to equation (7.31):

$$C_{0A} = \frac{0.05\ L\cdot 1\frac{g}{L}}{0.6\cdot 0.12\ \text{m}^2\cdot 0.04\ \text{m}} = 17.33\frac{\text{g}}{\text{m}^3}$$

$$C_{0B} = \frac{0.05\ L\cdot 2\frac{g}{L}}{0.6\cdot 0.12\ \text{m}^2\cdot 0.04\ \text{m}\cdot 0.12\ \text{m}^2\cdot 0.04\ \text{m}} = 34.67\frac{\text{g}}{\text{m}^3}$$

With the chosen ΔL of 4 cm = 0.04 m, a column length of 80 cm results in 20 sections. The plots of the time profiles of substances A and B in section 10 (middle of column) and section 20 (end of column and outlet) are shown in Figure 7.22. The plots show that at the column outlet substances A and B are well separated (and can be collected separately for further purification if desired). In the middle of the column the substances are not well separated and their peaks partially overlap. The plots also shows that substance A is weakly retained by the column and elutes early while substance B is retained more strongly and elutes later (the maximum adsorption capacity on the stationary phase, W_{max}, is lower for substance A than for substance B).

The check of the mass balances for substances A and B gives the following results, referring to equation (7.32) and taking, as an example, $t = 90$ s (all values are in g/m^2, i.e. masses per unit of cross sectional area of the column):

$C_{0A}\varepsilon\Delta L = 0.4159$ (substance A injected into the column)

$\sum_{t=0}^{n\Delta t} C_A\left(t, L = L_{bed}\right)u\varepsilon\Delta t = 0.4148$ (substance A that left the column)

$\sum_{L=0}^{L_{bed}} C_A\left(n\Delta t, L\right)\varepsilon\Delta L = 0.00057$ (substance A dissolved in the column)

$\sum_{L=0}^{L_{bed}} W_A\left(n\Delta t, L\right)(1-\varepsilon)\rho_{solid}\Delta L = 0.00058$ (substance A retained on the stationary phase)

$C_{0B}\varepsilon\Delta L = 0.8321$ (substance B injected into the column)

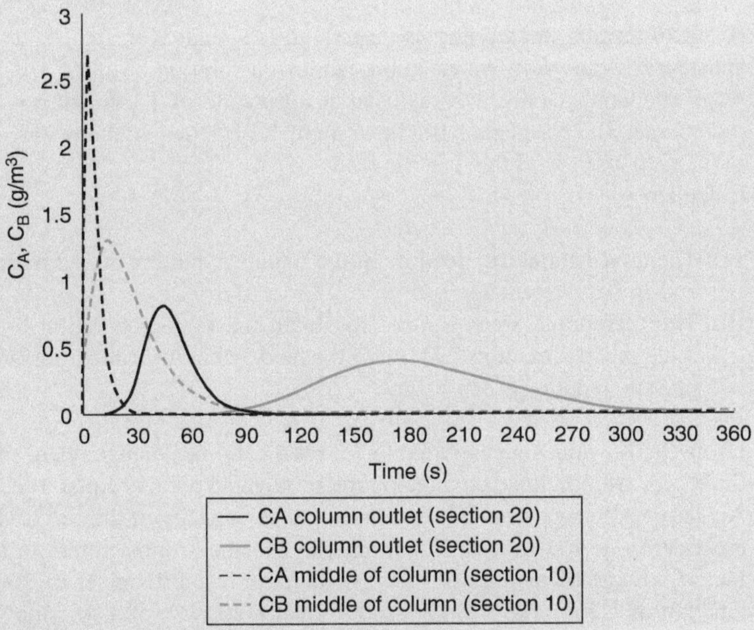

FIGURE 7.22 Plot with the solution of Example 7.9.

$\sum_{t=0}^{n\Delta t} C_B\left(t, L = L_{bed}\right)u\varepsilon\Delta t = 0.00689$ (substance B that left the column)

$\sum_{L=0}^{L_{bed}} C_B\left(n\Delta t, L\right)\varepsilon\Delta L = 0.1164$ (substance B dissolved in the column)

$\sum_{L=0}^{L_{bed}} W_B\left(n\Delta t, L\right)\left(1-\varepsilon\right)\rho_{solid}\Delta L = 0.7089$ (substance B retained on the stationary phase)

It can be easily verified that the mass balances match for both substances, i.e. for substance A:

$$0.4159 \cong 0.4148 + 0.00057 + 0.00058$$

and for substance B:

$$0.8321 \cong 0.00689 + 0.1164 + 0.7089$$

with errors of less than 1%.

The mass balances above also confirm the evidence of Figure 7.22, i.e. that at time $t = 90$ s, almost all of the substance A injected has left the column, while almost all of the substance B injected is still in the column, mostly on the stationary phase and some dissolved.

QUESTIONS AND PROBLEMS

7.1 A nanofiltration membrane is tested at lab scale for the separation of substances *A* and *B* in water. In the lab test, a solution of 5% *A* and 2% *B* in water (percentages are w/w) is used at a flowrate of 4 L/h using a 100 cm^2 membrane. The permeate has a flowrate of 3.5 L/h and contains 95% of the *B* in the feed and 1% of the *A* in the feed.
Calculate

(a) The total volumetric flowrate of the retentate and the mass fractions of *A* and *B* in the retentate.
(b) The membrane area required to obtain the same separation for a pilot scale process treating 500 L/h of the feed, using the same transmembrane pressure applied at lab scale.

7.2 Using the K' values given in Table 7.3 for a cation exchange resin, show that the K' for ammonium exchange on the protonated resin is equal to 2.01.

7.3 An anion exchange resin has an ion exchange capacity of 2 mol/L, where the moles are referred to monovalent anions and the volume refers to the geometric volume of the empty bed. The column has a diameter of 20 cm and a height of 1.2 m. The resin is used to treat a 200 L/h of a solution with the monovalent anion to be removed present at a concentration of 0.15 M.
Calculate how long the ion exchange process can last before the resin is fully saturated. Assume very fast (instantaneous) exchange.

7.4 It is desired to use adsorption to remove a contaminant (concentration 50 mg/L) from a fermentation mixture containing the product. The contaminant is adsorbed on activated carbon (Langmuir model with $W_{max} = 0.02$ g/g and $\alpha = 0.15$ m^3/g) while the product is not adsorbed. Assuming the column has a diameter of 20 cm, calculate the length of column required to remove the contaminant from a volume of 2 m^3 of fermentation mixture. Assume a void fraction of 0.5 and a density of the activated carbon of 2000 kg/m^3. Assume that the mass transfer rate in the column under the chosen operating conditions is such that the breakthrough time is 90% of the time for instantaneous adsorption.

7.5 Calculate the outlet adsorbate concentration over time and the breakthrough time to reach 5% of the feed concentration for an adsorption column fed with a stream at 4 m^3/h containing the adsorbate at 300 mg/L. Assume the column has a cross sectional area of 0.7 m^2, a void fraction of 0.4 and a length of 2 m. Assume that adsorption follows the Lamgmuir model with $W_{max} = 0.0035$ g/g and $\alpha = 0.08$. The mass transfer coefficient is 0.25 s^{-1} and the density of the solid phase is 2200 kg/m^3. For the numerical solution, it is recommended to use a time discretisation $\Delta t = 1$ s and to divide the column length in 10 sections.

7.6 Consider the need to separate three products dissolved in water from a fermentation mixture. The products can all be adsorbed on a suitable resin but have different adsorption isotherms and mass transfer coefficients. If the need is to recover and purify each of the three products, discuss whether adsorption or chromatography is the most appropriate process.

7.7 Compare the pros and cons of adsorption and chromatography for the separation and purification of fermentation mixtures.

8 Design Summary and Examples of Industrial Fermentation Processes

This chapter gives a short summary of the main design considerations covered in previous chapters and shows some examples of industrial fermentation processes. The global market size of chemicals obtained with fermentation processes has been estimated in the order of 110 billion US$ in 2024. While the market size of fermentation chemicals is large, large are also the challenges in the design and optimisation of these processes. The tools covered in previous chapters (kinetics and stoichiometry, mass balances, mass and heat transfer considerations) should be used to maximise process efficiency and minimise process costs.

8.1 SUMMARY OF DESIGN CONSIDERATIONS

We will summarise here the main steps for the design of a fermentation process for commercial scale. Here we assume that the aim of the fermentation is to produce a certain product to be sold on the market.

The main design steps can be summarised as follows:

- Step 1. Identify the microorganism that is able to produce the desired product.
- Step 2. Gain information from the literature on the metabolism of the microorganism: requirements for substrate and mineral elements, temperature, pH, type of metabolism (product produced as primary or secondary metabolite), suitability for growth under batch, continuous, fed-batch culture.
- Step 3. Decide the fermentation mode: batch, continuous, fed-batch. This can be decided based on literature evidence and/or on lab-scale experiments. Considerations on the productivity can underpin the choice of the fermentation mode, as shown in section 3.7.
- Step 4. Decide the operating parameters for the chosen fermentation mode. In all cases, the temperature, pH, concentration of substrate and of mineral elements should be chosen. When choosing the substrate concentration, considerations should be given to the desired productivity and microorganism concentration. Once the substrate concentration has been chosen, the requirements for mineral elements can be estimated from the fermentation stoichiometry (Chapter 3). For batch processes, the initial microorganism

DOI: 10.1201/9781003633822-8

concentration and the length of the batch need to be decided. For continuous and fed-batch processes, the initial microorganism concentration at start-up will not affect the steady state of the fermenter or the process performance. As long as the concentration chosen is high enough to avoid wash-out during start-up. For continuous and fed-batch process, the residence time, or the dilution rate, is the critical parameter to choose as it will greatly affect process performance and productivity. All these considerations, when choosing the operating parameters, should be based on experimental tests, guided by considerations on the reaction stoichiometry and kinetics and by the appropriate mass balances (Chapters 3 and 4).

- Step 5. Calculate the required reactor volume. The reactor volume should be calculated based on the desired product productivity and on the chosen residence time. If the required volume is too large, the process may be carried out in multiple fermenters in series.
- Step 6. If the fermentation is aerobic, the aeration system should be designed to maintain the desired oxygen concentration. Critical considerations in the design or the aeration system are choosing the agitator type and speed and of the aeration rate. Design of the aeration system should be based on the tools presented in Chapter 5 and underpinned by experimental measurements at lab-and/or pilot scale.
- Step 7. Design of the heat transfer system. The heat transfer system should be designed to maintain the desired temperature in the fermentation medium. This designing requires choosing the temperature inlet of the cooling/heating fluid, the type of jacket and the determination of the required heat transfer area. Design considerations for the heat transfer system are reported in Chapter 6.

If the vessel(s) to be used for the fermentation is (are) already available, the design steps above need to be adapted, considering that the reactor volume, agitation type and heat transfer system are available. If a new vessel needs to be purchased, it should be chosen, considering the design steps above, among the available designs from the main manufacturers, some of which are reported in the Bibliography section.

In the design steps above, we have not mentioned, for conciseness's sake, other important design choices to be made, e.g. pumps and compressors, tubing, filters. Clearly, the economics of the process, considering capital and operating costs, need to be considered to evaluate the viability of the proposed process.

8.2 BIOETHANOL

Bioethanol (Cardona et al., 2010) is the common name for the ethanol molecule (C_2H_5OH) when produced from biomass, rather than from fossil fuels as in conventional processes. Bioethanol is commercially produced using the anaerobic fermentation of the glucose derived from cereal crops or from sugar cane. Bioethanol is an important energy vector in the transition towards renewable fuels and it is mainly used in vehicles since it has similar combustion properties as conventional petrol.

Since bioethanol ultimately comes from biomass, which can be harvested at the same rate it is produced, it can be considered a renewable energy source. Furthermore, the carbon dioxide generated by the combustion of bioethanol is absorbed by the crops when they grow, in an ideally zero-carbon process. However, the energy consumption in the bioethanol production process from corn and in the growth of corn can contribute to carbon dioxide emissions, depending on the way this energy is produced.

Figure 8.1 shows the production volumes of bioethanol in different countries in years 2007–2018. The market size of bioethanol has been estimated as 33 billion US$ in 2020 and is expected to grow to more than 60 billion US$ by 2025.

The bioethanol production process is partly different depending on whether the feedstock is cereals or sugar cane.

Cereal feedstocks, e.g. corn, wheat, and barley, are mainly composed of starch (Table 8.1).

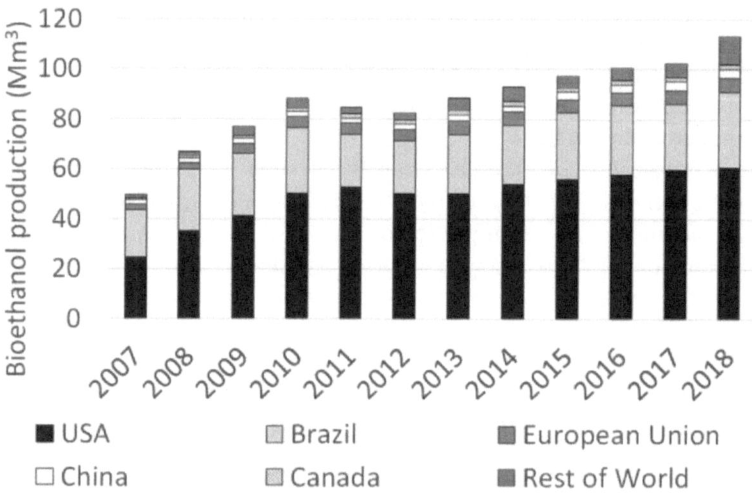

FIGURE 8.1 Production of bioethanol in years 2007–2018. (Adapted from www.afdc.energy.gov/data/.)

TABLE 8.1
Typical Chemical Composition of Cereals Used for Bioethanol Production (% w/w)

Feedstock	Available Carbohydrates (Mainly Starch)	Fibre	Protein	Lipids	Water
Wheat	59	13	11	2	15
Maize (corn)	65	10	9	4	12
Barley	63	10	11	2	12

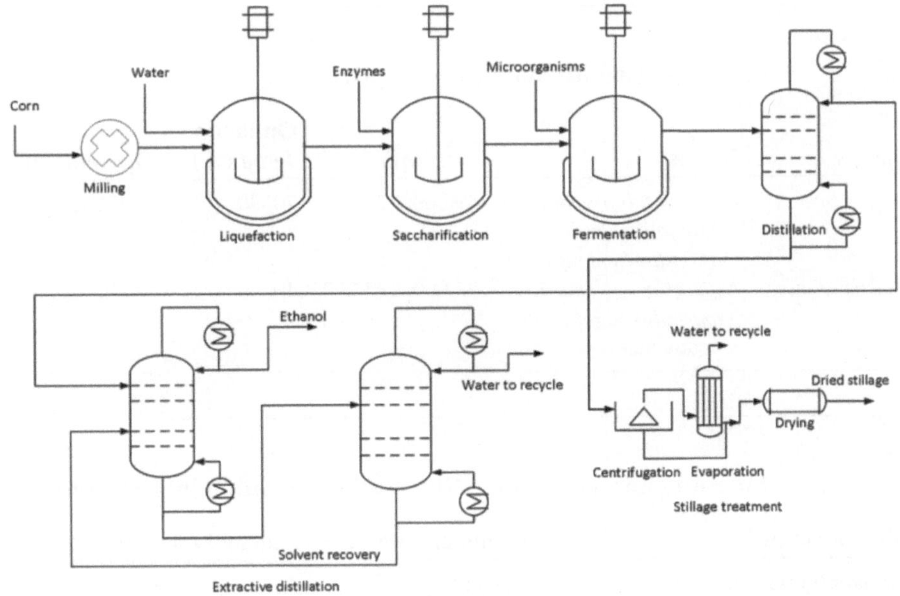

FIGURE 8.2 Simplified scheme of a process for the production of bioethanol from corn.

Starch is a polysaccharide made of glucose units. The main units in starch molecules are called amylose (linear polymer in which glucose molecules are joined by α-1,4 bonds) and amylopectin (branched polymer in which linear chains of glucose molecules are joined by α-1,6 bonds). The initial stages in the process (Figure 8.2) aim to generate the free glucose, which will be then fermented into ethanol. The first stage is milling, where the particle size of the corn grains is reduced to 3–5 mm. Milling is a mechanical process, usually carried out with hammer mills. In these mills, particle size reduction is obtained by using a hammer, which beats and breaks down the corn particles until they are small enough to pass through a screen. After milling, the cooking and liquefaction stages essentially consist of heating the milled grains to a temperature of 100–110 °C, usually with the addition of enzymes. This process breaks up the corn granules, making the starch freely available in the liquid medium and available for the following hydrolysis stage. In the starch hydrolysis stage, enzymes are added which hydrolyse starch to glucose. Table 8.2 reports some common enzymes used for starch hydrolysis, and their optimum conditions of temperature and pH. The reaction of starch hydrolysis can be written as in equation (8.1):

$$C_6H_{10}O_5 + H_2O \rightarrow C_6H_{12}O_6 \tag{8.1}$$

Once free glucose has been obtained, it is converted into ethanol in the fermentation stage. The fermentation is anaerobic, carried out by one of several species of microorganisms, e.g. the yeast *Saccharomyces cerevisiae*, or the bacterium *Zymomonas mobilis* (Table 8.3). In the fermentation, the microorganisms use ammonia

TABLE 8.2
Examples of Enzymes Used for Starch Hydrolysis

Enzyme	Source	MW	Optimum Temp (°C)	Optimum pH
α-amylases	*Bacillus subtilis, Bacillus amyloliquefaciens, Bacillus licheniformis*	49,000–62,000	70–90	5.5–6.5
Glucoamylases	*Aspergillus awamori, Aspergillus niger* and many others	38,000–112,000	40–60	4.5–5.0

TABLE 8.3
Examples of Microorganisms Used for Glucose Fermentation into Ethanol

Microorganisms	Optimum Temperature (°C)	Ethanol Tolerance (g/L)
Saccharomyces cerevisiae	30–37	160
Zymomonas mobilis	30	100
Clostridium thermocellum	55–65	10–30

as nitrogen source and generate, in addition to ethanol, new microorganisms, carbon dioxide and water. A typical stoichiometry for the fermentation process is reported in equation (8.2):

$$C_6H_{12}O_6 + 0.12NH_3 \rightarrow 0.12C_5H_7O_2N + 1.8C_2H_5OH + 1.8CO_2 + 0.36H_2O \qquad (8.2)$$

The maximum concentration of ethanol at the end of the fermentation is limited by the fact that ethanol is toxic for microorganisms at high concentrations. The maximum ethanol concentration that can be tolerated by microorganisms is different for different microorganisms, but it is typically in the range 100–160 g/L.

After the fermentation process, the bioethanol needs to be purified to at least 99 % w/w to be used as fuel. Separation of the ethanol from water is in general obtained using distillation. In distillation columns, ethanol, which is more volatile than water, is collected as the distillate at a concentration of up to approximately 90% w/w. At concentrations higher than this, ethanol and water form an azeotrope and further separation by distillation is no longer possible. In order to increase the ethanol purity to the desired 99% w/w, one of the following technologies is used: (a) vacuum distillation. The azeotrope tends to disappear at reduced pressures; (b) extractive or azeotropic distillation. By adding third components (often ethylene glycol for extractive distillation) the relative volatility between ethanol and water changes, allowing to reach a higher ethanol purity; (c) adsorption on molecular sieves. Molecular sieves are aluminosilicates that adsorb water but not ethanol, because water molecules are considerably smaller than ethanol molecules and can be trapped in the cage of the sieves, while ethanol is not retained. After the purification processes, ethanol of the desired purity is obtained. Any solids residues from the process are called stillage. Stillage contains all the inorganic and organic matter that is not starch, which is present in the feedstock (fibre, lipids, etc) and the produced microorganisms. Typically, stillage is concentrated by centrifugation, evaporation and drying and is used for animal feed.

The other common feedstock for bioethanol production is sugarcane. In sugarcane, most of the carbohydrates are present as sucrose (Table 8.4). Sucrose is a disaccharide composed of one molecule of glucose and one of fructose. Differently from starch, sucrose can be immediately metabolised by microorganisms, without previous hydrolysis. The pre-treatments of sugar cane for bioethanol production are therefore easier than the pre-treatments from cereal feedstocks. In the conventional process (Figure 8.3), bioethanol is produced from sugar cane as a by-product of the sugar industry. Sugarcane is collected manually or mechanically. Then, it is milled with addition of hot water and sugarcane juice, which contains sucrose and all the soluble species, is extracted. Bagasse is the by-product of sugarcane extraction and contains the lignocellulosic (fibre) fraction. Bagasse is typically burned on site to generate electricity. The sugarcane juice is concentrated, by means of evaporation and centrifugation, and in the end sugar crystals are obtained. The residual liquid after centrifugation is called molasses (Table 8.4) and can be used as substrate for fermentation (without any further pre-treatments) for ethanol production. The fermentation and separation/purification stages are the same as already described for the bioethanol production from cereals.

TABLE 8.4
Chemical Composition (% w/w) of Sugarcane and Molasses

Feedstock	Sucrose	Fibre	Protein	Lipids	Water
Sugarcane	14	13	<1	<1	70
Molasses	37	–	3.5	<1	19

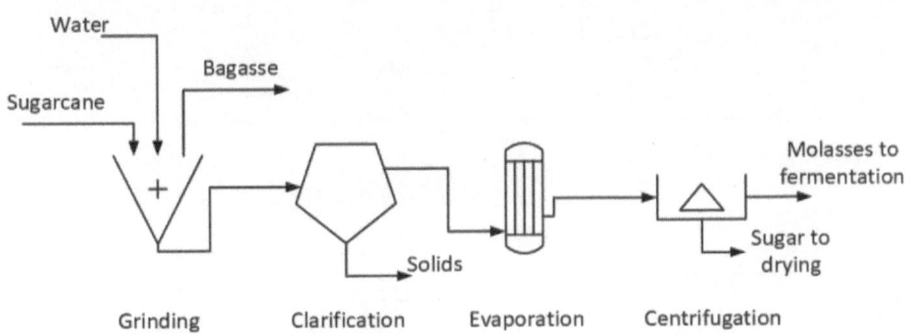

FIGURE 8.3 Simplified scheme of molasses production from sugarcane. Molasses are sent to fermentation for bioethanol production.

8.3 PENICILLIN

Penicillin is a group of antibiotics and historically is one of the first antibiotics produced at large scale. Today penicillin is still used in the treatment of many bacterial infections, e.g. those caused by *Staphylococci* and *Streptococci*. Penicillin is produced by certain species of the Penicillium microorganism under aerobic conditions. Current market size of penicillin is in the order of 2–3 billion US$.

Fermentation is carried out using sugars (typically glucose or lactose) as substrates, and the fungus Penicillium as microorganism. Also, nutrients containing nitrogen, phosphorus and other salts need to be added.

Penicillin (process flow sheet in Figure 8.4) is not produced during growth of the microorganism but is produced during the stationary phase after growth has stopped (secondary metabolite). When fermentation is complete, biomass is removed by filtration. A typical equipment used for biomass filtration is the rotary vacuum filter. After the biomass has been removed, the water phase is mixed with an organic solvent, typically amyl acetate or butyl acetate. The solvent dissolves penicillin but not the many other salts present in the fermentation medium, therefore purity of penicillin is increased. Solvent extraction is typically carried out in a centrifuge, which also allows for the separation of the water phase. The organic (solvent) phase is sent to the crystallisation stage, where several inorganic reagents are added (e.g. sodium carbonate). The reagents form a salt with penicillin. This allows the precipitation of penicillin (salts are not soluble in organic solvents) as a solid, which will be then further purified. The solid phase (penicillin salt) is separated by the solvent using centrifugation. The penicillin salt is collected from the centrifuge and sent to the dryer, where the residual solvent is removed using hot air.

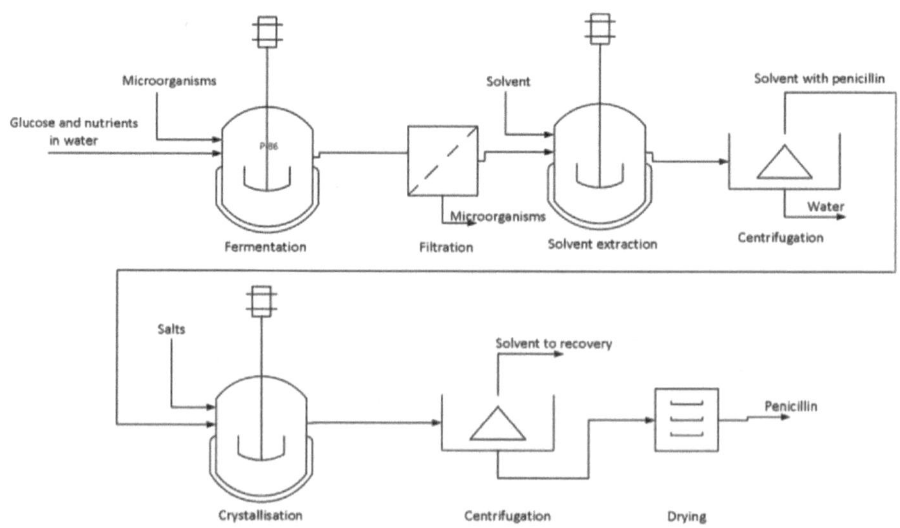

FIGURE 8.4 Simplified scheme of penicillin production.

8.4 WHISKY

Scotch whisky is one of Scotland's biggest exports, second only to oil and gas, as billions of litres are sold each year worldwide. Scotch whisky totals one quarter of all of food and drink exports from the UK making it a very important contributor to both the Scottish and the UK economies. The market size of Scotch whisky was estimated at £4.9 billion in 2019. In 2009, the Scotch Whisky Regulations defined Scotch whisky as a product "that has been distilled at a distillery in Scotland from water and malted barley (to which only whole grains of other cereals may be added)". Malt whisky is the whisky made using barley as only cereal, while grain whisky also uses other cereals (corn, wheat or rye) in addition to barley. The production processes of malt and grain whisky are significantly different.

The first stage in the production of malt whisky (process flowsheet in Figure 8.5) is called malting. The aim of malting is making the starch in the barley available for the next steps of the process. Barley is soaked in water for 2–3 days before being spread out on the floors of the malting house. In this way, natural enzymes are produced which will convert starch into glucose in the next step. The malting process is stopped by drying the barley (over peat fires in "kilns" or using air). After malting, the dried malted barley is ground down and mixed with hot water at a temperature of approximately 60–65 °C in a vessel known as a "mash-tun" (mashing). This process

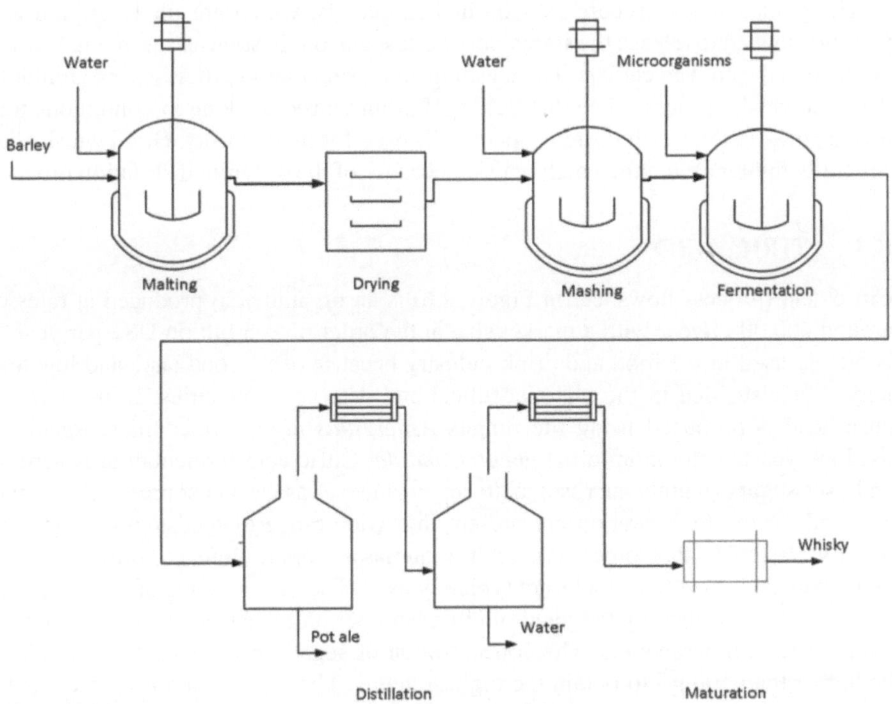

FIGURE 8.5 Simplified scheme of the malt whisky production process.

hydrolyses the starch from the barley and converts it into sugars, which dissolve and form a sugary liquid known as "wort". When the mashing process is complete, the liquid is removed from the spent barley and transferred to the fermenting vat where yeast (typically *Saccharomyces cerevisiae* strains) is added. The yeast converts glucose into ethanol and carbon dioxide. The fermentation is not sterile, so other microbial species may be present which convert glucose to other organic compounds which contribute to the taste, smell, and appearance of the final product. The final product of the fermentation is a solution ethanol/water (ethanol up to 10% w/w), plus all the other substances contained in barley. This product is known as "wash" and has a taste not too dissimilar from the taste of unhopped beer. The size of fermentation vessels is in the range 5–150 m^3. The fermentation is carried out in batch and typically lasts for 2–3 days. The stoichiometry of the fermentation is similar to the one of the bioethanol process.

The wash is then distilled in the characteristic "swan-necked" copper stills. Scottish whisky is traditionally distilled twice, with the second copper still being slightly smaller than the first. After the second distillation the ethanol content is 65–70% (spirit). The final stage in the process is called maturation. The spirit is put into oak casks and stored. The spirit must mature in casks for a minimum of three years before it is legally allowed to be called whisky in Scotland. During maturation, the flavours of the spirit combine with natural compounds in the wood cask and this gives the whisky its own characteristic flavour and aroma. During each year of maturation, about 2% of the spirit is lost through natural evaporation.

The grain whisky process uses un-malted cereals, which are finely ground and pressure cooked to release the starch into the suspension. In some cases, malted barley is added to the cooked cereals. The mashing and fermentation processes are similar to the malt whisky process. The distillation of grain whisky is done in continuous tray towers, instead than in the batch copper stills used for malt whisky. Grain whisky has generally lighter aroma than malt whisky, because of the different distillation process.

8.5 CITRIC ACID

Citric acid (process flowsheet in Figure 8.6) is an organic acid produced at rates of around 400,000 t/year with a market size in the order of 2–3 billion US$ per year. It is mainly used in the food and drink industry because of its good taste and low toxicity. It is also used in the pharmaceutical and detergent industries. In most cases, citric acid is produced using the fungus *Aspergillus niger*. Other microorganisms used are yeasts belonging to the genus *Candida*. Citric acid fermentation is aerobic and uses sugars (mainly sucrose, maltose, or glucose) as carbon sources. The sugars required for the fermentation are present in a wide range of feedstocks, which are usually waste of other processes, such as molasses, apple pomace, brewery waste and others and these feedstocks are typically used. The performance of the fermentation, mainly described by the yield of citric acid per unit mass of sugars, depends on many operation parameters. The concentration of sugars in the feedstocks needs to be higher than 100 g/l to obtain the highest yields. The concentration of trace metal ions is particularly important. It is important to provide trace metal ions (Fe, Cu, Mn,

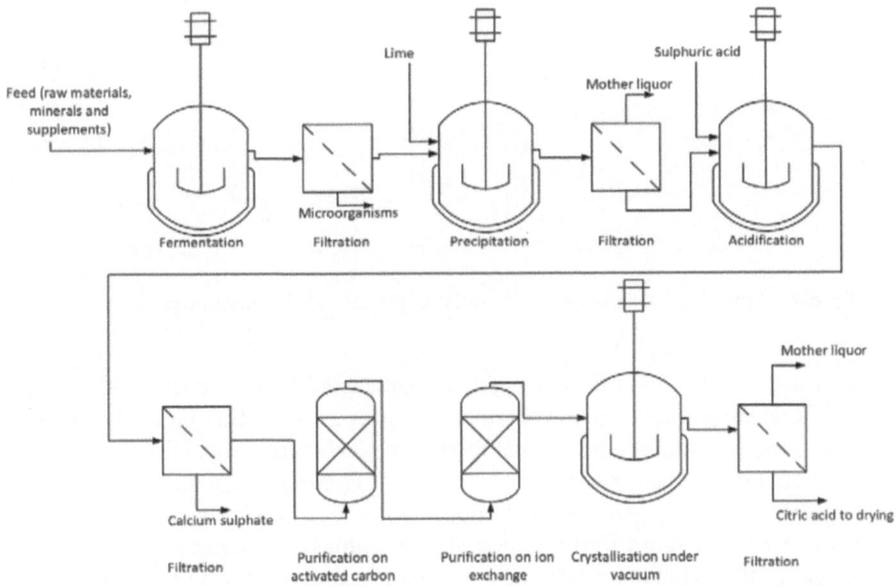

FIGURE 8.6 Simplified scheme of the citric acid production process.

Co, Ni) to support the growth of the microorganisms, however some of them (Mn^{2+}, Fe^{3+} and Zn^{2+}) need to be present at very low concentrations in order for citric acid to be produced. Therefore, many feedstocks used for citric acid production need to be purified by precipitation or ion exchange to decrease their concentration of trace metals. pH needs to be below 2.5 to obtain high citric acid yields, and this is easily achieved even in the absence of external acid addition because of the acidity of the produced citric acid. Aeration is also important and oxygen concentration needs to be maintained above 25 % of saturation to obtain high yields. After the fermentation, the microorganisms are removed by filtration and citric acid is precipitated by the addition of lime, forming tri-calcium citrate tetrahydrate, which has a low solubility in water. The precipitated calcium citrate is separated from water using filtration and the cake is washed several times with water to increase the purity of the salt. After filtration citric acid is formed again by adding sulphuric acid (this process also forms calcium sulphate which is filtered off). The resulting solution of citric acid is purified using activated carbon and ion exchange resins and citric acid is then crystallised under vacuum. Depending on the temperature of the crystallisation, anhydrous citric acid or citric acid monohydrate is obtained. Citric acid is then dried in the final stage of the process.

8.6 L-GLUTAMIC ACID

L-glutamic acid (or its salt L-glutamate) is the amino acid produced in the largest quantities, with a global production of over 1 Mt/year (process flowsheet in Figure 8.7) and a market size of over 10 billion US$ per year. It is used by a variety of industries: the

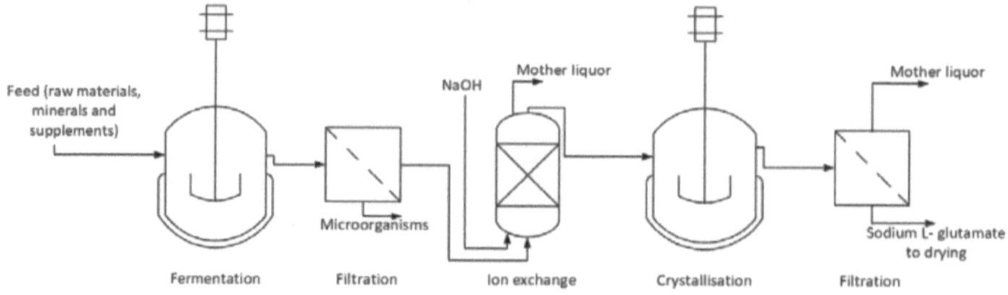

FIGURE 8.7 Simplified scheme of the production process for L-glutamic acid.

food industry uses L-glutamic acid as flavour enhancer, the pharmaceutical industry uses it as food supplement and the chemical industry as a building block. L-glutamic acid is also used as additive for animal feed. The production process is based on the fermentation with the bacterium *Corynebacterium glutamicum*, which, under appropriate conditions and using sugars as carbon sources, excretes large concentrations of L-glutamate in the medium. Various operating conditions can trigger in *C. glutamicum* the excretion of L-glutamate, the most common are as follows: (a) growth under biotin limitation; (b) addition of surfactants; (c) addition of penicillin. Furthermore, the concentrations of dissolved oxygen and ammonium and the pH (optimum range 7.0–7.7) need to be controlled in order to maximise the yields. The fed-batch configuration is often the preferred option in industrial processes, as it allows to vary the concentration of ammonium and triggering factors (e.g. surfactants) over the course of the fermentation. At the end of the fermentation, yields of L-glutamate of 60–70 % (based on the glucose used) can be obtained and glutamic acid is present mainly as ammonium L-glutamate. After fermentation, the microorganisms are separated by filtration and the liquid phase is passed through an anion exchange resin, where L-glutamate is retained on the resin. The resin is then eluted with sodium hydroxide, to form sodium glutamate which is released into solution and to regenerate the resin. Sodium glutamate is then crystallised from the solution and purified.

8.7 CELL-BASED VIRUS VACCINES

Vaccines are used to protect against diseases. In this section, we will see fermentation processes used to produce vaccines against diseases caused by viruses. Viruses are infective agents made of a nucleic acid molecule (DNA or RNA) coated by proteins and in some cases surrounded by a lipid envelope. Viruses are considered "borderline" between living and nonliving, as they can reproduce but only using living cells of other organisms. Viruses can infect any species, microorganisms, plants and animals. When viruses infect the human body, or any other organisms, they bind to and enter the cells of the host body, then they use the cell metabolic machinery to duplicate their genetic material and their protein coating, creating new virus particles, and finally leave the cells. Viruses cause diseases in many ways, e.g. by breaking and killing the cells of the host body. Organisms protect themselves against viruses with their immune response, with the use of certain cells and molecules, which can kill the infecting agents. The immune response is boosted by vaccination. Vaccines introduce inactivated or weakened viruses in the organism, so that, if the organism is infected with the active virus, it can quickly develop the appropriate immune response. These vaccine viruses are produced using either chicken egg cultures or animal (often mammalian) cells. In this section, we will see an example of a process flowsheet to produce influenza vaccine viruses from mammalian cells (Figure 8.8). Influenza is a very common disease that can impact hundreds of millions of people every year and is caused by various viruses belonging to the family of *Orthomyxoviridae*.

Influenza vaccines are inactivated or attenuated live viruses, sometimes containing adjuvants (typically aluminium salts), which help the body to develop a stronger immune response. The main mammalian cells used in the production of cell-based influenza vaccines are MDCK (Madine Darby canine kidney cells), Vero (Kidney epithelial monkey cells) or PER.C6 (human retina-derived cells). The cells are grown either attached to microcarriers (e.g. dextran beads, size in the order of magnitude of 100 μm), or as free-cell suspensions. The carbon source for cell culture is typically glucose supplemented with amino acids (glutamine) and pyruvate. Alternatively, complex animal-derived protein-rich media (serum) can also be used. The cell culture starts at small scale after thawing of the cell seeds and continues at larger scale in fermenters of size up to several m^3. When the cell concentration is high enough (this typically takes several days after inoculation of the fermenter) the virus is added and can propagate using the mammalian cells. Virus propagation lasts typically for 3–5 days. At the end of the process, the mammalian cells are separated by

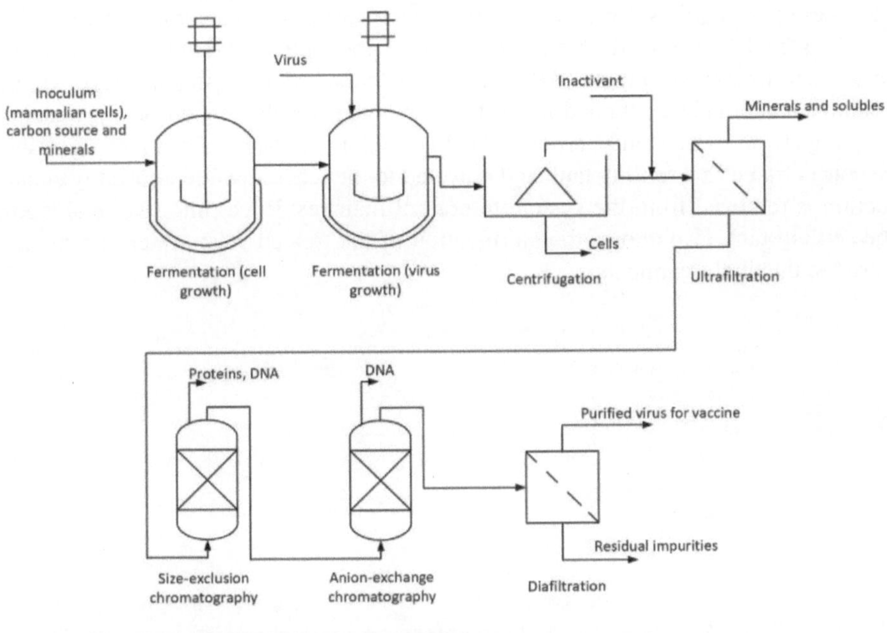

FIGURE 8.8 Example of a process flowsheet to produce influenza vaccine using mammalian cells. (Adapted from Kalbfuss et al., 2007a, 2007b.)

filtration or centrifugation. Mammalian cells are much larger than viruses (10–20 μm vs ~ 100 nm) and can be separated easily by mechanical means. The virus is then inactivated using chemicals. The suspension containing the inactivated virus is then flown through ultrafiltration membrane modules to remove small soluble molecules. The concentrated virus is purified using chromatography. Size-exclusion chromatography removes proteins (which have a much smaller size than viruses) and part of the DNA. The rest of the DNA is removed by anion exchange chromatography. Finally, the virus is purified by diafiltration to remove any residual impurities and salts.

8.8 PROCESSES WITH UNDEFINED MIXED CULTURES

Most processes considered so far use pure cultures of selected microbial species, chosen based on the desired product. In the sector of wastewater and organic waste treatment, however, processes which use undefined mixed cultures are more common. Undefined mixed cultures are consortia of many different species of microorganisms, which coexist in the same reactor. Mixed culture processes are usually started up with an inoculum composed of undefined microbial species from another similar plant or using the microorganisms, which are naturally present in wastewater or waste. Mixed culture processes are usually better suitable for the treatment of wastewaters and organic waste, because these waste streams contain many organic contaminants of different chemical nature, which need a consortium of different microorganisms for their degradation. On the other hand, pure cultures can only grow on a limited range of substrates. Mixed culture processes can be described with the same concepts and models used for pure culture processes, although the complexity of the substrates and processes often requires some modification in the terminology and rate equations.

An example of mixed culture process is the activated sludge process for wastewater treatment (Figure 8.9). In this process, microorganisms degrade the organic contaminants contained in wastewaters under aerobic conditions, producing microorganisms, water and carbon dioxide as final products. The microorganisms are concentrated in a settling tank and recycled to the reactor, while a (usually small) fraction is removed from the system to control their residence time (waste sludge). The settling tank also allows the clarification of the treated wastewater and its discharge to the final treatments.

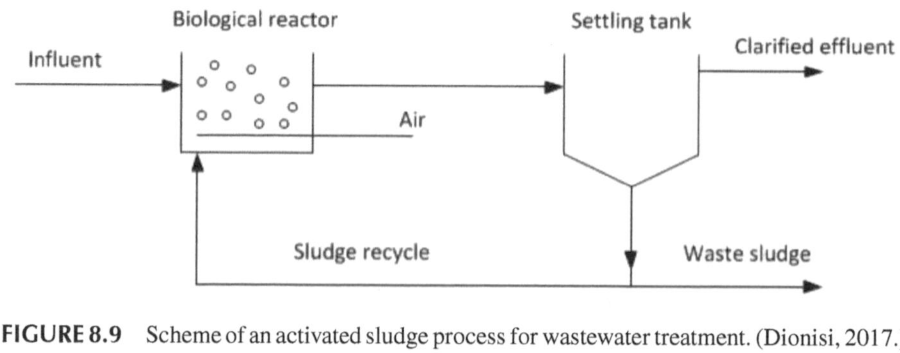

FIGURE 8.9 Scheme of an activated sludge process for wastewater treatment. (Dionisi, 2017.)

Another example of mixed culture process is anaerobic digestion of organic waste. In this process (Figure 8.10) a consortium of microorganisms converts the biodegradable components of organic waste into methane and carbon dioxide (biogas), using a complex series of chemical reactions (Figure 8.11), also producing, like in most

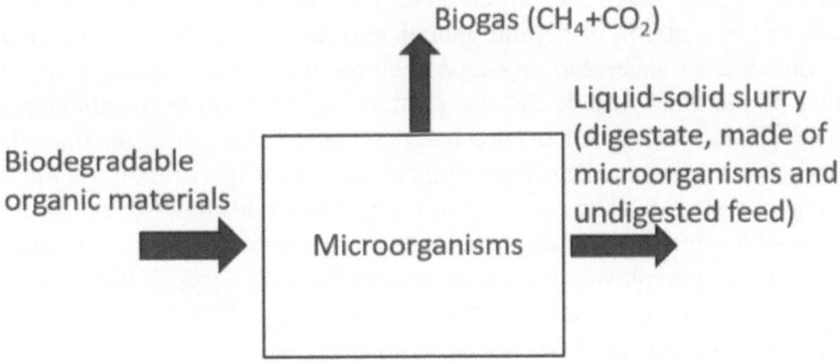

FIGURE 8.10 Scheme of an anaerobic digester for the treatment of organic waste.

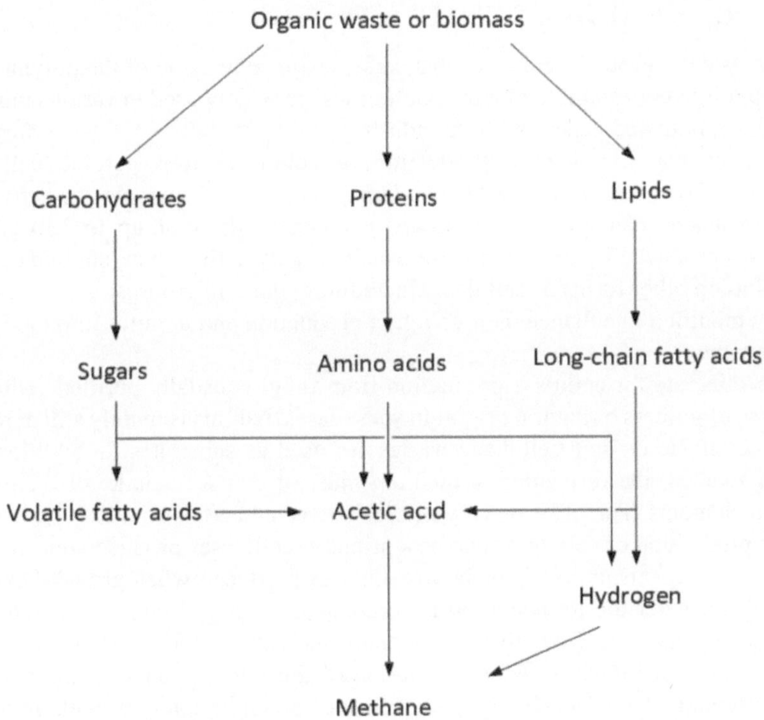

FIGURE 8.11 Simplified scheme of the complex series or reactions occurring in anaerobic digesters.

fermentation reactions, new microorganisms and water. The components of the waste (carbohydrates, proteins and fats) are first hydrolysed by extracellular enzymes into their monomeric constituents (sugars, amino acids and fatty acids), the monomers are then converted into organic acids and alcohols, in some cases with the production of hydrogen. Other reactions convert the organic acids and alcohols into acetic acid, which is then converted into methane. Hydrogen is also converted to methane by microorganisms, which use carbon dioxide as carbon source. Each of the steps in the anaerobic conversion of organic matter into methane is carried out by different microorganisms, so anaerobic digestion requires the coexistence of many different microbial populations, which use the products of each other metabolism as their substrates. The produced methane is usually combusted with air to generate electricity and/or heat, therefore anaerobic digestion can be seen as a process to produce renewable energy from waste. The solid-liquid slurry (digestate) that leaves the digester is composed of any undigested organics, the microorganisms, and the nutrients (particularly nitrogen and phosphorus) generated by the hydrolysis of the organic matter. The digestate can be used in agriculture, partially replacing the need for synthetic fertilisers. The organic acids produced as intermediates in the anaerobic digestion process have also been proposed as carbon source to produce biodegradable plastics (polyhydroxyalkanoates, PHAs) (Villano et al., 2014).

8.9 PRODUCTION OF CELLULASES

Cellulases are a group of enzymes that catalyse the conversion of the polymer cellulose into oligomers and into glucose. Cellulases are widely used in various industries, e.g. textile, pulp and paper and food industry.

In many cases, industrial production of cellulases uses aerobic cellulolytic fungi, e.g. *Hypocrea jecorina (Trichoderma reesei)* or *Humicola insolens*. These microorganisms can produce cellulases at concentrations of up to 130 g/L. The cellulases produced by these fungi are usually highly active on crystalline cellulose, as well as on other forms of cellulose. In addition, these microorganisms can be genetically modified to enhance their cellulase production and to tailor it to the industry requirements.

The substrate for cellulase production from fungi is usually purified cellulose or cellulose oligomers of natural origin. In some cases, but this is mainly still at research stage, cellulosic or lignocellulosic wastes are used as substrates. In addition to the carbon sources, the fermentation medium must of course include all the essential mineral elements. The pH is usually acidic, between 4 and 6.

The production of cellulases can be a primary (cellulases produced during growth of the microorganisms) or secondary (cellulases produced when growth has ended) metabolism. After the fermentation is completed, the liquid mixture containing the enzymes is separated from the microorganisms using mechanical separation, e.g. filtration or centrifugation. After separation of the microorganisms, the enzyme is usually separated and purified by a sequence of precipitation (e.g. with ammonium sulphate), re-dissolution of the enzyme in the precipitate (e.g. using sodium acetate buffer), dialysis with removal of low molecular weight substances and ion exchange chromatography. Depending on the required purity of the enzyme, not all the purification steps may be needed.

QUESTIONS AND PROBLEMS

8.1 Discuss and compare continuous and batch fermentation for the production of bioethanol and penicillin.

8.2 Propose a feeding strategy for the production of L-glutamic acid in a fed-batch reactor.

8.3 Which environmental factors trigger the production of citric acid?

8.4 The stoichiometry of glucose fermentation to ethanol with the microorganism *Saccharomyces cerevisiae* is written below:

$$C_6H_{12}O_6 + 0.12NH_3 \rightarrow 0.12C_5H_7O_2N + 1.8C_2H_5OH + 1.8CO_2 + 0.36H_2O$$

A simplified process scheme for ethanol production from glucose is reported in Figure 8.12. The scheme shows the fermenter (R1) and the first distillation column (S1).

The information available on the mass flow rates of some of the components in the various streams is reported in Table 8.5.

Calculate the mass flow rate of glucose in the feed (stream 1).

Calculate the mass flow rate of ethanol in stream 4.

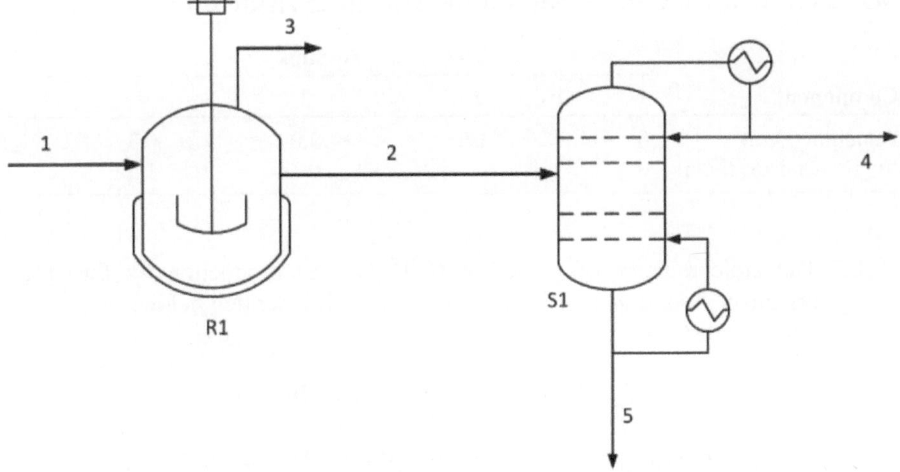

FIGURE 8.12 Simplified scheme of ethanol production for Problem 8.4.

TABLE 8.5
Mass Flow Rates of Components for the Scheme in Figure 8.12

	Streams				
Component	**1**	**2**	**3**	**4**	**5**
Glucose (kg/h)		50	0	0	
Ethanol (kg/h)	0		25		10
Carbon dioxide (kg/h)	0	0	200		

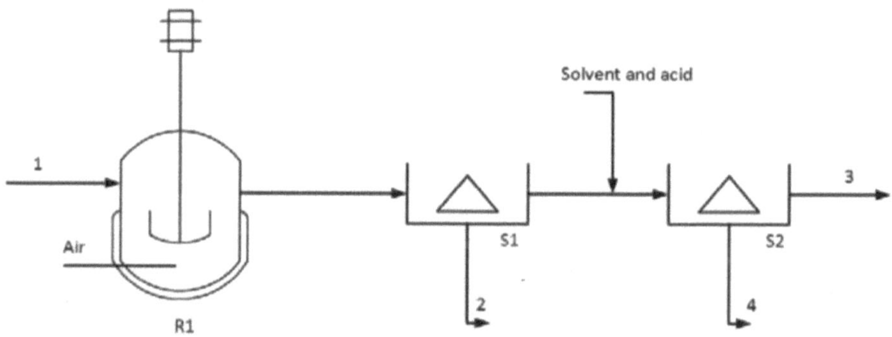

FIGURE 8.13 Simplified process flowsheet of the penicillin production process.

TABLE 8.6
Mass Flow Rates of Components for the Scheme in Figure 8.13

	Streams			
Component	**1**	**2**	**3**	**4**
Penicillin (kg/h)	0	0.1	3.0	0.3
Microorganisms (kg/h)	0		0	1.5

8.5 The stoichiometry of penicillin ($C_9H_{11}O_4N_2S$) production by the microorganism *Penicillium chrysogenum* ($C_5H_7O_2N$) is written below:

$$C_6H_{12}O_6 + 3.24O_2 + 0.55NH_3 + 0.037H_2SO_4 \rightarrow$$

$$0.48C_5H_7O_2N + 0.037C_9H_{11}O_4N_2S + 3.27CO_2 + 4.98H_2O$$

A simplified process flow sheet showing the reactor (R1), the filter (S1) and the solvent extractor-separator (S2) is shown in Figure 8.13.

The information available on the mass flow rates of some of the components in the various streams is reported in Table 8.6.

Calculate

- The oxygen consumption rate by the microorganisms in the reactor.
- The mass flow rate of microorganisms in stream 2.

8.6 Consider the process flowsheet in Figure 8.14, which shows a simplified scheme of whisky production from barley. The simplified stream table is in Table 8.7.

The reaction of starch hydrolysis is $C_6H_{10}O_5 + H_2O \rightarrow C_6H_{12}O_6$

The fermentation of glucose uses ammonia and produces microorganisms, ethanol, carbon dioxide and water. Assume that all the produced carbon dioxide leaves in stream 2.

Calculate

- The mass flow rate of ethanol in stream 5 (as kg/d).
- The mass flow rate (kg/d) of carbon dioxide in stream 2.

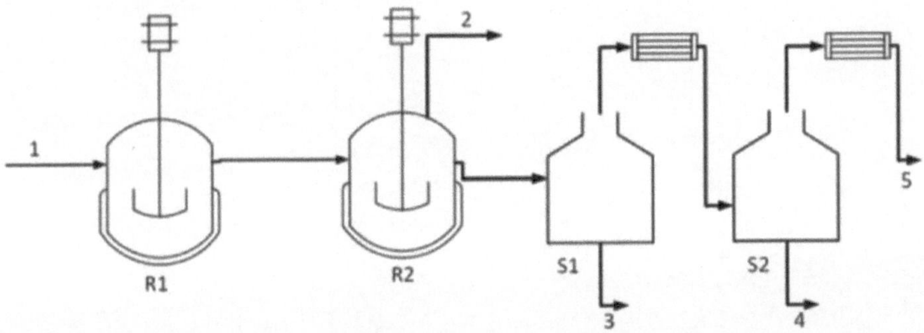

FIGURE 8.14 Simplified process flowsheet of the whisky production process. R1 = reactor for starch hydrolysis; R2 = reactor for glucose fermentation; S1, S2 = distillation stages.

TABLE 8.7
Mass Flow Rates (in kg/d) of Components for the Scheme in Figure 8.14

	Streams				
Component	**1**	**2**	**3**	**4**	**5**
Starch ($C_6H_{10}O_5$)	10,000	0	700	0	0
Glucose ($C_6H_{12}O_6$)	0	0	1,000	0	0
Ethanol (C_2H_6O)	0	0	113	200	
Yeast ($C_5H_7O_2N$)	0	0	1,890	0	0

Appendices

Appendix A: Numerical Solution of Differential Equations in Excel

The mass balances during batch enzymatic (Chapter 1) or fermentation (Chapter 4) reactions, as for any batch processes, are described by differential equations. The substrate balance inside a porous support containing enzymes (Chapter 1) is also described by a differential equation. This section describes a procedure to solve differential equations using Microsoft Excel.

Consider the generic differential equation:

$$\frac{dx}{dt} = f(x) \tag{A.1}$$

with the initial condition $x(t=0) = x_0$.

Solving this differential equation means finding the function $x(t)$ that satisfies equation (A.1). A numerical solution of the equation can be found by noting that, if the time increment Δt is very small, we have

$$x\big|_{t+\Delta t} = x\big|_t + \frac{dx}{dt}\bigg|_t \Delta t = x\big|_t + f(x)\big|_t \Delta t \tag{A.2}$$

In this way the values of the function $x(t)$ can be obtained as a series of data points starting from the initial condition $x(0) = x_0$, e.g.

$$\begin{aligned} x\big|_{\Delta t} &= x_0 + f(x)\big|_0 \Delta t, \\ x\big|_{2\Delta t} &= x_{\Delta t} + f(x)\big|_{\Delta t} \Delta t, \\ &\dots \\ x\big|_{n\cdot\Delta t} &= x_{(n-1)\Delta t} + f(x)\big|_{(n-1)\Delta t} \Delta t \end{aligned} \tag{A.3}$$

This numerical integration of the differential equation (A.1) gives correct results only if Δt is sufficiently small. Equations (A.3) can be solved using Excel.

Example A.1

Solve numerically the differential equation

$$\frac{dx}{dt} = 1.5 \cdot x^2 \tag{A.4}$$

over the time interval $t = [0 \quad 0.5]$ with the initial condition $x(0) = 1$. The units of time and x in this example are arbitrary.

Solution

In order to solve this equation numerically using Excel, we need to generate a column of time values using small increments Δt, e.g. Δt=0.01. Then we need to create columns for x and for $\frac{dx}{dt} = 1.5 \cdot x^2$. The first value in the x column will be $x(0) = 1$ and the following values will be given by $x(0.01) = x(0) + \left.\frac{dx}{dt}\right|_0 \cdot 0.01$, $x(0.02) = x(0.01) + \left.\frac{dx}{dt}\right|_{0.01} \cdot 0.01$ and so on. This is shown in the screenshot below (Figure A.1).

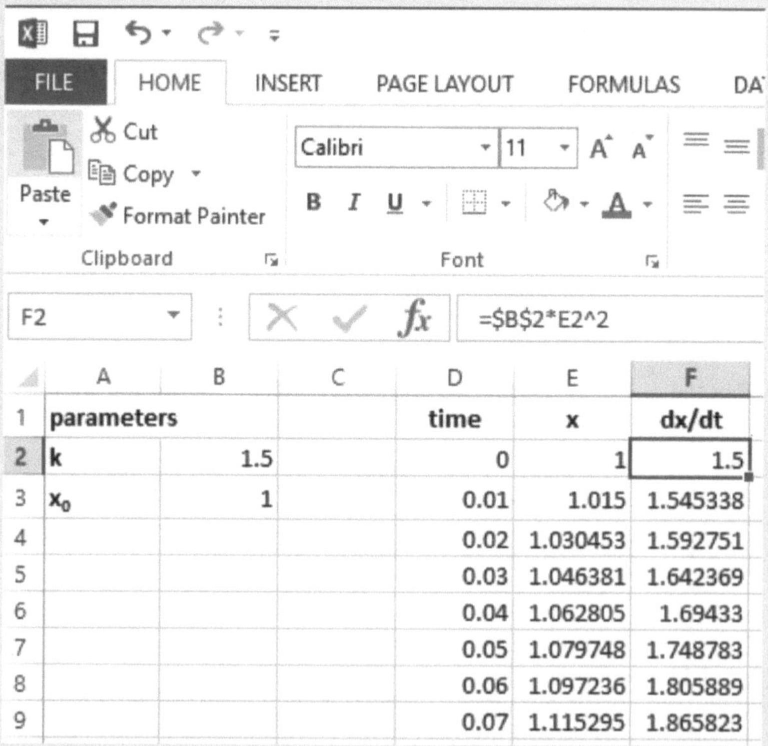

FIGURE A.1 Excel screenshot exemplifying the numerical solution of the differential equation in Example A.1.

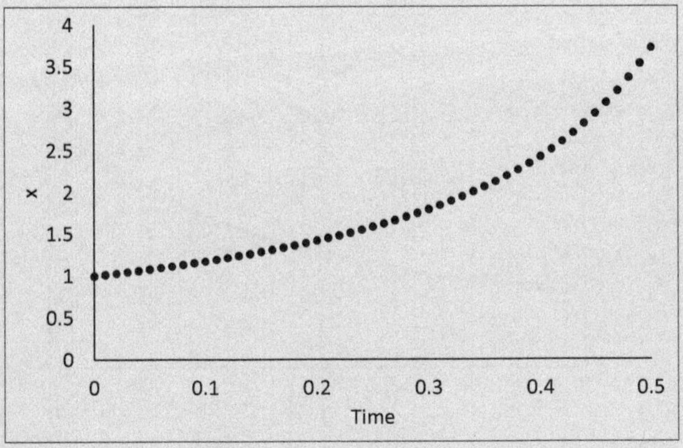

FIGURE A.2 Plot of $x(t)$ generated by numerical integration of the differential equation in Example A.1. Values of x obtained with $\Delta t = 0.01$.

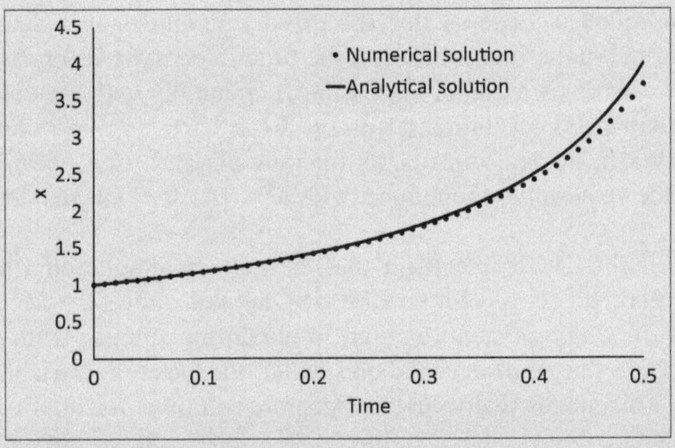

FIGURE A.3 Comparison of the numerical solution (obtained with $\Delta t = 0.01$) with the analytical solution.

The values of $x(t)$ obtained with this procedure are shown in Figure A.2.

The numerical solution obtained in this way can be compared to the analytical solution, which, in this particular case, can be obtained very easily:

$$x = \frac{x_0}{1 - 1.5 \cdot t \cdot x_0}$$

The comparison of the analytical and numerical solution is shown in Figure A.3. It can be seen that the agreement between the numerical and the analytical solution is very good, this means that the numerical error is very small.

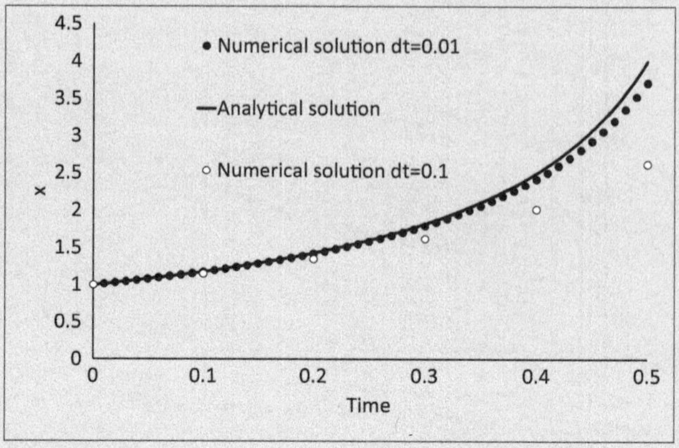

FIGURE A.4 Comparison of the numerical solution obtained with $\Delta t = 0.1$ with the numerical solution obtained with $\Delta t = 0.01$ and the analytical solution.

It is important to observe that the critical parameter that determines the accuracy of the numerical solution is the time interval used for discretisation, Δt, which needs to be as small as possible, compatibly with the computational power available. For example, let us see what happens if we choose a time interval $\Delta t = 0.1$, i.e. ten times larger than the *dt* value used so far. Figure A.4 compares the values of the function x with $\Delta t = 0.1$, $\Delta t = 0.01$ and the analytical solution.

From Figure A.4., it is evident that the numerical solution with $\Delta t = 0.1$ is largely incorrect (large numerical error) and this shows the need to choose a value of Δt, which is small enough to obtain the solution with the desired accuracy. The value of Δt needs to be smaller, the faster the variable x changes with time. This means that to obtain accurate solutions for high values of the derivative $\frac{dx}{dt}$ the value of Δt needs to be smaller than for low values of the derivative.

Appendix B: Numerical Solution of Systems of Equations in Excel

Systems of equations describe the steady state of continuous-flow processes like the chemostat. In many cases these equations can be solved manually by substitution, however this section presents a method to solve them using Excel.
Consider a system of equations:

$$\begin{cases} f_1(x_1, x_2, \ldots, x_n) = 0 \\ f_2(x_1, x_2, \ldots, x_n) = 0 \\ \ldots\ldots \\ f_n(x_1, x_2, \ldots, x_n) = 0 \end{cases} \tag{B.1}$$

This is a system of n equations in the n unknowns x_1, x_2, … ., x_n. Solving this system of equations means finding the values of x_1, x_2,…x_n that satisfy all equations (B.1).

The procedure to solve equations (B.1) in Excel is quite simple. Firstly, initial guesses have to be given to the unknowns. Then the equations have to be written in the same form as for equations (B.1), i.e. with the left-hand side equal to 0. The left-hand sides of the equations have to be written into Excel, as a function of the given initial guesses for the unknowns. Excel will calculate a numerical value for each equation (or rather for the left-hand side of each equation), which will be in general different than 0, because the initial guesses given are not the solution of the system of equations. We will call these calculated values "residuals". Then each residual is squared and all the squared residuals are added up together. At this point Solver is used to find the minimum of the sum of the squared residuals, by manipulating the values of the unknowns. If Solver is successful and the sum of squared residuals is very close to 0, the values found for the unknown are the solutions of the system of equations.

Note that the procedure is quite sensitive (or very sensitive in some cases) to the given values of the initial guesses. Therefore, it is important to have an idea at least of the order of magnitude of the values of the unknowns. The procedure is shown in the example below.

Example B.1

Solve the following system of equations:

$$\begin{cases} x+2y+z=3 \\ 2x+y-z=5 \\ x-y+2z=2 \end{cases}$$

Solution

This is a system of three equations in the three unknowns *x,y,z*. This is a simple system that can be easily solved manually by substitution, however we will use it as an example of the Excel method. First of all, initial guesses have to be given to *x,y,z*, we will give the value of 1 to all of them. Then we need to write the equations in Excel in the form

$$\begin{cases} x+2y+z-3 \\ 2x+y-z-5 \\ x-y+2z-2 \end{cases}$$

Excel will calculate a value for each equation (residual). We need to square each of the residuals and then add all the squared residuals together. Then use Solver to find the values of *x,y,z* than make the sum of the squared residuals minimum.

This procedure is shown in the screenshot below (Figure B.1).

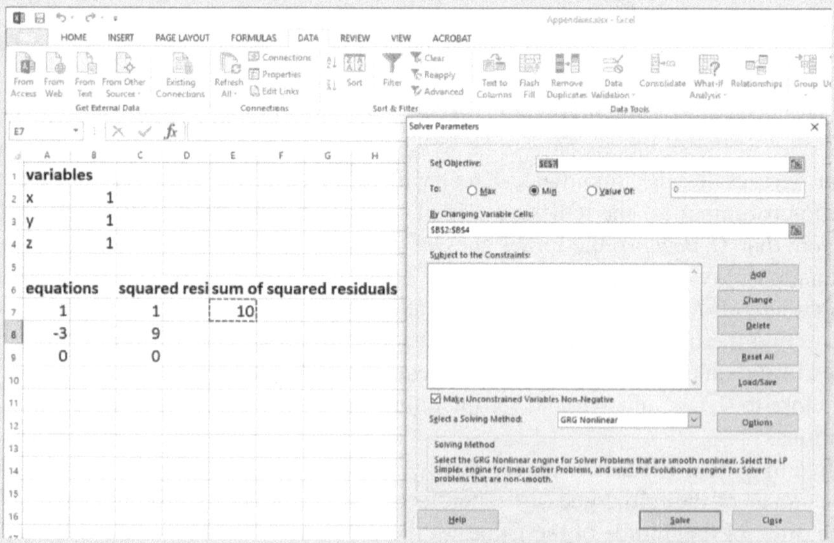

FIGURE B.1 Screenshot showing the procedure to solve Example B.1.

In this case solver gives the correct values of the unknowns:

$$\begin{cases} x = 2.33 \\ y = 0.33 \\ z = 0 \end{cases}$$

With these values the sum of the squared residuals is of the order of 10^{-12}, i.e. a very low value, which ensures we have found the solution of the system of equations.

Appendix C: Answers and Solutions

CHAPTER 1

1.1 Remembering that the Michaelis-Menten kinetics is $\left(-r_S\right) = k_2 \dfrac{\left[E_0\right]\left[S\right]}{K_m + \left[S\right]}$ we note that the units of k_2 depend on the units of r_S and of E_0, while the units of K_m must be the same as the units of S.

(a) $\left(-r_S\right)$ expressed in mol L^{-1} s^{-1}, E_0 in mol L^{-1} and S in kg m^{-3}. The units of K_m are kg m^{-3}. The units of k_2 are mol (of substrate) mol (of enzyme)$^{-1}$ s^{-1}.

(b) $\left(-r_S\right)$ expressed in g L^{-1} h^{-1}, E_0 in mol L^{-1} and S in mol m^{-3}. The units of K_m are mol m^{-3}. The units of k_2 are g (of substrate) mol (of enzyme)$^{-1}$ h^{-1}.

(c) $\left(-r_S\right)$ expressed in g L^{-1} h^{-1}, E_0 in unit L^{-1} and S in kmol m^{-3}. The units of K_m are kmol m^{-3}. The units of k_2 are g (of substrate) unit^{-1} h^{-1}.

(d) $\left(-r_S\right)$ expressed in kg m^{-3} d^{-1}, E_0 in units m^{-3} and S in g L^{-1}. The units of K_m are g L^{-1}. The units of k_2 are kg (of substrate) unit^{-1} d^{-1}.

1.2 We note that one unit of this enzyme converts the triglycerides in the oil at a rate of 1 mmol of fatty acid per minute under the assay conditions. 1 g of the commercial enzyme preparation contains 200 units, therefore the rate with the commercial enzyme is of 200 mmol of fatty acid per g of enzyme per minute.

(a) For 10 mmol/L of fatty acid in 2 L vessel per hour, it means the fatty acid production rate that we need is 20 mmol per hour, i.e. 0.33 mmol per minute = 330 mmol of fatty acid per minute. Since 1 g of enzyme converts 200 mmol of fatty acid per minute, we need 330/200 =1.65 g of enzyme in the 2 L vessel.

(b) For 5 mmol/L of fatty acid in 0.2 L flask per minute, it means the fatty acid production rate that we need is 1 mmol per minute. Since 1 g of enzyme converts 200 mmol of fatty acid per minute, we need 1/200 = 0.005 g of enzyme (=5 mg) in the 0.2 L flask.

(c) For 2 kg/m^3 of fatty acid in 5 m^3 vessel per hour, it means the fatty acid production rate that we need is 10 kg per hour, i.e. 39 mol per hour (the MW of palmitic acid is 256.43 g/mol) = 0.65 mol per minute = 650,000 mmol of fatty acid per minute. Since 1 g of enzyme converts 200 mmol of fatty acid per minute, we need 650,000/200 = 3250 g of enzyme (=3.25 kg) in the 5 m^3 vessel.

1.3 We note that the empirical formula for lactose is $C_{12}H_{22}O_{11}$(MW 342) and for both galactose and glucose is $C_6H_{12}O_6$ (MW 180). One molecule of lactose reacts with one molecule of water to give the products. Therefore, the reaction representing the conversion of lactose into galactose and glucose can be written as

$$C_{12}H_{22}O_{11} + H_2O \rightarrow 2C_6H_{12}O_6$$

(a) If the lactose rate is 10 mol L^{-1} h^{-1} the production rate of each of galactose and glucose is also 10 mol L^{-1} h^{-1}.

(b) If the lactose rate is 10 g L^{-1} h^{-1} the rate of water consumption is 10×18/342 = 0.53 g L^{-1} h^{-1} (because 18 g of water are consumed for every 342 g of lactose consumed). Therefore, the total production of galactose plus glucose will be (in mass): 10.53 g L^{-1} h^{-1} and the production rate of each of these sugars will be 10.53/2=5.265 g L^{-1} h^{-1}.

1.4 The Michaelis-Menten equation is

$$\left(-r_s\right) = \frac{k_2 E_0 S}{k_m + S}$$

We need to determine whether this equation fits the experimental data and, if so, the values of the parameters k_2 and K_m.

Several methods, described in Chapter 1, can be used to find the parameters k_2 and K_m. The experimental data are obtained in batch, this means that the methods described in section 1.2.3.1.4 and 1.2.3.6 can be used, while the method in section 1.2.3.5 cannot be used as it refers to data obtained in a continuous system. Essentially, the experimental data must be used to calculate $(-r_S)$, $1/S$, $(-r_S)/S$ and the other variables to be used for the linear regression methods. The values of the parameters should be found from the slope and intercept of the various plots. For the nonlinear parameter fitting method, the values of k_2 and K_m are found by solving the first order differential equation representing the mass balance for S, and then using Solver to find the k_2 and K_m values that minimise the objective function.

Summary of the parameters obtained with the different methods.

Method	k_2 (mmol/L/h)	K_m (mmol/L)
Lineweaver-Burk	36.71	0.615
Eadie-Hofstee	36.67	0.2236
Hanes-Woolf	22.07	0.329
Integration	19.76	0.194
Nonlinear fitting	20.03	0.2014

It is often not easy or not possible to determine which method gives the most accurate parameters. In this case the accuracy of parameter fitting can be evaluated by comparing the profiles of $S(t)$ obtained with the various sets of parameters. This can be done by solving the differential equation representing the mass balance of S with each of the parameter sets from the various methods and comparing the fitting visually and by looking at the objective function in each case. The results are shown in Figure C.1. Overall, most methods give very good and very similar fitting. The Lineweaver-Burk give larger discrepancies with the experimental data.

1.5 In order to find the parameters k_2 and K_m of the Michaelis-Menten model, the method requires linear regression of the experimental data considering that the data are obtained in a continuous flow reactor. Therefore, the method in section 1.2.3.5 applies in this case.

The linear regression for continuous reactors is

$$\frac{\tau}{S_{feed} - S} = \frac{K_m}{k_2 [E_0]} \frac{1}{[S]} + \frac{1}{k_2 [E_0]}$$

However, with this set of experimental data, $[E_0]$ is different in the various experiments. Therefore, the experimental data have to be linearised as follows:

$$\frac{\tau [E_0]}{S_{feed} - S} = \frac{K_m}{k_2} \frac{1}{[S]} + \frac{1}{k_2}$$

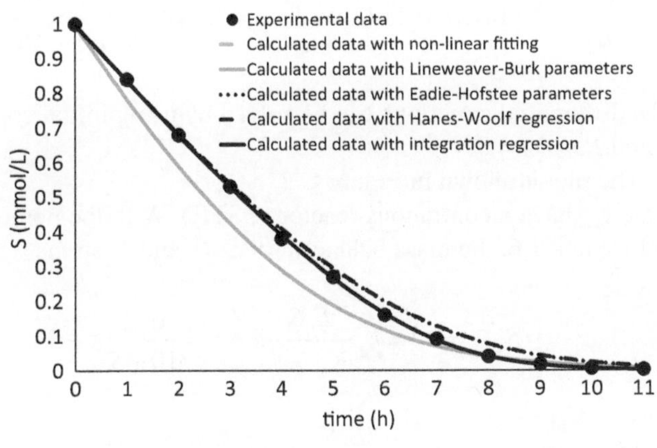

FIGURE C.1 Plot of different parameter estimation methods vs experimental data for Problem 1.4.

Therefore, plotting $\frac{\tau[E_0]}{S_{feed} - S}$ vs $\frac{1}{[S]}$ should give us a linear trend with slope equal to $\frac{K_m}{k_2}$ and intercept equal to $\frac{1}{k_2}$. The plot and the values of the parameters are reported in the Excel spreadsheet. Note the very good agreement between the model and the experimental data. However, only three data points are available, more experiments should be done for a full validation of the model.

1.6 The empirical formulas for tripalmitin (substrate, *S*), palmitic acid (product, P_1) and glycerol (product, P_2) are, respectively, $C_{51}H_{98}O_6$, $C_{16}H_{32}O_2$ and $C_3H_8O_3$. The reaction for tripalmitin hydrolysis catalysed by the enzyme lipase is:

$$C_{51}H_{98}O_6 + 3H_2O \rightarrow 3C_{16}H_{32}O_2 + C_3H_8O_3$$

The time profiles during the batch reaction for the substrate tripalmitin and for the products palmitic acid and glycerol are obtained by solving the differential equations, with all the concentrations in g/L and the time expressed in min (the units of r_S are therefore g L^{-1} min^{-1}):

$$\frac{dS}{dt} = r_S = -k_2\frac{E_0S}{K_m + S} = \frac{2 \cdot 3 \cdot S}{10 + S} = -\frac{6 \cdot S}{10 + S}$$

$$\frac{dP_1}{dt} = 0.953r_S = 0.953k_2\frac{E_0S}{K_m + S} = 0.953\frac{2 \cdot 3 \cdot S}{10 + S} = \frac{5.718 \cdot S}{10 + S}$$

$$\frac{dP_2}{dt} = 0.114r_S = 0.114k_2\frac{E_0S}{K_m + S} = 0.114\frac{2 \cdot 3 \cdot S}{K_m + S} = \frac{0.684 \cdot S}{10 + S}$$

The differential equations can be solved by using initial conditions $S_0 = 80$ g/L and $P_1 = P_2 = 0$.

The plot is shown in Figure C.2.

1.7 Here we have a continuous reactor (CSTR). With the numerical values given in Problem 1.6, the mass balances for *S*, P_1 and P_2 in the CSTR are

$$S_{feed} = 80 = k_2\frac{E_0S}{K_m + S}\tau + S = \frac{6 \cdot S}{10 + S}\tau + S$$

$$0.953k_2\frac{E_0S}{K_m + S}\tau = \frac{5.718 \cdot S}{10 + S}\tau = P_1$$

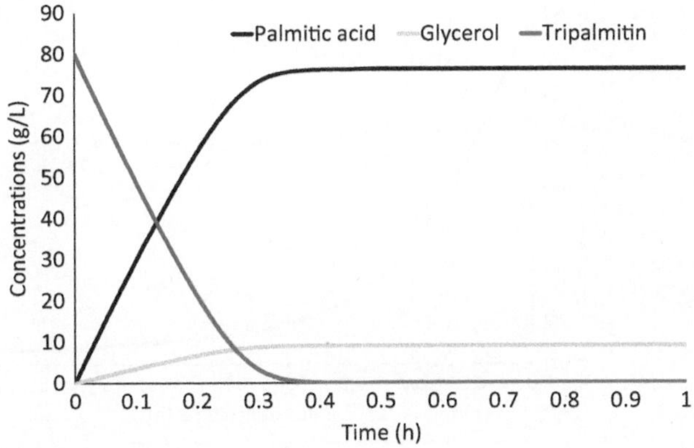

FIGURE C.2 Results of Problem 1.6.

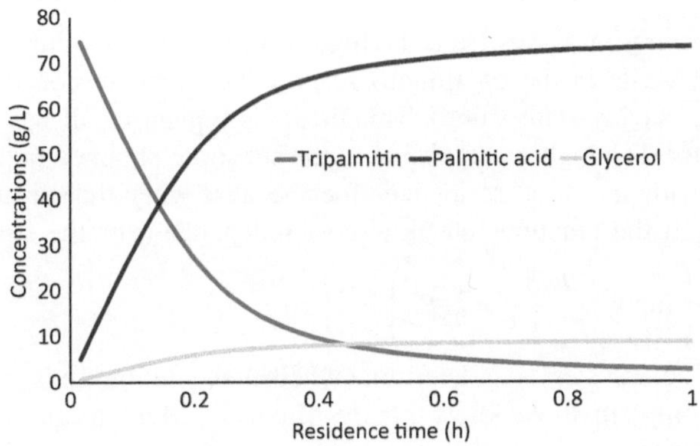

FIGURE C.3 Profiles of substrate and products vs residence time for Problem 1.7.

$$\frac{0.684 \cdot S}{10+S}\tau = P_2$$

The mass balance for the substrate can be used to calculate S as a function of t. This mass balance can be solved as a nonlinear equation in Excel or as a quadratic equation. Once S is found, P_1 and P_2 can be directly calculated from their respective mass balances. All calculations are shown in the Excel spreadsheet. The profiles of S, P_1 and P_2 vs the residence time are shown Figure C.3. The comparison of the tripalmitin concentration in the batch and in the CSTR as a function of the residence time (which is the reaction time for the batch reactor) is shown in Figure C.4. The figure shows that for relatively long residence times the concentration in the batch is lower than in the CSTR. This is because the reaction rate depends on the substrate concentration which on average is higher in a batch reactor (in the batch reactor the substrate

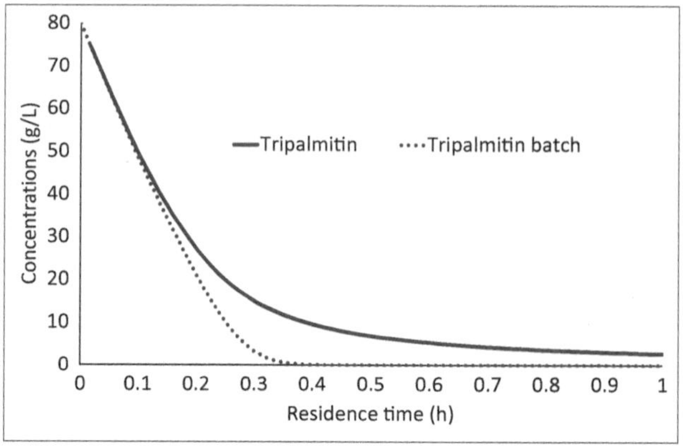

FIGURE C.4 Comparison of substrate profile vs residence time in CSTR and in batch for Problem 1.7.

concentration varies from the highest to the lowest value during the reaction time, while in the continuous reactor the substrate concentration is always at the steady-state value). This effect is typical of all batch and continuous reactor. The higher conversion of the substrate obtained in the batch reactor is not evident at short residence times because when the substrate concentration is high the reaction rate is almost independent of the substrate concentration $\left(\lim_{S\to\infty} k_2 \frac{E_0 S}{K_m + S} = k_2 E_0\right)$.

1.8 (a) The first step is to determine whether the inhibition is competitive or noncompetitive. We know that the Michaelis-Menten equation for competitive inhibition is

$$(-r_S) = \frac{k_2 E_0 S}{K_m\left(1 + \frac{I}{K_I}\right) + S}$$

while for noncompetitive inhibition we have

$$(-r_S) = \frac{k_2 E_0}{\left(1 + \frac{I}{K_I}\right)\left(1 + \frac{K_m}{S}\right)}$$

Therefore, for competitive inhibition the rate equation can be rearranged as

$$\frac{1}{(-r_S)} = \frac{1}{S}\frac{K_m\left(1 + \frac{I}{K_I}\right)}{k_2 E_0} + \frac{1}{k_2 E_0}$$

This shows that if the inhibition is competitive a plot of $\frac{1}{(-r_S)}$ $vs \frac{1}{S}$ gives the same intercept ($\frac{1}{k_2 E_0}$) both in the presence and in the absence of inhibitor.

On the other hand, for non-competitive inhibition the rate equation can be rearranged as

$$\frac{1}{(-r_S)} = \frac{1}{S}\frac{K_m\left(1+\frac{I}{K_I}\right)}{k_2 E_0} + \frac{\left(1+\frac{I}{K_I}\right)}{k_2 E_0}$$

and therefore, if the inhibition is non-competitive in a plot of $\frac{1}{(-r_S)}$ $vs \frac{1}{S}$ the intercept with and without inhibitor will be different.

Therefore, the problem can be solved by plotting $\frac{1}{(-r_S)}$ $vs \frac{1}{S}$ with and without inhibitor and observing the shape of the lines obtained. From the plots the values of k_2, K_I and K_m can be obtained.

The obtained plot is shown in Figure C.5. From the plot it is clear that the intercept with and without inhibitor is essentially the same (5003 vs 5169) while the slope changes. This indicates competitive inhibition.

The kinetic parameters k_2 and K_m can be obtained from the intercept and slope of the plot without inhibitor. Once k_2 and K_m are known, the parameter K_I can be obtained from the slope of the plot with the inhibitor.

The resulting parameter values are

$$k_2 = 0.0200 \text{ mol mol}^{-1} \text{ min}^{-1}$$
$$K_M = 8.6\times10^{-4} \text{ mol/L}$$
$$K_I = 1.37\times10^{-5} \text{ mol/L}$$

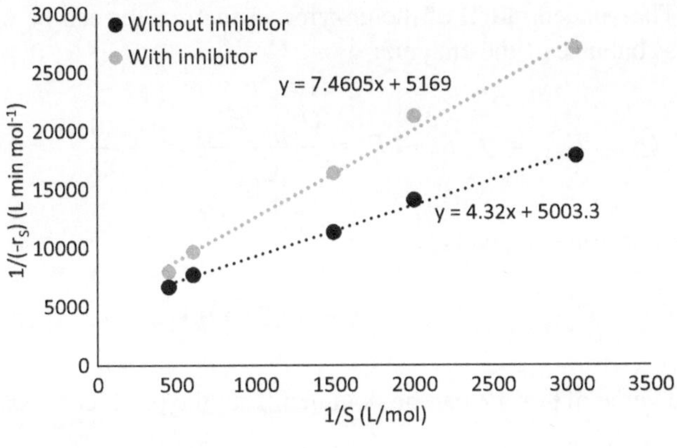

FIGURE C.5 Solution plot for Problem 1.8a.

(b) In the presence of a competitive inhibitor, as in this case, the substrate rate is $(-r_S) = k_2 \frac{[E_0][S]}{K_m\left(1+\frac{[I]}{K_I}\right)+[S]}$ The concentrations of S and E_0 can be calculated from the given data, $(-r_S)$ is given and the concentration of the inhibitor can be calculated by re-arranging and solving for $[I]$:

$$[I] = \frac{\left([S]\left(k_2[E_0]-(-r_S)\right)-(-r_S)K_m\right)K_I}{(-r_S)K_m}$$

The concentrations of S and E_0 can be calculated as follows. The total solution volume is the sum of the enzyme, substrate and inhibitor solutions, i.e. it is 125 mL. The concentration of enzyme in the enzyme solution (50 mL) is 0.01 mol/L, the concentration of the enzyme in the final solution is therefore 0.01×50/125 = 0.004 mol/L. The concentration of substrate in the final solution is 0.0008×50/125 = 0.00032 mol/L.
Therefore, the pesticide concentration is

$$[I] = \frac{\left(0.00032\left(0.020.004-18\cdot10^{-6}\right)-18\cdot10^{-6}0.00086\right)1.37\cdot10^{-5}}{18\cdot10^{-6}0.00086}$$
$$= 3.86\cdot10^{-6}\,\text{mol}/\text{L}$$

This is the concentration of inhibitor in the final solution of volume 125 mL. Since we used 25 mL of the pesticide solution, the concentration of inhibitor in the pesticide solution is $3.86\cdot\frac{10^{-6}125}{25} = 1.9310^{-5}\,\text{mol}/\text{L}$

1.9 We will use the units of mol, L and s for number of molecules, volume and time, respectively. The continuous reactor receives three influent streams and has one effluent stream. The effluent stream has a flow rate equal to the sum of the three influent streams, i.e. equal to 0.8 L/s.

The concentration of the enzyme in the reactor can be obtained from the mass balance of the enzyme:

$$Q_{enzyme}E_{feed} = Q_{tot}E_0 \rightarrow E_0 = \frac{Q_{enzyme}E_{feed}}{Q_{tot}} = \frac{0.2\cdot0.01}{0.8} = 0.0025\,\text{mol}/\text{L}$$

The substrate balance is

$$Q_{substrate}S_{feed} = (-r_S)V + Q_{tot}S \rightarrow 0.1\cdot2 = 0.2 = (-r_S)V + 0.8\cdot S$$

The value of $(-r_S)V$ can be obtained from the product balance:

$$Q_{total}P = 2(-r_S)V \rightarrow 0.8\cdot.2 = 0.16 = 2(-r_S)V \rightarrow (-r_S)V = 0.08\ \text{mol}/\text{s}$$

Inserting the obtained value of $(-r_S)V = 0.08$ in the substrate balance we obtain the value of S (substrate concentration in the reactor), $S = 0.15\,\text{mol/L}$. The equation for $(-r_S)$ comes from the equation for competitive inhibition, i.e.

$$(-r_s) = \frac{k_2 E_0 S}{K_m\left(1 + \frac{I}{K_I}\right) + S}$$

The concentration of inhibitor I in the reactor can be obtained from the inhibitor balance:

$$Q_{inhibitor} I_{feed} = Q_{tot} I \rightarrow I = \frac{Q_{inhibitor} I_{feed}}{Q_{tot}} = \frac{0.5 \cdot 0.1}{0.8} = 0.0625\,\text{mol/L}$$

Since the kinetic parameters k_2, K_m, K_I are given and the concentrations of S, E_0 and I have been calculated, we can now calculate $(-r_S)$ obtaining $(-r_S) = 0.000356 \frac{mol}{L \cdot s}$

Since we have calculated that $(-r_S)V = 0.08$ mol/s and $(-r_S) = 0.000356 \frac{\text{mol}}{\text{L} \cdot \text{s}}$ we can now calculate the required volume V as $V = \frac{(-r_S)V}{(-r_S)} = 222.22$ L.

1.10 Since in this case

$$\frac{dP}{dt} = k_2 E_0$$

we have

$$\frac{dP}{dt} = k_2 E_{0IN} e^{-k_d \cdot t}$$

which can be integrated

$$\int_0^P dP = k_2 E_{0IN} \int_0^t e^{-k_d \cdot t} dt$$

the integration gives

$$P = \frac{k_2 E_{0IN}}{k_d}\left(1 - e^{-k_d t}\right)$$

We know that after 12 h the reaction is complete, and the maximum concentration of product P has been reached. When $t \rightarrow \infty$ (we can assume that 12 h here is long enough for this assumption to hold), $P = \frac{k_2 E_0}{k_d}$. Therefore, the initial enzyme concentration is given by

$$E_{0IN}=\frac{k_dP}{k_2}=\frac{0.1\frac{1}{\min}\cdot 0.002\frac{\text{mol}}{\text{L}}}{104\frac{\text{mol}}{\text{mol}\cdot\min}}=1.92\cdot 10^{-6}\,\frac{\text{mol}}{\text{L}}$$

The enzyme concentration can be converted in mass units by using the given MW of the enzyme, obtaining $E_{0IN}=1.92\cdot 10^{-6}\,\frac{\text{mol}}{\text{L}}\cdot 50{,}000\frac{\text{g}}{\text{mol}}=0.096\frac{\text{g}}{\text{L}}$

The concentration of total crude protein in the initial reaction mixture is $\frac{10\text{ mg}}{1\text{ mL}}=10\frac{\text{g}}{\text{L}}$. Therefore, the fraction of the enzyme in the initial crude preparation is $\frac{0.096\frac{\text{g}}{\text{L}}}{10\frac{\text{g}}{\text{L}}}=0.96\%\text{ w/w}$

1.11 From the plot obtained with the free enzyme we can calculate the kinetic parameters of the enzyme.

Indeed, the Michaelis-Menten equation:

$$\left(-r_S\right)=\frac{k_2E_0S}{K_m+S}$$

can be rearranged as follows:

$$\frac{1}{\left(-r_S\right)}=\frac{K_m}{k_2E_0S}+\frac{1}{k_2E_0}$$

So, a plot of $\frac{1}{\left(-r_S\right)}\,vs\,\frac{1}{S}$, such as the one given in the figure, allows the determination of the kinetic parameters.

From the intercept, we have

$\frac{1}{k_2E_0}=6.97\text{ l min/g}$, which gives

$k_2=1.44$ g substrate/(g enzyme min)

and from the slope

$$\frac{K_m}{k_2E_0}=194.89\text{ min}$$

$k_m=27.96$ g substrate/l

In the commercial process the enzyme is immobilised on nonporous beads, but there are no diffusion limitations. This means that the kinetic parameters measured for the free enzyme also apply to the immobilised enzyme and that the concentration on the support surface is the same as the bulk concentration.

Since the desired product concentration P_{out} is expressed in mol/l, and we do not know the MW of species P, we will carry out the mass balance for the commercial process using moles. The first thing to do is to convert the values of the kinetic parameters in mol S from g S.

So we have

$k_2 = 0.02$ mol S/(g enzyme min)
$K_m = 0.40$ mol S/l

The mass balance for the product in the batch reactor is (where the concentrations are expressed in mol/l):

$$\frac{dp}{dt} = 2(-r_s) = 2\frac{k_2 E_0 S}{K_m + S} = -2\left(\frac{ds}{dt}\right)$$

Therefore,

$$dP = -2dS$$

and integrating $\int_{P_0}^{P} dP = (P - P_0) = -2\int_{S_0}^{S} dS = -2(S - S_0)$

so the desired product concentration P_{out} and the correspondent substrate concentration S_{out} are related as follows:

$$S_{out} = S_0 - \frac{(P_{out} - P_0)}{2}$$

where $P_0 = 0$ (no product at time 0) and $S_0 = 50/70 = 0.71$ mol/L. The required concentration of S is therefore $S_{\text{out}} = 0.21$ mol/l. We need to determine the enzyme concentration E_0 that is needed to achieve a substrate concentration S_{out} (starting from S_0) in a time t_{batch}. We integrate the mass balance for the substrate in the batch reactor:

$$-\frac{dS}{dt} = \frac{k_2 E_0 S}{K_m + S}$$

and we obtain

$$\frac{\ln\left(\frac{S_0}{S_{out}}\right)}{t_{batch}} = \frac{k_2 E_0}{K_m} - \frac{S_0 - S_{out}}{t_{batch}} \cdot \frac{1}{K_m} \rightarrow E_0$$

$$= \left[\frac{\ln\left(\frac{S_0}{S_{out}}\right)}{t_{batch}} + \frac{S_0 - S_{out}}{t_{batch}} \cdot \frac{1}{K_m} \right] \frac{K_m}{k_2}$$

$$= \left[\frac{\ln\left(\frac{0.71}{0.21}\right)}{2h} + \frac{(0.71-0.21)\frac{\text{mol}}{\text{L}}}{2h} \cdot \frac{1}{0.40\frac{\text{mol}}{\text{L}}} \right] \frac{0.40\frac{\text{mol}}{\text{L}}}{1.2\frac{\text{mol}}{\text{g}\cdot\text{h}}}$$

$$= 1.77\frac{\text{g}}{\text{L}}$$

The required number of support beads per unit volume can now be calculated since the enzyme concentration E_0 has to be equal to the enzyme attached on the surface of the support (the surface density of the enzyme on the support, $E_{surface}$, is known):

$$E_0 = \frac{n_{bead} \pi D_{bead}^2 E_{surface}}{V} \rightarrow \frac{n_{bead}}{V} = \frac{E_0}{\pi D_{bead}^2 E_{surface}}$$

$$= \frac{1.77\frac{g}{L}}{\pi (0.1cm)^2 \, 1 \cdot 10^{-2} \frac{g}{cm^2}} = 5.64 \cdot 10^3$$

beads per liter of reactor.

1.12 (a) At steady state the flux of urea to the surface of the beads must be equal to the rate of urea degradation due to the enzymatic reaction. This can be written as

$$J\left(\frac{\text{mgurea}}{\text{cm}^2 \cdot \text{s}}\right) = k_L\left(S_b - S_S\right) = k_2 E_{surf} \frac{S_S}{K_m + S_S}$$

This can be solved graphically and from the intersection of the curves the value of S_S is determined.

With the numerical values given in this problem, after conversion of units so that length is in m, mass is in kg, volume in m^3 and time in s:

$$0.002\left(1.0 - S_S\right) = 0.001\frac{S_S}{0.2 + S_S}$$

The plot with the calculation of S_S is shown in Figure C.6. From the plot or from the calculated values we see that $S_s \cong 0.62\frac{\text{kg}}{\text{m}^3}$

(b) Once S_S has been found, the rate of urea degradation is simply given by

$$J\left(\frac{\text{mg urea}}{\text{cm}^2 \times \text{s}}\right) = k_L\left(S_b - S_S\right)$$

or can be obtained directly from the plot at the intersection point. We obtain $J \cong 7.56 \cdot 10^{-4}$ kg m^{-2} s^{-1}.

1.13 (a) Assuming that the film resistance to mass transfer is negligible, i.e. $S_S = S_b$, the rate of urea hydrolysis can be expressed as

$$\left(-r_S\right) = \eta\frac{k_2 E_0 S_b}{K_m + S_b}$$

The Thiele modulus is given by

$$\varphi = R\sqrt{\frac{k_2 E_0 / K_m}{D_e}}$$

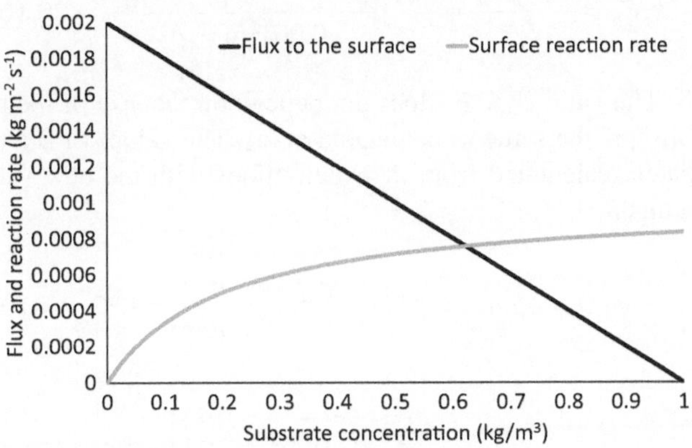

FIGURE C.6 Solution plot for Problem 1.12.

k_2E_0 can be eliminated combining the two equations above to give

$$(-r_S) = \frac{\eta\varphi^2 D_e K_m}{R^2}\frac{S_b}{K_m + S_b}$$

Under the hypotheses of this problem $\eta = \frac{3}{\phi}$

Therefore, we have

$$(-r_S) = \frac{3\varphi D_e K_m}{R^2}\frac{S_b}{K_m + S_b}$$

In this equation everything is known except φ, which therefore can be calculated as

$$\begin{aligned}\varphi &= \frac{R^2(-r_S)}{3D_e K_m}\left(\frac{K_m + S_b}{S_b}\right)\\ &= \frac{0.001^2\,\mathrm{m}^2\,10\dfrac{\mathrm{mol}}{\mathrm{m}^3\mathrm{h}}}{3\cdot 5.4\cdot 10^{-6}\dfrac{\mathrm{m}^2}{\mathrm{h}}0.2\dfrac{\mathrm{mol}}{\mathrm{m}^3}}\left(\frac{0.2\dfrac{\mathrm{mol}}{\mathrm{m}^3}+0.5\dfrac{\mathrm{mol}}{\mathrm{m}^3}}{0.5\dfrac{\mathrm{mol}}{\mathrm{m}^3}}\right)\\ &= 4.32\end{aligned}$$

From the Thiele modulus, the effectiveness factor and the maximum reaction rate can be calculated as:

$$\eta = \frac{3}{\varphi} = 0.69$$

$$k_2E_0 = \frac{\varphi^2 D_e K_m}{R^2} = \frac{4.32^2\cdot 5.4\cdot 10^{-6}\dfrac{\mathrm{m}^2}{\mathrm{h}}0.2\dfrac{\mathrm{mol}}{\mathrm{m}^3}}{0.001^2\,\mathrm{m}^2} = 20.16\frac{\mathrm{mol}}{\mathrm{m}^3\mathrm{h}}$$

(b) The value of k_2E_0 does not depend on the size of the particle and therefore it is the same as calculated at (a). The values of φ and η can be immediately calculated from their definitions with the new value of the particle radius R.

$$\varphi(D = 4\ \mathrm{mm}) = \varphi(D = 2\ \mathrm{mm})\frac{4\ \mathrm{mm}}{2\ \mathrm{mm}} = 4.32\cdot 2 = 8.64$$

$$\eta = \frac{3}{\varphi} = \frac{3}{8.64} = 0.35$$

1.14 The time profile of S_b can be obtained by solving the differential equation $\frac{dS_b}{dt} = -k_L\left(S_b - S_s\right)a$. For each time step and for each value of S_b the value of S_S is calculated by solving the equation $k_L\left(S_b - S_s\right) = \frac{k_2 E_{surf} S_s}{K_m + S_s}$. This is a second-order equation, which can be solved for S_S as follows. The equation can be re-arranged as

$$\left(-k_L\right)S_s^2 + \left(k_L S_b - k_L K_m - k_2 E_{surf}\right)S_s + k_L S_b K_m = 0$$

The solution of this second order equation is (remember that for the second order equation $ax^2+bx+c=0$ in the unknown x and with a, b, c as parameters, the solutions are $x = \frac{-b \mp \sqrt{b^2 - 4ac}}{2a}$, and we take the only solution that has physical sense):

$$S_s = \frac{-k_L S_b + k_L K_m + k_2 E_{surf} - \sqrt{\left(k_L S_b - k_L K_m - k_2 E_{surf}\right)^2 + 4 S_b K_m k_L^2}}{-2k_L}$$

This equation should be solved simultaneously to the differential equation $\frac{dS_b}{dt} = -k_L\left(S_b - S_s\right)a$ to obtain the profiles of S_b and S_s over time in the batch reactor. The calculations are plotted in Figure C.7.

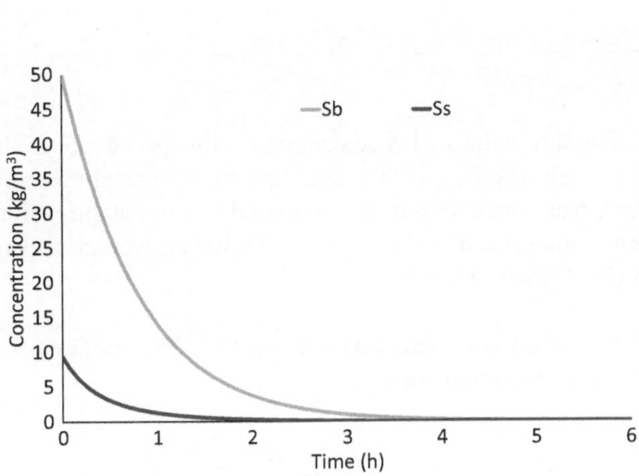

FIGURE C.7 Solution plot for Problem 1.14.

1.15 The problem can be solved by simultaneously solving the two equations that describe the mass balances for S_b and S_S:

$$S_{bfeed} = \frac{k_2 E_{surf} S_s}{K_m + S_s} a\tau + S_b \qquad k_L\left(S_b - S_s\right) = \frac{k_2 E_{surf} S_s}{K_m + S_s}$$

These equations can be solved in various ways. One way is to express S_b as a function of S_S from the mass balance for S_b and then substitute in the mass balance for S_S.

By isolating S_b from the first equation we obtain

$$S_b = S_{bfeed} - \frac{k_2 E_{surf} S_s}{K_m + S_s} a\tau$$

which can be substituted in the second equation to give

$$k_L\left(S_{bfeed} - \frac{k_2 E_{surf} S_s}{K_m + S_s} a\tau - S_s\right) = \frac{k_2 E_{surf} S_s}{K_m + S_s}$$

which can be re-arranged as a second-order equation in the unknown S_S

$$\left(-k_L\right)S_s^2 + \left(k_L S_{bfeed} - k_L K_m - k_2 E_{surf} - k_L^2 E_{surf} a\tau\right)S_s + k_L S_{bfeed} K_m = 0$$

The solution of this equation is

$$S_s = \frac{-k_L S_{bfeed} + k_L K_m + k_2 E_{surf} + k_L^2 E_{surf} a\tau - \sqrt{(k_L S_{bfeed} - k_L K_m - k_2 E_{surf} - k_L^2 E_{surf} a\tau)^2 + 4 S_{bfeed} K_m k_L^2}}{-2k_L}$$

For any values of S_S calculated with the equation above, the value of S_b can be calculated. The calculations are are plotted in Figure C.8. Note the lower conversion of S_b compared to the batch reactor, for a given residence time/reaction time, due to the lower average substrate concentration in CSTR than in batch.

1.16 The solution is obtained by solving the differential equation for enzymatic reactions in porous media:

$$D_e \frac{2}{r}\frac{dS}{dr} + D_e \frac{d^2S}{dr^2} = \frac{k_2 E_0 S}{K_m + S}$$

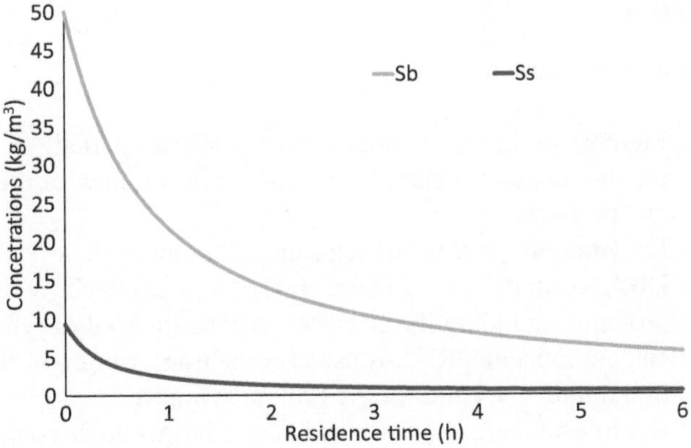

FIGURE C.8 Solution plot for Problem 1.15.

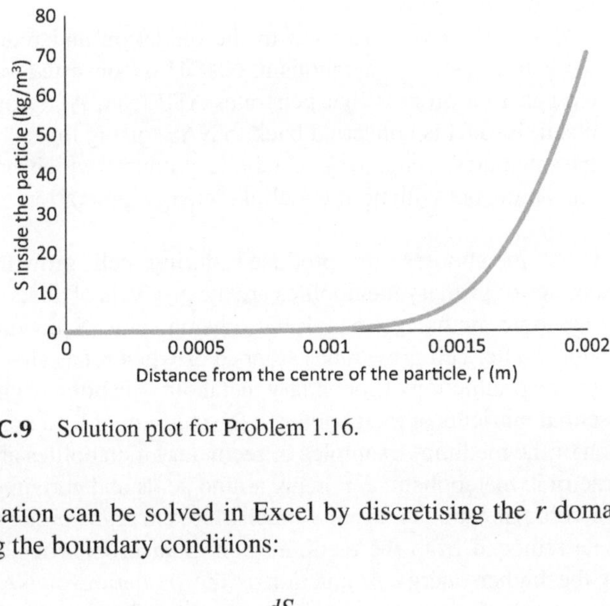

FIGURE C.9 Solution plot for Problem 1.16.

This equation can be solved in Excel by discretising the r domain and by imposing the boundary conditions:

$$r = 0 \quad \frac{dS}{dr} = 0$$

$r = R \quad S = S_b$ (ignoring the resistance of the film surrounding the particles)

We obtain the plot in Figure C.9 (by approximating $r = 0$ with $r = \Delta r$ in the numerical solution, as seen in Chapter 1), with the values of the effectiveness factor of $\eta = 0.55$ and of the Thiele modulus $\Phi = 16.3$. The profile of S, the relatively low value of the effectiveness factor and the relatively high value of the Thiele modulus indicate significant diffusional resistances inside the particles.

CHAPTER 2

2.1 Main cell components and their functions:

- Membranes are the external cell boundaries and allow for the retention of the intracellular materials and for the in-out flow of the substrate, nutrient and products.
- Enzymes are proteins which catalyse all the metabolic reactions in the cell.
- DNA contains the genetic information, which is used to synthesise proteins, including the enzymes, which are used to synthesise all the main cell components. RNA is used for the transcription of the genetic information in the DNA and for the protein synthesis.
- Cytoplasm includes water and all the intracellular components of the cell.
- ATP and ADP are used as energy vectors for chemical reactions in the cell. ATP transfers energy to the reagents to allow their conversion into products and in this process is converted to ADP. ADP is converted back to ATP by catabolic reactions.
- NADH and NAD^+ are used in the oxidation and reduction reactions in the cell. In aerobic metabolism, NADH is converted back to NAD^+ using oxygen in a process that generates ATP from ADP. In anaerobic metabolism, NADH is converted back to NAD^+ using the substrate or metabolic intermediates, generating metabolic products which have often commercial value, but with no intracellular energy generation.

2.2 Primary metabolites are produced during cell growth and duplication. Examples of primary metabolites are the products of anaerobic fermentations, for example methane, ethanol and organic acids. Secondary metabolites are produced after cell growth has stopped or when it has slowed down considerably. The production of secondary metabolites is often triggered by the lack of essential nutrients or metabolites in the medium or by a nonbalanced composition of the medium. Examples of secondary metabolites are organic products of aerobic metabolism, e.g. many amino acids and certain organic acids.

2.3 In aerobic metabolism usually much more energy is produced per unit of substrate removed from the medium than in anaerobic metabolism. The reason for the higher energy production is the oxidation of NADH molecules by oxygen, which generates up to 3 moles of ATP per mol of NADH converted to NAD^+ (this process is called oxidative phosphorylation). Under anaerobic conditions there is no oxygen and oxidative phosphorylation cannot happen. Therefore, the energy production in anaerobic metabolism is only due to substrate-level phosphorylation, and is much lower, per unit of metabolised substrate, than under aerobic conditions. The energy required for the synthesis of cell components is similar for aerobic and anaerobic metabolism. Therefore, under anaerobic conditions a higher fraction of the substrate needs to be used in the catabolic pathways than under aerobic conditions, giving a lower production of microorganisms per unit of substrate removed.

2.4 Batch processes are easy to operate and allow for an easy traceability of the product back to the production batch, which is good for safety and regulatory purposes, especially for the pharmaceutical industry. Continuous processes do not have downtime and have usually higher microorganism concentration than batch processes, giving therefore higher reaction rates. However, chemostat processes require a careful control of the fermentation conditions with risks of process instability and contamination. Fed-batch processes combine the easy traceability of batch processes with the high microorganism concentration of continuous processes. Furthermore, fed-batch processes are highly flexible because they allow for a change in the feed composition during the feeding, therefore they are particularly suitable for the production of secondary metabolites, whose production is triggered by a change in the medium composition. However fed-batch reactors have downtime due to the removal of the reaction volume and their volumetric productivity is lower than for continuous processes because part of the fermentation volume remains in the reactor and does not contribute to the productivity.

2.5 Stirred tank reactors have the advantages of maintaining the same composition in any points of the reactors. In addition, mass and heat transfer rates are enhanced by mixing. Packed- and fluidised-bed fermenters allow for microorganism retention in the reactor, which facilitates product purification and can give higher microorganism concentration and so higher reaction rates. However, mass and heat transfer rates may be lower than for stirred tank feremembers due to the absence of mechanical agitation. Airlift reactors give good mixing similar as for stirred tank fermenters and are better for shear-sensitive cultures, however the mass and heat transfer rates may be lower than in agitated vessels. Bubble columns are cheap to manufacture and are suitable for shear-sensitive cultures, however the mass and heat transfer rates are typically low because of the absence of mechanical agitation.

CHAPTER 3

3.1 The generic stoichiometry for this fermentation can be written as

$$C_2H_5OH + aO_2 + bNH_3 \rightarrow cC_5H_7O_2N + dCO_2 + eH_2O$$

We know that 0.20 kg of microorganisms are produced per kg of ethanol consumed, so

$$\frac{c \cdot 113}{46} = 0.20 \quad \Rightarrow \quad c = 0.08$$

The other coefficients can be calculated from the elemental balances for *C*, *H*, *N* and *O*:

$$\begin{cases} C: 2 = 5 \cdot 0.08 + d \\ H: 6 + 3b = 7 \cdot 0.08 + 2e \\ O: 1 + 2a = 2 \cdot 0.08 + 2d + e \\ N: b = 0.08 \end{cases}$$

The solution of the system of equations gives b = 0.08; d = 1.60; e = 2.84; a = 2.60. So the stoichiometry for this process is

$$C_2H_5OH + 2.60O_2 + 0.08NH_3 \rightarrow 0.08C_5H_7O_2N + 1.60CO_2 + 2.84H_2O$$

3.2 Glucose converted is 30,000·0.95=28,500 t/year.

- From the reaction stoichiometry 0.60 kg of glutamic acid are produced per kg of glucose, so the glutamic acid produced is 28,500·0.60 = 17,100 t/year.
- 0.45 kg of carbon dioxide are produced per kg of glucose, so 12,888 t/year of carbon dioxide are produced.
- 0.39 kg of oxygen are consumed per kg of glucose, so 11,115 t/year of oxygen are consumed.

3.3

- From the stoichiometry we need 0.1·14 = 1.4 g of N per 0.7·192 = 134.4 g of citric acid, so we need 0.0104 tonnes of N per tonne of citric acid. Per 10,000 tonnes of citric acid per year we need 104.17 tonnes of N, i.e. 398 tonnes of ammonium chloride.
- From the stoichiometry we need 0.56 tonnes of oxygen per tonne of citric acid produced, so to produce 10,000 tonnes of citric acid we need 5600 tonnes of oxygen.
- 0.43 tonnes of carbon dioxide are produced per tonne of citric acid produced, so 4300 tonnes of carbon dioxide will be produced for the production of 10,000 tonnes of citric acid per year

3.4 The generic stoichiometry for this fermentation can be written as

$$C_6H_{12}O_6 + aO_2 + bNH_3 \rightarrow cC_5H_7O_2N + dC_3H_5O_2N + eCO_2 + fH_2O$$

We know that 0.20 kg of microorganisms are produced per kg of glucose consumed, so

$$\frac{c \cdot 113}{180} = 0.20 \quad \Rightarrow \quad c = 0.32$$

We also know that 0.60 kg of protease are produced per kg of glucose consumed, so

$$\frac{d \cdot 87}{180} = 0.60 \quad \Rightarrow \quad d = 1.24$$

The remaining coefficients are calculated from the elemental balances:

$$\begin{cases} C: 6 = 5 \cdot 0.32 + 3 \cdot 1.24 + e \\ H: 12 + 3b = 7 \cdot 0.32 + 5 \cdot 1.24 + 2f \\ O: 6 + 2a = 2 \cdot 0.32 + 2 \cdot 1.24 + 2e + f \\ N: b = 0.32 + 1.24 \end{cases}$$

The solution of the system of equations gives b = 1.56; e = 0.68; f = 4.12; a = 1.30. So, the stoichiometry for this process is

$$C_6H_{12}O_6 + 1.30O_2 + 1.56NH_3 \rightarrow 0.32C_5H_7O_2N + 1.24C_3H_5O_2N + 0.68CO_2 + 4.12H_2O$$

3.5 The coefficient x can be determined using the C, H or O balance. The C balance is

$$6 = 5 \cdot 0.15 + 2x + 2.1$$

from which x=1.575.

3.6 The rate of lactic acid (the product) production is proportional to the rate of lactose (the substrate) removal according to

$$r_P = Y_{P/S} \cdot (-r_S)$$

With the data available, to determine $Y_{P/S}$ we need to calculate the stoichiometry of the fermentation. The general stoichiometry for this fermentation is

$$C_{12}H_{22}O_{11} + aNH_3 \rightarrow bC_5H_7O_2N + cC_3H_6O_3 + dH_2O$$

It is known that the initial lactose concentration is 0.80·50 = 40 g/L. The removal of 70% of the initial lactose corresponds to a removal of 28 g/L of lactose. This coincides with the formation of 2 g/L of microorganisms, so

$$\frac{b \cdot 113}{342} = \frac{2}{28} \quad \Rightarrow \quad b = 0.22$$

The remaining coefficients can be calculated from the elemental balances:

$$\begin{cases} C: 12 = 5 \cdot 0.22 + 3c \\ H: 22 + 3a = 7 \cdot 0.22 + 6c + 2d \\ O: 11 = 2 \cdot 0.22 + 3c + d \\ N: a = 0.22 \end{cases}$$

The solution of the system of equations gives a = 0.22; c = 3.63; d = –0.33 (this means that water is a reagent and not a product). So, the stoichiometry for this process is

$$C_{12}H_{22}O_{11} + 0.22NH_3 + 0.33H_2O \rightarrow 0.22C_5H_7O_2N + 3.63C_3H_6O_3$$

From this stoichiometry, $Y_{P/S}$ = 0.96 kg/kg. Therefore, the rate of lactic acid production when the rate of lactose removal is 0.4 g/L.h is

$$r_P = 0.96 \cdot 0.4 = 0.38 \frac{\text{g}}{\text{L} \cdot \text{h}}$$

Note that in this case the general fermentation equation has only four unknown coefficients, because there is no carbon dioxide production. However, it is not possible to determine the four coefficients using the elemental balances alone, because the balances of H and O are the combination of each other, so they are not independent equations. The information on the microorganism yield is therefore still needed. Related to this, also note that, in the solution of the four elemental balances the coefficient d can be obtained either from the H or from the O balance obtaining the same result.

3.7 According to equation (3.19):

$$\mu_{max} = Ae^{-\frac{Ea}{RT}}$$

Taking the natural logarithm of both sides:

$$\ln \mu_{max} = \ln A - \frac{E_a}{RT}$$

Therefore, if we generate a series of values of μ_{max} at different temperatures, the plot of $\ln \mu_{max}$ vs $1/T$ should give a straight line with intercept equal to ln A and slope equal to $-Ea/R$, allowing for the calculation of A and Ea. For the determination of each μ_{max} value at each temperature, experiments should be carried out by measuring the biomass and substrate concentration as a function of time, as shown in section 3.4.

3.8 We use the Monod model with maintenance metabolism. In this case,

$$r_S = -\frac{\mu \cdot X}{Y_{X/S}} - m_S X$$

$$r_X = \mu \cdot X$$

Therefore,

$$r_S = -\frac{r_X}{Y_{X/S}} - m_S X = -\frac{1.7\frac{\text{kg}}{\text{m}^3 \cdot \text{h}} \cdot 24\frac{\text{h}}{\text{d}}}{0.2\frac{\text{kg}}{\text{kg}}} - 0.4d^{-1} \cdot 2.5\frac{\text{kg}}{\text{m}^3} = -205\frac{\text{kg}}{\text{m}^3 \cdot \text{d}}$$

3.9 Equation (3.17), which describes the inhibition of microbial growth by a substance *I*, can be re-written as

$$\mu = \frac{\mu_{max} S}{K_S + S} \frac{K_I}{K_I + I}$$

The rate of substrate removal is therefore

$$r_S = -\frac{\mu \cdot X}{Y_{X/S}} = -\frac{\mu_{max} S}{K_S + S} \frac{K_I}{K_I + I} \frac{X}{Y_{X/S}}$$

If $S >> K_S$, this equation becomes

$$r_S = -\mu_{max} \frac{K_I}{K_I + I} \frac{X}{Y_{X/S}}$$

which can be re-arranged as

$$\frac{X}{(-r_S)} = \frac{Y_{X/S}}{\mu_{max}} + \frac{Y_{X/S}}{\mu_{max} K_I} I$$

Therefore, a series of batch experiments can be carried out, under the condition that at the start of each experiment $S >> K_S$, measuring the initial substrate removal rate at different concentrations of the inhibitor I. Then $X/(-r_S)$ should be plotted vs I, obtaining a linear plot with slope equal to $Y_{X/S}/\mu_{max}.K_I$, from which K_I can be calculated since $Y_{X/S}$ and μ_{max} are known.

3.10 According to equation (3.19):

$$\mu_{max} = Ae^{-\frac{Ea}{RT}}$$

This equation can be linearised as

$$\ln \mu_{max} = \ln A - \frac{E_a}{RT}$$

Therefore, the values of A and Ea can be obtained from a linear plot of $\ln \mu_{max}$ vs $1/T$. With the given experimental data the linear plot is in Figure C.10. From the regression line, we find that

$$A = e^{34.193} = 7.077 \cdot 10^{14}$$

and

$$\frac{E_a}{R} = 9{,}588K \quad \Rightarrow \quad E_a = 79.68 \frac{\text{kJ}}{\text{mol}}$$

At 28 °C the value of μ_{max} is

$$\mu_{max} = 7.077 \cdot 10^{14} e^{-\frac{79{,}670}{8.31 \cdot 301.15}} = 10.6\,\text{d}^{-1}$$

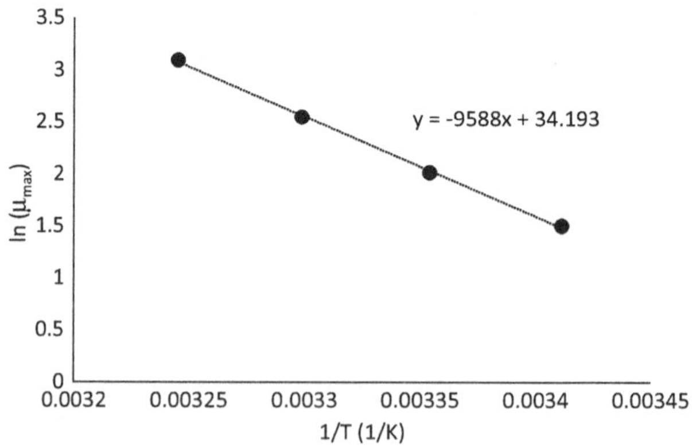

FIGURE C.10 Plot for Problem 3.10.

3.11 If the metabolism is described by the Monod model with endogenous metabolism, the linearization of the rate equations for substrate removal and biomass growth gives

$$\frac{X}{(-r_S)} = \frac{Y_{X/S}}{\mu_{max}} + \frac{Y_{X/S}}{\mu_{max}} K_S \frac{1}{S}$$

and

$$\frac{r_X}{(-r_S)} = Y_{X/S} - b\frac{Y_{X/S}}{\mu_{max}} - \frac{bK_S Y_{X/S}}{\mu_{max}} \frac{1}{S}$$

With the data of this problem, the plots of $X/(-r_S)$ and $r_X/(-r_S)$ vs $1/S$ are given in Figures C.11.

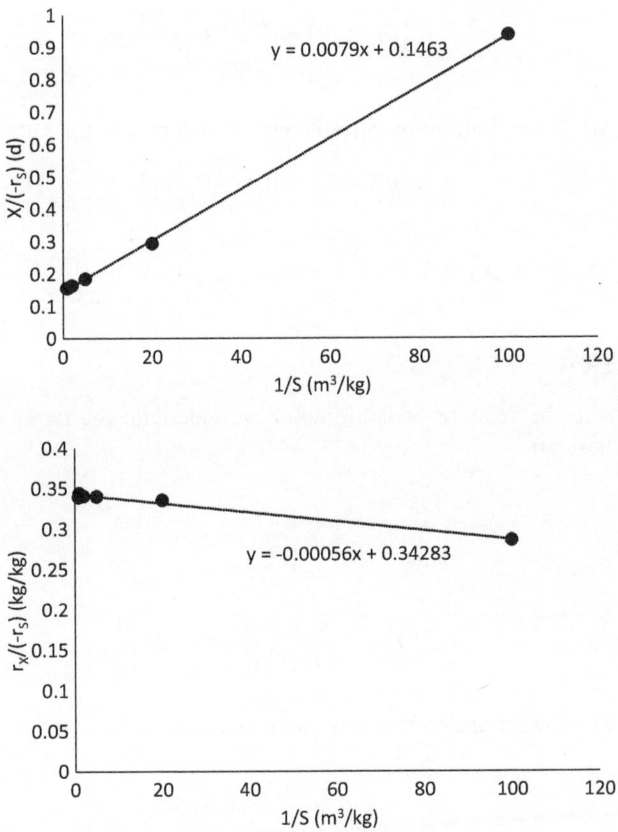

FIGURE C.11 Plots for Problem 3.11.

From the plot of $X/(-r_S)$ vs $1/S$ we get that

$$\frac{Y_{X/S}}{\mu_{max}} = 0.1463\,d; \quad \frac{Y_{X/S}}{\mu_{max}} K_S = 0.0079 \frac{\text{kg}\cdot\text{d}}{\text{m}^3}$$

Therefore $K_S = 0.054$ kg/m^3.
From the plot of $r_X/(-r_S)$ vs $1/S$ we get that

$$Y_{X/S} - b\frac{Y_{X/S}}{\mu_{max}} = 0.3428\frac{\text{kg}}{\text{kg}}; \quad \frac{bK_S Y_{X/S}}{\mu_{max}} = 0.00056\frac{\text{kg}}{\text{m}^3}$$

Therefore, $b = 0.07$ d^{-1}, $Y_{X/S} = 0.35$ kg/kg and $\mu_{max} = 2.4$ d^{-1}

3.12 The total rate of oxygen consumption by the microorganisms is given by

$$\left(-r_{O2biom}\right) = \frac{\mu X}{Y_{X/O2}} + \frac{bX}{Y_{X/O2,end}}$$

We know that in this case the net growth rate of the microorganisms is

$$r_X = \mu X - bX = 0.3\frac{\text{kg}}{\text{m}^3\cdot\text{h}} \text{ and } bX = 0.045\frac{\text{kg}}{\text{m}^3\text{d}}$$

Therefore $\mu X = 0.255\dfrac{\text{kg}}{\text{m}^3\text{d}}$ and $\left(-r_{O2biom}\right) = 0.38\dfrac{\text{kg}}{\text{m}^3\text{d}}$

CHAPTER 4

4.1 From the reaction stoichiometry we calculate the parameters $Y_{X/S}$ and $Y_{P/X}$. These are

$$Y_{X/S} = \frac{0.1\cdot 113}{180} = 0.063\frac{\text{kg}}{\text{kg}}$$

$$Y_{P/X} = \frac{0.66\cdot 192}{0.1\cdot 113} = 11.21\frac{\text{kg}}{\text{kg}}$$

The maximum growth rate can be calculated from the doubling time:

$$\mu_{\max} = \frac{\ln 2}{t_d} = 0.315 hr^{-1}$$

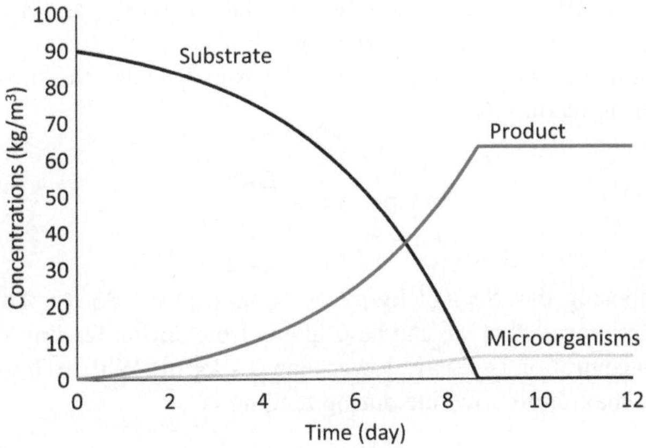

FIGURE C.12 Plot for Problem 4.1.

The differential equations that give the profiles for the substrate, microorganisms and product are, in this case (with concentrations in kg/m^3 and time in h):

$$\frac{dX}{dt} = \frac{0.198S}{0.07+S}X$$

$$\frac{dS}{dt} = -\frac{0.198S}{0.07+S}\frac{X}{0.063}$$

$$\frac{dP}{dt} = \frac{0.198S}{0.07+S}X \cdot 11.21$$

These equations can be solved with $S_0 = 90$ kg/m^3, $X_0 = 0.2$ kg/m^3 and $P_0 = 0$. The obtained profiles are reported in Figure C.12.

4.2 The effluent substrate concentration in a chemostat is given by

$$S = \frac{DK_S}{\mu_{max} - D}$$

This equation can be applied to the two operating conditions given in the problem (where the time is in d and the concentration in kg/m^3):

$$0.05 = \frac{1.0K_S}{\mu_{max} - 1.0}$$

$$0.15 = \frac{1.3K_S}{\mu_{max} - 1.3}$$

These are two equations in the two unknowns μ_{max} and K_S. The solution gives μ_{max} = 1.53 d^{-1} and K_S = 0.026 kg/m^3.

4.3 For a fed-batch reactor at the "quasi-steady state" the substrate concentration during feeding is

$$S = \frac{DK_S}{\mu_{\max} - D}$$

Imposing that S = 0.1 kg/m^3, we obtain D = 5.33 d^{-1}. This is the maximum dilution rate that we can have at any time during feeding so that the substrate concentration is always lower than 0.1 kg/m^3. With an initial volume of 1 m^3, the maximum flow rate during feeding is

$$Q = V \cdot D = 5.33 \frac{\mathrm{m}^3}{\mathrm{d}}$$

4.4 From the reaction stoichiometry:

$$Y_{P/X} = 9.50 \frac{\mathrm{kg}}{\mathrm{kg}}$$

$$Y_{X/S} = 0.063 \frac{\mathrm{kg}}{\mathrm{kg}}$$

$$Y_{X/O2} = 0.16 \frac{\mathrm{kg}}{\mathrm{kg}}$$

The concentrations of biomass, substrate and product and the oxygen consumption rate as a function of the dilution rate at steady state are given by

$$S = \frac{DK_S}{\mu_{\max} - D}$$

$$X = Y_{X/S}\left(S_0 - S\right)$$

$$P = X \cdot Y_{P/X}$$

$$r_{O2biom} = -D\frac{X}{Y_{X/O_2}}$$

The results are plotted in Figure C.13.

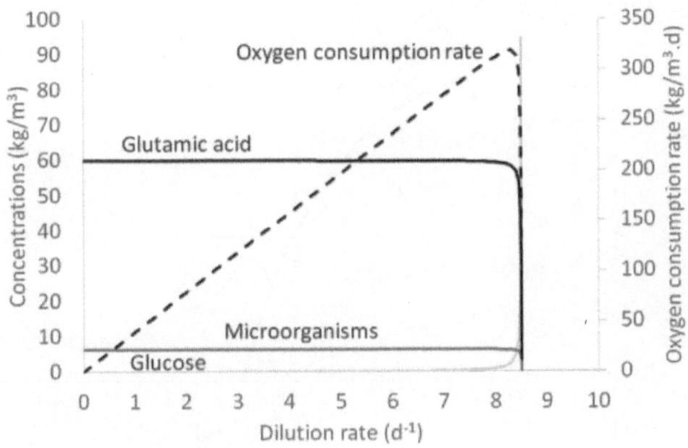

FIGURE C.13 Plot for Problem 4.4.

4.5 The yields $Y_{X/S}$ and $Y_{P/X}$ are in this case, from the reaction stoichiometry:

$$Y_{P/X} = 6.11\frac{\text{kg}}{\text{kg}}$$

$$Y_{X/S} = 0.075\frac{\text{kg}}{\text{kg}}$$

For the batch process, the time profiles of microorganisms, glucose and ethanol are given by the following equations, where time is in *d*:

$$\frac{dX}{dt} = \frac{12.1S}{0.1+S}X$$

$$\frac{dS}{dt} = -\frac{12.1S}{0.1+S}\frac{X}{0.075}$$

$$\frac{dP}{dt} = \frac{12.1S}{0.1+S}X \cdot 6.11$$

The initial conditions are $S_0 = 120$ kg/m³, $X_0 = 0.5$ kg/m³ and $P_0 = 0$. The time profiles are shown in Figure C.14.

From the graph, we see that the time required to obtain 99% glucose removal is 0.242 d. The downtime is 2 h = 0.083 d, so the total time for one batch is approximately 0.33 d. At the end of each batch the concentration of ethanol is 53 kg/m³. So, the ethanol productivity is 53/0.33 = 161 kg/m³·d = 58.6 t/m³·year. Therefore, the reactor volume required to produce 20,000 t of ethanol per year is 20,000/58.6 = 341.2 m³.

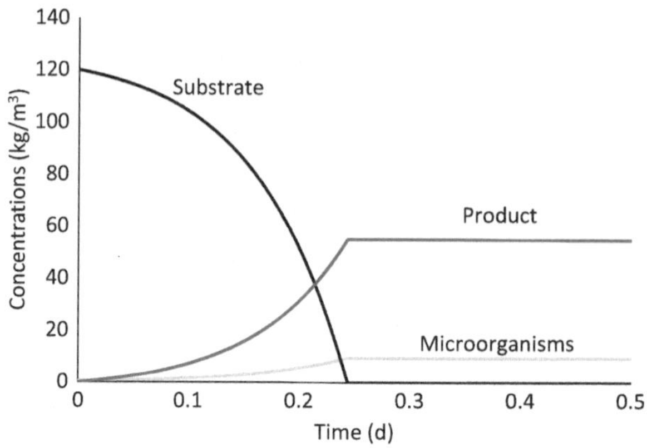

FIGURE C.14 Plot for Problem 4.5.

For the chemostat, the dilution rate required for a concentration of 1.2 kg glucose/m^3 (which corresponds to 99% conversion of the glucose in the feed), is given by

$$D = \frac{\mu_{max} S}{K_S + S} = 11.2\,d^{-1}$$

From the reaction stoichiometry we consume 2.17 t of glucose per 1 t of ethanol, so to produce 20,000 t of ethanol per year we need to feed 43,400 t of glucose per year. This allows us to calculate the required flow rate to the reactor:

$$Q = \frac{43{,}400\,\frac{\text{t}}{\text{year}}}{0.12\,\frac{\text{t}}{\text{m}^3}} = 361{,}700\,\frac{\text{m}^3}{\text{year}}$$

The required volume of the chemostat is therefore:

$$V = \frac{Q}{D} = \frac{361{,}700\,\frac{\text{m}^3}{\text{year}}}{\dfrac{11.2\,\frac{1}{\text{d}}}{365\,\frac{\text{d}}{\text{year}}}} = 88.5\text{m}^3$$

For the fed-batch reactor the annual feed flow rate is the same as calculated for the chemostat, i.e. 361,700 m^3/year. However, the reactor is fed over 11 h per cycle, with 1 h downtime between batches. Therefore, the actual feed flow rate over the 11 hr feeding time will be 361,700·12/11 = 394,582 m^3/year = 1081 m^3/d = 45.0 m^3/h. The maximum dilution rate which is required to maintain the glucose concentration always below 0.1 kg/m^3 is the same as calculated for the chemostat, i.e. 11.2 d^{-1}. The initial volume of the fed-batch reactor is therefore:

$$V_0 = \frac{Q}{D} = 96.5 \text{ m}^3$$

Starting with a volume of 96.5 m^3, with a flow rate of 45 m^3/h, the final volume after 11 h feeding will be 591 m^3.

4.6 With the yields reported in Problem 4.1, the equations for the batch reactor are

$$\frac{dX}{dt} = \frac{10.0S}{0.15+S}X$$

$$\frac{dS}{dt} = -\frac{10.0S}{0.15+S}\frac{X}{0.063}$$

$$\frac{dP}{dt} = \frac{10.0S}{0.15+S}X \cdot 11.21$$

The initial conditions are S_0 =100 kg/m^3, X_0 = 0.2 kg/m^3 and P_0 = 0. From the solution of the equations, the time required to obtain concentration of glucose of 0.5 kg/m3 is 0.35 d. By adding the downtime (2 h = 0.083 d), the total batch time will be 0.43 d. When the glucose concentration is 0.5 kg/m^3, the citric acid concentration is 70 kg/m^3 (this can be obtained from the graph or directly from the product yield or reaction stoichiometry). Therefore, the citric acid productivity in the batch process is 70/0.43 = 162.8 kg/m^3·d.

For the chemostat, the dilution rate required to achieve a glucose concentration of 0.5 kg/m^3 is given by

$$D = \frac{10.0 \cdot 0.5}{0.15+0.5} = 7.7\text{d}^{-1}$$

The effluent citric acid concentration is 70 kg/m^3 (same as for the batch reactor), therefore the productivity per unit volume of reactor is: 70·7.7= 539 kg/m^3·d.

For the fed-batch reactor, the maximum dilution rate at the start of the cycle is still 7.7 d^{-1}.

The product productivity is given by

$$\frac{dP_{TOT}}{dt}\left(\frac{\text{kgproduct}}{\text{d}}\right) = \frac{Y_{P/S}S_0D}{1+Dt_{feed}} = 119.0\frac{\text{kg}}{\text{m}^3\text{d}}$$

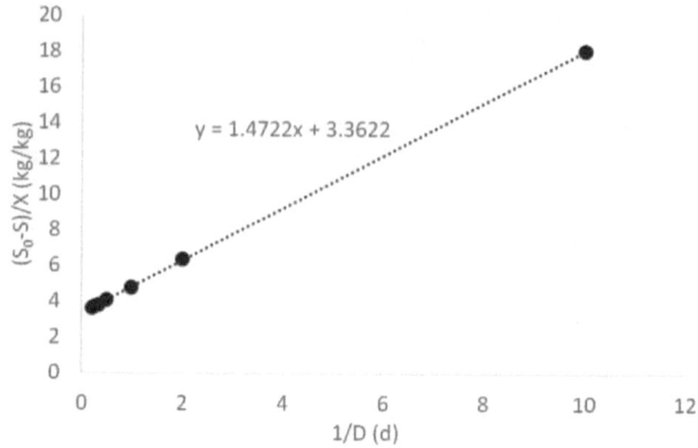

FIGURE C.15 Plot for Problem 4.7.

This is the citric acid productivity during the feed period. However we need to consider that there is no production during the downtime, so the actual productivity will be 119·11/12=109.1 kg/m³·d.

4.7 The parameters $Y_{X/S}$ and m_S can be obtained by plotting $\frac{(S_0 - S)}{X}$ vs $\frac{1}{D}$. With the data of this problem the plot is shown in Figure C.15.

From the slope, we obtain m_S = 1.47 d^{-1}, from the intercept we obtain $Y_{X/S}$ = 0.30 kg/kg.

4.8 At T = 30 °C, we calculate the μ_{max} from the doubling time:

From equation (3.19), the ratio between the μ_{max} at two temperatures T_1 (30 °C = 303.15 K) and T_2 (35 °C=308.15 K) is

$$\frac{\mu_{\max 2}}{\mu_{\max 1}} = e^{-\frac{Ea}{R}\left(\frac{1}{T_2}-\frac{1}{T_1}\right)} = e^{-\frac{90{,}000}{8.31}\left(\frac{1}{308.15}-\frac{1}{303.15}\right)}$$

We obtain μ_{max2} = 0.62 d^{-1}.

4.9 For each data series we can plot $\frac{1}{S}$ vs $\frac{1}{D}$ (Figure C.16) to calculate μ_{max}.

From the plot we obtain the μ_{max} at the three temperatures: T = 20 °C μ_{max} = 5.2 d^{-1}; T = 25 °C μ_{max} = 11.3 d^{-1}; T = 30 °C μ_{max} = 26.9 d^{-1}

To obtain the parameters A and Ea of equation (3.19), we need to plot $\ln(\mu_{max})$ vs $\frac{1}{T}$. The plot is reported in Figure C.17.

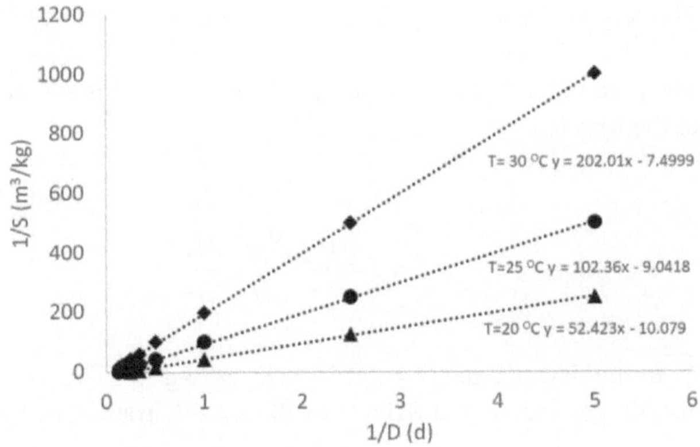

FIGURE C.16 Plot for Problem 4.9.

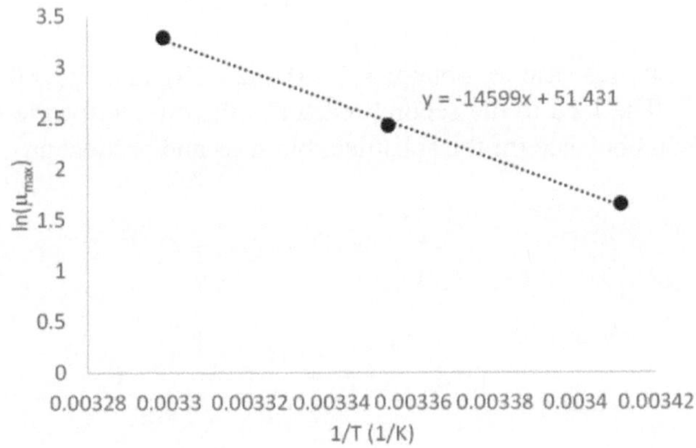

FIGURE C.17 Plot for Problem 4.9.

From the slope of the plot we calculate $E_a = 121{,}318$ J/mol and $A = 2.17\cdot10^{22}$.

4.10 From the microorganisms balance, the required dilution rate is

$$D = \frac{Q}{V} = \frac{\mu_{max} S}{K_S + S} = 1.59\,\text{day}^{-1}$$

The conversion of sucrose is 97%. We produce 0.8 kg of lactic acid per kg of sucrose converted. To produce 15 t/day of lactic acid we need to convert 15/0.8=18.75 t of sucrose and we need to process 18.75/0.97=19.33 t/d of sucrose = $Q{\cdot}S_0$

Since the concentration of sucrose in the feed (S_0) is 100 g/L = 0.1 t/m³, the required feed flow rate is 193.3 m³/d.

The required volume of the reactor is therefore $V = Q/D = 193.3\ m^3/d/1.59\ d^{-1} = 121.57\ m^3$.

4.11 The parameter K_S can be calculated from the effluent substrate concentration in the first reactor:

$$K_S = \frac{\mu_{max} S_1 - DS_1}{D_1}$$

In the first reactor $D = 2\ d^{-1}$, so $K_S = 9\ kg/m^3$. The parameters $Y_{X/S}$ and $Y_{P/S}$ can also be calculated from the effluent substrate concentration:

$$X_1 = Y_{X/S}\left(S_0 - S_1\right)$$

$$P_1 = Y_{P/S}\left(S_0 - S_1\right)$$

In this case we obtain: $Y_{X/S} = 0.052$ kg/kg and $Y_{P/S} = 0.36$ kg/kg.

The feed to the second reactor is the effluent of the first reactor, so the mass balance for the substrate, biomass and product are

$$\frac{\mu_{max} S_2}{K_S + S_2} X_2 \cdot V_2 + QX_1 = Q \cdot X_2$$

$$Q \cdot S_1 = Q \cdot S_2 + \frac{\mu_{max} S_2}{K_S + S_2} \frac{X_2}{Y_{X/S}} \cdot V_2$$

$$\frac{\mu_{max} S_2}{K_S + S_2} \frac{X_2}{Y_{X/S}} Y_{P/S} \cdot V_2 + QP_1 = Q \cdot P_2$$

We have that $Q/V_2 = D_2 = 1.33\ d^{-1}$. The equations can be therefore written as follows:

$$\frac{\mu_{max} S_2}{K_S + S_2} X_2 + D_2 X_1 = D_2 X_2$$

$$D_2 S_1 = D_2 S_2 + \frac{\mu_{max} S_2}{K_S + S_2} \frac{X_2}{Y_{X/S}}$$

$$\frac{\mu_{max} S_2}{K_S + S_2} \frac{X_2}{Y_{X/S}} Y_{P/S} + D_2 P_1 = D_2 P_2$$

This is a system of three equations in the three unknowns S_2, X_2, P_2, which can be solved analytically or numerically. The solution gives $S_2 = 0.045$ kg/m^3, $X_2 = 5.15$ kg/m^3 and $P_2 = 36.06$ kg/m^3. If we wanted to achieve the same final substrate effluent concentration (S_2 = 0.045 kg/m^3) using one single CSTR, with sterile feed and inlet substrate concentration of 100 kg/m^3, the required dilution rate would be

$$D = \frac{\mu_{max} S_2}{K_S + S_2} = 0.040\,\text{d}^{-1}$$

And the required volume would be $V = Q/D = 250$ m^3. Note that the volume of one single reactor would be much higher than the combined volumes of two reactors in series. This is because in this case the parameter K_S is larger than the substrate concentration in the reactors, and the reaction rate is limited by the substrate concentration. In two reactors in series, the reaction rate is larger than in one single reactor because the substrate concentration in the first reactor is larger than the one in one single reactor.

4.12 At the quasi-steady state the biomass balance in a fed-batch reactor can be written as

$$D = \frac{Q}{V} = \frac{\mu_{max}}{K_S + S}$$

In a fed-batch reactor at quasi-steady state the substrate concentration, although it can be assumed very low and much lower than the feed concentration, is the highest at the beginning of the feed and lowest at the end of the feed. Therefore, in a chemostat process, which wants to achieve the same effluent substrate concentration as the fed-batch, the dilution rate can be equal to the dilution rate of the fed-batch at the start of the cycle (this is the maximum dilution rate in the fed-batch cycle). This will ensure that the substrate concentration in the chemostat is always much lower than the feed while ensuring that the chemostat volume is kept to a minimum.

For the fed-batch, the dilution rate at the start of the cycle is

$$D = \frac{Q}{V} = \frac{1.6\frac{m^3}{h}}{2m^3} = 0.8\text{h}^{-1}$$

This is the required dilution rate for the chemostat. The residence time in the chemostat is therefore $\tau = \frac{1}{D} = 1.25\,\text{d}$.

To calculate the effluent substrate and biomass concentrations in the chemostat we need the coefficients $Y_{P/S}$ and $Y_{X/S}$, which can be obtained from the fed-batch data:

$$X = Y_{X/S} S_0$$

$$P = Y_{P/S} S_0$$

We obtain: $Y_{X/S} = 0.10$ kg/kg and $Y_{P/S} = 0.33$ kg/kg. Therefore, the effluent concentrations of a chemostat with 80 kg/m^3 feed will be $X = 8$ kg/m^3 and $P = 26.4$ kg/m^3.

4.13 The biomass concentration throughout the length of the feed phase is:

$$X = Y_{X/S} S_0 = 7.5 \frac{\text{kg}}{\text{m}^3}$$

At the end of the feed phase, after the pH shift, the rate of product generation is

$$r_P = 8.9X = 66.75 \frac{\text{kg}}{\text{m}^3 \cdot \text{d}}$$

The product concentration at the end of the cycle is

$$P = r_P \cdot t = 5.56 \frac{\text{kg}}{\text{m}^3}$$

In each cycle we generate 3 m^3 of effluent, so we produce 16.7 kg of product per cycle. The cycle consists of 3 h feeding, plus 2 h product production and 1 h downtime, so the total length is 6 h. In 8000 h/year we can run 1333 cycles, so we produce 22,261 kg of product per year or 61.0 kg/d.

4.14 The data on the two lab-scale chemostat experiments can be used to calculate the parameters μ_{max}, K_S, $Y_{X/S}$ and $Y_{P/S}$ for the fermentation.

The parameters μ_{max} and K_S can be obtained from the effluent substrate concentration in the two experiments:

$$D_1 = \frac{\mu_{max} S_1}{K_S + S_1} \quad D_2 = \frac{\mu_{max} S_2}{K_S + S_2}$$

where $D_1 = 3$ d^{-1}, $D_2 = 5$ d^{-1}, $S_1 = 0.5$ kg/m^3, $S_2 = 1.1$ kg/m^3. Solving this system of two equations in the two unknowns μ_{max} and K_S we obtain $\mu_{max} = 11.25$ d^{-1}, $K_S = 1.375$ kg/m^3.

From the biomass and product production in the first experiment we obtain $Y_{X/S}$ and $Y_{P/S}$:

$$X_1 = Y_{X/S}\left(S_0 - S_1\right)$$

$$P_1 = Y_{P/S}\left(S_0 - S_1\right)$$

We obtain $Y_{X/S} = 0.051$ kg/kg and $Y_{P/S} = 0.28$ kg/kg.

The dilution rate that gives the optimum product productivity is obtained by plotting the volumetric productivity DP vs D. The graph is shown in Figure C.18 (for a feed of 50 kg/m^3).

From this graph, we see that the optimum volumetric product productivity is 113.2 kg/m^3·d and is obtained for $D_{opt} = 9.4$ d^{-1}. The biomass concentration that corresponds to this dilution rate is 2.2 kg/m^3. If the vessel volume is 2 m^3, the product productivity is 226.2 kg/d.

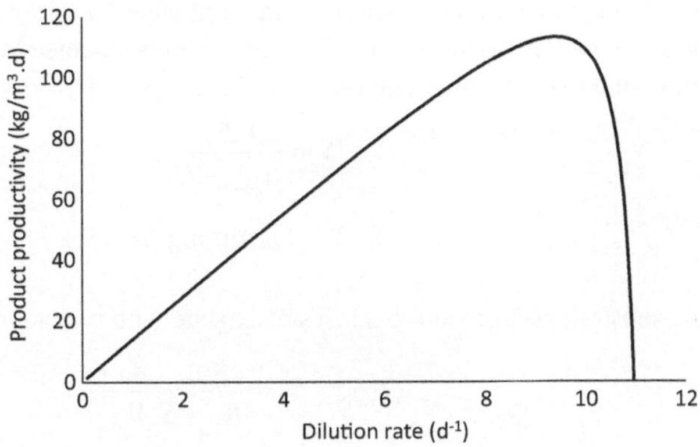

FIGURE C.18 Plot for Problem 4.14.

4.15 For a chemostat where microbial metabolism is described by the Monod equation with maintenance, the biomass and substrate concentrations at steady state are

$$X = \frac{Y_{X/S}\left(S_0 - S\right)}{1 + \frac{m_S Y_{X/S}}{D}}$$

$$D = \frac{\mu_{max}}{K_S + S}$$

Using the biomass and substrate concentration obtained at the two dilution rates, we have a system of four equations with the four unknowns μ_{max}, K_S, $Y_{X/S}$, m_s. After re-arrangements we obtain: μ_{max}=0.75 h^{-1}, K_S = 0.2 kg/m^3, $Y_{X/S}$ = 0.65 kg/kg, m_S = 0.11 h^{-1}.

4.16 In a fed-batch reactor at the quasi-steady state, the substrate concentration is linked to the dilution rate by the equation:

$$D = \frac{Q}{V} = \frac{\mu_{max}}{K_S + S}$$

The dilution rate, and so the substrate concentration, is highest at the start of the feed, when the volume is minimum. The substrate concentration that corresponds to 1% of the feed concentration is S = 0.2 kg/m^3. The maximum feed flow rate that can be used is therefore

$$Q = \frac{\mu_{max} S}{K_S + S} V = \frac{7.3 \cdot 0.2}{0.1 + 0.2} 0.2 = 0.97\,\mathrm{m}^3\,/\,\mathrm{d}$$

4.17 For maintenance metabolism, the mass balances for biomass and product are the same as for the Monod model without maintenance, and they lead to the same quasi-steady state equations for biomass and product, respectively:

$$S = \frac{DK_S}{\mu_{max} - D}$$

$$P = S_0 \cdot Y_{P/S} \text{(assuming } S << S_0\text{)}$$

The substrate balance at the quasi steady state with maintenance metabolism is

$$Q(S_0 - S) = \frac{\mu_{max}}{K_S + S} \frac{X \cdot V}{Y_{X/S}}$$

which gives (assuming $S << S_0$)

$$X = \frac{Y_{X/S} S_0}{1 + \frac{m_S Y_{X/S}}{D}}$$

For the endogenous metabolism, the mass balances for substrate and product are the same as for the simple Monod model, giving (assuming $S << S_0$):

$$X = Y_{X/S} S_0$$

and

$$P = S_0 \cdot Y_{P/S}$$

The biomass balance at quasi steady state with endogenous metabolism is

$$XQ = \frac{\mu_{max} S}{K_S + S} X \cdot V - bXV$$

which gives

$$S = \frac{K_S (D + b)}{\mu_{max} - (D + b)}$$

Note that the quasi-steady state equations for the fed-batch reactor with maintenance and endogenous metabolism are the same as for the corresponding chemostat models.

4.18 The maximum allowable substrate concentration in the fed-batch cycle is 0.3 kg/m^3 (1 % of 30 kg/m^3). This means that the maximum feed flow-rate is

$$Q = \frac{15 \cdot 0.3}{0.08 + 0.3} \cdot 1 = 11.8 \frac{\text{m}^3}{\text{d}}$$

We need to feed 9 m^3 per cycle, so the time required for each feed is 0.76 d = 18.2 h. Including the downtime, the total cycle time is 18.7 h.

The product concentration throughout all the cycle can be approximated by (assuming $S<<S_0$):

$$P = 30 \cdot 0.3 = 9 \frac{\text{kg}}{\text{m}^3}$$

Each batch produces 9 m^3 of effluent at a product concentration of 9 kg/m^3, so each batch produces 81 kg of product. In 8000 h/year we can have 427 batches, with a total product production of 34,587 kg/year or 94.8 kg/d.

Biomass concentration is given by

$$X = \frac{0.2 \cdot 30}{1+\frac{0.8 \cdot 0.2}{11.8}} = 5.9 \frac{\text{kg}}{\text{m}^3}$$

Each batch produces 53.1 kg of biomass, with a biomass productivity of 62.1 kg/d.

CHAPTER 5

5.1 Factors that affect the oxygen transfer rate in agitated vessels are as follows:

- Mixing conditions (type, speed and dimensions of the agitator, geometry of the vessel, presence of baffles, etc). The k_La, in particular the specific area a, increases as the turbulence in the vessel increases, because gas bubbles are broken into into smaller bubbles, with a larger area per unit volume of reactor. Increase in turbulence is usually obtained by increasing the size and stirring speed of the agitator. The geometry of the vessel and the presence of baffles also affect the mixing conditions and therefore the mass transfer coefficient.
- Temperature. Higher temperatures increase the film coefficient k_L, because the diffusivity of oxygen in water increases. However, the oxygen solubility in water decreases as the temperature increases, and this causes a reduction in the driving force.
- Gas flow rate. Increasing the gas flow rate increases the number of bubbles per unit volume of the vessel, this increases the area available for mass transfer and in turn increases the oxygen transfer rate;
- Composition of the inlet gas. Higher oxygen concentrations in the inlet gas correspond to higher saturation concentration of oxygen in water, increasing the driving force and the oxygen transfer rate.
- Concentration of solids and of microorganisms. Higher concentrations of solids or microorganisms increase the viscosity of the liquid phase, decreasing the coefficient k_L and decreasing the oxygen transfer rate;

- Total pressure. For a given mass flow rate of the inlet gas, increasing the pressure in the vessel, increases the oxygen saturation concentration in the liquid phase, increasing the driving force. However, higher pressures will cause the gas bubbles to be smaller, decreasing the area available for mass transfer per unit volume (parameter a).

5.2 The slope of the plot $\ln\left(CO_2^* - CO_2\right)$ vs t is equal to $-k_L a$. From the graph we see that $k_L a = 0.18\ \text{min}^{-1}$. If $CO_2^* = 9.1\frac{\text{mg}}{\text{l}}$ and $CO_2 = 2.1\frac{\text{mg}}{\text{l}}$ the oxygen transfer rate is

$$r_{O2transf} = 0.18\cdot(9.1-2.1) = 1.26\frac{\text{mg}}{\text{l}\cdot\text{min}} = 1.26\cdot 10^{-3}\frac{\text{kg}}{\text{m}^3\ \text{min}}$$

5.3 For geometrica lly similar vessels:

$$\frac{D_{small}}{H_{small}} = \frac{D_{large}}{H_{large}}$$

and

$$\frac{D_{ag,small}}{D_{small}} = \frac{D_{ag,large}}{D_{large}}$$

For the small vessel $\frac{D_{small}}{H_{small}} = 1$, so this ratio will be the same for the large vessel. For the large vessel the volume is

$$V_{large} = \frac{\pi}{4}D_{large}^2 H_{large} = \frac{\pi}{4}D_{large}^3$$

which gives

$D_{large} = 2.67$ m

Therefore $H_{large} = 2.67$ m.

For the small vessel the ratio between the agitator dimeter and the vessel diameter is 0.3, this ratio will be the same for the large vessel, which will have therefore an agitator of a diameter equal to $D_{ag,large} = 0.3D_{large} = 0.80$ m.

5.4 We need to find the dimensions of the large vessel. For the small vessel:

$$V_{small} = \frac{\pi}{4}D_{small}^2 H_{small} \quad \Rightarrow \quad H_{small} = \frac{4V_{small}}{\pi D_{small}^2} = 1.57m$$

Therefore $\frac{D_{small}}{H_{small}} = \frac{D_{large}}{H_{large}} = 0.57$

For the large vessel

$$V_{large} = \frac{\pi}{4} D_{large}^2 H_{large} = \frac{\pi}{4} \frac{D_{large}^3}{0.57} \Rightarrow D_{large} = \sqrt[3]{\frac{4 \cdot 0.57 V_{large}}{\pi}} = 1.94\,\text{m}$$

and $H_{large} = \frac{D_{large}}{0.57} = 3.40$ m

Furthermore, the ratio between agitator diameter and vessel diameter is

$$\frac{D_{ag,small}}{D_{small}} = \frac{D_{ag,large}}{D_{large}} = 0.33$$

Therefore $D_{ag,large}$=0.65 m

The ratio between the mass transfer coefficients in both vessels is

$$\frac{k_L a_{small}}{k_L a_{large}} = \left(\frac{P_{Gsmall}}{P_{Glarge}} \frac{V_{large}}{V_{small}}\right)^{0.62} \left(\frac{Q_{Gsmall}}{Q_{Glarge}} \frac{A_{large}}{A_{small}}\right)^{0.34}$$
$$= \left(\frac{N_{small}^3 D_{ag,small}^5}{N_{large}^3 D_{ag,large}^5} \frac{V_{large}}{V_{small}}\right)^{0.62} \left(\frac{Q_{Gsmall}}{Q_{Glarge}} \frac{D_{large}^2}{D_{small}^2}\right)^{0.34}$$

Replacing the known numerical values this becomes

$$1 = \left(\frac{2.5^3 0.3^5}{N_{large}^3 0.65^5} \frac{10}{1}\right)^{0.62} \left(\frac{250}{2{,}500} \frac{1.94^2}{0.9^2}\right)^{0.34}$$

From this we obtain $N_{large} = 1.29\ \text{s}^{-1} = 77$ rpm.

5.5 The oxygen mass balance for the chemostat is, ignoring the oxygen concentration in the feed:

$$k_L a\left(C_{O2}^* - C_{O2}\right)V = Q \cdot C_{O2} + \left(-r_{O2biom}\right) \cdot V$$

which becomes

$$k_L a\left(C_{O2}^* - C_{O2}\right) = DC_{O2} + \left(-r_{O2biom}\right)$$

The substrate consumption rate in this chemostat is

$$\left(-r_S\right)V = Q\left(S_0 - S\right) \Rightarrow \left(-r_S\right) = D\left(S_0 - S\right) = 0.5(40 - 4) = 18 \frac{\text{kg}}{\text{m}^3\text{d}}$$

From the fermentation stoichiometry we calculate the ratio between oxygen and substrate consumption:

$$Y_{O2/S} = \frac{32 \cdot 3}{180} = 0.53 \frac{\text{kg}}{\text{kg}}$$

Therefore, the rate of oxygen consumption is

$$\left(-r_{O2biom}\right) = \left(-r_S\right) Y_{O2/S} = 9.5 \frac{\text{kg}}{\text{m}^3\text{d}}$$

The $k_L a$ in this system is

$$k_L a = \frac{DC_{O2} + \left(-r_{O2biom}\right)}{\left(C^*_{O2} - C_{O2}\right)} = 1583 \text{d}^{-1}$$

5.6 In the small reactor (reactor 1):

$$\left(-r_{O2,max1}\right) V_1 = Q_{O2max1} = k_L a_1 \left(C^*_{O2} - C_{O2min}\right) V_1$$

From this equation we obtain $k_L a_1 = 7.1 \text{ min}^{-1}$.

In the large reactor (reactor 2):

$$\left(-r_{O2,max2}\right) V_2 = Q_{O2max2} = k_L a_2 \left(C^*_{O2} - C_{O2min}\right) V_2$$

$k_L a_2$ can be calculated from the given correlation for oxygen transfer rate:

$$\frac{k_L a_2}{k_L a_1} = \frac{\alpha \left(\frac{P_G}{V}\right)_2^{0.7} \left(\frac{Q_G}{A}\right)_2^{0.2}}{\alpha \left(\frac{P_G}{V}\right)_1^{0.7} \left(\frac{Q_G}{A}\right)_1^{0.2}} = \gamma^{0.7}$$

Therefore $k_L a_2 = 4.4 \text{ min}^{-1}$ and

$$Q_{O2max2} = k_L a_2 \left(C^*_{O2} - C_{O2min}\right) V_2 = 308 \frac{g_{O2}}{min}$$

Q_{O2max2} is the maximum oxygen transfer rate achievable in the large reactor.

We need now to calculate how much biomass is produced per unit of oxygen consumed, this can be obtained by calculation of the growth stoichiometry. In this case the general growth stoichiometry is

$$C_6H_{12}O_6 + aO_2 + bNH_3 \rightarrow cCO_2 + dH_2O + eC_5H_7O_2N$$

The coefficients a,b,c,d,e can be calculated from the elemental balances and from the given growth yield, $Y_{X/S}$. We obtain the following stoichiometry:

$$C_6H_{12}O_6 + 3.6O_2 + 0.48NH_3 \rightarrow 3.6CO_2 + 5.04H_2O + 0.48C_5H_7O_2N$$

From this stoichiometry we have

$$Y_{O2/X} = \frac{32 \cdot 3.6}{0.48 \cdot 113} = 2.12\frac{\text{kg}}{\text{kg}}$$

The maximum biomass production rate in the large vessel is

$$r_{Xmax}V_2 = (QX)_{max} = \frac{Q_{O2max2}}{Y_{O2/X}} = 209\frac{\text{kg}}{\text{d}}$$

5.7 From the batch test we can calculate the growth stoichiometry and the kinetic parameters. The generic growth stoichiometry in this case is

$$\text{C}_6\text{H}_{12}\text{O}_6 + a\text{O}_2 + b\text{NH}_3 \rightarrow c\text{C}_5\text{H}_7\text{O}_2\text{N} + d\text{H}_2\text{O} + e\text{CO}_2$$

We know that $Y_{X/S} = 0.3$ kg/kg. From the elemental balances we obtain all the coefficients:

$$\text{C}_6\text{H}_{12}\text{O}_6 + 1.2\text{O}_2 + 0.16\text{NH}_3 \rightarrow 0.16\text{C}_5\text{H}_7\text{O}_2\text{N} + 1.68\text{H}_2\text{O} + 1.2\text{CO}_2$$

From the stoichiometry we obtain $Y_{X/O2} = 0.47$ kg/kg.

The graph gives the kinetic parameters. From the linearization of the substrate removal rate according to the Monod equation we obtain

$$\frac{X}{(-r_S)} = \frac{Y_{X/S}}{\mu_{max}}K_S\frac{1}{S} + \frac{Y_{X/S}}{\mu_{max}}$$

From the graph the intercept is

$$\frac{Y_{X/S}}{\mu_{max}} = 0.088\frac{\text{kg}}{\text{kg}}\text{d}$$

From which we calculate $\mu_{max} = 3.41\ \text{d}^{-1}$.

The slope from the graph is

$$\frac{Y_{X/S}}{\mu_{\max}}K_S = 0.0111\frac{\text{kg}\cdot\text{d}}{\text{m}^3} \quad \Rightarrow \quad K_S = 0.13\frac{\text{kg}}{\text{m}^3}$$

The chemostat will work with $D = 0.9 \cdot \mu max = 3.07\ \mathrm{d}^{-1}$. Under these conditions the effluent substrate concentration will be

$$S = \frac{DK_S}{\mu_{max} - D} = 1.18 \frac{\mathrm{kg}}{\mathrm{m}^3}$$

And the biomass concentration will be

$$X = Y_{X/S}\left(S_0 - S\right) = 5.65 \frac{\mathrm{kg}}{\mathrm{m}^3}$$

The oxygen balance in the chemostat is

$$k_L a\left(C_{O2}^* - C_{O2}\right) = D \cdot C_{O2} + \frac{DX}{Y_{X/O2}}$$

The required $k_L a$ is therefore

$$k_L a = \frac{D \cdot C_{O2} + \dfrac{DX}{Y_{X/O2}}}{\left(C_{O2}^* - C_{O2}\right)} = 5678\ \mathrm{d}^{-1}$$

5.8 The mass balance for oxygen during the batch fermentation is

$$\frac{dC_{O2}}{dt} = k_L a\left(C_{O2}^* - C_{O2}\right) - \frac{\mu_{max}}{K_S + S} \frac{X}{Y_{X/O2}}$$

From the problem we know that it needs to be $dC_{O2}/dt = 0$, so the required $k_L a$ during the fermentation will be given by

$$k_L a = \frac{\mu_{max}}{K_S + S} \frac{X}{Y_{X/O2}} \frac{1}{\left(C_{O2}^* - C_{O2}\right)}$$

With the assumption that $S >> K_S$

$$k_L a = \mu_{max} \frac{X}{Y_{X/O2}} \frac{1}{\left(C_{O2}^* - C_{O2}\right)}$$

From the given stoichiometry: $Y_{X/O2} = 0.15$ kg biom/kg O_2

μ_{max} can be calculated from the doubling time:

$$t_d = \ln \frac{2}{\mu_{max}}$$

which gives $\mu_{max} = 0.76\ h^{-1}$

We need to calculate the biomass concentration after 4.5 hours:

$X = X_0 e^{\mu_{max} t}$, which gives $X = 3.06$ g/L.

Therefore, the $k_L a$ required after 4.5 hours will be equal to 3365 h^{-1} = 0.935 s^{-1},

From the given correlations we obtain

$k_L a = 0.020 \cdot 2.40^{0.8} N^{2.4}$ from which N can be calculated

$N = 3.67$ revs/s.

5.9 (a) From the reaction stoichiomtery 0.98 g of acetic acid are produced per g of ethanol converted. The concentration of acetic acid in the reactor is 48 g/L, so 48/0.98 = 49 g/L of ethanol are converted. The ethanol concentration in the feed is equal to 60 g/L so the effluent ethanol concentration will be 60-49 = 11 g/L.

Under the new conditions, an acetic acid productivity of 1500 kg/d is desired. With the same flow rate and reactor volume, this means that the acetic acid concentration in the reactor needs to be 1500/20 = 75 kg/m^3. Therefore, under the new conditions 75/0.98 = 76.5 kg/m^3 of ethanol need to be converted. The effluent ethanol concentration will remain equal to 11 g/L (the residence time does not change), so the glucose concentration in the feed under the new conditions needs to be equal to 76.5+11 = 87.5 kg/m^3.

b) The $k_L a$ for oxygen under the initial conditions can be calculated using a mass balance for oxygen:

$$r_{O2transf} V = Q \cdot C_{O2} + \left(-r_{O2biom}\right) \cdot V$$

r_{O2biom} is the oxygen consumption rate by the microorganisms and can be calculated from the stoichiometry of the reaction and from the given mass flow rates:

$$\begin{aligned}\left(-r_{\mathrm{O2biom}}\right) &= \frac{Q \cdot C_{\mathrm{CH3COOH}}}{V} Y_{\mathrm{O2/CH3COOH}} \\ &= \frac{20\ m^3/\mathrm{d} \cdot 48\ \mathrm{kg}/m^3 \cdot 0.71\ \mathrm{kgO2/kg\ acetic\ acid}}{10 m^3} \\ &= 68.2\ \mathrm{kgO2}/m^3.\mathrm{d}\end{aligned}$$

The equation for the rate of oxygen transfer is

$$r_{O2transf} \left(\frac{\mathrm{kg}_{\mathrm{O}_2}}{\mathrm{m}^3 \cdot \mathrm{day}} \right) = k_L a \cdot \left(C^*_{O2} - C_{O2} \right)$$

So, the only unknown in the oxygen balance is the k_La, which can be calculated as equal to 17,950 d^{-1}.

Under the new conditions we aim to produce 1500 kg of acetic acid per day. This means that the oxygen consumption rate by the microorganisms will be equal to

$$\left(-r_{O2biom}\right)=\frac{1500\,\text{kg}/\text{d}\cdot 0.71\,\text{kgO}_2/\text{kg acetic acid}}{10\text{m}^3}=106.5\,\text{kgO}_2/\text{m}^3.\text{d}$$

If the k_La stays the same we can use the oxygen balance to calculate the new oxygen concentration in the reactor. This will be equal to 2.1 mg/L, so, yes, the current aeration system can work well under the new conditions and can provide the required oxygen concentration.

5.10 From the oxygen balance in a chemostat:

$$k_La=\frac{Q\cdot\left(C_{O2}-C_{O20}\right)+D\dfrac{X}{Y_{X/O2}}\cdot V}{\left(C^*_{O2}-C_{O2}\right)V}$$

In this case we ignore the contributions of the inlet and outlet streams, so

$$k_La=\frac{D\dfrac{X}{Y_{X/O2}}\cdot V}{\left(C^*_{O2}-C_{O2}\right)V}$$

In the scale-up the residence time and the inlet substrate concentration are kept constant, this means that D and X are the same in the small and large reactors, therefore,

$$\frac{k_La_{large}}{k_La_{small}}=\frac{\left(C^*_{O2}-C_{O2}\right)_{small}}{\left(C^*_{O2}-C_{O2}\right)_{large}}=\frac{8.0-4.0}{8.0-2.0}=0.67$$

For geometrically similar vessels

$$\frac{A_{small}}{A_{large}}=\left(\frac{D_{small}}{D_{large}}\right)^2\quad\frac{V_{small}}{V_{large}}=\left(\frac{D_{small}}{D_{large}}\right)^3$$

Combining these equations, with the condition that the gas flow rate per unit volume is the same in the scale-up:

$$\frac{\left(\dfrac{Q_G}{A}\right)_{large}}{\left(\dfrac{Q_G}{A}\right)_{small}}=\left(\frac{V_{large}}{V_{small}}\right)^{0.333}$$

Using the correlation for k_La:

$$\frac{k_La_{large}}{k_La_{small}}=\frac{\left(\frac{P_G}{V}\right)_{large}^{0.75}\left(\frac{Q_G}{A}\right)_{large}^{0.45}}{\left(\frac{P_G}{V}\right)_{small}^{0.75}\left(\frac{Q_G}{A}\right)_{small}^{0.45}}$$

$$=\left(\frac{P_{Glarge}}{P_{Gsmall}}\right)^{0.75}\left(\frac{V_{small}}{V_{large}}\right)^{0.75}\left(\frac{V_{large}}{V_{small}}\right)^{0.33\cdot 0.45}$$

$$=\left(\frac{P_{Glarge}}{P_{Gsmall}}\right)^{0.75}\left(\frac{V_{small}}{V_{large}}\right)^{0.60}$$

which becomes

$$0.67=\left(\frac{P_{Glarge}}{P_{Gsmall}}\right)^{0.75}(0.20)^{0.60}$$

and so $P_{Glarge}/P_{Gsmall}=2.12$. Since the ratio P_G/P is the same for both vessels, the ratio between P_{large} and P_{small} will also be equal to 2.12.

Comparing the power numbers for the large and small vessels:

$$\frac{\left(\frac{P}{\rho_L N^3 D_{ag}^5}\right)_{large}}{\left(\frac{P}{\rho_L N^3 D_{ag}^5}\right)_{small}}=1 \;\Rightarrow\; \frac{P_{large}}{P_{small}}$$

$$=\frac{\left(N^3D_{ag}^5\right)_{large}}{\left(N^3D_{ag}^5\right)_{small}}\Rightarrow\left(\frac{N_{large}}{N_{small}}\right)=\left[\frac{\frac{P_{large}}{P_{small}}}{\left(\frac{D_{aglarge}}{D_{agsmall}}\right)^5}\right]^{0.33}$$

Since the vessels are geometrically similar $D_{ag,small}/D_{ag,large}=(V_{small}/V_{large})^{0.33}$ and so

$$\left(\frac{N_{large}}{N_{small}}\right)=\left(\frac{2.12}{5^{0.33\cdot 5}}\right)^{0.33}=0.53$$

Therefore, the agitation speed required in the large vessel is $N = 0.53 \cdot 180 = 95.4$ rpm.

5.11 The oxygen balance in the chemostat is

$$k_L a\left(C_{O2}^* - C_{O2}\right) = D \cdot C_{O2} + \frac{DX}{Y_{X/O2}}$$

Since the feed, residence time and oxygen saturation concentration are the same for the small and the large reactor, the steady-state biomass concentration will also be the same, so the requirement to have the same dissolved oxygen concentration implies that the $k_L a$ will have to be the same.

From the given correlation for $k_L a$:

$$\frac{k_L a_{large}}{k_L a_{small}} = \frac{0.03\left(\frac{P_G}{V}\right)_{large}^{0.69}\left(\frac{Q_G}{A}\right)_{large}^{0.39}}{0.03\left(\frac{P_G}{V}\right)_{small}^{0.69}\left(\frac{Q_G}{A}\right)_{small}^{0.39}}$$

Re-arranging and considering that in this case the superficial gas velocity remains the same in the two reactors:

$$\frac{k_L a_{large}}{k_L a_{small}} = \frac{\left(\frac{P_G}{V}\right)_{large}^{0.69}}{\left(\frac{P_G}{V}\right)_{small}^{0.69}} = \left(\frac{P_{G,large}}{P_{G,small}}\frac{V_{small}}{V_{large}}\right)^{0.69} = \left(\frac{N_{large}^3 D_{ag,large}^5}{N_{small}^3 D_{ag,small}^5}\frac{V_{small}}{V_{large}}\right)^{0.69}$$

Under conditions of geometric similarity:

$$\frac{D_{ag,large}}{D_{ag,small}} = \left(\frac{V_{large}}{V_{small}}\right)^{\frac{1}{3}}$$

Therefore, the condition of equal $k_L a$ in the large and small vessels becomes:

$$1 = \left(\frac{N_{large}^3}{N_{small}^3}\left(\frac{V_{small}}{V_{large}}\right)^{\frac{2}{3}}\right)^{0.69} = \frac{N_{large}^{2.07}}{N_{small}^{2.07}}\frac{V_{small}^{0.46}}{V_{large}^{0.46}} \Rightarrow N_{large} = 120 \cdot 6^{0.22} = 179 \text{ rpm}$$

5.12 The volumetric oxygen mass transfer rate is

$$r_{O2transf} = k_L a \cdot \left(C_{O2}^* - C_{O2}\right)$$

The correlation for k_l is

$$Sh_L = \frac{k_l \cdot d_{bubbles}}{D_{g,L}} = 2.0 + 0.31\left(\frac{d_{bubbles}^3 \rho_L g}{\mu_L D_{g,L}}\right)^{\frac{1}{3}}$$

From this correlation

$$k_l = \left[2.0 + 0.31\left(\frac{d_{bubbles}^3 \rho_L g}{\mu_L D_{g,L}}\right)^{\frac{1}{3}}\right]\frac{D_{g,L}}{d_{bubbles}}$$

And therefore

$$\frac{k_{l,35}}{k_{l,25}} = \frac{\left[2.0 + 0.31\left(\frac{d_{bubbles}^3 \rho_L g}{\mu_L D_{g,L}}\right)_{35}^{\frac{1}{3}}\right]\left(\frac{D_{g,L}}{d_{bubbles}}\right)_{35}}{\left[2.0 + 0.31\left(\frac{d_{bubbles}^3 \rho_L g}{\mu_L D_{g,L}}\right)_{25}^{\frac{1}{3}}\right]\left(\frac{D_{g,L}}{d_{bubbles}}\right)_{25}}$$

$$= \frac{\left[2.0 + 0.31\left(\frac{0.002^3 1000 \cdot 9.8}{0.00072 \cdot 3.0 \cdot 10^{-9}}\right)_{35}^{\frac{1}{3}}\right]\left(\frac{3.0 \cdot 10^{-9}}{0.002}\right)_{35}}{\left[2.0 + 0.31\left(\frac{0.002^3 1000 \cdot 9.8}{0.00089 \cdot 2.2 \cdot 10^{-9}}\right)_{25}^{\frac{1}{3}}\right]\left(\frac{2.2 \cdot 10^{-9}}{0.002}\right)_{25}} = 1.33$$

For the specific area a

$$a = 1.44\left[\left(\frac{P_G}{V}\right)^{0.4}\left(\frac{\rho_L}{\sigma^3}\right)^{0.2}\right]\left(\frac{v_g}{v_t}\right)^{0.5}$$

With the assumptions of this problem, all the physical properties and variables in this equation don't change between 25 °C and 35 °C, so $a_{25} = a_{35}$.

Therefore

$$\frac{k_L a_{35}}{k_L a_{25}} = 1.33$$

and

$$\frac{r_{O2transf,35}}{r_{O2transf,25}} = \frac{k_La_{35}\cdot(7.5-2.5)}{k_La_{25}\cdot(9.3-2.5)} = 0.98 \quad \Rightarrow \quad r_{O2transf,35} = 0.137\frac{\text{kg}}{\text{m}^3\text{s}}$$

Note that the rate of oxygen transfer at 35 °C is approximately the same as at 25 °C because the increase in the k_l is compensated by the decrease in the driving force.

5.13 The mass balance for oxygen in a chemostat is

$$k_La\left(C_{O2}^* - C_{O2}\right)V = QC_{O2} + \frac{r_XV}{Y_{X/O2}}$$

Since we know the maximum k_La that we can have in this vessel, we can calculate the maximum biomass production rate:

$$\begin{aligned} r_X &= \left[k_La\left(C_{O2}^* - C_{O2}\right) - DC_{O2}\right]Y_{X/O2} \\ &= \left[313.2(0.009-0.002) - 0.125\cdot 0.002\right]0.18 \\ &= 0.39\frac{\text{kg}}{\text{m}^3\text{h}} \end{aligned}$$

The maximum product volumetric production rate is therefore

$$r_P = r_XY_{P/X} = 2.0\frac{\text{kg}}{\text{m}^3\text{h}}$$

and the product productivity is

$$r_PV = 4.0\frac{\text{kg}}{\text{h}} = 48\frac{\text{kg}}{\text{d}}.$$

The maximum possible substrate concentration in the feed is obtained from the substrate balance, which, with the hypothesis that $S \ll S_0$, is

$$QS_0 = \frac{r_XV}{Y_{X/S}} \quad \Rightarrow \quad S_0 = \frac{r_X}{DY_{X/S}} = 31.2\frac{\text{kg}}{\text{m}^3}$$

5.14 The mass balance for oxygen in a chemostat is

$$k_La\left(C_{O2}^* - C_{O2}\right)V = QC_{O2} + \frac{r_XV}{Y_{X/O2}}$$

which can be re-arranged as

$$k_La\left(C_{O2}^* - C_{O2}\right) = DC_{O2} + \frac{DX}{Y_{X/O2}}$$

The value of $k_L a$ for this chemostat is

$$k_L a = \frac{DC_{O2} + \frac{DX}{Y_{X/O2}}}{\left(C^*_{O2} - C_{O2}\right)} = \frac{0.083\frac{1}{h}\cdot 0.0035\frac{kg}{m^3} + 0.083\frac{1}{h}\frac{4}{0.25}\frac{kg}{m^3}}{(0.0092 - 0.0035)\frac{kg}{m^3}} = 233\ \text{h}^{-1}$$

Under the new proposed conditions of residence time 10 h, which corresponds to D = 0.10 h^{-1}, the biomass concentration should stay the same, because the biomass concentration in a chemostat is given by

$$X = Y_{X/S}\left(S_0 - S\right)$$

and, with the assumptions of this problem, $S << S_0$ in all conditions.

The oxygen concentration with the new proposed residence time can be calculated again from the oxygen balance under the new conditions, which will have the same $k_L a$ and biomass concentration:

$$C_{O2} = \frac{k_L a C^*_{O2} - \frac{DX}{Y_{X/O2}}}{D + k_L a} = 0.0023\frac{\text{kg}}{\text{m}^3} = 2.3\frac{\text{mg}}{\text{L}}$$

Therefore, the residence time can be reduced to 10 h without reducing the oxygen concentration below the minimum acceptable level.

5.15 Experiments should be carried out in a range of agitation speed and inlet gas flow rate. For each condition the $k_L a$ should be measured with the oxygen absorption method (or with any other method).

In a first series of experiments the gas flow rate will be held constant while the agitation speed is varied. By varying the agitation speed the power draw will vary and we will collect a series of data of $k_L a$ vs P_G (we assume that adequate correlations are available for the calculation of the power draw of the agitator in the vessel, or that this can be measured directly). We will be able to calculate the parameter a of equation (5.22):

$$k_L a = k\left(\frac{P_G}{V}\right)^a v_g^b$$

Indeed, equation (5.22) can be linearised as

$$\ln\left(k_L a\right) = a\ln\left(\frac{P_G}{V}\right) + \ln\left(k v_g^b\right)$$

and the plot of ln(k_La) vs ln (P_G/V) should give a straight line with slope equal to a and intercept equal to $\ln\left(kv_g^b\right)$.

The next series of experiments should be carried out at constant agitation speed (i.e. constant P_G) and variable Q_G (i.e. variable v_g). The values of k_La collected in this series of experiments will be correlated with the linearised equation:

$$\ln\left(k_L a\right) = b\ln\left(v_g\right) + \ln\left(k\left(\frac{P_G}{V}\right)^a\right)$$

which will allow the calculation of b from the slope of the line ln(k_La) vs ln(v_g).

Once the values of a and b are known, the value of k can be calculated from either of the linearised form of the correlation [ln(k_La) vs ln (P_G/V) or ln(k_La) vs ln(v_g)]. The best procedure is to calculate k from both correlations and calculate the average of the two values obtained.

5.16 If we assume the model with maintenance, we need to take into account the oxygen consumed by the oxidation of the substrate for maintenance in the oxygen balance. The oxygen consumption due to maintenance is (equation (2.36)):

$$r_{O2biom,m} = \frac{-m_S X}{Y_{S/O2,m}}$$

Therefore, the oxygen balance for a chemostat becomes

$$Q \cdot C_{O20} + k_L a \cdot \left(C_{O2}^* - C_{O2}\right)V = Q \cdot C_{O2} + \frac{\mu_{max} S}{K_S + S}\frac{XV}{Y_{X/O2}} + \frac{m_S XV}{Y_{S/O2,m}}$$

Therefore equation (5.34) modifies into

$$k_L a = \frac{D \cdot \left[C_{O2} - C_{O20} + \frac{X}{Y_{X/O2}}\right] + \frac{m_S X}{Y_{S/O2,m}}}{\left(C_{O2}^* - C_{O2}\right)}$$

For the model with endogenous metabolism, the oxygen consumption due to endogenous metabolism is (equation (2.42)):

$$r_{O2biom,end} = \frac{-bX}{Y_{X/O2,end}}$$

and the oxygen balance in a chemostat becomes

$$Q \cdot C_{O20} + k_L a \cdot \left(C_{O2}^* - C_{O2}\right)V = Q \cdot C_{O2} + \frac{\mu_{max} S}{K_S + S}\frac{XV}{Y_{X/O2}} + \frac{bXV}{Y_{X/O2,end}}$$

Equation (5.34) modifies into

$$k_La = \frac{D \cdot \left[C_{O2} - C_{O20} + \dfrac{X}{Y_{X/O2}} \right] + \dfrac{bX}{Y_{S/O2,m}}}{\left(C^*_{O2} - C_{O2} \right)}$$

5.17 The mass balance for oxygen in a batch reactor is

$$\frac{dC_{O2}}{dt} = k_La\left(C^*_{O2} - C_{O2}\right) - \frac{\mu_{max}}{K_S + S}\frac{X}{Y_{X/O2}}$$

Under the requirement of this problem the oxygen concentration is constant ($\frac{dC_{O2}}{dt} = 0$), $S >> K_S$, therefore the required k_La during the batch can be calculated from

$$k_La = \frac{\mu_{max}\dfrac{X}{Y_{X/O2}}}{\left(C^*_{O2} - C_{O2}\right)}$$

The time profile of the biomass concentration X vs time is obtained from the biomass balance in the batch reactor:

$$\frac{dX}{dt} = \mu_{max}X \quad \Rightarrow \quad \int_{X0}^{X}\frac{dX}{X} = \mu_{max}\int_0^t dt \quad \Rightarrow \quad X = X_0e^{\mu_{max}t}$$

Therefore the time profile of the required k_La is given by

$$k_La = \frac{\mu_{max}\dfrac{X_0e^{\mu_{max}t}}{Y_{X/O2}}}{\left(C^*_{O2} - C_{O2}\right)}$$

From the correlation for k_La given in this problem, we calculate the following equation for the required gas flow rate as a function of the k_La:

$$Q_G = A\left(\frac{k_La}{k\left(\dfrac{P_G}{V}\right)^{0.60}}\right)^{\frac{1}{0.42}}$$

and introducing the previous equation for the time profile of k_La (note that in the equation for Q_G, k_La needs to be in s^{-1}, therefore the units of μ_{max} need to be converted accordingly):

$$Q_G = A\left(\frac{\mu_{max} X_0 e^{\mu_{max} t}}{Y_{X/O2}\left(C^*_{O2} - C_{O2}\right)0.045\left(\frac{P_G}{V}\right)^{0.60}}\right)^{\frac{1}{0.42}}$$

$$= 2.8\left(\frac{\frac{4.8}{86,400}\cdot 0.1e^{4.8t}}{0.25(0.009 - 0.003)0.045\left(\frac{1,000}{3}\right)^{0.60}}\right)^{\frac{1}{0.42}}$$

This equation gives the required time profile for the gas flow rate vs time in this batch process, ensuring that the oxygen concentration remains at 3.0 mg/L. The fermentation needs to be carried out until the substrate concentration drops to 0.2 kg/m^3, from the initial 20 kg/m^3. The substrate balance in this case is

$$\frac{dS}{dt} = -\frac{\mu_{max} X}{Y_{X/S}} \Rightarrow \int_{S_0}^{S} dS = -\frac{\mu_{max} X_0}{Y_{X/S}}\int_0^t e^{\mu_{max} t}dt \Rightarrow S_0 - S = \frac{X_0 e^{\mu_{max} t}}{Y_{X/S}}$$

Therefore the end time of the fermentation t_{final} is

$$t_{final} = \frac{\ln\left(\frac{\left(S_0 - S\right)Y_{X/S}}{X_0}\right)}{\mu_{max}} = 0.77\,\text{d} = 18.4\,\text{h}$$

The time profile of the required gas flow rate during the fermentation is shown in Figure C.19. Note that the required gas flow rate is very low at the beginning of the fermentation and then increases exponentially, because of the exponential increase in biomass concentration.

5.18. With the assumptions of ignoring the oxygen mass flow rates in the inlet and outlet streams, the mass balance for oxygen becomes

$$k_La\left(C^*_{O2} - C_{O2}\right) = \frac{DX}{Y_{X/O2}}$$

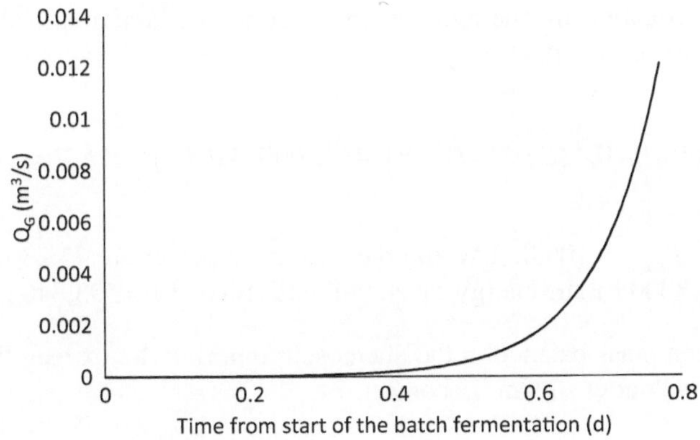

FIGURE C.19 Plot for Problem 5.17.

The required k_La to maintain the oxygen concentration C_{O2} in the chemostat is

$$k_La = \frac{DX}{Y_{X/O2}\left(C_{O2}^* - C_{O2}\right)}$$

In the initial conditions (subscript "old") the oxygen concentration is 2.0 mg/L, in the new conditions (subscript "new") the oxygen concentration is 4.0 mg/L. The ratio between the required k_La in the two conditions is:

$$\frac{k_La_{new}}{k_La_{old}} = \frac{\dfrac{DX}{Y_{X/O2}\left(C_{O2}^* - C_{O2new}\right)}}{\dfrac{DX}{Y_{X/O2}\left(C_{O2}^* - C_{O2old}\right)}} = \frac{\left(C_{O2}^* - C_{O2old}\right)}{\left(C_{O2}^* - C_{O2new}\right)} = \frac{8.5-4.0}{8.5-2.0} = 0.69$$

From the given correlation for k_La we obtain

$$\frac{k_La_{new}}{k_La_{old}} = \left(\frac{P_{Gnew}}{P_{Gold}}\right)^{0.70} = \left(\frac{N_{new}}{N_{old}}\right)^{2.1} \Rightarrow N_{new} = N_{old}\cdot\left(\frac{k_La_{new}}{k_La_{old}}\right)^{\frac{1}{2.1}} = 151 \text{ rpm}$$

The reduction in the required power draw can be calculated using the same equation:

$$\frac{P_{Gnew}}{P_{Gold}} = \left(\frac{k_La_{new}}{k_La_{old}}\right)^{\frac{1}{0.70}} = 0.59$$

From the equation for the agitator power draw as a function of the power number we can calculate P_{Gold}:

$$P_{Gold} = 0.43P = 0.43P_0\rho N_{old}^3 D_{ag}^5 = 0.43 \cdot 1{,}000 \cdot 4.2 \cdot \left(\frac{180}{60}\right)^3 1.3^5 = 181.05\ \text{kW}$$

Therefore P_{Gnew} = 106.82 kW and the savings in power are 74.23 kW. Over one year (8000 h), the energy saved will be $2.14 \cdot 10^{12}\ J$ = 593,840 kWh.

5.19 The oxygen mass balance in the chemostat, ignoring the oxygen flow with the inlet and outlet streams becomes:

$$r_{O2transf}V = k_L a\left(C_{O2}^* - C_{O2}\right)V = \frac{DXV}{Y_{X/O2}} = \frac{1.9d^{-1} \cdot 3.5\frac{\text{kg}}{\text{m}^3} \cdot 30\,\text{m}^3}{0.29\frac{\text{kg}}{\text{kg}}} = 687.9\frac{\text{kg}}{\text{d}}$$

This is the rate of oxygen transfer from the gas phase to the liquid phase. We need to calculate the rate at which oxygen is provided to the reactor with the inlet gas phase. The volumetric flow rate of the inlet gas phase is 3000 m^3/h, therefore the volumetric flow rate of inlet oxygen is 3000·0.21 = 630 m^3/h. The corresponding mass flow rate of inlet oxygen is

$$\frac{630\frac{\text{m}^3}{\text{h}} \cdot 1{,}000\frac{1}{\text{m}^3}\, 1\text{atm}}{0.0821\frac{1 \cdot \text{atm}}{\text{mol} \cdot \text{K}} \cdot 298.15\ \text{K}} = 25{,}737\frac{\text{mol}}{\text{h}} = 19{,}766\frac{\text{kg}}{\text{d}}$$

The efficiency of oxygen transfer is therefore 687.9/19,766=3.5 %.

5.20 The first step is to calculate the required dilution rate for a substrate concentration of 0.1 kg/m^3. This is

$$D = \frac{\mu_{max}S}{K_S + S} = 3.22\,\text{d}^{-1}$$

The influent flow rate is therefore: $Q = V{\cdot}D$ = 32.2 m^3/d.
The product productivity is

$$QP = Q\left(S_0 - S\right)Y_{P/S} = 578.3\frac{\text{kg}}{\text{d}}$$

The required $k_L a$ can be calculated from the oxygen balance:

$$k_L a\left(C_{O2}^* - C_{O2}\right)V = \frac{DXV}{Y_{X/O2}} \quad \Rightarrow \quad k_L a = \frac{DY_{X/S}\left(S_0 - S\right)}{Y_{X/O2}\left(C_{O2}^* - C_{O2}\right)} = 14{,}082\,\text{d}^{-1}$$

CHAPTER 6

6.1 In the scale-up of agitated vessels where fermentation reactions take place, the first criterion is to keep the residence time and the temperature the same in the small and in the large vessels, so that the performance of the process remains the same.

The condition that the residence time remains the same, means that the *Q/V* ratio remains the same for the small and large vessels.

The heat balance for a continuous process is described by the heat balances for the fluid in the reactor and in the jacket:

$$w_{feed}C_{Pfeed}\left(T_{IN}-T_R\right)+r_HV=UA\left(T_R-T_J\right)$$

$$UA\left(T_R-T_J\right)=W_J\cdot C_{PJ}\left(T_J-T_{JIN}\right)$$

The two heat balances can be re-arranged as

$$\frac{w_{feed}}{V}C_{Pfeed}\left(T_{IN}-T_R\right)+\frac{DX}{Y_H}=\frac{UA\left(T_R-T_J\right)}{V}$$

$$\frac{UA\left(T_R-T_J\right)}{V}=\frac{W_J\cdot C_{PJ}\left(T_J-T_{JIN}\right)}{V}$$

If the residence time and feed properties (density, composition and temperature) stay constant with scale-up, the heat balance for the fluid in the reactor indicates that, in order to maintain the reactor temperature T_R constant with scale-up, the ratio $\frac{UA\left(T_R-T_J\right)}{V}$ will have to be constant with scale-up. The heat balance for the fluid in the jacket indicates that, if the flow rate of the jacket fluid is scaled up to maintain W_J/V constant, the jacket temperature T_J will stay the same on scale-up.

The previous considerations indicate that, in order to maintain the reaction temperature constant with scale-up, the ratio *UA/V* needs to be kept constant. The ratio *A/V* is fixed by geometric considerations, i.e.

$$A=\frac{\pi D^2}{4}+\pi DH$$

$$V=\frac{\pi D^2}{4}H$$

Calling α the ratio *H*/*D*, which remains constant on scale-up for geometrically similar vessels:

$$A = \frac{\pi D^2}{4} + \pi\alpha D^2 = D^2\left(\frac{\pi}{4} + \pi\alpha\right)$$

$$V = \frac{\pi\alpha D^3}{4}$$

$$\frac{A}{V} = \frac{\left(4 + \frac{1}{\alpha}\right)}{D}$$

The ratio *A/V* decreases as the size of the vessel increases. Therefore, in order to keep the ratio *UA/V* constant with scale-up, *U* needs to be larger for the large vessel than for the small vessel. If the overall heat transfer coefficient *U* is controlled by the reactor-side coefficient, the heat transfer coefficient for the reactor $h_{reactor}$ needs to be larger for the larger vessel. Using the common correlation for $h_{reactor}$:

$$Nu = \frac{h_{reactor} D}{k_L} = a_1 \, \mathrm{Re}_{ag}^{a2} \, \mathrm{Pr}_{ag}^{0.33} \left(\frac{\mu_L}{\mu_{WL}}\right)^{a_3}$$

and noting that the physical properties of the fluid will stay the same on scale-up, we see that $h_{reactor}$ is proportional to $\frac{N^{a_2} D_{ag}^{2a_2}}{D}$, i.e. is proportional to $N^{a_2} D^{(2a_2-1)}$. Therefore, in the scale-up the agitation rate *N* needs to be adjusted to obtain an increased $h_{reactor}$, in order to obtain the same *UA/V* ratio with a lower *A/V*. In practice, depending on the scale-up ratio, this may or may not be possible, and it may be therefore needed, on scale-up, to use multiple vessels in parallel, rather than one big vessel, in order to have a more favourable *A/V* ratio.

6.2 The enthalpy of the reaction is

$$\Delta H_R\left(\frac{J}{mol\,glu\cos e}\right) = 0.12 H_{biom} + 1.8 H_{ethanol} + 1.8 H_{CO2} + 0.36 H_{H2O} - H_{GLu} - 0.12 H_{NH3}$$

For all the species the enthalpy (J/mol) is expressed as $H_{fi}(25\,°\mathrm{C}) + C_{pi} \cdot (40\text{–}25)$. So we have (all values in kJ/mol):

Biomass: –770+0.333*15 = –765
Ammonia: –79.6
Ethanol: –276
Carbon dioxide: –392
Water: –285
Glucose: –1270
And the enthalpy of reaction is –118.45 kJ/mol glucose.

To produce 1 kg of ethanol we need 2.17 kg of glucose to react. This corresponds to 12.05 mol of glucose. Therefore, the heat generated by the reaction when 1 kg of ethanol is produced is 1427 kJ.

6.3 By heat balance on reactor:

Heat generated = heat removed

Heat balance under condition *a*:

$$Q_{generated,\,a} = \frac{r_{Xa}V}{Y_H} = Q_{removed,a} = w_J.C_{PJ}.\left(T_{Ja} - T_{Jin}\right)$$

$$\frac{r_{Xa}V}{Y_H} = w_J.C_{PJ}.\left(T_{Ja} - T_{Jin}\right)$$

Now r_{Xa} can be calculated by mass balance for substrate in the chemostat:

$$QS_{0a} = \left(-r_{S,a}\right)V + QS_a$$

$$\left(-r_S\right) = \frac{r_X}{Y_{X/S}}$$

By putting value of rate, we get

$$QS_{0a} = \frac{r_{X,a}}{Y_{X/S}}V + QS_a$$

Now by biomass's mass balance:

$$r_{X,a}V = QX_a$$

$$r_{X,a} = \frac{QX_a}{V}$$

$$r_{X,a} = 0.001 \times \frac{4}{10} = 0.0004 \frac{\text{kg}}{\text{m}^3.\text{s}}$$

To calculate the value to $Y_{X/S}$ we obtain after rearranging:

$$Y_{X/S} = \frac{r_{X,a} V}{Q(S_{0a} - S_a)}$$

$$Y_{X/S} = \frac{X_a}{(S_{0a} - S_a)}$$

$$Y_{X/S} = \frac{4}{10-2} = 0.5 \frac{\text{kg biomass}}{\text{kg substrate}}$$

Now Y_H can be calculated using the heat balance for condition *a*:

$$Y_H = \frac{r_{Xa} V}{w_J . C_{PJ} \cdot \left(T_{Ja} - T_{Jin}\right)}$$

$$Y_H = \frac{0.0004 \times 10}{1.5 \times 4186 \times (15-5)}$$

$$Y_H = 6.37 \times 10^{-8} \text{ kg / J}$$

Now to calculate the heat transfer coefficient (U):

$$\frac{r_{Xa} V}{Y_H} = UA\left(T_{Ra} - T_{Ja}\right)$$

$$U = \frac{r_{Xa} V}{Y_H A\left(T_{Ra} - T_{Ja}\right)}$$

$$U = \frac{0.0004 \times 10}{6.37 \times 10^{-8} \times 30 \times (35-15)}$$

$$U = 104.65 \frac{\text{W}}{\text{m}^2.\text{K}}$$

Now for condition *b*, the heat balance for reactor:

$$\frac{r_{Xb} V}{Y_H} = w_J . C_{PJ} \cdot \left(T_{Jb} - T_{Jin}\right)$$

Now we need r_{Xb} and for that we will be doing the substrate mass balance:

$$QS_{0b} = \left(-r_{Sb}\right)V + QS_b$$

$$\left(-r_S\right) = \frac{r_X}{Y_{X/S}}$$

Therefore,

$$QS_{0b} = \frac{r_{X,b}}{Y_{X/S}}V + QS_b$$

$$r_{Xb} = \frac{\left(Q\left(S_{0b} - S_b\right)\right)}{V}Y_{X/S}$$

Since the residence time is the same, it will be

$$S_a = S_b$$

Now $r_{x,b}$ will be

$$r_{Xb} = \frac{\left(0.001(15-2)\right)}{10} \times 0.5 = 0.00065\frac{\text{kg}}{\text{m}^3.\text{s}}$$

From heat balance equation of condition *b*

$$T_{Jb} = T_{Jin} + \frac{r_{Xb}V}{Y_H.w_J.C_{PJ}}$$

$$T_{cool,out,b} = 5 + \frac{0.00065 \times 10}{6.37 \times 10^{-8} \times 1.5 \times 4186}$$

$$T_{Jb} = 21.25°C$$

Now to calculate the T_{Rb}, we will be using the other equation of heat balance, i.e.

$$\frac{r_{Xb}V}{Y_H} = UA\left(T_{Rb} - T_{Jb}\right)$$

$$T_{Rb} = T_{Jb} + \frac{r_{Xb}V}{Y_H UA}$$

$$T_{Rb} = 21.25 + \frac{0.00065 \times 10}{6.37 \times 10^{-8} \times 104.65 \times 30}$$

$$T_{Rb} = 53.75\ °\text{C}$$

Hence

$$T_{reactor,b} > T_{reactor,max}$$

which is not possible, therefore reactor cannot be operated with $S_0 = S_{0b}$.

6.4 The effluent substrate concentration is

$$S = \frac{DK_S}{m_{max} - D} = 0.50 \frac{\text{kg}}{\text{m}^3}$$

The biomass concentration is

$$X = Y_{X/S}\left(S_0 - S\right) = 9.75 \frac{\text{kg}}{\text{m}^3}$$

The rate of heat generation is

$$r_H V = \frac{DXV}{Y_H} = 188\,\text{kW}$$

With the assumptions of this problem, the heat balance for the fluid in the reactor is

$$r_H V = UA\left(T_R - T_J\right)$$

The heat balance for the fluid in the vessel is

$$w_J c_{PJ}\left(T_J - T_{JIN}\right) = UA\left(T_R - T_J\right)$$

Combining these equations we calculate $T_J = 27$ °C. T_R is obtained from

$$T_R = \frac{r_H V}{UA} + T_J = 48$$

6.5 The heat balance for the fluid in the reactor is, in this case:

$$w_{feed} C_{Pfeed} \left(T_{IN} - T_R\right) + r_H V = 0$$

The enthalpy of the reaction is $-1.56 \cdot 10^6$ J/mol glucose. We need to produce 15 t of citric acid per day. From the reaction stoichiometry we need to convert 1.34 kg glucose per kg of citric acid produced. So, we need to convert 20.1 t = 111.67 kmol of glucose per day. The rate of heat generation is therefore $r_H V = 1.56 \cdot 10^9$ J/kmol·111.67 kmol/d = $1.74 \cdot 10^{11}$ J/d. From the heat balance, the feed flow rate is

$$w_{feed} = \frac{-r_H V}{C_{Pfeed} \left(T_{IN} - T_R\right)} = \frac{-1.74 \cdot 10^{11}\,\mathrm{J/d}}{4.19 \cdot 10^3 J/\mathrm{kg.}^\circ\mathrm{C}\left(25 - 35\right)^\circ\mathrm{C}} = 4.15 \cdot 10^6 \frac{\mathrm{kg}}{\mathrm{d}}$$

Since the density of the feed is 1000 kg/m³, the flow rate of the feed is $Q = 4150$ m³/d. It has to be that $Q \cdot S_0 = 20{,}100$ kg of glucose/d, therefore $S_0 = 4.84$ kg/m³.

6.6 We need to write the enthalpy balances for the fluid in the reactor and for the fluid in the jacket. Under the hypotheses of the problem, these two equations are the following:

$$Q\rho c_P \left((Tfeed - T\right) + r_{react}\left(-\Delta H_R\right)V = UA\left(T - T_J\right)$$

$$w_J c_{PJ}\left(T_J - T_{JIN}\right) = UA\left(T - T_J\right)$$

These equations have the two unknowns T and T_J.

The term $r_{react}V$ represents the mol of glucose which react (kmol/h). The ethanol production rate is 150 kg/h = 3.26 kmol/h. From the reaction stoichiometry this corresponds to a glucose reaction rate of 2.17 kmol/h. Therefore, the heat generation rate by this reaction is equal to $r_{react}(-\Delta H_R)V$ = 256 MJ/h = 71.1 kW.

The flow rate through the reactor is $Q = D \cdot V = 44.8$ m³/d.

All the other terms in the heat balances are given and the two equations can be solved by substitution. The solution is $T_J = 17.6$ °C and $T = 37.3$ °C.

6.7 The heat balances for the fluid in the reactor and in the jacket can be re-arranged as

$$\frac{w_{feed}}{V} C_{Pfeed}\left(T_{IN} - T_R\right) + \frac{DX}{Y_H} = \frac{UA\left(T_R - T_J\right)}{V}$$

$$\frac{UA\left(T_R - T_J\right)}{V} = \frac{W_J \cdot C_{PJ}\left(T_J - T_{JIN}\right)}{V}$$

Since the residence time, feed properties and reactor temperature will be kept constant on scale up, the heat balance for the fluid in the reactor indicates that the ratio $\frac{UA(T_R - T_J)}{V}$ will be constant on scale up. Furthermore, T_J will need to stay the same on scale up, so the term *UA/V* will have to stay constant on scale-up.

The ratio *A/V* will change on scale-up:

$$\frac{A}{V} = \frac{\left(4 + \frac{1}{\alpha}\right)}{D}$$

In this case $\alpha = 1$, so $\frac{A}{V} = \frac{5}{D}$

The diameter of a cylindrical vessel is related to the volume by the equation:

$$D = \sqrt{\frac{4V}{\pi\alpha}}$$

Therefore, we have

$$\frac{D_{small}}{D_{large}} = \left(\frac{V_{small}}{V_{large}}\right)^{0.33}$$

The change in the *A/V* ratio on scale up will be

$$\frac{\left(\frac{A}{V}\right)_{large}}{\left(\frac{A}{V}\right)_{small}} = \frac{\frac{5}{D_{large}}}{\frac{5}{D_{small}}} = \frac{D_{small}}{D_{large}} = \left(\frac{V_{small}}{V_{large}}\right)^{0.33} = 0.46$$

So, in order for the *UA/V* ratio to be constant on scale-up, it will have to be

$$\frac{U_{large}\left(\frac{A}{V}\right)_{large}}{U_{small}\left(\frac{A}{V}\right)_{small}} = 1 \quad \Rightarrow \quad \frac{U_{large}}{U_{small}} = \frac{\left(\frac{A}{V}\right)_{small}}{\left(\frac{A}{V}\right)_{large}} = 2.15$$

Since we are assuming that $U \approx h_{reactor}$, this ratio is also equal to $h_{reactor,large}/h_{reactor,small}$.

With the given correlation for the heat transfer coefficients we get

$$\frac{h_{reactor,large}}{h_{reactor,small}} = \frac{N_{large}^{0.67} D_{large}^{0.34}}{N_{small}^{0.67} D_{small}^{0.34}} = \frac{N_{large}^{0.67}}{N_{small}^{0.67}} \cdot 1.30$$

Therefore, it will be

$$N_{large} = N_{small} \left(\frac{2.15}{1.30} \right)^{\frac{1}{0.67}} = 191 \text{ rpm}$$

6.8 The heat balances for the fluid in the chemostat and the fluid in the jacket are

$$w_{feed} C_{Pfeed} \left(T_{IN} - T_R\right) + \left(-r_S\right)\left(-\Delta H_R\right) V = UA\left(T_R - T_J\right)$$

$$UA\left(T_R - T_J\right) = W_J \cdot C_{PJ} \left(T_J - T_{JIN}\right)$$

which can be combined into

$$w_{feed} C_{Pfeed} \left(T_{IN} - T_R\right) + \left(-r_S\right)\left(-\Delta H_R\right) V = W_J \cdot C_{PJ} \left(T_J - T_{JIN}\right)$$

From this equation we can calculate the enthalpy of reaction:

$$\left(-\Delta H_R\right) = \frac{W_J \cdot C_{PJ} \left(T_J - T_{JIN}\right) - w_{feed} C_{Pfeed} \left(T_{IN} - T_R\right)}{\left(-r_S\right) V} =$$

$$\frac{100{,}000 \frac{\text{kg}}{\text{d}} \cdot 4{,}186 \frac{\text{J}}{\text{kg}} \cdot (17-10) \ ^\circ\text{C} - 80{,}000 \frac{\text{kg}}{\text{d}} \cdot 4186 \frac{\text{J}}{\text{kg}} \cdot (15-28) \ ^\circ\text{C}}{20 \frac{\text{m}^3}{\text{d}} \cdot 80 \frac{\text{kg}}{\text{m}^3}}$$

$$= 4.55 \cdot 10^6 \ \frac{\text{J}}{\text{kg}}$$

6.9 From the total mass balance for this reaction we calculate the amount of produced water per kg of reacted substrate. The water produced is 0.15 kg/kg. The enthalpy of reaction per unit mass of reacted substrate is

$$\Delta H_R \left(35 \ ^\circ\text{C}\right)\left(\frac{\text{J}}{\text{kg}}\right) = 0.12 H_{biom} \left(35 \ ^\circ\text{C}\right) + 0.40 H_{product} \left(35 \ ^\circ\text{C}\right)$$
$$+ 0.35 H_{CO2} \left(35 \ ^\circ\text{C}\right)$$

$$+0.15H_{water}\left(35\ ^\circ\mathrm{C}\right)-H_{subs}\left(35\ ^\circ\mathrm{C}\right)-H_{NH3}\left(35\ ^\circ\mathrm{C}\right)$$

where all the enthalpies are in J/kg. The enthalpies are calculated from the general equation:

$$H_i\left(35\ ^\circ\mathrm{C}\right)=\Delta H_{fi}\left(25\ ^\circ\mathrm{C}\right)+C_{Pi}\left(35-25\right)$$

Using this equation, the enthalpies of the reaction components in their reference states at 35 °C are (in J/kg):

$H_{subs}=-7.485\cdot10^6$
$H_{NH3}=-4.463\cdot10^6$
$H_{biom}=-7.97\cdot10^6$
$H_{product}=-1.0788\cdot10^7$
$H_{CO2}=-8.99\cdot10^6$
$H_{water}=-1.596\cdot10^7$

The enthalpy of the reaction is therefore

$$\Delta H_R=-3.238\cdot10^6\ \frac{\mathrm{J}}{\mathrm{kg}}$$

For an adiabatic system the heat balance for the fluid in the reactor becomes

$$w_{feed}C_{Pfeed}\left(T_{IN}-T_R\right)+\left(-r_S\right)\left(-\Delta H_R\right)V=0$$

The term $\left(-r_S\right)V$ is equal to QS_0, under the assumption that all the substrate is removed.

Therefore the heat balance becomes:

$$Q\rho C_{Pfeed}\left(T_{IN}-T_R\right)=-\left(-\Delta H_R\right)QS_0 \quad\Rightarrow\quad S_0=\frac{\rho C_{Pfeed}\left(T_R-T_{IN}\right)}{\left(-\Delta H_R\right)}=16.8\frac{\mathrm{kg}}{\mathrm{m}^3}$$

This is the maximum substrate concentration we can have in the feed with this fermentation reaction and these temperatures for the fluid in the reactor and for the feed.

6.10 The first step is the calculation of the required U when the reactor temperature is at its maximum value, 30 °C. The heat balances for the fluid in the vessel and the jacket are

$$w_{feed}C_{Pfeed}\left(T_{IN}-T_R\right)+\left(-r_S\right)\left(-\Delta H_R\right)V=UA\left(T_R-T_J\right)$$

$$UA\left(T_R-T_J\right)=W_J\cdot C_{PJ}\left(T_J-T_{JIN}\right)$$

With the information given in this problem, this is a system of two equations with the two unknowns T_J and U. With the given numerical values we have

$$A = \pi DH = 3.14 \cdot 3.2 \cdot 3.1 = 31.15 \text{ m}^2$$

$$(-r_S)V = 75 \cdot 70 = 5{,}250 \frac{\text{kg}}{\text{d}}$$

The two equations can be written as

$$75{,}000 \frac{kg}{d} 4{,}190 \frac{\text{J}}{\text{kg °C}} (20-30) \text{ °C} + 5250 \frac{\text{kg}}{\text{d}} \cdot 9.50 \cdot 10^5 \frac{\text{J}}{\text{kg}}$$
$$= U \cdot 31.25 m^2 (30 - T_J) \text{ °C}$$

$$U \cdot 31.25 \text{ m}^2 (30 - T_J) \text{°C} = 50{,}000 \frac{\text{kg}}{\text{d}} \cdot 4{,}190 \frac{\text{J}}{\text{kg°C}} (T_J - 5) \text{°C}$$

Solving these two equations we get $T_J = 13.8$ °C, $U = 89.1$ W/m^2.°C.
We need now to calculate the actual U that can be achieved in this vessel. U can be obtained from the combination of the heat transfer coefficients for the reactor and for the jacket:

$$\frac{1}{U} = \frac{1}{h_{jacket}} + \frac{D_{ext}}{D_{int} h_{reactor}} \cong \frac{1}{h_{jacket}} + \frac{1}{h_{reactor}}$$

$h_{reactor}$ is obtained from the correlation:

$$\frac{h_{reactor} D_{vessel}}{k_L} = 0.74 \, \text{Re}_{ag}^{0.67} \, \text{Pr}_{ag}^{0.33} \quad \Rightarrow \quad h_{reactor} = \frac{k_L 0.74 \left(\frac{N D_{ag}^2 \rho_L}{\mu_L} \right)^{0.67} \left(\frac{c_{PL} \mu_L}{k_L} \right)^{0.33}}{D_{vessel}}$$

The rotational speed of the agitator can be calculated from the available power (15 kW) and from the power number $P_0 = \frac{P}{\rho N^3 D_{ag}^5}$ which gives

$$N = \left(\frac{P}{\rho P_0 D_{ag}^5} \right)^{0.33} = 1.62 s^{-1} = 97 \text{ rpm}$$

Substituting this value for N and the other geometrical parameters and physical properties in the equation for $h_{reactor}$ we obtain

$$H_{react} = 697 \text{ W/m}^{2\text{o}}\text{C}$$

The heat transfer coefficient for the fluid in the jacket is given by

$$Nu_{jacket} = \frac{h_{jacket} D_e}{k_J} = 0.023 \, \text{Re}_J^{0.8} \, \text{Pr}_J^{0.33} \quad \Rightarrow \quad h_{jacket}$$

$$= \frac{k_J 0.023 \left(\dfrac{\rho_J v D_e}{\mu_J} \right)^{0.8} \left(\dfrac{c_{PJ} \mu_J}{k_J} \right)^{0.33}}{D_e}$$

Each spiral has a rectangular section with sides of 0.05 and 0.04 m, therefore the cross sectional area is $0.05 \cdot 0.04 = 0.002 \, \text{m}^2$ and the wetted perimeter is $2 \cdot (0.05 + 0.04) = 0.18 \, \text{m}$. The equivalent diameter is

$$De = \frac{4 \times 0.002}{0.18} = 0.044 \text{ m}$$

The fluid velocity in the jacket is

$$v = \frac{\dfrac{50}{86,400}}{0.002} = 0.29 \frac{\text{m}}{\text{s}}$$

Replacing the numerical values, we obtain for the fluid in the jacket:

$$h_{jacket} = 1217 \text{ W/m}^2.°\text{C}.$$

Combining the values obtained for h_{jacket} and $h_{reactor}$ we obtain

$$U = 443 \text{ W/m}^2.°\text{C}$$

This value of U is considerably higher than the value of U required for this process, so the fermentation can be carried out in the proposed vessel without exceeding the maximum allowable temperature for the reaction. The process can be carried out with an agitation rate lower than the maximum value or with a lower value for the flow rate of the jacket fluid.

6.11 The heat balances for the batch fermentation with a jacketed vessel are

$$M_R C_{PR} \frac{dT_R}{dt} = r_H \cdot V - UA(T_R - T_J) \Rightarrow 0$$

$$= \frac{\mu_{max} S}{K_S + S} \frac{X}{Y_{X/S}} (-\Delta H_R) V - UA(T_R - T_J)$$

$$M_J C_{PJ} \frac{dT_J}{dt} = W_J \cdot C_{PJ} (T_{JIN} - T_J) + UA(T_R - T_J)$$

These must be solved together with the mass balances for the substrate and microorganisms in the reactor:

$$\frac{dS}{dt} = -\frac{\mu_{max}}{K_S + S} \frac{X}{Y_{X/S}}$$

$$\frac{dX}{dt} = \frac{\mu_{max}}{K_S + S} X$$

The volume of the fluid in the reactor is

$$V = \frac{\pi D^2 H}{4} = 3.51 \text{ m}^3$$

The mass of fluid in the jacket is

$$M_J = \pi D H w \rho = 471 \text{ kg}$$

The area for heat transfer is

$$A = \pi D H = 9.42 \text{ m}^2$$

The heat balance for the fluid in the vessel can be re-arranged to calculate the jacket temperature T_J

$$T_J = T_R - \frac{\mu_{max} S}{K_S + S} \frac{X}{Y_{X/S}} (-\Delta H_R) \frac{V}{UA} = T_R - \left(-\frac{dS}{dt}\right)(-\Delta H_R) \frac{V}{UA}$$

This equation allows for the calculation of T_J from the substrate profile. Once the time profile of T_J is calculated, $\frac{dT_J}{dt}$ can be calculated numerically and the required time profile for w_J can be calculated from the heat balance of the fluid in the jacket:

$$W_J = \frac{M_J C_{PJ} \frac{dT_J}{dt} - UA(T_R - T_J)}{C_{PJ}(T_{JIN} - T_J)}$$

In the numerical solution, the first step is to calculate the time profiles of the substrate and biomass from the following equations:

$$\frac{dS}{dt} = -\frac{3.8S}{0.06+S}\frac{X}{0.14}$$

$$\frac{dX}{dt} = -\frac{3.8S}{0.06+S}X$$

Once the time profiles of S and X are calculated, the time profile of T_J is obtained from

$$T_J = 30 - \frac{3.8S}{0.06+S}\frac{X}{0.14}9.67\cdot 10^5 \frac{3.51}{100\cdot 86{,}400\cdot 9.42}$$

From the time profile of T_J vs t, $\frac{dT_J}{dt}$ can be calculated and finally the profile of w_J vs time is obtained from

$$W_J = \frac{471\cdot 4{,}190\frac{dT_J}{dt} - 100\cdot 86{,}400\cdot 9.42(30 - T_J)}{4{,}190(15 - T_J)}$$

The plots of the calculated variables are shown in Figure C.20.

6.12 For an adiabatic batch fermentation the heat balance is

$$M_R C_{PR}\frac{dT_R}{dt} = r_H \cdot V \quad \Rightarrow \quad \frac{dT_R}{dt} = \frac{\frac{\mu_{max}}{K_S + S}\frac{X}{Y_{X/S}}(-\Delta H_R)}{M_R C_{PR}}$$

This equation can be solved simultaneously to the mass balances for X and S:

$$\frac{dS}{dt} = -\frac{\mu_{max}}{K_S + S}\frac{X}{Y_{X/S}}$$

$$\frac{dX}{dt} = \frac{\mu_{max}}{K_S + S}X$$

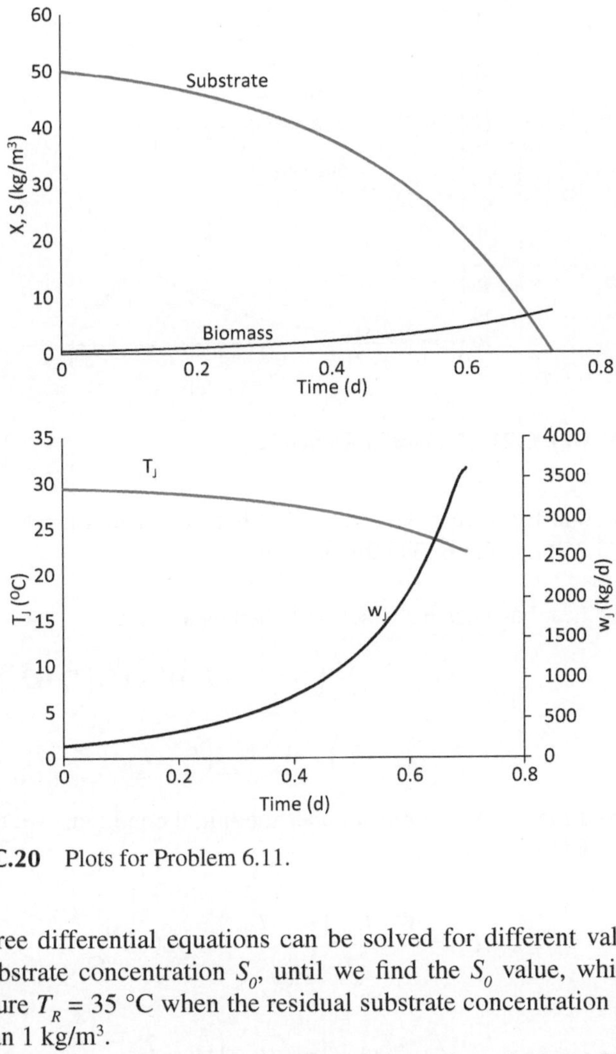

FIGURE C.20 Plots for Problem 6.11.

These three differential equations can be solved for different values of the initial substrate concentration S_0, until we find the S_0 value, which gives a temperature $T_R = 35$ °C when the residual substrate concentration is equal or lower than 1 kg/m^3.

The product concentration during the batch can be calculated from

$$\frac{dP}{dt} = \frac{\mu_{max}}{K_S + S} \frac{X}{Y_{X/S}} Y_{P/S}$$

Figure C.21 shows the profiles of S and T_R for two initial concentrations S_0. The maximum initial concentration is approximately 111 kg/m^3. This corresponds to a final product concentration of 38.9 kg/m^3, achieved in 0.95 d = 22.8 h. Adding the downtime the total time for one batch is 24.8 h. The volume of one batch is 10 m^3, so one batch produces 389 kg of product.

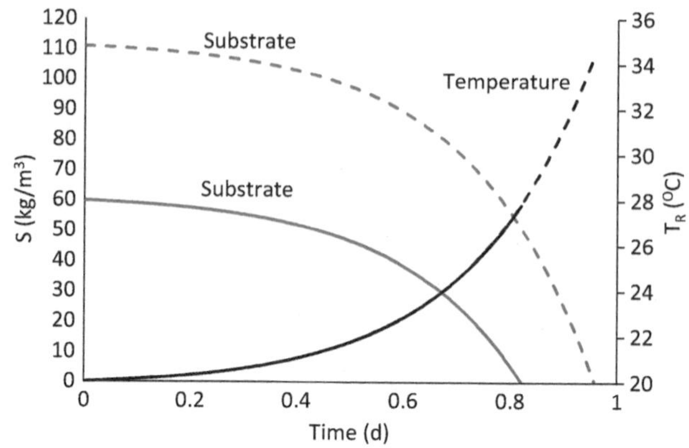

FIGURE C.21 Plot for Problem 6.12.

In 8000 h per year we can do 322 batches, with a product productivity of 125,258 kg/year or 343.2 kg/d.

6.13 The heat balances for a jacketed chemostat are

$$w_{feed}C_{Pfeed}\left(T_{IN}-T_{R}\right)+\left(-r_{S}\right)\left(-\Delta H_{R}\right)V=UA\left(T_{R}-T_{J}\right)$$

$$UA\left(T_{R}-T_{J}\right)=W_{J}\cdot C_{PJ}\left(T_{J}-T_{JIN}\right)$$

From these two equations under the initial conditions we can calculate the terms UA and $(-\Delta H_R)$.

$$UA=\frac{W_{J}\cdot C_{PJ}\left(T_{J}-T_{JIN}\right)}{\left(T_{R}-T_{J}\right)}=9699\frac{\text{W}}{\text{°C}}$$

$$\left(-\Delta H_{R}\right)=\frac{UA\left(T_{R}-T_{J}\right)-w_{feed}C_{Pfeed}\left(T_{IN}-T_{R}\right)}{\left(-r_{S}\right)V}=6.36\cdot 10^{6}\frac{\text{J}}{\text{kg}}$$

Under the new conditions of increased substrate concentration, U will have to increase because the substrate reoval rate $(-r_S)$ increases. We can use the two heat balances under the new conditions to calculate the new values of UA and T_J.

T_J can be obtained from

$$T_{J}=T_{JIN}+\frac{w_{feed}C_{Pfeed}\left(T_{IN}-T_{R}\right)+\left(-r_{S}\right)\left(-\Delta H_{R}\right)V}{W_{J}C_{PJ}}=21.9\text{ °C}$$

And the new value of *UA* from

$$UA = \frac{W_J \cdot C_{PJ}\left(T_J - T_{JIN}\right)}{\left(T_R - T_J\right)} = 14{,}249\frac{W}{m^2\ °C}$$

Since *A* doesn't change under the new conditions, *U* will have to increase

$$\frac{U_{new}}{U_{initial}} = 1.47 = \frac{h_{reactor,new}}{h_{reactor,initial}}$$

Using the given correlation for $h_{reactor}$, assuming that the only change is for the agitation speed:

$$\frac{h_{reactor,new}}{h_{reactor,initial}} = \left(\frac{N_{new}}{N_{initial}}\right)^{0.67} \Rightarrow N_{new} = N_{old}\left(\frac{h_{reactor,new}}{h_{reactor,initial}}\right)^{\frac{1}{0.67}} = 142\ \text{rpm}$$

To calculate the product productivity, we need the parameter $Y_{P/S,}$ which can be calculated under the initial conditions:

$$Y_{P/S} = \frac{P}{\left(S_0 - S\right)} = 0.43\frac{\text{kg}}{\text{kg}}$$

The product productivity under the new conditions (70 kg/m^3 of substrate in the feed) is

$$QP = Q\left(S_0 - S\right)Y_{P/S} = 731\frac{\text{kg}}{\text{d}}$$

(increased from 624 kg/d under the initial conditions).

The increase in the power can be calculated from the power number:

$$P = P_0 \rho N^3 D_{ag}^5 \Rightarrow \frac{P_{new}}{P_{initial}} = \left(\frac{N_{new}}{N_{initial}}\right)^3 = 5.59$$

The agitator power required under the new conditions is 5.59 times higher than the initial one. This is a very considerable increase in power requirements, which is likely to negatively affect the economics of the process.

CHAPTER 7

7.1 The feed has a flowrate of 4 L/h which corresponds to 4 kg/h (we can assume the density of water since this is a diluted water solution). The mass flowrates of *A* and *B* with the feed are, respectively, 0.05·4 = 0.2 and 0.02·4 = 0.08 kg/h. The permeate contains 1% of the *A* and 95% of the *B* in the feed, therefore the mass flowrates of *A* and *B* in the permeate are, respectively, 0.01·0.2 = 0.002 and 0.95·0.08 = 0.076 kg/h.

(a) The volumetric flow rate of the retentate is 4-3.5 = 0.5 L/h, which corresponds to 0.5 kg/h (again we assume the density of water). The mass flowrates of *A* and *B* in the retentate are, respectively, 0.2–0.002 = 0.198 and 0.08–0.076 = 0.004 kg/h. The mass fractions of *A* and *B* in the retentate are therefore, respectively, 0.198/0.5·100 = 39.6% w/w and 0.004/0.5·100 = 0.8% w/w. The process is therefore successful in concentrating species *A*, which is likely to be the high molecular weight species (while *B*, which mostly permeates through the membrane, is likely to have lower molecular weight than *A*).

(b) In the lab experiment the transmembrane flux is 4L/h/100 cm^2 = 400 L m^{-2} h^{-1}. To process 500 L/h with the same transmembrane pressure applied at lab scale, the required area of the membrane is 500 L h^{-1}/400 L m^{-2} h^{-1} = 1.25 m^2.

7.2 From the table, we see that the K' are referred to the Li^+ ion (K_{Li}' = 1.00) and are K'_H=1.27 and K'_{NH4} = 2.55.

The required K' for ammonia exchange with H+ ions on the resin is the equilibrium constant of the reaction:

$$NH_4^+ + HR \leftrightarrow NH_4R + H^+$$

and the required K' is

$$K'_{NH4-H} = \frac{C_{NH4R}C_{H+}}{C_{NH4+}C_{HR}}$$

In the table, the K_H' and K_{NH4}' are defined as

$$K'_{NH4} = \frac{C_{Li+}C_{NH4R}}{C_{NH4+}C_{LiR}} = 2.55$$

It is

$$\frac{K'_{NH4}}{K'_H} = \frac{C_{Li+}C_{NH4R}}{C_{NH4+}C_{LiR}} \cdot \frac{C_{H+}C_{LiR}}{C_{Li+}C_{HR}} = \frac{C_{NH4R}C_{H+}}{C_{NH4+}C_{HR}} = K'_{NH4-H} = \frac{2.55}{1.27} = 2.01$$

7.3 The column has a volume of $\frac{\pi 0.2^2}{4} 1.2 = 0.0377\,\text{m}^3 = 37.7\,L$. Therefore, the total moles of anions that the column can exchange before saturation is 2 mol/L·37.7 L = 75.4 mol. The cation to be exchanged enters the column at a rate of 200 L/h·0.15 mol/L = 30 mol/h. Therefore, the length of column operation is 75.4 mol/30 mol/h = 2.51 h= 2 h 31 min.

In this example we assumed very fast (instantaneous) ion exchange. This means that at breakthrough the ion profile is vertical (the concentration of the ion is zero before the column reaches saturation, then becomes instantaneously equal to the feed concentration once the column is saturated). In reality, ion exchange is not instantaneous and some ion will be released in the effluent before the column is fully saturated, which reduces the length of time the column can be used compared to the assumption of instantaneous exchange.

7.4 For a volume of fermentation mixture of 2 m^3 and a concentration of contaminant of 500 mg/L (= 500 g/m^3), the mass of contaminant to be adsorbed is $500 \frac{\text{g}}{\text{m}^3} \cdot 2\,\text{m}^3 = 1000$ g. According to the given adsorption isotherm, for a concentration inside the column equal to the concentration of the feed, which corresponds to saturation of the activated carbon, the concentration of contaminant on the solid is $W = 0.02 \frac{\text{g}}{\text{g}} \frac{0.15 \frac{\text{m}^3}{\text{g}} \cdot 500 \frac{\text{g}}{\text{m}^3}}{1 + 0.15 \frac{\text{m}^3}{\text{g}} \cdot 500 \frac{\text{g}}{\text{m}^3}} = 0.0197 \frac{\text{g}}{\text{g}}$ (note that this value is very close to W_{max}). Therefore, the mass of solid to adsorb 1000 g of contaminant is $\frac{1{,}000\, g_{contaminant}}{0.0197 \frac{g_{contaminant}}{g_{solid}}} = 50{,}761\, g_{solid}$. This mass corresponds to a volume of solid of $V_{solid} = \frac{50{,}761\, g_{solid}}{2.0 \cdot 10^6 \frac{g_{solid}}{m^3}} = 25.38\,L$.

This is the volume of dry solid without considering the empty spaces. The total volume of column, including the volume occupied by the fluid is: $V_{column} = A \cdot L = \frac{V_{solid}}{1-\varepsilon} = \frac{2.84\,L}{1-0.5} = 0.0508\ m^3$ (where A is the cross sectional area of the column and L is its length). With $A = \frac{\pi D^2}{4} = \frac{\pi 0.2^2}{4} = 0.0312\,\text{m}^2$. Therefore, the column length will be $L = \frac{0.0508\,\text{m}^3}{0.0312\,\text{m}^2} = 1.63\,\text{m}$. However, when breakthrough happens, not all the column has been saturated, and operation has to be stopped when 90% of the maximum theoretical flow has passed through the column. This means that the column must be longer than what calculated above for the ideal case of instantaneous adsorption and the required column length will be $\frac{1.63}{0.9} = 1.81\text{m}$.

7.5 The problem requires the solution of the mass balances of the adsorbate in an adsorption column, equations (7.28) and (7.29). In these equations, with the data of this problem:

$$u = \frac{Q}{\varepsilon A} = \frac{4\frac{\text{m}^3}{\text{h}}}{0.4 \cdot 0.7\ \text{m}^2} = 14.29\frac{\text{m}}{\text{h}} = 0.00397\frac{\text{m}}{\text{s}}$$

In the numerical solution, using 10 sections corresponds to $\Delta L = \frac{2m}{10} = 0.2m$.

The adsorption is described by the Langmuir isotherm, which gives in the mass transfer rate an equilibrium concentration C^* described by Equation (7.26), with W in g/g and C^* in g/m^3:

$$C^* = \frac{W}{0.08(0.0035 - W)}$$

The plot of the outlet concentration (section 10) vs time is reported in Figure C.22. The breakthrough time is defined in this case as the time when the outlet concentration reaches 5% of the feed concentration, i.e. $0.05 \cdot 300 = 15\,\text{g}/\text{m}^3$. From the data used for Figure C.22, the breakthrough time is equal to 5224 s.

7.6 If the aim of the process is the separation and purification of each of the three substances, chromatography may be suitable while adsorption will not work. Adsorption uses a continuous feed of the fermentation mixture and in this

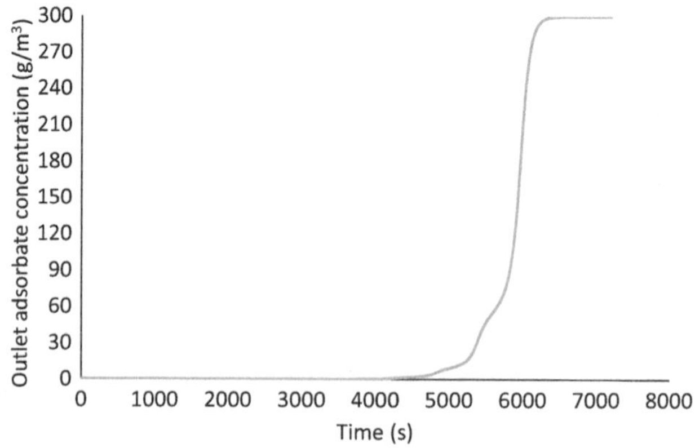

FIGURE C.22 Plot for Problem 7.5.

case would produce a solid phase with the three substances adsorbed, but the effluent would never contain the individual separated substances. The effluent from the adsorption column would initially only contain water (plus possibly other components of the fermentation broth different from the three products) then once the column saturates the three products would all end up in the effluent. Instead, in chromatography, because of the pulse injection of the feed, the three substances will be separated and can be collected separately with the effluent.

7.7 The advantage of chromatography is that it allows the separation and recovery of individual components in multicomponent mixtures, which is not possible to achieve by adsorption. However, chromatography has the disadvantages of the dilution of the products with the eluent (which is not present in an adsorption process), of the limited capacity and of the high cost. Instead, adsorption allows the removal of contaminants from mixtures, therefore achieving purification of the nonadsorbed substances, however it does not allow separation and recovery of the contaminants.

CHAPTER 8

8.1 Bioethanol is a primary metabolite, produced during the main metabolism of the microorganisms. A chemostat is suitable for its production and would generally give higher productivities than a batch process. Penicillin is a secondary metabolite, mainly produced when growth has stopped or has significantly slowed down. Production of penicillin is a two-phase process and therefore the use of a chemostat, characterised by steady-state conditions, is not recommended. It is better to use batch processes, where microroganisms can grow when substrate is in excess and can produce penicillin when the substrate is (almost) completely removed.

8.2 The most typical factors that trigger L-glutamic acid production with the microorganism *Corynebacterium glutamicum* are biotin limitation, presence of surfactants or of penicillin. A fed-batch process for the production of L-glutamic acid can be carried out starting with a balanced feed (with enough biotin, without surfactants or penicillin) in order to increase the number and mass of microorganisms, followed by an unbalanced feed with biotin limitation (or with penicillin or surfactants) to trigger the extraction of the amino acid.

8.3 Citric acid production is favoured by low concentrations of certain metals (Fe, Mn, Zn) and by relatively high oxygen concentrations in the medium (at least 25% of saturation).

8.4 We know from the data that 200 kg/h of carbon dioxide are produced. According to the stoichiometry 0.44 kg carbon dioxide are produced per kg of glucose that reacts. Therefore, the glucose reaction rate is 455 kg/h. We have

50 kg/h of glucose leaving the reactor, therefore the mass flow rate of glucose in the feed is 505 kg/h.

With the reaction of 455 kg/h of glucose we produce, according to the stoichiometry, 209 kg of ethanol. Of these, 25 are lost with stream 3 and 10 with stream 4, therefore the mass flow rate of ethanol in stream 4 is 174 kg/h.

8.5 The penicillin production rate is in total 3.4 kg/h. From the reaction stoichiometry 0.087 kg penicillin are produced per kg of oxygen consumed. So, the oxygen consumption is 3.4/0.087 = 39.1 kgO_2/h
From the reaction stoichiometry 0.166 kg penicillin/kg microorganisms are produced. So, for the production of 3.4 kg/hr of penicillin, 20.5 kg/h microorganisms are produced. Of these, 1.5 leave the system with stream 4, so 19.0 kg/h of microorganisms leave with stream 2.

8.6 From the stream table we see that 9,300 kg of starch per day are converted. From the stoichiometry of starch hydrolysis this corresponds to a production of glucose of 10,333 kg/d. 1000 kg/d of glucose leave the system unreacted, so 9333 kg/d of glucose are converted. We produce 0.20 kg of microorganisms per kg of glucose converted.

We use this information to calculate the stoichiometry of glucose fermentation to ethanol:

$$C_6H_{12}O_6 + aNH_3 \rightarrow bC_5H_7O_2N + cC_2H_6O + dCO_2 + eH_2O$$

$b = a = 0.32$ (from the N balance)

combining the elemental balances for O, C and H we obtain

$c = 1.47$, $d = 1.46$, $e = 0.97$ and the balanced stoichiometry is

$$\begin{aligned} C_6H_{12}O_6 + 0.32NH_3 \rightarrow\ & 0.32C_5H_7O_2N + 1.47C_2H_6O \\ & + 1.46CO_2 + 0.97H_2O \end{aligned}$$

From this stoichiometry 0.38 kg of ethanol are produced per kg of glucose converted, so in total we produce 9333·0.38=3,547 kg of ethanol per day. The total loss of ethanol with streams 3 and 4 is 313 kg/d, so the ethanol in steam 5 is 3234 kg/d.

The production of carbon dioxide is 0.36 kg/kg glucose, so the total production of carbon dioxide (all in stream 2) is 3360 kg/d.

Bibliography

JOURNALS AND BOOKS

Armstrong, D.C., M.R. Johns. Culture conditions affect the molecular weight properties of hyaluronic acid produced by *Streptococcus zooepidemicus*. *Applied and Environmental Microbiology*, 63, 2759–2764, 1997.

Azhar, S.H.M., R. Abdulla, S.A. Jambo, H. Marbawi, J.A. Gansau, A.A.M. Faik, K.F. Rodrigues. Yeasts in sustainable bioethanol production: A review. *Biochemistry and Biophysics Reports*, 10, 52–61, 2017.

Bailey, J.E., D.F. Ollis. *Biochemical Engineering Fundamentals*. Second Edition. McGraw-Hill, USA, 1986.

Benz, G.T. Piloting bioreactors for agitation scale-up. *CEP* 104, 32–34, 2008.

Cardona, C.A., O.J. Sanchez, L.F. Gutierrez. *Process Synthesis for Fuel Ethanol Production*. CRC Press, USA, 2010.

Dionisi, D. *Biological Wastewater Treatment Processes-Mass and Heat Balances*. CRC Press, USA, 2017.

Dionisi, D., I.M.O. Silva. Production of ethanol, organic acids and hydrogen: An opportunity for mixed culture biotechnology?. *Reviews in Environmental Science and Bio/Technology*, 15, 213–242, 2016.

El-Mansi, E.M.T., J. Nielsen, D. Mousdale, R.P. Carlson (Eds.). *Fermentation Microbiology and Biotechnology*. CRC Press, Boca Raton, FL, US, 2018.

Garcia-Ochoa, F., E. Gomez. Bioreactor scale-up and oxygen transfer rate in microbial processes: An overview. *Biotechnology Advances*, 27(2), 153–176, 2009.

Humbird, D., R. Davis, J.D. McMillan. Aeration costs in stirred-tank and bubble column bioreactors. *Biochemical Engineering Journal*, 127, 161–166, 2017.

Kalbfuss, B., M. Wolff, R. Morenweiser, U. Reichl. Purification of cell culture-derived human influenza A virus by size-exclusion and anion-exchange chromatography. *Biotechnology and Bioengineering*, 96(5), 932–944, 2007a.

Kalbfuss, B., Y. Genzel, M. Wolff, A. Zimmermann, R. Morenweiser, U. Reichl. Harvesting and concentration of human influenza A virus produced in serum-free mammalian cell culture for the production of vaccines. *Biotechnology and Bioengineering*, 97(1), 73–85, 2007b.

Kwiatkowski, J.R., A.J. McAloon, F. Taylor, D. Johnston, Modeling the process and costs of fuel ethanol production by the corn dry-grind process. *Ind Crops Prod* 23, 288–296, 2006.

McCabe, W.L., J.C. Smith, P. Harriott. *Unit Operations of Chemical Engineering*. McGraw-Hill Chemical Engineering Series, New York, NY, USA, 1993.

Oberoi, H.S., P.V. Vadlani, R.L. Madl, L. Saida, J.P. Abeykoon. Ethanol production from orange peels: Two-stage hydrolysis and fermentation studies using optimized parameters through experimental design. *Journal of Agricultural and Food Chemistry*, 58, 3422–3429, 2010.

Okpokwasili, G.C., C.O. Nweke. Microbial growth and substrate utilization kinetics. *African Journal of Biotechnology*, 5, 305–317, 2005.

Pérez Rubio, A., J.M. Eiros. Cell culture-derived flu vaccine: Present and future. *Human vaccines Immunothera* 14, 8, 2018.

Pollard, D.J., G. Hunt, T. Kirschner, P. Salmon. Rheological characterization of a fungal fermentation for the production of pneumocandins. *Bioprocess and Biosystems Engineering* 24(6), 373–383, 2002.

Ratledge, C., B. Kristiansen (Eds). *Basic Biotechnology*. Second Edition. Cambridge University Press, UK, 2001.

Schulze, K.L., R.S. Lipe. Relationship between substrate concentration, growth rate, and respiration rate of *Escherichia coli* in continuous culture. *Archiv für Mikrobiologie* 48(1), 1–20, 1964.

Shuler, M.L., F. Kargi, M. DeLisa. *Bioprocess Engineering. Basic Concepts. Third Edition.* Prentice Hall, USA, 2017.

Sinnott, R., G. Towler. *Chemical Engineering Design*. Fifth Edition. Coulson & Richardson's Chemical Engineering Series, IChemE. Butterworth-Heinemann, Burlington, MA, USA, 2009.

Stanbury, P., A. Whitaker, S. Hall, *Principles of Fermentation Technology*. 3rd Edition. Butterworth-Heinemann, Exeter, UK, 2016.

Taguchi, H., A.E. Humphrey. Dynamic measurement of the volumetric oxygen transfer coefficient in fermentation systems. *Journal of Fermentation Technology*, 44, 881–889, 1966.

Todaro, C.C., H.C. Vogel (Eds). *Fermentation and Biochemical Engineering Handbook*. Elsevier Science & Technology Books, The Netherlands, 2014.

Van Leeuwen, M.R., P. Krijgsheld, T.T. Wyatt, E.A. Golovina, H. Menke, A. Dekker, J. Stark, H. Stam, R. Bleichrodt, H.A.B. Wösten, J. Dijksterhuis. The effect of natamycin on the transcriptome of conidia of *Aspergillus niger*. Studies in mycology 74, 71–85, 2013.

Villano, M., F. Valentino, A. Barbetta, L. Martino, M. Scandola, M. Majone Polyhydroxyalkanoates production with mixed microbial cultures: From culture selection to polymer recovery in a high-rate continuous process. New biotechnology 31, 289–296, 2014.

Whytes, M.J., N.L. Morgan, J.S. Rockey, G. Higton. *Industrial Microbiology: An Introduction*. Blackwell Science, Oxford, UK, 2001.

INTERNET REFERENCES

https://commons.wikimedia.org/wiki/File:20101212_200110_LactobacillusAcidophilus.jpg

https://bacdive.dsmz.de/strain/7096

https://commons.wikimedia.org/w/index.php?curid=52254246

https://commons.wikimedia.org/wiki/File:Cell_membrane_detailed_diagram_4.svg

Molview.org

https://commons.wikimedia.org/wiki/File:Hexokinase_induced_fit.svg

https://en.wikipedia.org/wiki/Nucleotide#/media/File:0322_DNA_Nucleotides.jpg

https://commons.wikimedia.org/wiki/File:RNA_chemical_structure.GIF

http://2010.igem.org/Team:UCL_London/Fermenter_Mechanics

https://sites.google.com/site/fermentersin/types

www.technologynetworks.com/cell-science/articles/prokaryotes-vs-eukaryotes-what-are-the-key-differences-336095

COMMERCIAL PROVIDERS OF LAB, PILOT AND FULL-SCALE FERMENTERS AND FERMENTATION EQUIPMENT

Many companies provide fermenters and fermentation equipment at lab, pilot and/or full scale. The list below, not exhaustive, mentions some providers active worldwide.

Applikon Biotechnology, applikon-biotechnology.com
De Dietrich Process Systems, dedietrich.com
Gea, gea.com
GPC Bio, gpc-bio.com
INFORS HT, infors-ht.com
Pall, pall.com
Paul Mueller Company, paulmueller.com
Pierre Guerin Technologies, pierreguerin.com
Sartorius, sartorius.com
Solida Biotech, solidabiotech.com

Index

A

Acetic acid, concentration of, 389
Adenosine di-phosphate (ADP), 93, 94
Adenosine triphosphate (ATP), 93, 94
  energy-rich molecule, 93
  enzymatic reaction, 92
Adiabatic batch fermentation, 414
  heat balance, 414
  maximum product productivity, 416
Adiabatic reactors, scale-up/scale-down of, 256
Adsorbate, column outlet concentration, 306
Adsorption equilibrium, 299–300
  applications of, 298
  Langmuir/Freundlich isotherms, 299
  mass balance equations, 304
Adsorption isotherm, 419
Adsorption, mass balances, 298
  equilibrium, 299–300
  rate of mass transfer, 300–306
  substances removal, 298
Aerated fermenter, conceptual scheme of, 170
Aerobic fermentations, 95–96, 103, 107, 135, 140, 147, 166, 169, 192, 198, 224, 237, 247
Aerobic metabolism, 98, 362
Agitation speed, 186
Agitator, for fermentation process, 189
Agitator power, 417
  draw, 205
  scale-up/ scale-down of, 258
Agitator speed, 178
  gassed power draw, 208
  scaled-up/ scaled-down vessels, 207
Agitator type, 178
Air flow rate, 178
Airlift fermenter, scheme of, 105
Alkylation method, 5
Amino acids, 1
  production, 325
  side chain, 2
Ammonia, concentration of, 238
Ammonium sulphate, 154
Anaerobic digesters, 329
Anaerobic digestion processes, 217
Anaerobic fermentations, 147, 224, 229, 317, 362
Animal-derived protein-rich media, 327
Anion exchange resin, equilibrium constant, 281
Anion exchangers groups, 276
Antifoam agents, 107
Arrhenius model, with activation energy, 51
Aspergillus Niger, 87, 99, 325

B

Bacillus subtilis, 181, 194
Batch fermentations
  average reaction rate, 162
  differential equations, 415
  disadvantages of, 101
  dissolved oxygen profile, 192
  growth-limiting substrate, 101
  heat balances, with jacketed vessel, 413
  with jacketed vessel, 413
  microorganisms, 192, 246
  process of, 246, 251
  profiles, 156, 158
    comparison of, 157
    substrate concentration, collecting data of, 160
  qualitative profiles of, 102
  qualitative time profiles, 246
Batch profiles, endogenous metabolism, 159
Batch reactors
  continuous reactors estimation, 160–162
  endogenous metabolism model, 155
  enzymes
    concentration, 36
    conceptual scheme of, 58
  fermentation, 247
  gas flow rate vs time, 398
  graphical comparison, 65
  graphical determination, 63, 64
  heat balance, 245
  mass balances, 140–143, 355
    equations, 58, 59
    for oxygen, 397
    for substrate, 32
  microorganisms' concentration, 163
  nonlinear parameter fitting method, 28
  nonporous support, 59
    enzymes immobilised, 58
  process of, 363
  reaction parameter, 60
  second-order equation, 63
  substrate concentration, 36
    conversion rate, 64, 65
    time profile of, 28
  substrate profile, 27, 32
    enzyme concentration, 33

enzymes, cost of, 33
mass balance, 34
Batch test, generic growth stoichiometry, 387
Biochemical engineering industry
crossflow filtration, scheme of, 272
fermentation effluent, 273
Biodegradable carbon source, 100
Bioethanol production, 1, 5, 217, 317–321, 421
Biofuels, 5, 106
Biomass
concentration, 156, 236, 246, 380, 388, 398, 406
time profile of, 397
heat balance, 248
mass balances, 157, 159, 403
maximum biomass production rate, 387
metabolism, 155
productivity, 146
typical profiles of, 147
Biomass profiles, mass balances, 157
Bubble column fermenter
scheme of, 105
stirred tank reactors, 105

C

Capital costs, 79
Carbon dioxide
mass flow rates, 332
production of, 422
Catabolism, aerobic fermentation processes, 96
Cations exchange resin, 284
equilibrium constant, 281
Li^+, 281
Cell disruption, nonmechanical methods, 108
Cell membrane, 91, 96, 108
Cellular enzymes, denaturation of, 119
Cellular metabolism, 91, 96
Chemostat, 374, 380, 381
dilution rate, 375
heat balances, 409
for jacketed chemostat, 416
jacketed chemostat, heat balances, 416
mass balance
for oxygen, 394
for substrate, 403
microbial metabolism, 381
oxygen balance, 388, 390, 392, 396
oxygen concentration, 399
oxygen mass balance, 385, 388, 390, 394, 400
substrate consumption rate, 385
Chromatography, 108, 307
advantage of, 421
conceptual schemes of, 308
mass balances, 307–313
conceptual scheme, 310
mass transfer, 307–313
overview of, 307
process description, 309
Citric acid, industrial fermentation processes, 324–325
Citric acid production process, 325, 376, 421
Commercial-scale fermentation processes, 86
Continuous fermentation process, 234, 237
Continuous fermenter, 101, 143–144, 146, 151, 153, 155–156, 163–164, 198, 201, 203
endogenous metabolism, 161
profiles, 158
Continuous-flow reactors (CSTRs) method, 79, 162, 348–350, 360, 379
batch, comparison of, 162–164
conceptual scheme of, 60
enzymatic reaction, 24, 25, 35, 66
enzyme concentration, 36
enzymes immobilised, on nonporous supports, 60
enzyme/substrate concentration, 37
experimental data, 25, 26
first-order kinetics, 145
kinetic parameters
estimation, 160–162
from experimental data, 23–25
mass transfer coefficient, 62
nonfermentation reaction, 145
plot of, 26
reaction rate parameter, 62
re-arranging equation, 26, 37
residence time, 36, 37, 61
second-order equation, 67
substrate concentration, 23, 37
substrate mass balance, 75
Thiele modulus, 75
tripalmitin concentration, 349
Cooling fluid, flow rate of, 239
Corynebacterium glutamicum, 326, 421
Cylindrical vessel, diameter of, 408
Cytoplasm, 91–92, 362

D

Dairy industry, lactose-rich by-product, 106
Deoxyribonucleic acid (DNA), 92
Design fermentation processes, 110
Detergents, 5
Differential equations, using Microsoft Excel, 337–340
DNA double helix, 93
Downstream processes, 266
enzymatic reaction, 266
Penicillium chrysogenum, 266

E

Eadie-Hofstee method, 16, 17, 31
  integration method, 21
Eadie-Hofstee plot, 16, 17
  experimental data, 18
  re-arrangements of, 17
Effluent substrate concentration, 37, 144, 163, 371, 406
Electron acceptor, 95–96, 98–99
Endogenous metabolism
  batch profiles, 159
  continuous fermenter, 161
  continuous reactor, 156
  fermenter profiles, 160
  mass balances, modifications of, 154–160
Enthalpies
  added/ removed, 229
  biomass concentration, 231
  biomass/substrate, combustion of, 224
  of chemical species, 217–221
  components per unit mass, 240
  diluted water-based mixtures, 233
  fermentation reaction, 222, 223, 224
  fluid in jacket, 243
  glucose, 402–403, 407
    formation of, 223
  heat balance, 245, 410
  inlet stream, 241, 243
  jacket fluid, 256
    heat balance, 401, 407, 410
  mass balance, 409
  outlet streams, 242, 244
  physical properties, 219, 238
  reaction stoichiometry, 233
  reactor temperature, 243
  right-hand side, 233
  steady-state mass balances, 234
  substances, 263
Enzymatic hydrolysis, acid-base mechanism, 4
Enzymatic reactions, 40, 78, 79
  adsorption applications, 298
  amino acids, side chain, 2
  applications of, 5–6
  catalysed reactions, kinetics of, 1, *See also* Enzyme catalysis
  catalysis/mechanisms, 2–3
  competitive inhibitors, kinetics, 39–40
  continuous-flow reactors (CSTRs) method, 24
  cost analysis of, 79
  CSTR (continuous reactor), 35
  design of, 77–79
  differential equation, 360
  economics, 77–79
  effectiveness factor vs Thiele modulus, 74
  immobilisation/materials, 3–6
    nonporous supports, 54–62
    porous supports, 68–75
  inhibitors, effect of, 38–39
  integration method, 22, 23
  kinetic parameters, from experimental data, 55
    continuous-flow reactors (CSTRs) method, 23–25
    Eadie-Hofstee method, 16–17
    Hanes-Woolf method, 18–19
    integration method, 21–22
    Lineweaver-Burk method, 13–14
    nonlinear parameter fitting method, 27–29
  kinetic parameters, with inhibitors, 43–45
  kinetics reactions, 7
  limitations of, 3–4
  mass balances
    for batch reactors, 32–33
    for continuous-flow reactors, 35–37
  mass transfer coefficient, 56
  Michaelis-Menten kinetics, 38
    analysis of, 10–11
    derivation of, 7–10
    substrate removal rate, 12
  nature/chemistry/structure, 1–2
  vs. nonbiological heterogeneous catalysts, 4
  noncompetitive inhibitors, kinetics, 41–42
  nonporous support/covalent bonding, 5
  on nonporous supports, 54
  overview of, 1
  pH effect, 48–50
  reaction parameter, 57
  scheme of, 92
  surface reaction rate, 57
  temperature, effect of, 51–52
  volumetric reaction rate, 57
Enzyme balance, 41
Enzyme catalysis
  application, 6
  conceptual sketch of, 5
  heterogeneous catalysts, 4
  immobilisation methods, 5, 6
  limiting reagent, 7
  lock-and-key model, 3
  mechanisms of, 2
  nonbiological catalysts, 3
  reaction progression, 3
Enzyme chymotrypsin, acid-base mechanism, 4
Enzyme concentration, 352, 353, 354
  batch reactor, on substrate profile, 33
  first-order model, 52
  inhibition parameter, 47
  maximum noninhibited rate, 45

pH effect
optimum value, 49, 51
plot of reaction rate, 50
rate of substrate transfer, 54
reaction rate vs temperature, 52, 53
suspect inhibitor, 46
Enzyme deactivation, 51
Enzyme glucoamylase, 11
Enzyme inhibition, 38
Enzyme/inhibitor complex, 39
Enzymes, 92
substrate complex, 39
to triglycerides, 345
Enzyme-substrate (ES) complex, 2, 7
Equilibrium constant, 8, 173, 279, 281–283, 287, 292
Escherichia coli, 139, 169
Ethanol
anaerobic fermentation, 229
concentration of, 149, 373
glucose, anaerobic fermentation of, 147
mass balances, 157, 159
mass flow rates, 230, 332, 333
production, from glucose, 124, 149, 224
Eukaryotic/prokaryotic cells, 90

F

Fed-batch processes, 363
batch fermentation, 103, 139, 154, *See also* Fed-batch reactor
fermentation process, 102
substrate concentration, 382
Fed-batch reactor, 139, 149, 372, 374, 375
batch, comparison of, 162–164
biomass balance, 379
biomass concentration, 151
design flow rate, 163, 164
lower reaction rate, 163
operational sequence of, 150
product productivity, 152
quasi-steady state, 150, 151, 152, 372, 379, 381
Saccharomyces cerevisiae, 153
substrate concentration, 151
Feed flow rate
fermentation stoichiometry, 153
quasi-steady state concentrations, 153
Fermentation constant, scale-up/scale-down, 256
Fermentation processes
advantage of, 88
agitated vessels, scale-up of, 401
cell growth, stoichiometry/metabolism of, 95–99
conditions, 220, 221
culture cultivation/maintenance, 107
enthalpy, of chemical species, 217–221
industrial, *See* Industrial fermentation processes
kinetics/stoichiometry
endogenous metabolism, 127–135
growth stoichiometry, 123–127
inhibitors/temperature/pH, effect of, 117–120
kinetic parameters, experimental measurement of, 120–123
overview of, 110–117
pH, on growth rate, 117–120
rate equations, for secondary metabolites, 127–135
yield coefficients, 123–127
maintenance/endogenous metabolism, 100
media/equipment, sterilisation of, 108–109
metabolic products, 100
metabolites, extraction/separation of, 107–108
microbial cells
growth/product formation, qualitative time profiles of, 100–101
structure/components of, 90–95
microorganisms, 86–88
classification of, 88–90
lactic acid production, using chemical synthesis, 89
overview of, 86
processes/reactors, 101–106
starting materials, 106–107
vessels, geometry/materials, 109
Fermentation reaction, kinetics of, 162
Fermentations consume oxygen, *See* Oxygen transfer, fermentations
Fermentation stoichiometry, 124, 153, 218, 222, 316, 386
Fermentation temperature, 217, 239, 256, 258, 259
Fermentation vessels, sterilisation of, 109
Fick's law of diffusion, 69, 170
Film resistance, to mass transfer, 357
Fluid, heat balance, 406
Fluidised bed fermenters, 104
packed, 369
scheme of, 104
Fluid velocity, for jacket, 254
Foam, 107
Forming tri-calcium citrate tetrahydrate, 325
Free enzymes, 79
concentration of, 8
lock- and- key model, 3
Freundlich model, 300–301

G

Gas chromatography (GC), 307
Gas flow rate, 178–181, 199, 203–204, 383, 390, 397

Gel permeation chromatography (GPC), 307
Generic stoichiometry, 364
  for fermentation, 363
  microorganisms, on acetic acid, 123
Genetic information, transcription of, 362
"Global" mass transfer coefficient, 172
Glucose, 364
  aerobic catabolism of, 98
  aerobic fermentation of, 95
  anaerobic fermentation of, 147, 224
  mass balances, 157, 159
  mass flow rates, 332, 333
  plot of, 148
  profiles of, 148
Glucose 6-phosphate, 96
Glucose concentration, 147, 154, 195, 238, 247, 375
Glucose consumption rate, 201
Glucose, fermentation of, 126
L-Glutamate, yields of, 326
Glutamic acid, 88, 364
L-Glutamic acid production, 325, 326, 421
Glycolysis reaction, 98
Grain whisky process, 324
Guanosine diphosphate (GDP), 98
Guanosine triphosphate (GTP), 98

H

Haldane/Andrews equation, 117
Hanes-Woolf linearisation, 19
  data, re-arrangements, 20
  experimental data, re- arrangement, 19
Hanes-Woolf methods, 18, 21, 31
  enzymatic reaction, 19
  integration method, 21
  parameters determination, 19
  plot, 20
Heat balances, 217
  adiabatic batch fermentation, 414
  for batch fermentation, 245–250
  for continuous fermentation, 228–245
  equation, 405
  on reactor, 403
  scale-up/scale-down problem, 256–260
  variable temperature, effect of, 250–251
Heat generation, rate of, 217, 218, 222–228
Heat-sensitive media, 109
Heat transfer coefficients, 251, 253, 255, 257, 408
  in agitated vessels, 251–255
  material, conductivity of, 251
Heat transfer resistance, 257, 259
High-performance liquid chromatography (HPLC), 307
Humicola insolens, 330
Hypocrea jecorina, 330

I

Ideal gas law, 171, 202
Immune response, by vaccination, 327
Industrial fermentation processes
  bioethanol, 317–321
  cell-based virus vaccines, 326–328
  cellulases production, 330–331
  citric acid, 324–325
  design considerations, 316–317
  ethanol production, 331
  L-glutamic acid, 325–326
  mass flow rates, 332, 333
  penicillin, 322
  undefined mixed cultures, 328–330
  whisky, 323–324
Industrial ion exchange processes, 274
Influenza vaccine, using mammalian cells, 327
Inlet gas
  composition of, 383
  mass flow rate of, 237, 247
  phase, volumetric flow rate of, 400
  total mass flow rate of, 237
Ion exchange
  adsorption, 309
  concentration-based activity coefficient, 279
  conceptual scheme of, 274, 275
  equilibrium constant, 281
  equilibrium reactions, 279–283
  fermentation, 275
  generic *A*+ ion, 292
  generic exchangeable ion, 278
  mass balance equations, 283–286, 288, 290, 297
  mass transfer, 283–286
    qualitative effect of, 279
    rate of, 277, 292, 293
  overview of, 274
  packed column, application of, 277
  polystyrene resins, 276
  reaction, 282
  regeneration process, 274, 277–280
  resins/column operation, 274–277
  strong-acid/strong-base resins, 277
  weak-acid/weak-base resins, 277
Iron-deficient media, 100

J

Jacket
  batch reactor, scheme of, 245
  fermenter, 230
  fluid flow rate, 259
  fluid velocity, 254
  heat balances, 248, 401, 407, 410
  heat transfer coefficients, 411

Prandtl umber, 254
Reynolds number, 254
transfer coefficient, 412
vessels, scheme of, 253
Jacket temperature
profiles of reactor, 250
reactor temperature, 235

K

Kidney epithelial monkey cells, 327
Kinetic parameters, 147
estimation, continuous-flow reactor, 160–162
from experimental data
continuous-flow reactors (CSTRs) method, 23–25
Eadie-Hofstee method, 16–17
Hanes-Woolf method, 18–19
integration method, 21–22
Lineweaver-Burk method, 13–14
nonlinear parameter fitting method, 27–29
with inhibitors, 43–45
mass balance, 355
species, MW of, 355
without inhibitor, 351

L

Lactic acid
mass flowrate, 273
nanofiltration process, 273
Lactic acid production
process scheme, 89
rate of, 365
Lactobacillus acidophilus, 87
Lactose, empirical formula, 346
Langmuir equation, 300–302
Langmuir isotherm, 300–302, 309–310, 420
Lineweaver-Burk method, 13, 21, 31
data analysis, 13
experimental data, 18
limitation of, 14
linearisation, 15, 26
parameters determination, 14
plot for, 15, 17
re-arrangements of, 15
substrate concentration, 14
Lineweaver-Burk plot, 16, 18, 24, 44
Liquid/gas films, viscosity, 178
Liquid phase, temperature, 178

M

Madine Darby canine kidney cells (MDCK), 327
Malt whisky, 323–324
Mammalian cells, 88, 327–328
Mass balances, 146, 197
accumulation, 139
of adsorbate, 420
adsorption, 298
equilibrium, 299–300
rate of mass transfer, 300–306
substances removal, 298
for batch reactors, 140–143
biomass in fermenter, 232
chromatography, 307–313, *See also* Chromatography
continuous reactors, 143–149
with endogenous metabolism, 155
design enzymatic processes, 78
endogenous metabolism, modifications of, 154–160
equations, 250
for adsorption, 304
integration of, 294–295
fed-batch reactors, 149–154
fermentation processes, 151
general equation, 139
immobilised, on porous supports, 75
ion exchange, 283–286
kinetic parameters, 355
metabolism, 157, 382
oxygen, during batch fermentation, 388
porous supports, conceptual scheme, 69
process parameters, effect of, 297
reaction processes, 71
secondary metabolites, modifications of, 154–160
second-order differential equation, 70, 71
for substances, 312
conversion rate, 77
substrate flux, 70
Mass transfer
chromatography, 307–313
coefficient of, 179, 180, 181, 184, 186, 198, 200, 297, 302, 305
ion exchange, 283–286
oxygen in fermenter, 174
rate of, 61, 174, 286, 287
value of, 178
Mass transfer theory, 169, 179
Membranes, 362
Messenger RNA (mRNA), 93
Methanol, 106
Michaelis-Menten equation, 13, 346–347, 350, 354
Michaelis-Menten kinetics, 7, 10, 11, 13, 18, 21, 38, 43, 52, 54, 345
for competitive inhibition, 39
enzymatic reactions, 38
analysis of, 10–11
derivation of, 7–10

substrate removal rate, 12
linearisation, 16
substrate concentration, 14
substrate rate vs substrate concentration, 11
substrate removal rate, 12
Michaelis-Menten model, 7, 78, 347
Michaelis-Menten parameters
data collected, in CSTR, 24
determination of, 24
experimental data, 31
Michalis-Menten equation, 25
integration method, 21
parameters determination, 14, 16, 19
Microbial metabolism, 87, 106, 161, 176
Microorganisms
acclimation phase, 100
acetic acid (CH_3COOH), 123
Acetobacter, 193
Aspergillus Niger conidia, 87, 125
Bacillus subtilis, 194
batch fermentation process, 192, 246
batch reactor
mass balances, 140
substrate, time profiles of, 141
binary combination, 89
biomass concentration, 127
biomass growth, 110, 115
rate of, 113
substrate removal, 128
biomass, stoichiometric coefficient, 227
$C_5H_7O_2N$, 135
cake resistance, 269
carbon dioxide/water, 131, 132
characteristics of, 89
chemical/physical nature, 266
chemical stoichiometry, 95
classification of, 88
composition of, 383
continuous fermenter, scheme of, 143
crossflow filtration, 268–274
chemical structures of, 270
conceptual scheme, 269, 270
membrane, 269
crossflow membrane
conceptual scheme of, 271
modules, 271
differential equations, 371
dilution rate, 377
downstream processing, 107, 108
endogenous metabolism
model, 132
rate of, 131
ethanol concentration, 193
Eukarya domain, 89
experimental data, plot of, 122
fermentation fluid vs time, 247
fermentation medium, 267
fermentation processes, 86–88, 226
classification of, 88–90
lactic acid production, using chemical synthesis, 89
fermentation products, 90
filtrate, mass flow rate of, 268
generic stoichiometry, 123
glucose
aerobic conversion of, 227
anaerobic conversion, 227
fermentation of, 126
growth kinetics, 127
growth rate, 110, 111
heat balances, continuous fermenter, 228
inhibit biomass growth, 118
inhibiting substrate, 118
kinetic parameters, 112, 117, 128
determination of, 120
limiting substrate, 112, 113
linearization of, 123, 130, 133, 134
mass flow rates, 268, 332
mass of biomass, 114
metabolism of, 86
microbial metabolism, conceptual scheme, 87
Monod model, 129
multicellular organisms, 86
net growth rate, 370
nitrogen consumption, volumetric rate of, 115
nitrogen source, 153
organic substrate, 121, 129
oxidation of, 132
oxygen consumption, 115, 128, 370, 389, 390
packed bed reactors, 104
pH inhibition factor, 119
pH range, 118
pressure filtration/centrifugal filtration, 266
production of, 146
prokaryotic/eukaryotic cells, 90
qualitative profiles of, 101
rate of heat generation, 227
reaction stoichiometry, 248
in reactor, 413
re-arranging equation, 120
separation of, 266–268
shifting order rate equation, 111
stoichiometric coefficients, 125, 126, 196
stoichiometry of, 153
substrate, 116, 117, 122, 128
consumption, 127
removal, 143
typical profiles, 146
yield coefficients, 120, 124, 125
Microsoft Excel, numerical integration, 293

Model accuracy, 305
Molecular weights (MW), 1
Monod equation, 111, 112, 191, 387
Monod kinetics, 144
Monod model, 367, 369, 382

N

NADH molecules, 362
NAD^{+}, oxidation/reduction reactions, 94, 95
Nanofiltration, 269, 273
Neolithic China, 86
Newtonian viscosity, 180
Nicotinamide adenine dinucleotide (NADH), 94, 95
Nitrogen
  concentration, in fermentation media, 106
  fermentation stoichiometry, 124
Noncompetitive inhibition
  kinetics model, 41
  linearisation of, 43, 44
  maximum substrate rate, 42
  rate equation, 42
  substrate concentrations, 43
Noninhibited enzyme, 38
Nonlinear parameter fitting method, 28, 29, 30, 32, 346
Nucleic acid molecule, 326
Nusselt number, 251, 252

O

Organelles, 91
Organic matter, anaerobic conversion of, 330
Organic waste, anaerobic digester, 329
Oxidative phosphorylation, 98, 362
Oxygen
  absorption method, 395
  balance, in continuous reactors, 198
  consumption, *See* Oxygen consumption
  diffusivities, 171
  experimental data, 177
  mass transfer coefficient, 176
  saturation concentration of, 182
  sensors, 176
  solubility, 191
Oxygen concentration, 199, 395
  continuous fermenter, 200
  mass balances, 197
Oxygen consumption, 192, 386, 396
  fermentation stoichiometry, 386
  microorganisms per unit volume, 115
Oxygen fermenter, mass transfer coefficient, 184
Oxygen flow rate, 178, 201
  mass flow rates, 398
  oxygen transfers, 400
Oxygen transfer, fermentations
  aeration system, 169
  mass transfer coefficients
    correlations for, 178–182
    measurement of, 175–178
    power draw, in agitated vessels, 182–190
  oxygen balances
    in batch reactors, 191–197
    in continuous reactors, 198–202
  scale-up/scale-down, 203–208, 256–260
  theory of, 169–175
Oxygen transfers, 169
  efficiency of, 201, 400
  fermentations, *See* Oxygen transfer, fermentations
  mass flow rate, 400
  mass transfer, 237
    coefficient of, 284
    rate, 174, 175, 176
  for oxygen mass balance, 198
  rate of, 191, 383, 384, 386
  scaled-up/ scale-down vessel, 203, 205

P

Packed bed reactors
  microorganisms, 104
  scheme of, 104
Packed column, application of, 277
Parameter estimation methods *vs.* experimental data, 347
Penicillin production
  mass flow rates, 332
  rate, 422
  simplified scheme of, 322
Penicillium microorganism, 322
phosphorus, 106
pH sensor, 107
Polyhydroxyalkanoates (PHAs), 99
Polymeric resins, 298
Polysulfone, 269–270
Porous supports
  enzymatic reactions
    conceptual model, 68
    effectiveness factor vs Thiele modulus, 74
  enzymes immobilised, 69
    conceptual scheme of, 68
  intrinsic reaction rate, 73
  mass balances, conceptual scheme, 69
  substrate profile, 72, 73
    conceptual scheme, 69
Potassium, exchange of, 282
Power per unit volume, 188
Prandtl number, 252
  fluid in reactor, 251
  for jacket, 254
Preparative chromatography, 307

Protein synthesis, 362
Pyruvate, from glucose, 96, 97
Pyruvate oxidation, to acetyl-CoA, 99

Q

Quasi-steady state, 372, 379

R

Reaction rates, variable temperature effect, 250–251
Reactor-side coefficient, 402
Reactor temperature, continuous process, 250
Resin, H+ ions, 418
Reynolds number, 182, 183, 190, 203, 252, 258
  fluid in reactor, 251
  for jacket, 255
Ribonucleic acid (RNA), 93
RNA nucleotide, with ribose sugar, 94

S

Saccharomyces cerevisiae strains, 139, 153, 217, 324
Scotch whisky, 323
Secondary metabolites, 99
  mass balances, modifications of, 154–160
  penicillin, 322
Sherwood number, 185
  for liquid phase coefficient, 179, 185
  mass transfer coefficient, 185
Solid-liquid slurry, 330
Solvent extraction, 108
Solver add-in in Excel, 28, 29
Starch
  hydrolysis, 422
  mass flow rates, 333
Steady-state biomass concentration, 155, 392
Steady state mass balance, 69, 234
Sterilisation, 108
Stirred tank fermenter, components of, 103
Stirred tank reactors, 103, 105–106, 363
Stoichiometry reaction, 364, 370, 373
Substance
  enthalpy properties, 263
  thermodynamic properties, 260
Substrate concentration, 371
  batch reactor, 349
  limitation, 29
Substrate consumption, fermentation stoichiometry, 386
Substrate, continuous fermentation feed, 236
Substrate mass balance, 34, 405
  enzyme reactions, 72
  mass transfer rate, 72
  numerical solution of, 71
Sucrose, conversion of, 377
Sugars, purified substrates, 106
Sulfonic acid (SO_3H), 275
Sulphite oxidation method, 176
Sulphur, 106
Superficial gas velocities, 185, 186, 202
"Swan-necked" copper stills, 324
Systems of equations, in Microsoft Excel
  boundary conditions, 361
  numerical solution, 341–343

T

Thermodynamic properties, substance, 260
Thiele modulus, 74, 357, 358
Tricarboxylic acid cycle, 98
Triglycerides
  enzyme, 345
  in oil, 345
Tripalmitin, empirical formulas, 348
"Two-film" model, 169, 172
  mass transfer theory, 169
  scheme of, 170

U

Urea degradation, 356, 357
Urea hydrolysis, 357

V

Vaccination, immune response, 327
Variable temperature, reaction rates effect, 250–251
Vessel diameter
  agitator diameter, 385
  agitator of, 384
Vessels
  diameter of agitator, 194
  gassed power, 188
  geometry, 178
  heat transfer rate, 236
  mass transfer correlation, 187, 189
Virus propagation, 327
Volumetric mass transfer rate, 173
Volumetric oxygen mass transfer rate, 392

W

Wastewater treatment, activated sludge process, 328
Water-based fermentation medium, 108
Water, physical properties of, 242
Whisky production process, 333

Y

Yeast, mass flow rates, 333